FOUNDATIONS OF
APHASIA REHABILITATION

Related Pergamon journals

Journal of Neurolinguistics*
Editor: John Marshall

Language Sciences*
Editor: Paul Hopper

Neuropsychologia*
Editors: Giovanni Berlucchi and Malcolm Jeeves

*Free specimen copy available on request

FOUNDATIONS OF APHASIA REHABILITATION

edited by

MICHEL PARADIS

McGill University

INTERNATIONAL ASSOCIATION
OF LOGOPEDICS AND PHONIATRICS

PERGAMON PRESS
OXFORD · NEW YORK · SEOUL · TOKYO

U.K.	Pergamon Press Ltd, Headington Hill Hall, Oxford OX3 0BW, England
U.S.A.	Pergamon Press, Inc., 660 White Plains Road, Tarrytown, New York 10591-5153, U.S.A.
KOREA	Pergamon Press Korea, K.P.O. Box 315, Seoul 110-603, Korea
JAPAN	Pergamon Press Japan, Tsunashima Building Annex, 3-20-12 Yushima, Bunkyo-ku, Tokyo 113, Japan

Copyright © 1993 Pergamon Press Ltd

All Rights Reserved. No part of this publication may be reproduced, stored in a retrieval system or transmitted in any form or by any means: electronic, electrostatic, magnetic tape, mechanical, photocopying, recording or otherwise, without permission in writing from the publishers.

First edition 1993

ISBN 0-08-041940-2

Library of Congress Cataloging in Publication Data
A catalogue record for this book is available from the Library of Congress.

British Library Cataloguing in Publication Data
A catalogue record for this book is available from the British Library.

Printed and bound in Great Britain by BPCC Wheatons Ltd, Exeter

Contents

List of Contributors		vii
Preface		ix
I.	Classification and assessment of rehabilitation methods	1
	1. Inventory and classification of aphasia rehabilitation methods *S. Methé, W. Huber and M. Paradis*	3
	2. Efficacy of various methods of therapy *R. T. Wertz*	61
	3. Efficacy of language systematic learning approaches to treatment *L. Springer and K. Willmes*	77
II.	Linguistic foundations of aphasia rehabilitation methods	99
	4. Linguistic foundations of rehabilitation methods *J. Wilk and M. Paradis*	101
	5. Linguistic principles underlying two aphasia tests *M. Meth, L. Obler and P. Walsh*	195
	6. Pragmatics applied to aphasia rehabilitation *L. Perkins and R. Lesser*	211
III.	Cognitive foundations of aphasia rehabilitation methods	247
	7. Cognitive neuropsychology and practitioner-researchers in aphasia therapy *R. Lesser*	249
	8. Therapeutic approaches to naming disorders *H. Kremin*	261
	9. Reading and writing: Cognitive therapies of written language *H. Kremin*	293
	10. A review of therapy at the level of the sentence in aphasia *S. Byng and R. Lesser*	319
IV.	Neurological foundations of aphasia rehabilitation methods	363
	11. Neurophysiological foundations of aphasia rehabilitation *A. Kertesz*	365
	12. Interhemispheric participation in recovery from aphasia *J. Vendrell, P. Vendrell and D. Ibáñez*	379

V. Foundations of aphasia rehabilitation for special populations   411

13. Bilingual aphasia rehabilitation   413
    M. Paradis
14. Aphasic adults with premorbid heroin and alcohol abuse   421
    C. Bushell, L. Obler and P. Kerman-Lerner

Conclusion   427

Author Index   431

Subject Index   443

List of Contributors

Camille Bushell, Manhattan Veterans Administration Medical Center, and Department of Speech and Hearing Sciences, CUNY Graduate School, New York, U.S.A.
Sally Byng, Department of Clinical Communication Studies, City University, London, United Kingdom.
Walter Huber, Neurologische Klinik, RWTH, Aachen, Germany.
Daniel Ibáñez, Section of Neuropsychology, Department of Neurology, Hospital de la Santa Creu i Sant Pau, Barcelona, Catalonia, Spain.
Patricia Kerman-Lerner, Speech Pathology and Audiology Service, Goldwater Memorial Hospital, NYU Medical Center.
Andrew Kertesz, Department of Clinical and Neurological Sciences, The University of Western Ontario, and Saint Joseph's Hospital, London, Ontario, Canada.
Helgard Kremin, INSERM U. 302, Pavillon Clérambault, Hôpital de la Salpétrière, Paris, France.
Ruth Lesser, Department of Speech, University of Newcastle upon Tyne, Newcastle upon Tyne, United Kingdom.
Margaret Meth, Department of Speech and Hearing Sciences, CUNY Graduate School, New York, U.S.A.
Susan Methé, Department of Linguistics and School of Human Communication Disorders, McGill University, Montreal, Quebec, Canada.
Loraine K. Obler, Department of Speech and Hearing Sciences, CUNY Graduate School, New York, and Division of Communication Disorders, Emerson College, Boston, Massachussetts, U.S.A.
Michel Paradis, Department of Linguistics, McGill University, and Laboratoire de neuroscience de la cognition, Département de psychologie, Université du Québec à Montréal, Montreal, Quebec, Canada.
Lisa Perkins, Department of Speech, University of Newcastle upon Tyne, Newcastle upon Tyne, United Kingdom.
Luise Springer, School of Logopedics, Medical Faculty, RWTH, Aachen, Germany.
Josep Vendrell, Section of Neuropsychology, Department of Neurology, Hospital de la Santa Creu i Sant Pau, Barcelona, Catalonia, Spain.
Pere Vendrell, Department of Neurology, Hospital de la Santa Creu i Sant Pau, Barcelona, Catalonia, Spain.

Patricia Walsh, Department of Speech and Hearing Sciences, CUNY Graduate School, New York, U.S.A.

Robert T. Wertz, Audiology and Speech Pathology, Veterans Administration Medical Center, Martinez, California, and Department of Neurology, School of Medicine, University of California, Davis, California, U.S.A.

Joanne Wilk, Department of Linguistics, McGill University, Montreal, and Department of Communicative Disorders, The University of Western Ontario, London, Ontario, Canada.

Klaus Willmes, Neurologische Klinik, RWTH, Aachen, Germany.

Preface

At its August 1989 meeting in Prague, Czechoslovakia, on the occasion of the XXIst Congress of the International Association of Logopedics and Phoniatrics, the Aphasia Committee undertook to prepare a report on the theoretical foundations of aphasia rehabilitation to be presented at the next IALP congress in 1992. Each member set out to study a specific aspect of the principles underlying aphasia rehabilitation practices. A first draft of these investigations was circulated among the members of the Aphasia Committee prior to a meeting in Montreal in June, 1991, hosted by the *Laboratoire de Neuroscience de la Cognition* of the *Université du Québec à Montréal*. After a thorough discussion of each contribution, authors wrote a final report which was presented at the IALP XXIInd Congress in Hannover, Germany, in August, 1992, and is published in this volume. Hence all chapters in the present book have been written specifically for this volume by members of the IALP Aphasia Committee and their collaborators.

The purpose of this book is to provide students of language rehabilitation and professional language pathologists with an overview of the theoretical foundations of their field of endeavor, i.e., with the rationale behind what speech therapists have been doing over the past decades, and with pointers indicating the direction in which current theoretical principles are steering the field. Only when one understands why one is doing what one does, can one do what one does— effectively.

In this volume, the reader will find the rationale that has been invoked for, or is inferred from, every intervention procedure reviewed (Chap. 1); the assessment of the efficacy of the various methods as investigated so far (Chaps. 2 and 3); the linguistic presuppositions, whether explicit or implicit, of these various methods (Chap. 4), and those of two widely used tests, the Boston Diagnostic Aphasia Examination and the Minnesota Test for the Differential Diagnosis of Aphasia (Chap. 5); as well as a discussion of the role of pragmatics in language rehabilitation (Chap. 6). In addition, the contribution of cognitive neuropsychology (Chap. 7), its impact on therapeutic approaches to naming disorders (Chap. 8) and to reading and writing (Chap. 9), as well as implications for therapy at the level (Chap. 10) are explored. Neurophysiological foundations of, and interhemispheric participation in recovery from aphasia are presented (Chaps. 11 and 12) and, finally, two special populations, bilingual aphasic patients (Chap. 13) and aphasic adults with premorbid heroin and alcohol abuse (Chap. 14) are considered.

Most chapters contain a detailed review of the relevant literature. In addition, the chapters on the classification of rehabilitation methods (Chap. 1) and on their linguistic foundations (Chap. 4) are followed by an extensive appendix that summarizes respectively 42 and 60 studies in terms, where applicable, of subject/therapy area, therapy description, hypothesis tested, rationale, research method, results, and conclusion/discussion.

x   Preface

The book is not written from one particular theoretical perspective but rather presents an account of the various theoretical positions that have guided practice over the past quarter of a century and that continue to influence language therapy up to this day. As will be stressed in the following chapters, theoretical beliefs about the nature of language representations, the nature of language processing, and the nature of aphasia, change over time and, consequently, so do aphasia rehabilitation procedures.

If, as Hécaen(1972) proposed, aphasias are deficits in the use of specific rules of the various structural levels of language, therapy will depend on the speech-language pathologist's view of the very nature of these rules, which themselves are determined by the linguistic model selected. The various available linguistic models are obviously not definitive, as there are not only numerous competing types of grammars (e.g., structural, functional, stratificational, tagmemic, transformational, and relational), but also within the same general framework, there are fundamental principles that change from year to year if not from month to month (e.g., from transformational grammar to government binding, to the latest abandonment of deep and surface structure in favour of only logical and phonological form [Chomsky, 1991]). Hence none of these characterizations of linguistic structure can be taken as the actual representation that can serve as a permanent basis for intervention. Yet, the classification of language deficits and the grammatical description of a type of aphasia, which serve as the basis for any method of language retraining, crucially depend on, and are inherently limited by, the explanatory and predictive power of the selected linguistic theory.

Because language rehabilitation must be built on multidisciplinary foundations, linguistic analyses are complemented by cognitive, neurofunctional, neurophysiological and neuroanatomical considerations. As our knowledge of these improves, so too does our theory of language rehabilitation.

Thanks are due to Molly Mack for a thorough critical review of the manuscript, and to John Matthews for expert technical assistance throughout.

References

Chomsky, N. 1991. Course in Linguistics, Department of Linguistics and philisophy, Massachussetts Institute of Technology, October.
Hécaen, H. 1972. Neurolinguistique et neuropsychologie. *Langages*, **25**: 3-5

Part I Classification and assessment of rehabilitation methods

Improvement seems to be a function of both the severity of aphasia and the amount of time that has elapsed post-onset. The less severe the aphasia and the earlier the intervention, the better the chances of full or partial recovery. The more severe the aphasia and the later the intervention, the less successful the treatment. Indeed, very little improvement has been reported in cases of severe aphasia and/or when therapy was provided after the period of spontaneous recovery (e.g., over six months post-onset). This may be taken as providing support for two hypotheses. The first is that recovery of language depends on undamaged portions of the language areas and/or adjacent areas of the *left* hemisphere; and the second is that either (a) improvement is attributable to spontaneous recovery and likely to occur with or without therapy or (b) therapy is only effective during the spontaneous recovery period, but when provided during that period, it does increase the extent of recovery. It is very difficult to tease apart the contributions of therapy and of spontaneous recovery since a comparison between treated and non-treated groups can never be fully valid because of the lack of homogeneity inherent in patient groups.

One variable that may not have received the attention it deserves is the degree of metalinguistic knowledge pre-onset. The generally better prognosis in highly educated individuals might be attributable to the availability of metalinguistic knowledge in the form of explicit knowledge which can be consciously applied to the controlled production of utterances. Since it is only procedural memory (Cohen, 1984) underlying the automatic use of language (i.e., implicit linguistic competence) that is affected by the aphasia, conscious knowledge of various aspects of grammar is not necessarily impaired, and hence can be used as a compensatory strategy. The ability to use metalinguistic knowledge in the absence of procedural competence may also explain the recovery of a foreign language over the premorbidly more fluent native language in some bilingual patients who have learned their foreign language in a formal manner (Paradis, in press).

There seem to be five major compensatory strategies available to rehabilitate aphasic patients, short of retraining the automatic use of their linguistic competence, and assuming that reacquiring linguistic competence is extremely difficult in adults beyond the period of spontaneous recovery. They are as follows: (1) avoidance of problematic linguistic material by the use of simpler forms; (2) use of pragmatic aspects of language (whether or not they turn out to be subserved by the right hemisphere); (3) use of communicative strategies other than language (e.g., pantomime, Blissymbols, Amerind sign, etc.); (4) use of other forms of support believed to be dependent on the intact right hemisphere (e.g., Melodic Intonation Therapy); and (5) substitution of the controlled use of metalinguistic declarative knowledge for the unavailable automatic linguistic competence.

This section starts with an inventory of the various aphasia therapy methods that have been used over the past decades. Methods have been grouped together on the basis of their explicit or implicit theoretical presuppositions. The following two chapters examine the efficacy of various methods. Chapter 2 discusses the methodological problems inherent in various designs for evaluating the effectiveness of treatment. Chapter 3 investigates the efficacy of systematic learning approaches and reports on two experimental studies, each comparing two pairs of methods—namely, a linguistically structured approach with a stimulation approach, and the traditional PACE approach with a modified PACE approach.

References

Cohen, N. 1984 Preserved learning capacity in amnesia: Evidence for multiple memory systems. In L. R. Squire and N. Butters (eds.), *The neuropsychology of human memory*. New York: Guilford Press. Pp. 83-103.

Paradis, M. forthcoming. Differential involvement of cerebral memory structures as a consequence of whether languages are learned formally or acquired incidentally. In N. Ellis (ed.), *Implicit and explicit learning of languages*.

1 Inventory and classification of rehabilitation methods

Susan Methé, Walter Huber and Michel Paradis

INTRODUCTION

The purpose of this chapter is to provide the reader with an inventory and classification of the various language therapy methods described over the past 25 years, with an emphasis on the rationale behind their methodology.

Aphasia is not easily defined. The most that can be agreed upon is that it is a disorder which in some way affects the normal processing of language in one or more of the language modalities. For each theory of cerebral language function, there is a respective theory or approach which attempts to explain how to remediate language dysfunction.

The rationales behind the various aphasia therapies can be grouped into two main areas. These depend on what aspects of language are assumed to have remained intact after cerebral injury. One school of thought assumes that, after brain damage, the areas subserving language are still intact but are not easily accessible. The basis of therapy is the assumption that language can be accessed. In other words, through retraining, the left hemisphere can relearn what has been lost, or the inhibited language functions can be reactivated. Within this approach there are those who believe that intensive language stimulation alone should be sufficient for language to return. Others believe that the brain can be retrained for language through awareness, self-cueing or programmed instruction.

An alternative approach assumes that if certain language functions are lost because of a lesion to cerebral tissue, they will not be recoverable. Any language which remains is assumed to be subserved by undamaged brain tissue. The patient must therefore be taught to compensate for what has been lost. This can be done by tapping into the intact right hemisphere, by teaching the patient to use alternative (non-linguistic) means of communication or by focusing on effective verbal communication, stressing pragmatic aspects of language use rather than grammaticality.

Aphasia not only affects language in the patient but disrupts social interactions and affective states which may lead to depression. A full therapy program includes not only language or communication therapy but therapy aids aimed at coping with aphasia in affective and social contexts. In the following sections, various therapeutic approaches will be described and evaluated.

LANGUAGE REACTIVATION AND RETRAINING

Stimulation Approaches

Since World War II, the field of aphasia therapy (in the strict neuropsychological sense) has developed rapidly, resulting in a variety of contents and methods which, when taken together, make rather conflicting theoretical assumptions on the nature of aphasia and its consequence for therapy planning (Chapey,1986; Code & Müller, 1989).

In the 50's and 60's, theoretical work was mainly concerned with global goals and methodology (cf. Schuell et al., 1955; Wepman, 1955). Activation of any verbal behaviour was thought to be essential. The standard procedure was auditory stimulation. Enough motivation in the patient would suffice to reactivate impaired language functions. Overall severity and time post-onset were seen as important intervening factors. But neither a specific pattern of aphasic symptoms nor the possible underlying mechanisms or strategies were systematically considered for therapy planning.

STIMULATION THERAPY

Stimulation therapy treats aphasia as a general language deficit which crosses all language modalities and which varies from patient to patient because of other cerebral areas affected by brain damage. (Schuell et al., 1964). It is therefore assumed that there is only one type of aphasia but it can be exhibited in different ways due to severity level or to the presence of other concomitant cognitive or sensori-motor dysfunctions.

The rationale behind stimulation therapy is that aphasia interferes with proper language processing but does not result in a loss of language function. If sensory stimulation affects the brain, then through intensive stimulation the frequency of neurons firing and the number of fibers activated increases. This activation is claimed to be essential for the organization, storage and retrieval patterns in the brain (Duffy, 1986). If certain patterns of excitation were established during the acquisition of language, then appropriate stimulation should permit retrieval of these patterns.

One characteristic of stimulation therapy is that it focuses primarily on language and not on communication. Some argue that language forms only part of the communication process and that complete therapy must focus on pragmatic aspects as well (Aten, 1982, 1986; Davis and Wilcox, 1985; Helm-Estabrooks et al., 1982). However, a problem which has prompted many researchers to seek out new methods is the limited improvement resulting from traditional stimulation therapy (Helm-Estabrooks et al., 1987; Prins et al., 1989; Aten et al., 1982; Aten, 1986; Duffy, 1986). Schuell reports that stimulation therapy is effective except with severe aphasics. Others report similar findings (e.g., Broida, 1977). For example, one study showed that stimulation therapy was effective for all aphasic populations but the amount of success was

proportional to the amount of time after onset and the severity of the aphasia (Basso et al., 1979).
One study has taken stimulation therapy one step further. Kushner and Winitz (1977) assessed the effect of comprehension training on a patient. Comprehension training is consistent with stimulation therapy in that it uses auditory stimulation but also includes components of a pairwise relationship between sound and meaning. The comprehension approach stresses comprehension before production. Once a concept is understood, the patient should be able to produce the sound associated with the meaning. This method involved having the patient point to the correct picture following a single word stimulus. The same pictures were presented repeatedly until all could be named. At baseline the patient showed deficiencies in verbal production and comprehension and showed word-finding difficulties. Following treatment, the patient produced all of the lexical items correctly on demand.

Sentence-Level Auditory Comprehension (SLAC) Treatment Program

A recent application of stimulation therapy is SLAC. Since the auditory comprehension modality is impaired to some extent in most aphasic patients, SLAC is designed to improve auditory comprehension using controlled auditory stimulation (Naeser et al., 1986). SLAC uses the retained phoneme discrimination ability found in mixed, global and Wernicke's aphasics and uses it to facilitate sentence-level auditory comprehension.
SLAC uses sets of cards with prerecorded stimuli attached. Each card plays back a single stimulus item. Patients can reinsert the stimulus card as often as they wish until they have heard the card and can consistently give a correct response choice. Since the purpose of SLAC is to train the auditory modality to focus on verbal information without visual cues, the stimuli are never written on the front of the cards. Chronic aphasic patients were found to improve with treatment initiated more than six months post-onset. SLAC can be used with mild or more severe aphasics with comprehension deficits and can be adjusted to the patient's own level.

LURIA'S THEORY

An outstanding exception to stimulation therapy during the 60's is Luria's approach which anticipated many later developments of cognitive neuropsychology. However, in Luria's view, too, facilitation and reactivation were the main goals for language rehabilitation. He postulated that multiple facilitation would eventually result in a functional reorganization of the impaired language system (Luria et al., 1969).
An application of Luria's theory can be found in Language Enrichment Therapy (LET) (Lesser et al., 1986). LET is based on Luria's theory that the reacquisition of language in aphasic patients depends on the reintegration of linguistic systems and the restoration of the interaction between language and other mental functions. Relatives of the aphasic patients were trained to

administer LET and the patient could receive the treatment at home whenever convenient. LET is a pedagogical approach rather than a therapy aimed at functional communication. LET consists of a small-step hierarchical program which involves training the patient in vocabulary, comprehension, repetition, naming, constructing sentences, reading and writing. Although the patients enjoyed the independence the therapy provided, it was found that LET did not increase the amount of time spent at home doing therapy. The improvement found in the patients tested could not be attributed to the therapy because the treatment was administered during the period of spontaneous recovery.

Behaviourist Approaches

PROGRAMMED INSTRUCTION

Under the influence of behaviouristic learning theory, the stimulation approach gradually changed into a methodological framework for behaviour modification in aphasic patients. This is usually referred to as the programmed instruction approach (e.g., Holland, 1970; see also the review by Seron, 1987). Several standards were set for therapy procedures. Each therapy program should contain defined success criteria, which must be related to the baseline performance of the patient. The course of therapy must be planned and conducted in small steps. For each step, intervention techniques are specified as well as mastery criteria for the patient's performance. Furthermore, a hierarchy of difficulties is introduced. The specific hierarchy is defined for task demands and/or training material.

Response-Contingent Small-Step Treatment

This treatment approach assumes aphasia involves a decrease in the cognitive ability to organize covert and overt responses. The Response Contingent Small Step Treatment (RCSST) teaches the patient to adopt strategies designed to enhance covert and overt symbol manipulation using reinforcement theory (Bollinger and Stout, 1976).

Used for a variety of aphasia types, this therapy provides flexibility for task difficulty depending upon the patient's success rate. RCSST accurately defines the communication tasks presented during treatment and maintains an accurate measure of the patient's performance and current communication status.

The small steps are due to the identification of task hierarchies. Each task can be increased or decreased in difficulty by manipulating a single variable. For example, to increase or decrease the difficulty of a task, cues can be removed or added. When the patient fails at a level, the task is decreased by one step in difficulty. The target step should be within the patient's range but too difficult to yield consistently correct responses. This method allows for very little recurrence of errors, as these may have a negative effect on the patient's motivation.

Audiovisual Stimulation and Direct-Production Treatment

One study compared programmed instruction with stimulation therapy. Thompson and McReynolds (1986) performed a study which attempted to overcome the previous problems in assessing the effects of the two approaches. Many of the previous studies were not controlled. Independent variables were poorly defined, making replication difficult. Thompson and McReynolds also felt that generalization effects were an important outcome of treatment and needed to be examined. The linguistic feature trained for was Wh-interrogatives using alternate methods of audio-visual stimulation and direct-production treatment. Four non-fluent agrammatic patients were tested using drawings which elicited Wh-interrogative responses for both methods of hearing. It was found that direct-production treatment was consistently more effective than audiovisual stimulation in eliciting production of Wh-interrogatives. The responses were generalized to similar contexts but not to other interrogatives or untrained stimulus situations.

Filmed Language Instruction

Another example of programmed instruction involves systematic filmed language instruction. Di Carlo (1980) tested 14 aphasic patients assigned randomly to an experimental and a control group. The experimental group was presented with training films including ten perceptual films, five thinking films and thirty language films. The control group received traditional therapy. It was found that neither the experimental nor the control group demonstrated significant gains after instruction. All patients continued to produce sequencing errors in structure and showed syntactic and semantic errors. The experimental group did improve their general communication by learning how to use situationalcues, gestures and simple lexicon substitutions.

Sentence Repetition

Frequent error correction tends to cause aphasic patients to "lose track" of their sentences during production. If speech fluency is stressed through avoidance of error corrections then the patient should become more fluent. Kohn et al. (1990) propose a treatment in which the patient is trained to repeat sentences differing along variables known to affect sentence repetition, such as the number of words, the number of syllables per word and the richness of semantic content. One patient with conduction aphasia was trained to produce sentences just beyond his ability level. Sentence difficulty increased gradually upon successful completion of sentences at the previous level. Significant improvement was found in sentence repetition and in the number of content words. The patient's ability to carry on a conversation also improved.

Loose Training

Kearns (1985) proposed a treatment technique which reinforces creative language use instead of demanding specific target responses. Results from other studies show that overly structured didactic language training, does not effectively alter communication in the natural environment. Kearns therapy combines pragmatic approaches to language rehabilitation with intervention techniques. Highly structured approaches may inhibit verbal output. Therefore, a more semantic approach which loosens response requirements may be beneficial to the aphasic patient. One Broca's aphasic was treated using "loose training". The technique of forward-chaining was used. with this procedure, the clinician presents a picture and asks the patient to describe it. For each output, the clinician expands the output from the patient and gives reinforcement. Any information produced by the patient relevant to the picture is marked as correct since response elaboration and creative language is what is encouraged. In the patient examined, moderate degree of generalization was exhibited. The program was effective in facilitating patient-initiated response elaboration.

Group therapy

In another application of the behaviourist theory, Bloom (1962) suggested that rehabilitation of aphasic patients can be effectively attained through the dynamics of group stimulus and operant reinforcement. According to Skinnerian theory, verbal behaviour occurs as a function of dependent and independent variables in an environment which includes speaker and audience behaviour. This form of group therapy does not involve dialogue learning, play acting or role taking nor has it been used to provide socialization or recreation. Instead, the patients are taught to recognize daily useful concepts, such as ordering in a restaurant, handling money etc. Reinforcement occurrs when the appropriate response is obtained. The therapy was found to be most successful for global aphasics, although all patients showed some degree of improvement in comprehension and social effectiveness. Mild aphasic patients were found to benefit more from individual therapy. Carryover to untrained situations was minimal.

Syndrome/Symptom Approach

It was not until the seventies that specific types of linguistic content and communicative goals became important for therapy planning. This went hand in hand with the neoclassical revival of the notion of syndromes and symptoms. Approaches to aphasia therapy were based on varying rationales depending on which aspect of the aphasic disorder was deemed to be the debilitating factor.

In present clinical practice, therapy planning commonly starts from the aphasic syndrome found in an individual patient (Albert et al., 1981; La Pointe, 1990; Springer, 1986). There are two main reasons for doing this, one practical and one theoretical. First, syndromes are widely used as "abbreviations" for general performance characteristics to be expected in various patients. Second, syndromes reflect combinations of symptoms that may have common underlying pathological mechanisms at which therapy should be directed.

Syndrome classification requires systematic clinical observation of the patient's verbal behaviour across several linguistic modalities and with respect to units and regularities at different linguistic levels of description (phonological, lexical, morphological, and syntactic).

SYMPTOM THERAPY

Aphasic syndromes are described as characteristic combinations of symptoms. By *symptoms* is meant either the quantification of level of performance, such as poor or good comprehension, or the qualification of erroneous utterances in production, such as recurring utterances, automatized elements, stereotypies, word-finding difficulties, semantic or phonemic paraphasias, neologisms, agrammatism (omission of grammatical elements in obligatory contexts), paragrammatism (substitution of grammatical elements), etc.

The Preventative Method

Telegraphic speech is a symptom which often develops after the onset of aphasia. Most therapies usually start after the deficit has been manifested. Beyn and Shokhor-Trotskaya (1966) believe that by recognizing the primary speech deficit it is possible to prevent the emergence of secondary symptoms. The preventative method attempts to prevent telegraphic speech from developing in nonfluent aphasia by training functional categories before spontaneous lexical words occur. Once the patients have reacquired a grammatical basis they should be able to fit in lexical categories. Twenty-five patients with deprived expressive speech but good comprehension were trained. As soon as possible after the onset of the aphasia the patients were taught the simplest possible words which could express whole ideas—i.e."no", "oh", "there", "give". Later the utterances were increased in length using pronouns, adverbs, and auxiliaries. These utterances included "I want", and "I shall", combined with predicates such as "eat" and "sleep". When spontaneous words appeared over the course of the aphasia, names of objects were gradually introduced. Telegraphic speech did not emerge in any of the 25 patients trained.

METALINGUISTIC AWARENESS OF SYMPTOMS

During the late 60's and 70's, metalinguistic tasks began to be used widely in aphasia therapy (e.g., Springer and Weniger, 1980). Through the understanding of aphasic deficits, symptoms became more predictable and therapies were designed to help the patients in becoming aware of the symptoms they were exhibiting. These tasks required the patients to make conscious decisions about phonemic, graphemic, lexical and grammatical structures, thereby activating what was left of their linguistic knowledge.

Treatment of Aphasic Perseveration (TAP) is one retraining method which treats perseveration in moderate to severe aphasics. Perseveration is often observed in aphasic patients and has been regarded as an important part of aphasic syndromes. It is proposed that treatment of perseveration rather than other linguistic errors may improve language performance (Helm-Estabrooks, et al., 1987).

The rationale behind this approach is the assumption that perseveration can be raised to a conscious level and subsequently be actively inhibited by the patient. The therapy includes a discussion with the patients as to why they tend to repeat certain words. The clinician then trains the patients not to perseverate but to give no response or to ask for help in response to a stimulus.

In a study by Helm-Estabrooks et al. (1987), three stroke patients were trained using this method. All three had moderate to severe aphasia showing perseveration. After the treatment each of the patients showed significant improvement with TAP although it may have been difficult to separate the treatment effect from spontaneous recovery, since the time post-onset was only two months.

Self-cueing

Word-finding deficits are very common in aphasic syndromes. However, cues given by the therapist can often trigger a forgotten word. One of the many problems with training a patient in the clinic is the difficulty the patient has in transferring the new skills to novel situations. Self-cueing tries to bridge the gap between the clinic and the outside world and has been shown to be quite effective in word-finding deficits.

Aphasic patients make errors which they correct when given a cue by the therapist. If patients are taught to provide themselves with linguistic cues, their spontaneous speech may improve. Clinicians must determine the errors associated with the intended response for the patient. From this a decision must be made as to which associative cues trigger the correct response. The clinician can then train the patient to produce the cues that are helpful (Berman and Peele, 1967).

A study by Huntley et al. (1986) examined various types of cues in an effort to determine if certain cues were more effective than others. Cues such as presentation of the initial syllable of a target word or sentence-completion tasks were tested. No particular cues proved better than any others but Huntley et al. found that severe aphasics benefitted from simultaneous cues while mild aphasics did not. In another study, Whitney and Goldstein (1989) tested a program aimed at training mild aphasics to monitor their speech. The results showed that even mild aphasics improved significantly by producing fewer dysfluencies.

Recently, cueing for word-finding difficulties has been achieved through the use of microcomputers. These devices have been tested as both prostheses and treatment methods. When patients show difficulties in remembering a word, they may prompt a microcomputer to provide the forgotten word (Katz, 1987).

For example the computer may display a cue such as, "Do you remember the first letter of the word?" The computer analyzes the responses to the various questions and compiles a list of probable words. The patient chooses the appropriate word and the computer synthesizes the speech sounds corresponding to the chosen word. Not only has this device been effective for word-finding difficulties but, after several weeks, patients have been able to ask themselves the same questions as those presented by the computer —without having to use the computer.

A computer can also be used to generate synthesized sound cues (Bruce and Howard, 1987). Patients who responded to phonemic cues but who had difficulty producing them spontaneously were trained using a microcomputer. When patients experienced difficulty in producing a sound they were instructed to press the corresponding letter of the keyboard. The computer would generate the appropriate sound, and the sound cue could trigger the forgotten word. This method proved to be a promising approach to the treatment of anomia. However, a few drawbacks exist with this method. The English orthographic system does not consist of a simple letter-to-phoneme correspondence. Therefore patients can make errors by choosing the incorrect letter (e.g., the phonemes /s/ or /k/ can both be represented by the letter "c"). It was also found that the computer was not as friendly as a therapist and that therapists could probably provide more cues than a microcomputer could (e.g., transition to the following vowel after initial sound cueing).

Language-Oriented Approaches

STRUCTURE-BASED

Under this rubric, several approaches are addressed. All start from descriptions and theories of normal language processing. This provides a framework in which aphasic symptoms and syndromes can be specified and consequences for therapy interventions can be formulated.

The therapeutic goal is to make the patient indirectly aware of linguistic units and regularities that are specifically lost or are available for one item but not for another. The objective here is to induce relearning processes (Poeck et al., 1977; Kotten, 1981). These techniques are based mainly on the selection of linguistic problems and material to be practiced. If necessary, they are supplemented by explicit grammatical rules. Most patients must be asked to perform the same or similar items over and over again for the performance to stabilize.

It was in the late 60's that research on aphasia increasingly made use of methods taken from descriptive linguistics and that aphasic symptoms were interpreted in terms of grammatical theories (e.g., Goodglass, 1968). Studies of linguistic behaviour in aphasia tried to determine which linguistic structures were affected and which were preserved. Based on clinical and experimental findings, hierarchies of inherent linguistic difficulties were proposed.

This structural approach soon influenced the content and method of aphasia therapies especially in Europe (Howard and Hatfield, 1987), that bore a resemblance to grammatical exercises used in textbooks for second language learning (e.g., Engl et al., 1982). Some authors call this approach didactic. And indeed, the patient is typically presented with training material that exemplifies basic units and regularities of each component of the language system.

The Sentence Construction Board

The sentence construction board (SCB) is based on transformational and structural linguistics. Proposed by Davis (1973), it is a type of language therapy directed toward discrimination, identification and production of speech sounds for the comprehension and expression of ideas. It is a small-step hierarchical program which trains patients to string words together with the help of visual cues.

The sentence construction board contains coloured lights, used as cues representing grammatical classes of words. The complexity of syntactic patterns is gradually increased as the patient improves. The light cues can be introduced and faded as needed depending on the level of complexity for which the patient is being trained. This method can be used as a flexible adjunct to the training or practicing of syntactic constructions.

Systematic Therapy Program for Auditory Comprehension Disorders (STACDAP)

Prins et al. (1989) developed a method called Systematic Therapy Program for Auditory Comprehension Disorders (STACDAP). This therapy method, although based on Schuell's stimulation approach, is based upon a strict interpretation of linguistic theory. Specifically, it is believed that oral comprehension is almost always present in aphasia. STACDAP trains subjects with auditory comprehension disorders to relearn language in a hierarchical manner starting from phonology and progressing through morphology and syntax. Responses are scored in a detailed and objective manner. For these reasons the therapy is useful for a wide variety of patients with oral comprehension deficits.

The therapy is conducted in four steps. The first step involves non-verbal tasks which train some basic cognitive and perceptual functions (e.g., picture matching). Once the patient has mastered this level, phonological tasks are introduced (e.g., matching a stimulus with one of four pictures differing in one phoneme). The next step contains lexico-semantic tasks that train the patient to understand different words grouped into semantic categories. The final step is morpho-syntactic and trains verb tenses, aspect, gender, etc.

It was found that no treatment proved to be more effective than another and that neither the STACDAP treatment nor stimulation treatment had any clear effect on recovery from aphasia.

Training Formal Structure of Language

Cubelli et al. (1988) proposed a pilot study which would teach patients with conduction aphasia to control their phonemic production by orienting their attention to the formal structure of words and sentences. Therapy proceeded in a small-step hierarchical manner. Word structure was introduced by comparing visually similar words (i.e., syllable metathesis, letter substitutions, etc.) and choosing letters which belong to particular words. Syntax was introduced by assembling sentences with words in the phonological section, comparing pictures with their corresponding sentences and giving the patient anagram sentences which had to be unscrambled.

All patients improved in linguistic skills, although this could be attributed to spontaneous recovery. (None of the patients was more than three months post-onset.)

Mapping thematic relations

This therapy, proposed by Byng (1988), has a semantic basis. It aims at improving sentence production and comprehension when thematic relations must be mapped (i.e., agent and patient) Aphasia may disrupt the ability to map thematic relations. Formal training of thematic roles may improve language abilities in aphasic patients. Two subjects with moderate to severe comprehension deficits were trained.

Therapy consisted of practicing reversible sentences using a meaning card which indicated the meaning diagrammatically using pictures and corresponding colours which indicated the reverse-role relationships and prepositions used in the sentences.

Improvement was observed in tasks where mapping of thematic roles was crucial. However, the patients learned the principles but not the strategies. Therefore the method was seen as inappropriate for the disorder.

DEBLOCKING

For each aphasic syndrome, and possibly even for each patient, significant deviations from the general hierarchy of inherent task difficulties may be expected. Some task components can be more impaired than others. Some input or output routes or central processing components may even appear to be blocked. Weigl (1961) suggested systematic deblocking as a therapy method. When a patient performed poorly on one particular task the therapist would try to find related tasks in which the patient performed substantially better. The therapist would then arrange the tasks in a deblocking chain with the relatively preserved task preceding the impaired one. This led to facilitation, i.e., the patient performed well on what he could not do before. For example, some aphasic patients are unable to name objects. However, naming becomes possible after these patients have successfully activated the target name in some other task, such as reading the written name or comprehending the spoken name by pointing to the corresponding object. Deblocking effects can also be obtained when the training includes not just one but several words from the same semantic field with randomized order from task to task. Facilitation is, however, less likely to occur when semantically unrelated words are used (Springer, 1979; Weigl, 1979). Parallel similarity effects were reported for deblocking syntactic and morphological structures (Weigl and Bierwisch, 1970). From the perspective of modern neurolinguistics, it appears that these effects are based on preactivation of units/regularities of central processing components rather than on a "deblocking" of the impaired output or input routes.

Syntax Stimulation Program

This approach is based on the assumption that patients with a syntactic disorder have impaired and inconsistent access to syntactic knowledge rather than a loss of that knowledge. Proper cues can remove blocks that interfere with the accessing process of syntactic information. This method, proposed by Helm-Estabrooks et al. (1981) was called Syntax Stimulation Program (SSP); it was later renamed the Helm Elicited Language Program for Syntax Stimulation (HELPSS) (Helm-Estabrooks, 1982).

With this method, agrammatic patients are trained to produce various sentence types. As the patients improve, the difficulty of the sentence type is increased. For example, imperative constructions are easier than transitive ones which are easier than passive ones, etc. Therapy is conducted in two steps. During the first step, the patients are asked to produce a delayed repetition of a target response. During the second step, the patients were asked to complete a story with a self-retrieved response.

After completion of the syntax stimulation program, patients showed improvement in the intonational contours of the sentences produced in spontaneous speech, and they showed some untrained variations.

Compensatory Approaches

Many researchers have attempted to rehabilitate aphasic patients by compensating for language losses rather than by retraining lost or inhibited language functions. This approach assumes that damaged brain areas responsible for particular aspects of language cannot be retrained. And, although the brain does show a certain amount of plasticity, it is assumed that in adult patients, other areas cannot relearn what the damaged area subserved. Within this framework, the only way to rehabilitate speech and language in an aphasic patient is to take advantage of what functions remain.

Communication in which language is an important (but not the only) means of communication is stressed. Non-linguistic means of communication and functional communciation using pragmatics are also used. One category of such techniques is known as Augmentative and Alternative Communication

(AAC). Historically, AAC techniques have three important roles in aphasia treatment. Firstly, these techniques can be used as a compensatory means of communication or as an alternative to spoken language. Secondly, these techniques may be used to facilitate the re-acquisition of spoken language. Thirdly, the techniques may serve as an associative "link" to enable spoken language skills to occur. During the 1970's, signs and gestures, Blissymbols, and computerized and augmentative devices were used for aphasic patients. During the 1980's, speech prostheses were included in the list of AAC devices. It soon became apparent that patients could not learn large vocabularies of alternative systems and the rate of acquisition was very slow. Generalization to natural contexts was minimal. The use of these methods declined throughout the 80's. It is now believed that, in general, AAC techniques should be used as tools for and as adjuncts to traditional language therapy (Kraat, 1990).

RIGHT-HEMISPHERE INVOLVEMENT

Many reports point to the likelihood that the right hemisphere subserves many functions including pragmatic aspects of language use, humour and visuo-spatial processing (Perkins and Lesser, this volume). Some believe that the right hemisphere also subserves latent language abilities suppressed by the dominant left hemisphere. It is assumed that if aphasia has disrupted language functions in the left hemisphere, language abilities can be tapped by compensatory communicative methods based on functions subserved by the intact right hemisphere. Recently, there has been a focus on rehabilitating aphasic patients by improving their ability to communicate rather than by emphasizing the need to produce grammatical sentences (Perkins and Lesser, this volume; see also Davis and Wilcox, 1985).

PRAGMATICS

Pragmatics is the study of the relationship between language behaviour and the contexts in which it is used. A precise description of the domain of pragmatics is still evolving (Davis and Wilcox, 1985). It has been shown that pragmatic skills are well preserved in aphasia and are less vulnerable than disruptions in the lexicon, syntax, phonology, etc. Pragmatic approaches are flexible and provide the patient with an alternative to communicating linguistically, but it must be noted that linguistic change is not to be expected. These methods should not be used alone, but should be combined with counseling and linguistic training (Holland, 1991).

Visual Action Therapy

It is believed that pantomime training rather than linguistic training may also increase communicative effectiveness by circumventing linguistic deficits (Helm-Estabrooks, 1982). Visual Action Therapy (VAT) is specifically aimed at improving the communicative abilities of global aphasics (Helm-Estabrooks,

1982). Gestures are encouraged because hand gestures require less refined motor control than do the articulatory movements required for speech. Non-orthographic stimuli are used because severe auditory comprehension deficits are usually present in global aphasia. Patients are taught to produce symbolic gestures for a visually absent stimulus. Therapy is conducted in a series of hierarchically arranged steps. The patient is taught that a line drawing represents a real object and that the object can be represented by a gesture.

Patients improved significantly in their ability to perform pantomimes with untrained objects. An unexpected improvement was also found in auditory comprehension. This can be explained by the possibility that patients may employ internal verbal monitoring during the training program. VAT also appears to improve general attentional, visuo-spatial and visual search skills. VAT may also reintegrate some conceptual systems necessary for linguistic performance. In effect, right-hemisphere functions appear to be stimulated, thereby increasing access to the latent pathways that subserve language (Fazzini et al., 1986).

Promoting Aphasics' Communicative Effectiveness

The assumptions behind the approach called Promoting Aphasics' Communicative Effectiveness (PACE) are that any aphasic patient can communicate in some way (Davis and Wilcox, 1985). Hence patients are taught to use their pragmatic knowledge and to draw on a variety of channels of communication in order to communicate effectively.

During therapy, the clinician and patient are seated at a table with a stack of stimulus cards placed on the table. The stimulus cards include picture or written descriptions and serve as topics for a series of brief conversations between the patient and clinician. Each person takes a turn selecting the topic and conveying the message to the other person. The patient is encouraged to use any means of communication (e.g., writing, gestures, etc.) to convey the message.

Patients have shown an increase in communicative effectiveness using PACE. The authors claim that the communicative focus of PACE causes patients to think less about talking while becoming more linguistically adequate in the process.

Functional Communication Therapy

Functional Communication Therapy (FCT) is another pragmatic approach which attempts to improve the patient's use of information in conducting daily activities, interacting socially and expressing current physical and psychological needs (Aten, 1986).

During FCT, the clinician alternates between listener/speaker and creates role-playing situations which require an equal amount of participation from the patient. The clinician provides cues and modeled responses while encouraging unaided responses (Aten et al., 1982). Patients are encouraged to use gestures whenever possible. The main goals are to maximize comprehension during informational exchanges and to improve the communicative effectiveness of the patient's message using several possible channels of which language is only one. The treatment setting is as natural as possible, stimulating real-life scenes and situations. Stimuli may include menus, road maps, calendars, etc. Gains have been modest but patients have shown significant improvements in functional communication abilities, although they ceased to improve once the treatment was withdrawn.

Communicative Strategies

Communicative strategies are devices that are useful for communicating a message effectively despite grammatical or lexical errors. These devices include asking non-aphasic speakers to speak slower or to simplify complicated messages. Compensatory strategies, including the use of non-verbal messages and writing, may also be used by the aphasic person. The clinician observes the communicative success of patients' interactions with people in their lives. The findings are then shared with the patients and the people important to them. Specific strategies are then trained and eventually extended to natural conversation (Holland, 1991).

Behaviour Modification to Train Requesting

Behaviour methods have been shown to produce changes in pragmatic skills in aphasic patients. Doyle et al. (1987) showed significant changes in a Broca's aphasia patient in talkativeness, inquisitiveness and conversational success. Three conversational topics were chosen, including those involving information about the individual, leisure activities and health. The clinician prompted the patient to ask for information on the topics and subtopics. Correct requests included a question morpheme, a content word and rising intonation. Grammatical accuracy was not required. Pragmatically, it is inappropriate to share information which is already known by the speakers involved. Hence the topics Doyle et al. trained the patient for contained information that was of communicative value to strangers.

Conversational Coaching

A method developed by Holland (1991) involves training the patient to use pragmatic and augmentative strategies. Conversational coaching approximates conversation and communication in the outside world. A short monologue which is slightly too difficult for the patient is prepared by the clinician. The patient, in communicating the script, is required to use the methods trained for previously. The clinician then suggests alternate ways that the patient could communicate the script. The patient then communicates this to a family member who has not heard the information in the script. The clinician coaches the patient throughout the script and encourages family members to practice the methods previously trained. The procedure is then repeated with a stranger.

TAPPING INTO THE LANGUAGE CAPACITIES OF THE RIGHT-HEMISPHERE

The right hemisphere, although non-dominant for language in most people subserves (among other abilities) knowledge and use of melody patterns, rhythm and speech prosody. It is also believed that the language capacities of the right hemisphere are suppressed by the dominant left hemisphere in normal conditions. Indeed, following left-hemisphere damage, the right hemisphere may assist and perhaps diminish the language dominance of the left hemisphere (Goldfarb and Bader, 1979). Therefore, latent language capacities of the right hemisphere may be stimulated and a reorganization of the interhemispheric processes may be increased with participation of the right hemisphere (Sparks and Deck, 1986).

Melodic Intonation Therapy

Melodic Intonation Therapy or MIT (Albert, Sparks and Helm, 1973) consists of a gradual progression of length and difficulty of melodic and rythmic tasks. Melody patterns of simple useful sentences are hummed by the clinician while the rhythm is tapped onto the hand of the patient. The patient is asked to repeat the melody patterns while hand tapping. As the level of difficulty progresses, the clinician eventually fades the stimulus, decreases the number of cues and the amount of hand tapping. The patient is then asked to delay the response. If the patient is unable to accomplish the task, the clinician introduces the sentences with the appropriate words, rhythm, melody pattern and points of stress using *sprechgesang* (spoken song) with hand tapping and cues. The patient should ultimately articulate sentences with the proper prosody.

After being treated with MIT, there has been an improvement in performance among patients who already have reasonably good expressive ability but who show restrained output. According to Albert, Sparks and Helm (1973), MIT is not effective for global and Wernicke's aphasic patients. Goldfarb and Bader (1979) claim that MIT has limited effectiveness. Since the responses are presented under artificial circumstances, it has been suggested that they should be transferred to a more relevant and unstructured context.

Laughter Therapy

Laughter therapy is compatible with the acquisition of pragmatics (Potter and Goodman, 1983). For a majority of individuals, both humour and pragmatics are assumed to be subserved by the right hemisphere; hence these aspects are presumed not to be damaged by a lesion in the left hemisphere. Since the right hemisphere is believed to be involved in the processing of affective information, it is not unreasonable to speculate that laughter may serve to heighten right-hemisphere activity in the language relearning process. The right hemisphere

Inventory and classification 19

may be performing a mediational task in language encoding by activating a new pathway to the left hemisphere (Potter and Goodman, 1983).

Since laughter is contagious and humour is known to relieve tension and is relaxing, having a relaxed and tension-free patient may also be useful in regular therapy. In one study, two subjects were given laughter therapy (Potter and Goodman, 1983). Subject A exhibited a mild to moderate expressive deficit and word-finding difficulty. Subject B exhibited severe receptive and expressive aphasia, auditory comprehension and memory deficits. The clinician introduced a laugher box claiming that it would help the patient feel happier and perform better. The patient was encouraged to laugh upon the presentation of 18 seconds of taped laughter. If ever the patient became discouraged, the tape was played and the activity continued. Regardless of the performance of the patient the tape was played again with positive reinforcement. Substantial gains were found when the laughter tape was introduced. When the tape was withdrawn, responses fell to baseline.

It was concluded that laughter therapy could be used as an effective device for mediating progress by helping to access abstract and concrete reasoning functions and by creating cognitive bases for information processing and problem solving.

NON-VERBAL COMMUNICATION

It is presumed that visual symbol processing involves the right hemisphere. It is therefore hypothesized that it may be possible to bypass the verbal modality by substituting spoken words with visual symbols. The following methods differ from those previously mentioned in that they stress non-verbal means of communication.

Blissymbolics

Blissymbols are line-drawn symbols representing basic nouns, verbs, adjectives and function words. The lexicon is large enough to allow for the specific needs of the patient, and it contains some syntactic structures. Small symbol boxes which contain both the written word and the Blissymbol are ordered according to word class, and located on a portable board. The symbols are stable in time (i.e., they must be pointed to) and are visuo-spatial. The symbols are easily intelligible for non-aphasic communication partners.

Blissymbols were used by Johannsen-Horbach et al. (1985) to rehabilitate aphasic patients when there was little chance of improvement in oral and written language. The subjects were global aphasics who showed no significant improvement in their expressive speech during conventional therapy. The patients were taught to use the symbols in appropriate situations. Relatives were encouraged to become acquainted with the system.

Each patient acquired the symbol lexicon and understood the meaning of the function word symbols. Although Blissymbols can be used as a non-verbal means of communication, the patients did not generate any linguistic structures.

Sign Language

Sign languages have been believed to be a possible alternative to spoken languages because they bypass the vocal modality. It is now known that sign languages, those used by the deaf (e.g., ASL), are structured languages and are processed in the left-hemisphere just as spoken languages are (Poizner, Bellugi and Klima, 1989). Therefore it is understandable that true sign languages have consistently failed as a replacement for language in aphasic patients (Benson, 1979).

On the other hand, a communicative system derived from American Indian Sign Language (Amerind) has been proposed as a non-linguistic communicative system. Amerind is composed of symbolic gestures and does not have a grammar. The main goal of rehabilitation is to permit the patient to communicate needs and wants. Benson (1979) claims that Amerind was accepted by severely disabled aphasic patients who failed to respond to traditional language therapies.

In a study by Coelho (1987), a severe aphasic with acceptable auditory comprehension was taught basic signs either fabricated or taken from Amerind. The patient was first taught to imitate a signed model. She was then instructed to point to a picture of the concept signed. Finally she was asked to produce the sign when presented with the pictured concept. The patient eventually acquired 47 signs but used only 16 during social interactions. The authors concluded that the total number of signs which can be acquired is related to the severity of the aphasia. It was found that sign acquisition does not guarantee communicative effectiveness even though a patient can acquire a variety of signs when a structured program is presented. One study in particular (Coelho, 1990) looked at whether signs can be generalized to untrained situations and whether combinations of signs can be made into simple grammars. It was concluded that even severe aphasic patients can acquire some single signs. Less severely impaired aphasics can acquire and learn simple grammars. Overall performance appears strongly related to the severity of the aphasia. The author also found that generalization within the experiment was not comparable to communicative use.

Pharmacotherapy

Pharmacotherapy has been used to treat many other acquired syndromes so it was thought possible that pharmacotherapy could have a use in aphasia therapy. Although past results vary in effectiveness, pharmacotherapy looks promising as knowledge of cerebral physiology becomes more complete.

MEPROBAMATE

West and Stockel (1965) studied the effects of meprobamate on recovery from aphasia. Meprobamate has tranquillizing effects, and is a muscle relaxant and anti-convulsant. Previously meprobamate studies showed improved reaction time in learning. It was thought that meprobamate could improve language skills. Patients with left cerebro-vascular lesions were tested in a double-blind procedure. No significant differences were found in the rate of progress of the two groups.

BROMOCRIPTINE

Bromocriptine is a neurotransmitter that acts as a dopamine agonist which influences limbic system function. The limbic system has been found to partially mediate language function. If some aphasic features disrupt certain neurotransmitter systems then these functions may be extremely sensitive to pharmacological manipulation.

According to Bachman and Albert (1990), dopamine may be an important neurotransmitter in language function. For example, if transcortical motor aphasia is due to reduced dopamine then pharmacotherapeutic agents may reduce the disruption caused by this type of aphasia (Albert et al., 1987).

In a preliminary study, Bachman and Morgan (1988) studied three patients and the effect that various doses of Bromocriptine had on their language performance. Each patient was given a low dose and a high dose of Bromocriptine. The patients were given a pre-test and were subsequently tested at low maintenance, high maintenance and at one month after treatment.

All patients tolerated the medication and showed improvement in naming. Subjects deteriorated to baseline after cessation of drug therapy. It was concluded that Bromocriptine does play a limited role in language improvement. A more extensive study must be conducted with a larger number of subjects and in double-blind testing.

L-DOPA

Very recently 10 patients with aphasia showed improvement of phonation and articulation after L-dopa was administered (Hocki et al. 1990).

Pharmacotherapy is a relatively new field in aphasia therapy. Future research will likely use combined or hierarchical therapeutic designs, not only with dopamine agents but also with drugs affecting other neurotransmitter systems. Only placebo-controlled, double-blind studies can adequately address the true efficacy of pharmacological agents in the treatment of aphasia. Pharmacology should be considered a potential supplement to existing aphasia therapies (Bachman and Albert, 1990).

THERAPY AIDS

Two aspects of aphasia therapy must be considered in clinical practice. These are complex factors of associated with aphasia (etiology, severity, site of lesion, prognosis) and the complex of factors associated with the patient's affective and social context (premorbid personality, support by families, friends and self-help groups, social status and profession, financial resources, etc.) The degree to which a new personal perspective can be built up and social integration can be achieved will depend on the individual's specific history of illness and handicap. Clinical regimen should consider the factors associated with the aphasia and the factors associated with the patient's social and affective context.

When a person is struck with aphasia it is not only his or her language which is affected. Other associated deficits due to brain damage include motor-sensory deficits, visual-field abnormalities, unilateral attention, apraxia, amnesia and visuo-spatial disorientation (Benson, 1979). Because a brain lesion often leaves a patient incapacitated in some way, feelings of depression and frustration are common. Proper rehabilitation takes into account more than language deficits, and there are a variety of aids which can be used as adjuncts to regular therapies.

The Social Context

The following therapy aids aim at improving the patient's social interaction abilities and promotes understanding of the aphasic syndrome and effects for both the patient and the patient's immediate social entourage.

FAMILY THERAPY

Aphasia is not just a linguistic handicap; it also interferes with the patient's communicative abilities in general. Partial sensory deprivation can be linked to reduced communicative ability. A patient's family life can be substantially changed due to disruption caused by the aphasia. Rehabilitative progress may be stunted by the disruptions at home. Family therapy improves family communication skills in order to avoid emotional problems and psychological deterioration (Wahrborg, 1989). Family therapy assumes that normal brain function depends on constantly changing stimulation. A severe reduction in either level of sensory input or its variability can disturb normal brain function and can affect behaviour. The method involves having the aphasic patient and his or her family members act out the emotional feelings connected with the trauma. With therapies aiming at developing and creating communicative skills, family therapy may prove to be very important.

LINGUISTIC ROLE PLAYING

Linguistic role-playing (Schlanger and Schlanger, 1970) as a therapy attempts to stimulate residual language and develop compensatory communicative abilities. The patients are taught to use gesture and pantomime in varying situations. For example, the aphasic patients may play themselves under nonstressful and stressful situations or play role-oriented activities in which they play someone else. Specific problems that a patient may encounter while affected with the aphasic deficit can be dealt with using psychodrama. Linguistic role playing has shown to provide some relief of frustration and anxiety apparent when communication is disrupted. The patients showed a loss of inhibition and an improvement in their ability to cope with stressful situations. Schlanger and Schlanger claim that although linguistic role playing may not necessarily improve the aphasic's linguistic structure it may act as a catalyst in releasing linguistic skills.

The Affective Context

The following therapeutic aids aim at helping the patient cope with the affective aspects of aphasia such as depression, anxiety and frustration. Laughter therapy, mentioned above, can also be used as a therapy aid by relaxing the patient and improving the mood of the therapy session.

HYPNOTHERAPY

Hypnosis has been used to sharpen memory and integrate skills learned during therapy. According to Gildston and Gildston (1986), normal consciousness is the central axis of the self which ties together and orders life experiences. The assumption behind hypnotherapy is that brain damage that results in aphasia alters the patient's concept of self which is significantly mediated by language. Confusion, anxiety and depression can interfere with language rehabilitation progress. Hypnotherapy may decrease mental anxiety because it probes the unconscious, which may be subserved by areas in the central nervous system not damaged by cerebral accident.

Very recent studies have begun to look at hypnosis and its benefits. MacFarlane and Duckworth (1990) circulated a questionnaire to clinicians asking which client groups were given hypnosis and what were the benefits. It was found that hypnosis was used to achieve deep relaxation and to reduce physical tension and anxiety regardless of the communication disorder. Hypnosis was used primarily with patients with voice and fluency problems but also with aphasic and apraxic patients.

Difficulties in implementing a hypnotherapeutic program included the client's misconception of hypnosis, other professional's minimal support for the program and the client's difficulty in attaining the hypnotic state. It is still not clear what advantages hypnotherapy may have over non-hypnotic therapies.

THE NATURAL COURSE OF APHASIA

From a neuropsychological point of view, there are three basic mechanisms of functional recovery in the brain: restitution, substitution and compensation (Singer, 1982). Each one appears to correlate in different ways with changes in linguistic behaviour at different time periods post-onset (cf. Rothi and Horner, 1983). Consequently, contents and methods of aphasia therapy vary depending on the phase of recovery the patient is in. The Aachen clinical regimen distinguishes three phases of aphasia therapy: activation, symptom specific training and consolidation (Springer and Weniger, 1980; Huberet al., 1991).

Restitution of impaired language function typically occurs during the first four weeks after the trauma and leads to full recovery in about a third of the patients. This third includes patients in whom the lesion affects regions of the brain in the neighbourhood of the perisylvian language areas. Transitory aphasia arises due to edema or bleeding into the brain tissue of the language areas. While resorption takes place, the impaired language functions become normalized. Initially, it is language comprehension which improves rapidly. Expressive language improves more gradually, changing from mutism to short passages of paraphasic and jargon-like-speech with long interruptions characterized by effortful and often unsuccessful attempts at formulating the intended message. From day to day, the language output becomes more fluent with word-finding difficulties remaining as a residual symptom. The goal of aphasia therapy in these patients is to enhance the evolution of temporarily impaired language functions. The classical stimulation approach as well as modality oriented facilitation techniques seem to be most appropriate here. In some patients with severely reduced speech output, melodic intonation therapy (MIT) works surprisingly well (Sparks et al., 1974).

Stimulation is not the only therapy method used during the acute phase. In patients with evolving global or severe Wernicke's aphasia, the therapist will try to block automatisms, perseverations, jargon and logorrheic speech from the very beginning. The overall therapeutic goal of the acute phase is to stimulate the patient by all available means to respond communicatively as appropriately as possible.

The next phase of aphasia therapy starts when stable physical conditions are reached and when an extensive clinical examination of the aphasic syndrome is possible. Now symptom specific training is provided, aiming primarily at the relearning of degraded linguistic knowledge, at reactivating impaired linguistic modalities, and at learning compensatory linguistic strategies. In other words, methods of the language-oriented approaches are applied.

During the phase of linguistic learning, a gradual functional reorganization of the impaired language system is assumed (Luria et al., 1969). This may be brought about by both substitution and compensation of impaired brain functions. It is not clear to what extent functional reorganization presupposes some intact language function of the left dominant functions (e.g., Moore, 1989).

Conscious linguistic relearning during aphasia therapy is not an absolutely necessary prerequisite for language improvement. Up to twelve months post-onset, there is also spontaneous recovery (Kertesz, 1984; Reinvang, 1984), i.e., the severity of the aphasic disturbances decreases even if there is no professional language therapy. Of course, there is almost always some amount of communicative stimulation by families, friends, physical therapists, etc.

The phase of symptom-specific training is carried through as long as comparison of post- and pre-treatment testing demonstrate substantial improvement. In many patients, treatment can be extended well into the chronic course of recovery beyond twelve months post-onset until a learning plateau is reached (see also Hanson et al., 1989).

The symptom-specific training must be complemented by a phase of consolidation. Consolidation aims at enhancing the transfer of practiced linguistic skills to everyday communicative situation (Davis, 1986; Prutting, 1982). Standard techniques are linguistic role playing (Schlanger and Schlanger, 1970; De Bleser and Weismann, 1981) and PACE therapy (Davis and Wilcox, 1985). Present therapy research tries to specify contents, scope and efficacy of these pragmatic approaches to aphasic therapy (e.g., Glindemann and Springer, 1989; Springer et al., 1991).

Another important aspect of aphasia therapy involves a total rehabilitation rather than strictly language or communicative therapy. It has been shown that a lesion to the brain in the language areas rarely affects only language. It is therefore important to implement adjuncts to aphasia therapy which focus on motor-sensory deficits, apraxia, amnesia, depression, frustration, anxiety and the maintenance of supportive and personal relationships.

CONCLUSION

A classification of the various therapeutic methods has been presented. It has been shown that there are two approaches to aphasia therapy depending on what language capacities are believed to remain after a cerebral injury. One approach is the retraining approach. It is an attempt at retraining what is lost. This field includes stimulation therapy, programmed instruction, preventative methods, self-cueing and linguistic approaches. The other approach is compensatory. By providing the patient with alternative means of communication the language functions which have been lost can be compensated for. These methods include functional communication, right-hemisphere involvement and non-verbal forms of communication.

Aphasia affects not only the patient's language but also the social and affective aspects of everyday life. Aids that are directed at these areas are thus also important, since depression and frustration are frequently concomitant effects of aphasia.

References

Albert, M.L., Bachman, D., Morgan, A., and Helm-Estabrooks, N. 1987. Pharmacotherapy for aphasia, *Neurology*, **37** (Suppl.1): 175.
Albert, M.L., Goodglass, H., Helm, N.A., Rubens, A.B. and Alexander, M.P., 1981. *Clinical Aspects of Dysphasia*. Wien: Springer
Albert, M.L., Sparks, R.W., and Helm, N.A. 1973. Melodic intonation therapy. *Archives of Neurology*, **29**: 130-131.
Aten, J.L. 1986. Functional Communication Treatment. In R. Chapey (ed.) *Language Intervention Strategies in Adult Aphasia*, Baltimore: Williams and Wilkins, 266-276.

Aten, J.L., Caligiuri, M.P., and Holland, A.L. 1982. The efficacy of functional communication therapy for chronic aphasic patients. *Journal of Speech and Hearing Disorders*, **47**: 93-96.
Bachman, D.L., Albert, M.L. 1990. The pharmacotherapy of aphasia: Historical perspective and directions for future research. *Aphasiology*, **4**: 407-413.
Bachman, D.L., and Morgan, A. 1988. The role of pharmacotherapy in the treatment of aphasia: preliminary results. *Aphasiology*, **2**: 225-228.
Basso, A., Capitani, E.,and Vignolo, L.A. 1979. Influence of rehabilitation on language skills in aphasic patients. *Archives of Neurology*, **36**: 190-196.
Benson, D.F. 1979. Aphasia Rehabilitation (editorial), *Archives of Neurology*, **36**: 190-196.
Berman, M., and Peele, L.M. 1967. Self-generated cues: a method for aiding aphasic and apractic patients, *Journal of Speech and Hearing Disorders*, **32**: 372-376.
Beyn, E.S. and Shokhor-Trotskaya, M.K. 1966.The preventative method of speech rehabilitation in aphasia, *Cortex*, **2**: 96-108.
Bollinger, R.L., and Stout, C.E. 1976. Response-contingent small-step treatment: Performance-based communication intervention. *Journal of Speech and Hearing Disorders*, **41**: 41-51.
Bloom, L.M. 1962. A rationale for group treatment of aphasic patients, *Journal of Speech and Hearing Disorders*, **27**: 11-16.
Broida, H. 1977. Language therapy effects in long term aphasia. *Archives of Physical Medicine and Rehabilitation*, **58**: 248-253.
Bruce, C. and Howard, D. 1987. Computer-generated phonemic cues: an effective aid for naming in aphasia. *British Journal of Disorders of Communication*, **22**: 191-201.
Byng, S. 1988. Sentence processing deficits: Theory and therapy, *Cognitive Neuropsychology*, **5**: 629-676.
Chapey, R., (ed), 1986. *Language intervention strategies in adult aphasia*. 2nd edition. Baltimore: Williams and Wilkins.
Code, C. and Müller, D.J. (eds.), 1989. Aphasia therapy, Sudies in disorders of communication. 2nd edition. London: Whurr.
Coelho, C.A. 1987. Sign acquisition and use following traumatic brain injury: A case report. *Archives of Physical Medicine and Rehabilitation*, **68**: 229-231.
Coelho, C.A. 1990. Acquisition and generalization of simple manual sign grammars by aphasic subjects. *Journal of Disorders of Communication*, **23**: 383-400.
Cubelli, R., Foresti, A., and Consolini, T. 1988. Reeducation strategies in conduction aphasia. *Journal of Communication Disorders*, **21**: 239-249.
Davis, G.A., 1973. Linguistics and language therapy: The sentence construction board. *Journal of Speech and Hearing Disorders*, **38**: 205-214
Davis, G.A. 1986. Pragmatics and treatment. In R. Chapey (ed), *Language Intervention Strategies in Adult Aphasia*. 2nd edition. Baltimore: Williams and Wilkins.

Davis, G.A. and Tan, L.L. 1987. Stimulation of sentence production in a case with agrammatism. *Journal of Communication Disorders*, 20: 447-457
Davis, G.A. and Wilcox, M.J. 1985. *Adult aphasia rehabilitation: Applied pragmatics*. San Diego, California: College-Hill Press.
De Bleser, R. 1987. From agrammatism to paragrammatism. German aphasiological traditions and grammatical disturbances. *Cognitive Neuropsychology*, 4: 187-250.
Di Carlo, L.M. 1980. Language recovery in aphasia: Effect of systematic filmed programmed instruction. *Archives of Physical Medicine and Rehabilitation*, 61: 41-44.
Doyle, P.J., Goldstein, H. and Bourgeois, M.S. 1987. Experimental analysis of syntax training of Broca's aphasia: A generalized and social validation study. *Journal of Speech and Hearing Disorders*, 52: 143-155.[cited in Holland, 1991]
Duffy, J.R. 1986. Schuell's stimulation approach to rehabilitation. In R. Chapey (ed.) *Language Intervention Strategies in Adult Aphasia*. Baltimore: Williams and Wilkins. Pp. 187-214.
Engl, E., Kotten, A., Ohlendorf, I. and Poser, E. 1982. Sprachübungen zur Aphasiebehandlung: Ein linguistisches Übungsprogramm mit Bildern. Berlin: Marhold.
Gildston, H. and Gildston, P. 1986. Hypnotherapy adult rehabilitation (abstract). *Folia Phoniatrica*, 38: 301
Glindemann, R. and Springer, L., 1989. PACE-Therapie und sprachsystematische Übungen - Ein integrativer Vorschlag zur Aphasietherapie. *Sprache-Stimme-Gehör*, 13: 188-192.
Goldfarb, R. and Bader, E. 1979. Espousing melodic intonation therapy in aphasia rehabilitation: A case study. *International Journal of Rehabilitation Research*, 2: 333-342.
Goodglass, H. 1968. Studies on the grammar of aphasics. In S. Rosenberg and J. Koplin (eds.) *Developments in applied psycholinguistics research*. New York: Macmillan
Hanson, W.R., Metter, E.J. and Riege, W.H. 1989. The course of chronic aphasia, *Aphsiology*, 3: 19-29.
Helm-Estabrooks, N., Emery, P. and Albert, M.L. 1987. Treatment of aphasic perseveration (TAP) program: A new approach to aphasia therapy. *Archives of Neurology*, 44: 1253-1255.
Helm-Estabrooks, N., Fitzpatrick, P.M. and Barresi, B. 1981. Response of an agrammatic patient to a syntax stimulation program for aphasia. *Journal of Speech and Hearing Disorders*, 46: 422-427.
Helm-Estabrooks, N., Fitzpatrick, P.M. and Barresi, B. 1982. Visual action therapy for global aphasia. *Journal of Speech and Hearing Disorders*, 47: 385-389.
Helm-Estabrooks, N. and Ramsberger, G. 1986. Aphasia treatment delivered by telephone. *Archives of Physical Medicine and Rehabilitation*, 67: 51-53.

Holland, A. 1970. Case studies in aphasia rehabilitation using programmed instruction. *Journal of Speech and Hearing Disorders*, 35: 377-390.
Holland, A. 1991. Pragmatic aspects of intervention in aphasia. *Journal of Neurolinguistics*, 6: 197-211.
Hocki, T., Kenhlies, M., Hofmann, R., and Haferkamp, G. 1990. Pharmacotherapie van-Stimm-und Artikulatiostörungen bei Aphasie (Pharmological Treatment of Phonatory and Articulatory Disorders), *Folia Phoniatrica*, 42: 283-287.
Howard, D. and Hatfield, F.M. 1987. *Aphasia Therapy: Historical and Contemporary Issues*. Hove: Lawrence Erlbaum Associates.
Huber, W., Poeck, K., and Springer, L. 1991. *Sprachstörungen*. Stuttgart: TRIAS.
Huber, W., Poeck, K. and Weniger, D., 1989. Aphasie. In K. Poeck (ed.), *Klinische Neuropsychologie*, 2. Auflage. Stuttgart; Theime.
Huntley, R.A. and Gonzalez, L.J., 1987. Treatment of verbal akinesia in a case of transcortical motor aphasia, *Aphasiology*, 2: 55-66.
Huntley, R.A., Pindzola, R.H., Weidner, W.E., 1986. The effectiveness simultaneous cues on Naming Disturbance in Aphasia. *Journal of Communication Disorders* 19: 261-270
Johannsen-Horbach, H., Cegla, B. and Moger, U. 1985. Treatment of chronic global aphasia: with a nonverbal communicative system. *Brain and Language*, 24: 74-82.
Katz, R.C. 1987. Efficacy of aphasia treatment using microcomputers. *Aphasiology*, 2: 141-149.
Kertesz, A. 1984. Recovery from aphasia. In L.L. La Pointe (ed.), *Aphasia and related neurogenic language disorders*, New York: Thieme.
Kearns, K.P. 1985. Response Elaboration Training for Patient Initiated Utterances. In R.H. Brookshire (ed) *Clinical Aphasiology*, 15: Minneapolis: BRK.
Kohn, S.E., Smith, K.L. and Arsenault, J.K. 1990. The remediation of conduction aphasia via sentence repetition: A case study. *British Journal of Disorders of Communication*, 25: 45-60.
Kotten, A. 1981. Aphasietherapie: Linguistisch gesteuerter Wiedererwerb der Muttersprache. In G. Peuser (ed.), *Methoden der angewandten Sprachwissenschaft*. Bonn: Bouvier.
Kushner, D. and Winitz, H. 1977. Extended comprehension practice applied to an aphasic patient, *Journal of Speech and Hearing Disorders*, 42: 296-306.
La Pointe, L.L. (ed.), 1990. *Aphasia and related reurogenic language disorders*, New York: Thieme.
Lesser, R., Bryan, K., Anderson, J. and Hilton, R. 1986. Involving relatives in aphasia therapy: An application of language enrichment therapy. *International Journal of Rehabilitation Research*, 9: 259-267
Luria, A.R., Naydin, V.L., Testvetkova, L.S. and Vinarskaya, E.N. 1969. Restoration of higher cortical function following local brain damage. In

P.J. Vinken and G.W. Bruyn (eds). *Handbook of Clinical Neurology*, **3**: 368-433.
MacFarlane, F.K. and Duckworth, M. 1990. The use of hypnosis in speech therapy: A questionnaire study, The *British Journal of Disorders of Communication*, **25**: 227-246.
Moore, W.H. 1989. Language recovery in aphasia: a right hemisphere perspective. *Aphasiology*, **3**: 101-110.
Naeser, M.A., Haas, G., Mazurski, P. and Laughlin, S. 1986. Sentence level auditory comprehension treatment program for aphasic adults. *Archives of Physical Medicine and Rehabilitation*, Pp. 393-399.
Perkins, L. and Lesser, R. 1992. Pragmatics applied to aphasia rehabilitation. In M. Paradis (ed.), *Foundations of Aphasia Rehabilitation*.
Poeck, K., Huber, W., Stachowiak, F.-J. and Weniger, D. 1977. Therapie der Aphasien. *Nervenarzt*, **48**: 119-126.
Poizner, H., Bellugi, U. and Klima, E.S. 1989. In F. Boller and J. Grafman (eds.), *Handbook of Neuropsychology*, Vol. 2. Amsterdam: Elsevier. Pp. 157-172.
Potter, R.E. and Goodman, N.J. 1988. The implementation of laughter as a therapy facilitator with adult aphasics. *Journal of Communication Disorders*, **16**: 41-48.
Prins, R.S., Schoonen, R. and Vermeulen, J. 1989. Efficacy of two different types of speech therapy for aphasic stroke patients, *Applied Psycholinguistics*, **10**: 85-123.
Prutting, C.A., 1982. Pragmatics as social competence. *Journal of Speech and Hearing Disorders*, **47**: 123-134.
Reinvang, I., 1984. The natural history of aphasia. In F.C. Rose (ed.), *Progress in aphasiology*. New York: Raven Press.
Rothi, L.J., Horner, J. 1983. Restitution and substitution: Two theories of recovery with application to neurobehavioural treatment, *Journal of Clinical Neuropsychology*, **5**: 73-81.
Schlanger, P.H. and Schlanger, B.B. 1970. Adapting role-playing activities with aphasic patients, *Journal of Speech and Hearing Disorders*, **35**: 229-234.
Schuell, H., Carrol, V. and Street, B.S. 1955. Clinical treatment of aphasia. *Journal of Speech and Hearing Disorders*, **20**: 43-53.
Schuell, H., Jenkins, J.J. and Jimenez-Pabon, E. 1964. *Aphasia in adults: Diagnosis, prognosis and treatment*. New-York: Hoeber Medical Division, Harper and Row Publishers.
Seron, X. 1987. Operant procedures and neuropsychological rehabilitation. In M.J. Meier, A.L. Benton, and L. Diller, (eds). *Neuropsychological Rehabilitation*. Edinburgh: Churchill Livingstone.
Singer, W. 1982. Recovery mechanisms in the mammalian brain. In J.G. Nicholls (ed). *Repair and regeneration of the nervous system*. Berlin: Springer.

Sparks, R.W. and Deck, J.W. 1986. Melodic intonation therapy in R. Chapey (ed.) *Language Intervention Strategies in Adult Aphasia*, Baltimore: Williams and Wilkins. Pp. 320-332.
Sparks, R., Helm, N. and Albert, M. 1974. Aphasia rehabilitation resulting from melodic intonation therapy. *Cortex*, **10**: 303-316.
Springer, L. 1979. Zur Anwendung der Deblockierungsmethode in der Sprachtherapie. In G. Peuser (Hrsg). *Studien zur Sprachtherapie.* München: Fink.
Springer, L. 1986. Behandlunsphasen einer syndromspezifischen Aphasietherapie. *Sprache-Stimme-Gehör*, **10**: 22-29.
Springer, L., Glindemann, R. Huber, W. and Willmes, K. 1991. How efficacious is PACE-therapy when language systematic training in ioncorporated?, *Aphasiology* (to appear).
Springer, L. and Weniger, D. 1980. Aphasietherapie aus logopädischlinguistischer Sicht. In Bohme, G., (ed.), *Therapie der Sprach-Sprech- und Stimmstrungen*, Stuttgart: Fischer.
Thompson, C.K. and McReynolds, L.V. 1986. Wh-Interrogative production in agrammatic aphasia: an experimental analysis of auditory-visual stimulation and direct-production treatment. *Journal of Speech and Hearing Research*, **29**: 193-206.
Währborg, P. 1989. Aphasia and family therapy, *Aphasiology*, 3, 479-482.
Weigl, E. 1961. The phenomenon of temporary deblocking in aphasia. *Zeitschrift für Phonetik, Sprachwissenschaft und Kommunikationsforschung*, **14**: 337-364.
Weigl, E. and Bierwisch, M. 1970. Neuropsychology and linguistics: Topics of common research. *Foundations of Language*, **6**: 1-18.
Weigl, E. 1979. Neuropsychologische und neurolinguistische Grundlagen eines Programms zur Rehabilitation aphasischer Störungen. In Peuser, G. (Hrsg). *Studien zur Sprachtherapie.* München: Fink.
Wepman, J.M. 1953. A conceptual model for the processes involved in recovery from aphasia. *Journal of Speech and Hearing Disorders*, **18**: 4-13.
West, R. and Stockel, S. 1965. The effect of meprobamate on recovery from aphasia. *Journal of Speech Research*, **8**: 57-62.
Whitney, J.L. and Goldstein, H. 1989. Using self-monitoring to reduce disfluencies in speakers with mild aphasia. *Journal of Speech and Hearing Disorders*, **54**: 576-586.

APPENDIX

STIMULATION APPROACHES

Basso, A., Capitani, E. and Vignolo, L.A. 1979. Influence of rehabilitation of language skills in aphasia patients: A controlled study. *Archives of Neurology*, **36**, 190-196.

Main topic: This study looks at what variables affect recovery from aphasia.
Related issues: Rehabilitation is based on two clinically observable facts: 1) Automatic-voluntary dissociation triggers deeply rooted verbal habits.
2) If patients cannot bring forth a response then they should be taught to do via an elicited response, first elicited in a more automatic way and then in more voluntary ways.
Rationale: All intentional verbal behaviour if appropriately reinforced through training, leaves a trace in the cerebral structures capable of carrying out linguistic activity. This trace increases the probability of occurrence of subsequent correct verbal behaviour.
Method: Looked at four variables: time between onset and first examination, type of aphasia (fluent versus non-fluent), overall severity of aphasia, rehabilitation between the first and second examination.
Subjects: 281 aphasic patients (162 reeducated, 119 controls), 199 men, 82 women, all but 7 right-handed.
Stimuli: Stimuli and responses involve sentences and meaningful words while the use of isolated syllables or phonemes is avoided.
Procedure: A number of 45-50 minutes of stimulation therapy.
Step 1: Assessment of the four main language modalities (oral expression, auditory verbal comprehension, reading and writing).
Step 2: Train patient using stimulus-response language exercises.
Findings: Formal language rehabilitation was not significantly different in the several groups of aphasics. Gains were more frequent in treated than in non-treated patients. Influence of treatment counterbalanced by negative effect of time since onset and overall severity of aphasia.
Claims: Relationship of type of aphasia to improvement was not significant. Type of aphasia does not play a role in prognosis; therefore all aphasic patients treated increase their chance of improvement as compared to their "spontaneous" recovery.

Broida, H. 1977. Language therapy effects in long-term aphasia, *Archives of Physical Medicine and Rehabilitation*, **58**, 248-253.

Main Topic: This study looks at whether communicative ability of aphasic patients improves when language therapy is introduced one to six years post-onset.

Rationale: The patients used in this study were well past the period of spontaneous recovery; therefore any changes in language status after therapy can be attributed to the therapy used. All language processes are interrelated neurologically at various levels and, by reactivating certain processes through treatment, reactivation of other modalities can occur.
Method: Therapy individually planned for each patient.
Subjects: 14 male patients, aged 43 to 79. All but one had lesions in the left posterior hemisphere; 10 patients showed apraxia of speech.
Procedure: 50 minutes, 3 to 45 times weekly.
Patients with apraxia of speech given audio and visual stimulation of phonemes, syllables and words. Useful words and phrases introduced and duo singing and chanting were used.
Patients with auditory comprehension deficits were required to respond gesturally to a variety of commands. Complexity was increased gradually.
Patients with deficits in ability to pantomime were given therapy to improve actions.
Treatment for word finding difficulties consisted of working at the breakdown level of training tasks.
Findings: All 14 subjects improved in at least one aspect of their communicative ability after therapy was initiated. 11 patients made notable improvement in 3 aspects.
Claims: Language therapy can improve the communicative abilities of aphasic patients even after spontaneous recovery.

Duffy, J.R., 1986. Schuell's stimulation approach to rehabilitation, In R. Chapey (Ed.), *Language Intervention Strategies in Adult Aphasia*, Baltimore: Williams and Wilkins. Pp. 187-214.

Main Topic: Outline of Schuell's stimulation approach to aphasia therapy
Rationale: Sensory stimulation affects brain activity, increases stimulation strength and frequency of firing neurons and the number of fibers activated. Repeated sensory stimulation is essential for the organization, storage and retrieval patterns in the brain. The language retrieval unit may work through patterns of excitation laid down during original learning. Nearly all aphasic patients exhibit deficits in auditory modalities. Multi-modal impairments stem from auditory functions.
Method: Employs strong, controlled and intensive auditory stimulation of the impaired symbol system as the primary tool to facilitate and maximize the patient's responses.
Procedure: Stimulation must be intensive, repetitive and adjusted to the patient's level. Each stimulus should elicit a response. The material should be varied and relevant to the patient's lifestyle. Feedback is provided when beneficial and motivating.

Findings: Cautious to confident conclusions that therapy helps most but not all patients. (Schuell has reported effectiveness for all but severe aphasics.) Spontaneous recovery may account for improvement.
Claims: Intensive controlled, auditory stimulation results in significant multimodality improvement.

Helm-Estabrooks, N., Fitzpatrick P.M. and Barresi, B. 1981. Response of an agrammatic patient to syntax stimulation program for aphasia. *Journal of Speech and Hearing Disorders*, **46**: 422-427.

Main topic: Syntax stimulation program (SSP) as a treatment for agrammatism.
Related issues: Few formal studies have measured the effects of specific syntax training program on speech production skills of agrammatics. Some neurolinguistic studies show patients have impaired and inconsistent access to syntactic knowledge rather than loss of that knowledge.
Rationale: Agrammatic patients should produce longer and syntactically more complex statements after treatment with SSP. With proper cues the blocks that interfere with the accessing process can be removed using a deblocking method.
Method: Use an audiolinguistic pattern drill and test at three intervals, pre-treatment, mid-treatment and post-treatment.
Subject: One patient, 35-year-old male, gunshot wound, 3 years post-onset, 9 months of therapy, agrammatism.
Stimuli: Sentence types in order of difficulty: imperative, transitive, imperative-transitive, Wh-interrogative, declarative-transitive, declarative- intransitive, comparative, yes-no questions.
Procedure: One half-hour daily session for 5 weeks. Each construction had multiple exemplars presented in story completion format at 2 levels of difficulty, A and B.
Level A: Patient must produce delayed repetition of target response.
Level B: Patient must complete a story with self- retrieved target response.
If patient can self-correct at 90% or better at level A then move to level B.
Findings: The patient showed greater improvement of intonational contours, more complex verb forms and a better ratio of functor to substantive words. The patient generalized, used full sentences with a close grammatical relationship to the constructions taught using SSP. Much of the patient's speech remained telegraphic especially in tension-producing situation.
Claims: Agrammatic patients can benefit from treatment using SSP.

Helm-Estabrooks, N. and Ramsberger, G. 1986. Aphasia treatment delivered by telephone. *Archives of Physical Medicine and Rehabilitation*, **67**, 51-53.

Main Topic: HELPSS delivered by telephone as an effective therapy for non-fluent agrammatic patients.

Rationale: It is sometimes difficult to establish outpatient treatment programs for patients geographically inaccessible or with transportation problems. Therapy can be administered by telephone. HELPSS is based on syntactic hierarchy of difficulty shown by agrammatic patients in controlled experiments.
Related Issues: Therapies that are highly structured and which primarily use auditory input and verbal output appear to have potential in aphasia therapy.
Method: Uses story completion format to elicit 11 sentence types accompanied by simple line drawings
Subject: One subject, age 41, 11 months post-onset, global aphasia
Stimuli: notebook containing line drawings corresponding to stories which train the sentence types.
Procedure: Clinician calls patient 3 to 4 times a week for 20 to 30 minutes. Level A: Short story ending with target sentence read by clinician, then reread omitting target sentence which must be supplied by patient. Level B: story does not contain target sentence, which must be supplied by patient
Findings: The patient completed the program successfully in 113 sessions over 34 weeks. Language skills improved as did ability to produce and understand grammatical instruction.
Claims: Appropriate aphasia rehabilitation can be conducted with success over the phone, the method is cost effective.

Kushner, D. and Winitz H. 1977. Extended comprehension practice applied to an aphasic patient, *Journal of Speech and Hearing Disorders*, **42**, 296-306

Main Topic: The effect of comprehension training on language production by an aphasic patient.
Rationale: Training in language comprehension, in contrast to training in language production may be the primary means by which language is acquired. Language production should follow once the foundations of language comprehension are acquired.
Related Issues: Stimulation training is usually defined as listening tasks. Comprehension training is distinguished from stimulation training in that it utilizes a pairwise relationship between sound and meaning.
Method: Training lexical item comprehension and production. Stimulus items systematically increase in difficulty. The patient was tested 4 times: at base-line, during treatment, when treatment was stopped at 3 months post-onset and when treatment resumed at 4 months post-onset.
Subjects: One patient, a 47-year-old male, deficiencies in verbal production and in comprehension, word-finding difficulties, limited production, first tested at one month post-onset.
Stimuli: Nineteen different common nouns represented and repeated frequently in 106 frames. Four frames on a single page; frames were gradually filled with line

drawings until all quadrants were filled. There is a possibility that the results reflect spontaneous improvement.
Procedure: Patient pointed to the correct picture after hearing the therapist produce a single isolated noun. Patient produced word when presented with picture.
Findings: At the end of treatment the patient produced short sentences. Treatment showed an increased ability to name items.
Claims: This therapy is consistent with treatment principles of stimulation therapy. Training the comprehension of specific single lexical items may generalize to the comprehension of other words and to complex structures.

Lesser, R. et al. 1986. Involving relatives in aphasia therapy: An application of language enrichment therapy, *International Journal of Rehabilitation Research*, **9**, 259-267.

Main Topic: Language Enrichment Therapy (LET) as a means of rehabilitating language in aphasic patients who have retained good comprehension skills.
Rationale: Based on theory by Luria, reacquisition of language in aphasia depends on reintegration of linguistic systems and restoration of interaction between language and other mental functions.
Related Issues: Only program aimed at providing complete and coordinated progression of language stimulation for aphasia.
Hypothesis being tested: LET may be a possible therapy for aphasic patients with good comprehension. LET may be efficacious when administered by a trained relative rather than a therapist.
Method: Pedagogic approach rather than functional communication approach. Small step hierarchical program.
Subjects: 9 patients, average age 61 years, 6 Broca's aphasics, 4 conduction, 1 Wernicke's, 1 anomic, 1 global. All had relatively good comprehension.
Procedure: 10 to 12 weekly sessions. Repeat exercise 20 times, varying the vocabulary. Each exercise involves comprehension, repetition, naming, constructing sentences, reading and writing tasks.
Findings: All therapists found LET materials suitable, liked the gradual progression for increasing the difficulty of the task and praised the multimodal nature or the therapy. Seven out of 9 showed improvement but were below 5 months post-onset; therefore could be due to spontaneous recovery.
Claims: LET is adaptable to different types of aphasia and severity level. LET administered at home does not appear to increase the amount of time spent at home doing language work. Results inconclusive because patients were less that 5 months post-onset.

Naeser, M.A., Haas, G., Mazurski, P. and Laughlin, S. 1986. Sentence level auditory comprehension treatment program for aphasia adults, *Archives of Physical Medicine and Rehabilitation*, 67, 393-399.

Main Topic: Sentence Level Auditory Comprehension (SLAC) as a means of facilitating improved sentence level auditory comprehension in mixed, global and Wernicke's aphasics.
Rationale: Auditory comprehension is impaired to some extent in most aphasic patients. It may be possible to improve auditory comprehension using controlled auditory stimulation.
Method: Train the auditory modality to focus on verbal information without visual cues.
Group 1 received SLAC for two to six months. Group 2 received SLAC plus other treatment for 1 to 8 months. Group 3 received SLAC plus other treatment for 2 to 6 months. Test before and after treatment.
Subjects: Group 1: chronic cases, 5 men, 2 women, aged 51 to 64 years, 7 months to 11 years post-onset.
Group 2: chronic cases, 5 men, aged 33 to 59 years old, 4 months to 3 years post-onset.
Group 3: acute cases, 4 men, aged 41 to 64 years old, 1.5 to 2 months post-onset.
Stimuli: Language master tape recorder and sets of cards with prerecorded stimuli attached.
Procedure: 30 minutes, 3 to 5 times weekly. Patients inserted the stimulus card as often as necessary until they could consistently give a correct response. Level 1: Phoneme discrimination
Level 2: Target word identification
Level 3: Target word identification within a sentence
Findings: Group 1: 5 out of 7 showed a good response to SLAC. Group 2: 3 out of 5 showed a good response to SLAC and other treatment programs. Group 3: 4 out of 4 showed a good response to SLAC and other treatment programs.
Claims: Chronic aphasic patients can improve with treatment initiated greater than six months post-onset. SLAC can be used with mild or more severe aphasic patients with comprehension deficits and can be adjusted to the patient's level.

BEHAVIORIST APPROACHES

Bloom, L.M. 1962. A Rationale for Group Treatment of Aphasic Patients, *Journal of Speech and Hearing Disorders*, 27: 11-16

Main Topic: Group therapy as a means of increasing functional communication ability in severly impaired aphasic patients.
Rationale: Aphasia rehabilitation can be effectively approached through the dynamics of group stimulation and operant reinforcement as provided by Skinnerian theory. Correct verbal behaviour can be reinforced through mediation of other people.
Related Issues: This group therapy is not used to provide socialization or recreation. Patients rated on their ability to use functional language rather then structural langauge.
Method: In group, recreate situation requiring appropriate verbal behaviour. The method does not involve dialogue learning, play-acting or role-taking.
Subjects: Patients with various levels of severity of aphasia were divided into groups according to their ability to communicate as estimated by their use and understanding of language in everyday situations.
Stimuli: Situations requiring daily useful concepts e.g ordering in a restaurant, handling money, etc.
Procedure: When inappropriate response was generated, negative behaviour is expected by the audience serves as reinforcement of that behaviour.
Findings: All patients showed some degree of improvement in comprehension and social effectiveness
Claims: The therapy was most successful for global aphasics. The patients could learn a behaviour in therapy but carryover is into daily situations was minimal. Mild aphasics benefit more from individual therapy.

Bollinger, R.L. and Stout, C.E. 1976. Response-contingent small-step treatment: performance-based communication intervention, *Journal of Speech and Hearing Disorders*, 41: 40-51

Main Topic: A structured behavioural method using small step programmed instruction designed to improve the speech and language of a variety of aphasic patients.
Rationale: The application of certain behavioural methodologies is effective in producing desired changes in linguistic and nonlinguistic performance. Aphasic patients require external structure through response constraints prior to the development of efficient covert programming of speech and language behaviours.
Method: Clinician guides patient through a series of task hierarchies varying the level of difficulty by manipulating a single variable. Feedback is used as operant reinforcement.
Subjects: A wide variety of brain-injured communication-disturbed patients.

Stimuli: Picture cards
Procedure: Clinician establishes a treatment hierarchy. Patient is presented with the lowest number of stimuli with the highest number of cues. The clinician increases the number and removes cues one by one as the patient improves. When patient experiences failure, the clinician drops to previous step and maintains that level until the patient is successful
Findings: Applicable to a wide range of communication disorders.
Claims: Good tool for obtaining an accurate baseline of patient performance. The development of treatment hierarchies has yielded important treatment information applicable to patients with dysarthria, apraxia and aphasia.

Cubelli, R., Foresti, A. and Consolini, T. 1988. Reeducation strategies in conduction aphasia, *Journal of Communication Disorders*, **21**: 239-249

Main Topic: An oriented reeducation strategy which teaches patients with conduction aphasia to control phonemic production.
Rationale: Conduction aphasia patients should learn to control the phonological expressive deficits without resorting to "lexical strategies" which often result in the verbal paraphasias.
Method: Metalinguistic judgement of phonological and syntactical structure of words and sentences.
Subjects: FC: 59 year old male, 1 month post-onset, phonemic jargon in all verbal productions.
PB: 50-year-old male, 3 months post-onset, phonemic paraphasias.
IS: 59-year-old female, 1 month post-onset, phonemic paraphasias.
Procedure:
4 times a week for 45 minutes
Method is the same for each step. Patient chooses from multiple choices of visually similar alternatives, patient must read and repeat choice aloud then read and repeat all three stimuli. The complexity of the task is gradually increased.
Step 1: Given picture, must match corresponding written word among visually simular alternatives.
Step 2: Patient must match scenes to written sentences.
Step 3: Patient must compose sequences of words by putting letters together.
Step 4: Given an anagram sentence, patient must guess meaning, arrange and real aloud, then repeat without reading.
Findings: All patients improved their linguistic skills and increased their self-confidence. Spontaneous recovery not the only factor in improvement of the patients since improvement only found in what was trained for.
Claims: The study was not adequately controlled and failed to account for all variables that influence recovery from aphasia. The study does provide a framework for an oriented reduction strategy.

Di Carlo, L. 1980. Language recovery in aphasia: effect of systematic filmed programmed instruction, *Archives of Physical Medicine and Rehabilitation*, **61**: 41-44

Main Topic: Systematic, graded filmed language instruction based on modern linguistic principles and theory as a means to improve aphasic patient's communicative behaviour.
Rationale: Programmed instruction which is based on modern learning theory is designed to modify the aphasic person's communicative ability.
Method: Subjects given a pre-test, mid-test and post-test. Subjects moved on to next film once 80% success was achieved.
Subjects: 14 male veterans between the ages of 32 and 69, CVA, randomly assigned to a experimental and a control group, 7 in each group. All received speech therapy but the experimental group in addition participated in filmed programmed instruction.
Stimuli: 10 perceptual training, 5 thinking training and 30 language training films. The language films were constructed for language development for the deaf based on a framework of modern linguistic learning principles and theory.
Procedure: Perceptual and thinking training films used for orientation and practice. The language films trained language using modern linguistic learning principles and theory.
Findings: No significant gains in either group in the formal parameters of communication, lexicon, syntax or semantics.
All patients continued to produce sequencing errors in phrase structure, syntax errors of omission, telegraphic productions and errors in semantic specificity.
Patients did improve general communication by learning how to capitalize on and use situational cues, gestures and lexicon substitutions.
Claims: Modification of the number and category of subjects and the material used may contribute to more positive results in the future.

Kearns, K.P. 1985. Response elaboration training for patient initiated utterances. In R.H. Brookshire (ed.), *Clinical Aphasiology Conference Proceedings*, Minneapolis: BRK.

Main Topic: "Loose Training" as a reinforcer for creative language in aphasic patients.
Rationale: Treatment techniques which reinforce creative language use instead of demanding specific target responses may facilitate generative responding for aphasic patients.
Related Issues: Overly structured, didactic language training does not effectively alter communication in the natural environment. Pragmatic approaches to language rehabilitation can be combined with intervention techniques as an effective communication therapy.
Method: 30 line drawings divided into three sets of 10 items each. Two sets used for training, one set for probing generalization. Content words used by the

patient in describing the picture do not have to be depicted on the stimulus item to be counted, as long as the information conveyed was clearly relevant to the stimulus picture.
Subject: One 50-year-old male, 3 years post-onset, severe Broca's aphasia with mild to moderate apraxia.
Stimulus: 30 black and white line drawings depicting unambiguous transitive and intransitive verbs.
Procedure: Training conducted 3 times weekly. Clinician presented stimulus and asked patient to describe the picture. The technique of forward-chaining used. For each output, the clinician expanded the output from the patient and gave reinforcement. Clinician asked as Wh-question relevant to the action. Patient gives response. Clinician asked patient to repeat the full sentence and offers reinforcement.
Findings: A moderate degree of generalization was exhibited by the patient. The program was effective in facilitating patient-initiated response elaboration
Claims: Highly structured approaches may inhibit verbal output. A more semantic approach which loosens response requirements appears to be beneficial.

Kohn, S.E., Smith, K.L. and Arsenault, J.K. 1990. The remediation of conduction aphasia via sentence repetition: A case study. *British Journal of Disorders of Communication*, **25**: 45-60.

Main Topic: Sentence repetition as an approach to remediating conduction aphasia.
Rationale: By basing treatment on sentence repetition, hope to reinforce the prosodic experience of producing complete sentences. Attempts at error correction causes tendency to "derail". If one focusses on speech fluency rather than on accuracy of word production, the patient should become more fluent.
Method: Gradual increase in difficulty of sentence input to patient.
Subject: CM, 74 years old, conduction aphasia, numeric paraphasias, 7 months post-onset.
Procedure: Pre- and post- test of sentence repetition using 30 sentences differing along variables known to affect sentence repetition such as the number of words, the number of syllables per word and the richness of semantic content.
New set of sentences introduced each week for daily repetition.
Sentences used just beyond patient's initial repetition ability, difficulty increased gradually.
Findings: Significant improvement in sentence repetition and number of content words. The patient generalized to other speech contexts. The patient's ability to carry on a conversation improved.
Claims: Sentence repetition appears to be effective in remediating conduction aphasia.

Thompson, C.K. and McReynolds, L.V. 1986. Wh-interrogative production in agrammatic aphasia: An experimental analysis of auditory-visual stimulation and direct-production treatment. *Journal of Speech and Hearing Research*, **29**: 193-206.

Main Topic: Comparison of auditory-visual stimulation treatment versus direct production-treatment examining the acquisition and generalization of Wh-interrogatives in agrammatic aphasic patients.
Rationale: Agrammatic patients may lack the linguistic behaviours necessary to initiate conversation and to request information. Training wh- interrogatives may allow patients to express their intentions more effectively in natural situations.
Method: Alternating treatment design with a multiple-baseline design across behaviours and across subjects. Four Wh-interrogatives (what, where, who and why) trained using one or both treatments.
Subjects: Four patients, 1 woman and 3 men, 35 to 58 years old, non-fluent agrammatic aphasia with good comprehension, all unable to produce grammatically complete questions.
Stimuli: 30 4 x 5 inch, black and white drawings used to elicit the four Wh-interrogative constructions.
Procedure: Direct-production treatment: therapist presented a picture stimulus and asked the patient to ask a question about the picture. Positive reinforcement used if patient gave correct response. Auditory-visual treatment: A picture stimulus card and an orthographic stimulus card presented to the patient, patient asked to listen to the therapist repeat the stimulus word slowly, then the patient is given the opportunity to produce the target word.
Findings: Direct-production treatment was consistently more effective than auditory-visual stimulation in facilitating Wh-interrogatives. Once a pattern for question production is learned, it can successfully be used in untrained contexts. Training effects did not generalize to untrained interrogative forms. Stimulus generalization to elicited-language samples was not significant.
Claims: Direct-production treatment is more effective than auditory-visual stimulation. Generalization may not be a natural process resulting in behaviour change.

SYNDROME APPROACHES

Beyn, E.S. and Shokhor-Trotskaya, M.K. 1966. The preventative method of speech rehabilitation in aphasia, *Cortex*, **2**: 96-108.

Main Topic: Preventing telegraphic speech before it emerges in patients with motor aphasia.
Related issues: Most other therapies usually start after the deficit has been manifested. There is a definite interdependence between different aspects of

speech which is formed in the course of ontogenesis. Aphasia may cause deprivation of the dynamic verbal function at early stage.
Rationale: Might be able to prevent the appearance of some secondary speech defects of aphasia which up to now seemed inevitable by recognizing the primary speech deficit. Should regulate the use of content words by excluding nominative words because they would only aggravate the pathology of inner speech. Since content words appear automatically training the patient for function words should lead to less telegraphic speech.
Method:
Subjects: 25 patients, motor aphasia, cerebral vascular disease, soon after stroke, all deprived of expressive speech, comprehension relatively intact, no apraxia, gross disturbances of reading and writing.
Procedure: Select simplest possible words which can function as a whole sentence—e.g., "oh", "ugh", "no", "there!" etc. Complicated lexical composition of sentences with pronouns, adverbs, auxiliaries, modals, adverbs of time, compound predicates etc. As soon as spontaneous words begin to appear, introduce names of objects gradually. Applied stimulation methods and disinhibited speech function by using automized forms of speech (singing, counting, reciting verse, etc.).
Findings: Telegraphic style of speech, inevitable with other methods, did not emerge.
Claims: Not only does the preventative method stimulate the development of speech but helps patients to recover their comprehension of the speech of others.

Bruce, C. and Howard, D. 1987. Computer-generated phonemic cues: an effective aid for naming in aphasia. *The British Journal of Disorders of Communication*, 22: 191-201.

Main Topic: Microcomputers as an aid to generate phonemic cues either as prosthesis or treatment.
Rationale: Patient's naming does not permanently improve but is only better in the presence of a cue provided by a therapist. Should be able to teach patient to overcome this problem through self-generated cues.
Method: Computer converts letter into sounds.
Stage 1: Train patient to use the aid.
Stage 2: Evaluate the success of the aid either as prosthesis or as a treatment. The patient is tested twice, once using the aid, once without the aid.
Subjects: 5 subjects with Broca's aphasia, non-fluent with word-finding difficulties, at least 6 months post-onset.
Stimuli: 150 drawings from "Cambridge Pictures"; each word's initial phoneme represented by a single letter.
Computer: Apple II digitized speech sound of nine consonants.
Procedure: Patient must name 100 pictures from a set of 50 training drawings and 50 test drawings.
5 sessions: patient to taught to press letters on a keyboard to generate a cue.

Findings: All subjects learned and enjoyed using the aid. Four or the patients showed better naming and were also able to generalize to untrained pictures.
Claims: This is a promising approach to treatment of word-retrieval difficulty in aphasia. Treatment which may cause long-term improvement. Could complement other approaches of word-retrieval problems among aphasic subjects.
Drawbacks: The computer is not as friendly as a therapist and the sound quality was low causing confusion between phonetically similar sounds. The English orthographic system is more complex than the system allows. Certain sounds require sequences of letters ("ch") or certain letters may represent more than one sound ("c" could be /s/ or /k/).

Helm-Estabrooks, N., Emery, P. and Albert, M.L. 1987. TAP: A new approach to aphasia therapy, *Archives of Neurology*, **44**: 1253-1255.

Main topic: Using TAP (Treatment of Aphasic Perseveration) as a means of preventing perseveration in aphasic patients.
Related issues: No previous study has focused on treatment of perseveration as an approach to language rehabilitation.
Rationale: Perseveration is frequently observed in aphasic patients and may be a major deterrent to evaluation and rehabilitation of the aphasic patient. If perseveration can be raised to a conscious level then it may be actively inhibited by the patient allowing for a correct, non-perseverative response and improved language performance.
Method: Single subject experimental design. Alternate TAP with traditional program
Subjects: 3 subjects, all moderate to severe aphasics.
Case 1: 55-year-old male, transcortical sensory aphasia, moderate persevera- tion.
Case 2: 58-year-old male, non-fluent with impaired repetition and naming, severe perseveration, auditory comprehension moderately impaired.
Case 3: 49-year-old male, right-handed, conduction aphasia, verbal output of overlearned phrases, poor repetition and naming, severe perseveration, auditory comprehension moderately impaired
Procedure: Therapist ordered semantic categories from weakest to strongest depending stimuli which elicit less perseveration to determine the order of presentation during TAP treatment. The stimulus pictures were shown. If patient showed no response he or she was given up to three cues. The therapist discussed with patient why he or she perseverated. The therapist encourages the patient to give no response or to ask for help instead of perseverating.
Findings: All 3 patients improved with TAP; standard treatment did not result in improvement.
Claims: May be difficult to separate treatment effects from spontaneous recovery (only 2 months post-onset) but when TAP is compared with standard aphasia therapy, TAP proved to be considerably better.

Huntley, R.A., Pindzola, R.H. and Weidner, W.E. 1986. The effectiveness simultaneous cues on naming disturbance in aphasia. *Journal of Communication Disorders*, **19**: 261-270.

Main Topic: Examines different combinations of cues and evaluates the influence of the severity of the naming disturbance on naming.
Rationale: Simultaneous prompts may stimulate naming pathways. Redundancy arouses the cortical connections in such a way as to exceed the threshold of lexical access.
Method: Confrontation naming task, two conditions included (1) no cues presented, (2) simultaneous cues presentation.
Condition 1 provided a baseline measure.
Patients divided into mild and severe groups.
Subjects: 12 males, 4 females with left anterior lesions, average age 63 years, average time post-onset 29 months. All had adequate comprehension but word-finding difficulties.
Stimuli: Visual cues arranged on a single page with corresponding photo.
96 black and white photos used (more realistic than line drawings)
Cue types: 1) initial syllable 2) sentence completion 3) printed word revealed letter by letter 4) word spelled aloud 5) multiple choice of three semantically or phonologically related words (including target word).
Procedure: Presented random photos, each cue type presented 12 times per subject. All photos randomized and paired with a cue type.
Findings: Severe aphasics improved after simultaneous cueing. Small percentage of improvement for mild aphasics. No significant difference between mild aphasia with no cueing and severe aphasia when prompted.
Claims: No particular simultaneous cue proved better than another. Simultaneous cue combinations are facilitative of word recall in aphasic patients. Severe anomic patients would benefit from concurrent cue presentations early in treatment process, rather than consecutive combinations of single prompt. Simultaneous cues are to be used in initial stages of clinical management to enhance patient success in naming. Self-cueing is essential in facilitating independence in naming process.

Katz, R.C. 1987. Efficacy of aphasia treatment using microcomputers, *Aphasiology*, **1**: 141-149

Main Topic: Using microcomputers as an aid for verbal production.
Rationale: A programmable computer may be used as a compensatory communicative device and have a therapeutic function for patients with word-finding difficulties.
Method: Patients used a small, portable device and use it in actual communicative situations.
Subjects: Patients with dysnomia.

Procedure: The computer provided the patient with a series of questions designed to help the patient identify the forgotten word. The computer analyzed the patients responses and displayed a list of the possible words. When the patients recognized the word, a key was pressed and the computer produced the word via synthesized speech.
Findings: After several weeks, the patients asked themselves the same questions without using the computer prompts.
Claims: This microcomputer is an effective device which trains patients to provide their own cues when word-finding difficulties arise.

Whitney, J.L. and Goldstein, H. 1989. Using self-monitoring to reduce dysfluencies in speakers with mild aphasia. *Journal of Speech and Hearing Disorders*, **54**: 576-586

Main Topic: Treatment program designed to train mildly aphasic speakers to monitor dysfluencies in connected speech.
Rationale: What breaks the grammatical strings of aphasic speakers is not grammatical errors but multiple interruptions and revisions. A delay strategy would provide time for word-retrieval and improve connected speech.
Related Issues: Patients with mild aphasia have more hesitations, are less fluent.
Method: Baseline measure using a memorable experience description. From a baseline description, the most frequent dysfluency became the first target behaviour for self-monitoring training and treatment.
Subjects: 3 males with single unilateral left-hemisphere lesion causing mild aphasia, 3 months post-onset.
Controls: 3 normal men matched for each subject for age, education level and language.
Procedure: 14 sessions of 30 to 45 minutes in clinician's office
Step 1: Patient given picture and asked to describe it. The clinician identifies each occurrence of the target behaviour for the patient.
Step 2: Patient is instructed to listen to his or her speech on tape and press the counter each time a dysfluency is heard.
Step 3: Clinician asks patient to self-monitor during an actual picture description.
Stimuli: 40 colour reproductions of Norman Rockwell posters.
Findings: Treatment effect found although many dysfluencies not accurately monitored by patients. All patients showed immediate reduction in frequency of target behaviour when self-monitoring introduced. Subjects 1 and 2 showed simultaneous reduction in untargeted dysfluencies and a clear generalization to other tasks. All patients promoted a "delay strategy" where the rate of syllables per minute dropped and silent pauses increased. Other listeners noticed a difference in the patients' speech.
Claims: Self-monitoring treatment program was simple, effective and efficient in reducing dysfluencies. Self-monitoring may allow for the development of a

more effective word-retrieval strategy. Further investigation is needed for efficiency claims.

LANGUAGE ORIENTED APPROACHES

Byng, S. 1988. Sentence processing deficits: theory and therapy, *Cognitive Neuropsychology*, 5: 629-676

Main Topic: A cognitive neuropsychological approach which trains mapping thematic relations in aphasic patients.
Rationale: Deficits are observed stemming from impairments in the procedures which map thematic relations in sentence comprehension and production. One specific mechanism which maps thematic roles and grammatical relations may be a central mechanism common to all modalities of input and output. When the comprehension deficit is due to impaired mapping, the hypothesis predicts parallel deficits in comprehension and production.
Method: Treatment for Subject 1: Performance is monitored and practiced at home by the patient. Training and practice of prescribed exercises for thematic relations in written locative sentences.
Treatment for Subject 2: Since impairment more severe than BRB, cues were of no use to JG. Therapy instead aimed at improving sentence comprehension and encouraging perception of thematic roles related to their position in the sentence.
Subjects: Subject 1: BRB, 41 years old, 5 years post-onset, moderate to severe expressive dysphasia, mild receptive dysphasia, poor at comprehending abstract words.
Subject 2: JG, 55 years old, 4 1/2 years post-onset, Broca's aphasia, severe auditory comprehension impairment.
Stimuli: 60 sets of pictures, meaning cards with and without colour cues. Diagrams on meaning cards explain the meaning of the preposition using colour correspondences.
Procedure: 4 locative prepositions trained.
Treatment 1 for BRB: 2 pictures presented, one correct, one reverse role. Patient must select correct picture, clues given through a "meaning card" which diagrammatically (e.g. through colour cues) explains the meaning of the sentence containing the preposition corresponding to the pictures.
Treatment 2 for BRB: a) picture word matching therapy written word and four pictures, selects picture then checks response. b) dictionary therapy: given synonyms in a list, must make up synonyms using a dictionary.
Generalization tested with a synonym judgement task (deciding whether two words were synonyms or not).
Treatment for JG: given pretraining to learn how to use cards in isolation. Then trained with treatment 1 given to BRB.
Findings: BRB improved significantly. BRB could produce two and three argument structures where previously could produce none. Untreated aspects did

not improve therefore recovery could not be accounted for by spontaneous recovery.
JG showed results no better than chance after 6 weeks of therapy. JG did show improvement on comprehension of simple reversible active sentences but not reversible locatives.
Claims: The patients learned the principles, not just the strategies of mapping thematic roles onto grammatical relations.
BRB appears to have reactivated a mapping mechanism available but not being used fully until recalled by therapy. Dictionary therapy more useful than picture matching therapy.
JG appears to have had to relearn how to map thematic relations thus accounting for the amount of time needed to accomplish the tasks.

Davis, G.A. 1973. Linguistics and language therapy: The sentence construction board, *Journal of Speech and Hearing Disorders*, **38**: 205-214

Main Topic: The Sentence Construction Board (SCB) as a means to facilitate therapy involving sentence structure.
Rationale: Linguistics helps the clinician know what to teach. Developed from basic linguistic principles (structural and transformational grammar), the SCB is based on Fitzgerald's method for teaching language to deaf children.
Method: Small step training of grammatical features. Cues are introduced and faded by the clinician.
Subjects: Patients with a variety of language pathologies including hearing-impaired children and aphasic adults.
Stimuli: The SCB is structured like a suitcase. Contains six vertical columns which represent grammatical classes of words. Rows of lights above the columns serve as cues. Lights below the columns indicate further differentiation within a category. Key words, pictures and word cards and vocabulary cards can be placed in the columns.
Procedure: The arrangement of lights indicate a particular syntactic pattern to be practiced. Lights serve as cues for grammatical classes arranged in prescribed sequences. Lessons evolve from simple to complex syntactic patterns depending on the level of success from the previous session.
Findings: Anecdotal evidence of improvement in aphasic patients.
Claims: The SCB should be used as a flexible supplement for practising structures already known by the patient.

Prins, R.S., Schoonen, R. and Vermeulen, J. 1989. Efficacy of two different types of speech therapy for aphasic stroke patients, *Applied Psycholinguistics*, 10: 85-123

Main Topic: Systematic Therapy Program for Auditory Comprehension Disorders (STACDAP) versus stimulation therapy as effective therapy for aphasic stroke patients.
Rationale: Auditory comprehension deficits are always present in aphasia. Rehabilitation of these deficits has a better prognosis than rehabilitation of production disorders because comprehension precedes production in language acquisition.
Related Issues: Few studies on aphasia therapy are controlled.
Method: Each group tested before and after treatment. Each therapy group administered the respective treatment for 5 months. The control group did not receive any treatment for that period.
STACDAP: 4 linguistic levels of difficulty: non-verbal, phonological, lexico-semantic, morphosyntactic.
Subjects: 32 subjects: 16 men, 16 women, average age 67 years, 3 months to 17 years post-onset, disorder of auditory language comprehension.
3 groups: STACDAP group, stimulation group and the control group.
Procedure: Each level contains exercises of 2 or 3 tasks increasing in complexity. Each task has 10 or 20 items, covering a wide range of linguistic elements and constructions.
Non-verbal: picture and sound matching.
Phonological: match spoken stimulus with pictures of phonologically similar words.
Lexico-semantic: comprehension tasks of stimuli grouped in semantic categories.
Morphosyntactic: match spoken sentences with pictures.
Stimulation therapy described in other sources.
Findings: No significant difference in the language gains in any one group as a result of the treatment administered.
Claims: Treatment with STACDAP yields no better results that stimulation therapy. Neither therapy has nay clear effect on recovery from aphasia. Patient selection criteria may have been inadequate. The sample sizes were small. The duration and frequency of treatment was insufficient and may be too varied to be learned in five months.

RIGHT-HEMISPHERE INVOLVEMENT

Albert, M.L., Sparks, R.W. and Helm, N. 1973. Melodic intonation therapy for aphasia, *Archives of Neurology*, 29: 130-131.

Main topic: Melodic Intonation Therapy through stimulation of the right-hemisphere as a treatment for non-fluent aphasia.

Related issues: Musical ability controlled by the right-hemisphere. Preserved singing ability has been noted in patients who are severely language impaired. Child aphasics do not relearn language, instead the right-hemisphere assumes responsibility for language functions.
Rationale: Training with MIT may stimulate latent language capacities of the right-hemisphere.
Method: Small-step hierarchical training of prosody. Stimulation of the right-hemisphere's latent language capacities.
Subjects: Case 1: 67-year-old male, 18 months after suffering from a stroke, good comprehension, limited output.
Case 2: 65-year-old male, 14 months after suffering from a stroke, auditory comprehension fair, output limited.
Case 3: 48-year-old female, occlusion of the left middle cerebral artery, fair auditory comprehension, limited output.
Findings: MIT produced significant improvement in the expressive ability of the patients treated.
Claims: Gains were too rapid to be accounted for by spontaneous recovery. Patients with production difficulties benefit from MIT. Patients with severe comprehension deficits do not benefit.

Aten, J.L. 1986. Functional communication treatment in R. Chapey (ed.), *Language Intervention Strategies in Adult Aphasia*, Baltimore: Williams and Wilkins. Pp. 266-276

Main Topic: Functional Communication Therapy (FCT) as a means to improve the patient's daily communicative needs.
Rationale: Aphasia is not just a disruption of language processes but rather involves disturbances of numerous para-linguistic and non-linguistic processes that are vital to successful, daily communication activities.
Related Issues: There have been limited gains using traditional language-based treatment.
Traditional Therapy:
1) Major focus is on language.
2) Clinician is role stimulator/facilitator and initiates and directs language based exchanges.
3) Goal is to improve language through stimulation of input modalities and to Improve spoken and written language by improving propositional word-retrieval and increasing length/completeness of utterances.
4) Uses convergent confrontative approaches to increase available vocabulary. 5) Treatment conducted in individual clinical settings using pictorial and/or printed stimuli, reduces emphasis on personal relevancy of lexicon or grammar.
6) Treatment is based on an analysis of individual patient's profiles of language strengths and deficits from results.
Functional Communication Therapy:
1) Major focus is on communication.

2) Clinician is alternative listener/speaker and creates situations requiring equal participation from the patient.
3) Goal is to maximize comprehension of informational exchange by enriching context of natural conversation accompanied by writing and gesture, language is only one of several possible communication channels.
4) Encourages divergent utterances or circumlocutions stimulated by general situational cues and interrogative probes, accepting related holophrastic and prosodic productions.
5) Treatment conducted in settings as natural as possible including groups and when contrived situations are employed, they stimulate real-life scenes/context and stress functional communication content.
6) Treatment based upon formal language test results but also stresses use of functional measures of communicative ability.

Aten, J.L., Caligiuri, M.P. and Holland, A.L. 1982. The efficacy of functional communication therapy for chronic aphasic patients, *Journal of Speech and Hearing Disorders*, **47**: 93-96

Main topic: This study examines the efficacy of functional communication therapy for aphasic patients.
Related issues: Currently used measures for assessing the aphasic patient appear more sensitive to speech-language performance than to changes in functional communication. Functional communication therapy is not modality bound like language-based therapy is.
Hypothesis is here that a period of functional communication therapy should produce changes in a chronic aphasic patient's performance on a test measuring functional communication ability. Traditional language test scores should not increase after a period of functional communication therapy. Improvement should be maintained after functional communication therapy is withdrawn.
Method: Role-playing tasks in selected communicative situations. Gestural responses are primary mode of communication modeled and selectively reinforced. Tested before and after treatment.
Subjects: 7 male patients, mean age of 56.0 years, left middle cerebral artery occlusion, average 97.9 months post-onset, non-fluent agrammatic aphasia with functional auditory comprehension.
Stimuli: everyday items such as road maps, menus, calendars, etc.
Procedure: 1-hour sessions twice weekly over 12 weeks. Clinician provides cues and modeled responses, encourages unaided responses. Patient is asked to facilitate communication.
Findings: Improvements in functions communication abilities as tested using the CADL test. No improvement on traditional language tests. No improvement after treatment withdrawn.
Claims: Patients trained with Functional Communication Therapy show improved functional communication abilities. Patients improved despite varying

times of onset of the aphasic syndrome; therefore improvement cannot be attributable to spontaneous recovery.

Goldfarb, R. and Bader, E. 1979. Espousing melodic intonation therapy in aphasia rehabilitation: A case study, *International Journal of Rehabilitation Research*, **21**: 333-342

Main Topic: A training procedure which enables severely affected aphasic patients to respond to questions bearing relevance to daily life.
Related issues: there has been a consistent failure of operant conditioning and language stimulation methods when treating severe aphasia.
Rationale: The right-hemisphere is dominant for music and may assist in diminishing the language dominance of the left-hemisphere. Training useful sentences through melody patterns may be an effective therapy method.
Method: Embeded short phrases or sentences in a simple, distinct melody pattern
Subjects:1 50-year-old patient, CVA, global aphasia, comprehension less impaired than production.
Stimuli:: 52 target sentences based on everyday needs.
Procedure: described previously.
Findings: Considerable improvement noted in imitation and context related responses.
Claims: Effective but limited, responses are presented under artificial circumstances, should be transferred to a more relevant, unstructured context.

Helm-Estabrooks, N., Fitzpatrick, P.M. and Barresi, B. 1982. Visual action therapy for global aphasia, *Journal of Speech and Hearing Disorders*, **47**: 385-389.

Main topic: Visual Action Therapy as an approach to facilitating gestural communication.
Rationale: Pantomime training rather than linguistic training may increase communicative effectiveness of severely impaired patients by circumventing the linguistic deficits.
Method: Hierarchically structured, 3 level program. Compared pre-treatment and post-treatment scores.
Subjects: 8 patients, mean age of 56.3 years, unilateral left-hemisphere damage, mean 46.8 weeks post-onset, marked impairment on all tasks except visual matching.
Procedure: one half hour sessions five times weekly for a mean of 6.63 weeks.
Level 1: Patient taught that line drawings represent real objects.
Level 2: Patient taught that objects and drawing can be represented by gesture.
Level 3: Patient taught to reproduce gestures in response to presentation of objects. All directives and reinforcements given non-verbally.

Stimuli: Eight uni-manual objects, coloured drawings of the objects, small drawing of each object, drawings of a person appropriately manipulating each object.
Findings: VAT patients improved significantly in their ability to perform pantomimes with untrained objects and in their ability to respond to auditory comprehension tests.
Claims: Patients may employ internal verbal monitoring during the training program. VAT may improve general attentional skills, visuo-spatial and visual search skills. VAT may also reintegrate some of the conceptual systems necessary for linguistic performance.

Potter, R.E. and Goodman, N.J. 1983. The implementation of laughter therapy as a therapy facilitator with adult aphasics, *Journal of Communication Disorders,* **16**: 41-48.

Main Topic: Humour therapy incompatible with the learning framework that the clinician attempts to establish in the language rehabilitation process, namely the acquisition of pragmatics.
Related Issues: Humour therapy used for other syndromes such as schizophrenia. Humour relieves tension and is relaxing. Causing the client to laugh by employing an uncomplicated humour-provoking device could be a very useful therapy adjunct. Laughter is contagious therefore it is possible to amuse the individual by playing an audiotaped recording of laughter.
Rationale: If humour is an adaptive mechanism, it could then be utilized as an effective device to mediate progress by opening up abstract and concrete reasoning functions and by creating cognitive bases for information processing and problem solving in persons with organic cerebral dysfunction.
Method: Comparison A-B-A single-subject design tape used as a reinforcer during therapy task.
Subjects: Subject A: 84 years old, 6 months post-onset, mild to moderate expressive language deficit, word-finding difficulty, high-level language formulation difficulties, mild dysarthria.
Subject B: 79 years old, 4 months post-onset, severe receptive and expressive language impairment, auditory comprehension and memory deficits, severe oral apraxia, unable to follow simple directives.
Stimuli: 18 seconds of taped laughter sound, recording used in order to avoid subjective interpretations of possibly humorous situations.
Procedure: Subject A: Expressed by gestures the use of ten common objects without prompting or modelling.
Subject B: Correctly imitate plosive sounds in monosyllabic words.
Clinician introduced laugh box saying it will help the patient feel happier and make the patient perform better. Plays tape and encourages the patient to laugh. If ever the patient becomes discouraged, the laughter tape was played and the activity was completed. Regardless of performance by the patient, the tape was

played again with positive reinforcement. Response data recorded during the first ten minutes of each 30 minute session.
Findings: Both patients showed substantial gains over baseline when humour/laughter was introduced into the therapy, when withdrawn, the target behaviour decelerated to baseline.
Claims: Laughter had a positive effect contributing to a pleasurable therapy atmosphere. It is a good "ice-breaker". Right hemisphere is involved almost exclusively in the processing of affective information therefore it is not unreasonable to speculate that humour/laughter stimuli may serve to heighten right-hemisphere activity in the language relearning process. Therapy is similar to MIT, the right-hemisphere may be activating a new pathway to the left-hemisphere.

Sparks, R.W. and Deck, J.W. 1986. Melodic intonation therapy. In R. Chapey (ed.), *Language Intervention Strategies in Adult Aphasia*. Baltimore: Williams and Wilkins. Pp. 320-332.

Main Topic: Melodic Intonation Therapy used for non-fluent aphasic patients.
Rationale: The right hemisphere subserves prosody. Reorganization of interhemispheric process can be achieved by increasing the participation of the right hemisphere.
Method: MIT trains the patient to sing based on the melody pattern, rhythm and points of stress of spoken language. Humming, spoken song (*sprechgesang*), unison repetition, hand-tapping used as cues to proper prosody of sentence.
Subjects: To be used with severely non-fluent aphasic patients with a distinct potential for recovery of language.
Procedure: Gradual progression and length and difficulty of tasks. "Backing up" (repeating the previous level after a failed level) used to correct errors.
Level 1: Patient and clinician simultaneously hummed and hand tapped. Clinician faded cues.
Level 2: Clinician intoned sentence with hand-tapping. Patient repeats.
Level 3: Clinician intoned sentence with hand-tapping. Patient joins in, clinician fades. Later patient repeatd after delay.
Level 4: Clinician intoned sentence with hand-tapping and sprechgesang, patient joins in. Later patient repeated after delay without hand-tapping.
Findings: MIT appears to be effective for a quarter of the aphasic population.
Claims: Recommended with no other language therapy. Other therapies may confuse the patient.

NON-VERBAL COMMUNICATION

Coelho, C.A. 1987. Sign acquisition and use following traumatic brain injury: A case report, *Archives of Physical Medicine and Rehabilitation*, **68**, 229-231.

Main Topic: Amerind as a useful tool for communication in aphasic patients.
Rationale: When other therapies fail and oral communication skills are nonfunctional, manual signing may be an alternative to oral communication.
Method: Single case training of signs.
Subject: 21-year-old female, severe aphasia, oral verbal apraxia, severe attentional and memory deficits, 8 months post-onset.
Stimuli: 20 basic signs chosen from Amerind and fabricated.
Procedure: 3-stage training procedure: imitation of sign model, pointing to picture of concept signed, producing sign when presented with pictured concept.
Findings: Subject acquired 47 signs but used only 16 in interaction.
Claims: The total number of signs learned is related to the severity of the aphasia.
Possible to acquire a variety of signs when a structured program is presented.
There is a vast gap between learning signs in response to stimuli and using it for functional communication.

Coelho, C.A. 1990. Acquisition and generalization of simple sign grammars by aphasic subjects, *Journal of Disorders of Communication*, 23: 383-400.

Main topic: Manual sign grammars as a non-verbal means of communication for aphasic patients.
Hypothesis being tested: With treatment, aphasic patients should be able to combine signs to make simple grammars and generalize these signs and grammars to untrained situations.

EXPERIMENT 1
Method: 4-stage, step-by-step training program.
Training progressed through imitation to recognition to production of signs.
Patient taught to combine the signs acquired to form a simple grammar.
Subjects: Subject 1: 67-year-old male, moderate to severe aphasic with limited production, 15 months post-onset.
Subject 2: 57-year-old male, moderate to severe non-fluent aphasia, 3 years post-onset.
Subject 3: 52-year-old male, severe non-fluent aphasia, 21 months post-onset.
Subject 4: 52-year-old male, severe non-fluent aphasia, 29 months post-onset.
Stimuli: 14 signs selected for vocabulary from AmerInd or ASL, 8 objects, 4

actions, 2 agents. Only signs which were iconic and easily adapted to one-handed production were selected.
Procedure: Training program of three 45-60 minute sessions per week. Asked to present sign combinations when presented with a picture.
Findings: Subjects 1 and 2 acquired and generalized all of the sign combinations. Subject 3 acquired and generalized only 40% of the sign combinations. Subject 4 acquired 71 % of the signs and no sign combinations.
Claims: Severe aphasic patients can acquire some single signs. Less severe aphasics can acquire and generalize simple grammars. Manual sign acquisition is strongly correlated with severity of aphasia.
EXPERIMENT 2
Rationale: Less severe aphasic subjects may be able to produce sign combinations in less structured, more spontaneous tasks.
Method: Same as in Experiment 1
Subjects: Subjects 1 and 2 from Experiment 1.
Stimuli: 18 signs from AmerInd and ASL, actions, agents, objects and adjectives (same requirements as in Experiment 1).
Procedure: 2 sessions of 45 minutes per week.
Baseline established for 3 picture scenarios, conducted generalization problems after each session. Progression from imitation, recognition and production, all signs trained until acquired.
Findings: Subjects 1 and 2 rapidly acquired vocabulary and single signs and 9 combinations. Neither subjects showed generalizations to untrained pictures. Maintenance of sign combinations was weak and inconsistent.
Claims: Generalization of signs to untrained settings does not occur without additional training or somehow structuring the environment.

Johannsen-Horbach, H., Cegla, B. and Moger, U. 1985. Treatment of chronic global aphasia with a non-verbal communication system, *Brain and Language*, **24**: 74-82

Main Topic: Blissymbolics used as a non-verbal communication system for aphasia patients with global aphasia.
Rationale: Blissymbols are visuo-spatial, stable in time and probably involve right-hemisphere processing. They may be an effective means of communication for global aphasics when the chance of oral and written communication is small.
Related Issues: Blissymbols are easily intelligible for non-aphasic communication partners. The lexicon is large enough to allow for specific needs of patients. Blissymbols contain some syntactic structures.
Method: No reinforcements or restrictions. Teach patients basic nouns, verbs, adjectives, adverbs and function words. Train patients to produce and understand simple sentences in Blissymbolics.
Relative should be acquainted with Blissymbolics

Subjects: Global aphasics who did not show significant improvement of expressive speech during conventional therapy,
Stimuli: Line symbols and equivalent words written in individual boxes,
Procedure: The patients are taught to point to the corresponding box of the word wished to be communicated.
Findings: All patients acquired the symbol lexicon, all understood the meaning of the function words symbols. Tendency to perseverate was milder with the symbols than with the spoken modality. Patients overcame the perseveration more easily with the symbols than with the words.
Claims: Blissymbolics seen as a useful tool of communication.

PHARMACOTHERAPY

Albert, M.L., Bachman, D., Morgan, A., and Helm-Estabrooks, N. 1987. Pharmacotherapy for aphasia, *Neurology*, 37: 130-131.

Main Topic: Pharmacotherapy in aphasia treatment.
Rationale: Aphasic features might disrupt certain neurotransmitter systems. Impaired initiation of speech production in transcortical motor aphasia may be due to reduced dopamine. A pharmacotherapeutic agent may reduce the disruptions.
Method: Clinical trial of pharmacotherapy using Bromocriptine.
Subject: 62-year-old male, severe transcortical motor aphasia, perseveration, naming and comprehension moderately impaired, 2 1/2 years post-onset.
Procedure: Patient received standard aphasia therapy and plateaued, Bromocriptine administed in slow increasing doses.
Findings: Language improved substantially due to reduced response latency, improvement in naming and decrease in paraphasias.
Patient deteriorated to baseline after cessation of drug therapy.
Claims: Pharmacotherapy appears to be a promising treatment for aphasia.

Bachman, D.L. and Albert, M.L. 1990. The pharmacotherapy of aphasia: historical perspective and directions for future research, *Aphasiology*, 4: 407-413.

Main Topic: Pharmacotherapy in aphasia research.
Rationale: If speech and language therapy manifest their effects through physiological changes within the central nervous system, (i.e., change in neuronal sprouting or neurotransmitter receptor density) the manipulation of the biochemical environment within which these changes occur may substantially affect the clinical outcome.
Findings: Although only a small number of scientific studies have been conducted to evaluate the efficacy of pharmacological agents in treatment of aphasia, several conclusions emerge:

1) Certain features of aphasia may be amenable to pharmacological manipulation.
2) Dopamine system may be an important neurotransmitter.
3) Only placebo-controlled, double-blind studies can adequately address the true efficacy of pharmacological agents in the treatment of aphasia.
4) Future research will likely use combined or hierarchical therapeutic designs, not only with dopamine agents but also drugs affecting other neurotransmitter systems.
Claims: Pharmacology should be considered a potential supplement to existing aphasia therapies.

Bachman, D.L. and Morgan, A. 1988. The role of pharmacotherapy in the treatment of aphasia: Preliminary results, *Aphasiology*, **2**: 225-228

Main topic: Bromocriptine as a treatment for aphasia.
Rationale: Bromocriptine (dopamine agent) is a neurotransmitter that influences limbic system function; the limbic system partially mediates language function.
Related issues: The limbic system is extremely sensitive to pharmacological manipulation.
Method: Bromocriptine was tested three times at low maintenance, high maintenance and at one month after treatment.
Subjects: 1) 62-year-old male, left frontal intracerebral haemorrhage, 3 1/2 years post-onset, transcortical motor aphasia.
2) 73-year-old male, multiple bilateral thrombic cerebral infarct, 3-4 years post-onset, mixed anterior aphasia.
3) 62-year-old male, severe Broca's aphasia, thromboembolic infarct, 1 1/2 years post-onset.
Findings: All patients tolerated the medication; patient 1 improved the most. All 3 patients showed an improvement in confrontational naming and a reduction of pauses within and between utterances.
Claims: Bromocriptine may play a limited role in language improvement. Best results are seen in transcortical motor aphasia.
Comments: The number of subjects tested was small and the tresting was not double blind.

West, R. and Stockel, S. 1965. The effect of meprobamate on recovery from aphasia, *Journal of Speech Research*, **8**: 57-62

Main Topic: Meprobamate as a medication for aphasia recovery.
Rationale: In the past Meprobamate has improved reaction time in learning and may improve langauge skills.
Method: 2 groups (medication group and placebo group), double-blind procedure.

Subjects: 29 subjects, 21 male, 8 female suffering from a cerebral vascular accidents resulting in right hemiplegia.
Stimuli: Meprobamate is a tranquilizing muscle-relaxant and anti-convulsant.
Findings: Predictions were inconsistent; no significant difference found between the experimental group and the control group.
Claims: Rate of progress found in the patients not dependent on medication.

THE SOCIAL CONTEXT

Schlanger, P.H. and Schlanger, B.B. 1970. Adapting role-playing activities with aphasic patients, *Journal of Speech and Hearing Disorders*, **35**: 229-235.

Main Topic: Role playing as a compensatory communicative strategy for aphasic patients.
Rationale: Aphasic patients become frustrated and tend to withdraw from linguistic interactions. Training patients to fully use their residual language and to develop compensatory communicative skills may help them to adjust to their changed language status.
Method: Build up patient's self-confidence through communicating in groups of other aphasic patients. Role playing used to deal with emotions of self and others. Psychodrama is used to vent frustrations.
Procedure: 1) Patient used gesture and pantomime to communicate within a group.
2) Patients acted out situation-oriented situations in which they played themselves under nonstress and stress situations.
3) Patients acted out role-oriented activities in which they plays someone else.
4) Patient led through a problem-oriented situation in which psychodrama is used.
Findings: Over the several years in which role-playing has been used, patients have felt some relief of frustration and anxiety concerning deficient communication, have experienced a loss of inhibition, have felt a strong sense of successful accomplishment and have gained insight into the problems of self and the feelings and actions of others.
Claims: Role playing does not necessarily improve the aphasic's linguistic structure but does act as a catalyst in releasing these skills.

Warhborg, P. 1989. Aphasia and family therapy, *Aphasiology*, **3**: 479- 482

Main topic: Family therapy as a therapy aid for aphasic patients.
Related issues: Aphasia is a linguistic handicap; also interferes with person's communicative ability in general, partial sensory deprivation linked to reduced communicative ability. Aphasia disrupts family life; if family life is dramatically changed it may stunt any rehabilitative progress in therapy.
Rationale: Therapy has potential to restore aphasic's relational life. 1)

Assumes the family operates as a system. 2) Self is created through relationships. 3) Identity is biologically represented in the brain. 4) Biological representations of identity depend on exchange through communication. 5) Communication breakdown interferes with ability to uphold relationship thus causing emotional disorders.
Family therapy, with systematic striving for development and creation of communicative skills, might offer an important rationale in the treatment of this obstacle.
Goal: Improve family communicative skills. Avoid emotional disorders and psychological deterioration.
Method: Therapist analyses the communicative status in the family and introduces therapeutic interventions available to the aphasic person.
Have the aphasic patient and other members act out the emotional trauma connected with the aphasic disorder.
Claims: Promising results have been reported using family therapy and related approaches.

THE AFFECTIVE CONTEXT

Gildston, H. and Gildston, P. 1986. Hypnotherapy adult rehabilitation (abstract), *Folia Phoniatrica*, **38**: 301.

Main Topic: Hypnotherapy used to encourage rapid, extensive and stable progress in selected adult patients with aphasia.
Rationale: Associated with normal consciousness is the central axis of self which ties together and orders life experiences. Brain damage associated with aphasia disrupts the integrating core, generating an altered mental space wherein the patient's concept of self (which is significantly mediated by language) is likely to be seriously fragmented.
Accompanying confusion, anxiety and depression can interfere markedly with progress.
Holistic intervention—i.e., hypnotherapy, by trained clinicians may reverse the process (at least in part) because it probes the unconscious, which may derive from locations in the central nervous system not damaged by cerebral accident.
Subliminal resources, untapped by standardized tests can often be retrieved through hypnotherapy and then ushered into conscious awareness to abet the struggle for better control over propositional symbol manipulation.
Case studies illustrate supportive use, in trance, of individualized metaphors, imagery and suggestions to sharpen memory, integrate skills and enhance the capacity for more accurate decoding and encoding.

MacFarlane, F.K. and Duckworth, M. 1990. The use of hypnosis in speech therapy: A questionnaire study, *The British Journal of Communication Disorders*, **25**: 227-246.

Main Topic: The use of hypnosis in speech therapy determined through a questionnaire study.
Rationale: No clear theoretical rationale exists although there has been a recent interest in using hypnosis with adult acquired neurological disorders.
Related Issues: Few studies look at long-term maintenance of hypnotherapy.
Method: Questionnaire asked clinicians to identify client groups with whom hypnosis was used and the main benefits of use.
Procedure: 86 questionnaires were distributed; 60% were completed, 32 contained usable information.
Findings: Few clinicians use hypnosis as a direct means of obtaining fluent speech, more use it to encourage the client's personal development.
Clinicians tend to use hypnosis to encourage clients to take more responsibility for their own treatment.
Clients complain of difficulty of attaining a hypnotic state, lack of discipline and difficulty in finding time for hypnosis.
Both clients and clinicians have misconceptions of hypnosis.
Not clear what advantages hypnotic over non-hypnotic forms of relaxation are.
Claims: The extent to which therapists perceive communication problems as influencing the client's self-evaluation may play a significant part in leading them to decide upon use of hypnosis.
Hypnosis has been very effective for adults with dysphasia and voice and fluency problems by inducing deep relaxation and providing the client with insight into positive attitude towards self and situation.
Autohypnosis helps to ensure that improvements are transferred outside the clinic.
Hypnotherapy should be used as an adjunct to other therapies.

2 Efficacy of various methods

Robert T. Wertz

DEFINING THE QUESTION

The question, "Are various methods of therapy efficacious for various types of aphasic patients?," requires elaboration and some definitions. For example, one must assume that language treatment for aphasia is, indeed, efficacious. Further, one must assume that there are "various methods" of therapy. Finally, one must assume that aphasic patients can be classified, validly and reliably, into "various types." The extent to which these assumptions can be supported will determine whether the query is an appropriate research question.

Efficacy of Treatment for Aphasia

The accummulated evidence (Basso, Capitani and Vignolo, 1979; Shewan and Kertesz, 1984; Wertz et al., 1986) indicates that language treatment for aphasia is efficacious for patients who are aphasic subsequent to a single, left hemisphere thromboembolic infarct and who receive at least three hours of treatment each week for at least five months (Wertz, 1987). For other patients, there is no empirical evidence to demonstrate language treatment for aphasia is efficacious.

Various Methods of Therapy

Certainly, aphasia treatments have been described as different. There is "traditional treatment," typically ascribed to that introduced by Schuell (Schuell, Jenkins and Jimenez-Pabon, 1964) in which all communicative modalities — auditory comprehension, reading, oral expressive language, and writing— are "stimulated and facilitated" in a stimulus- response paradigm. There is "functional" (Holland, 1977) and the closely allied "pragmatic" (Davis and Wilcox, 1985) aphasia therapy in which emphasis is placed on communicative context rather than communicative content. There are a variety of treatments designed for specific types of aphasic patients, for example, Melodic Intonation Therapy (Sparks, Helm and Albert, 1974) for Broca's aphasia and Visual Action Therapy (Helm-Estabrooks, Fitzpatrick and Barresi, 1982) for global aphasia. And, more recently, there are a variety of computerized aphasia treatments (Katz, 1986, 1987).

Types of Aphasic Patients

Whether aphasic patients can be classified into various types has been argued since the earliest interest in aphasia. Currently, there are those (Darley, 1982) who insist on aphasia without adjectives and maintain differences among aphasic patients result solely from differences in severity or the presence of a nonaphasic coexisting disorde:. Others (Goodglass and Kaplan, 1983; Kertesz, 1979) suggest that the variety in symptoms among aphasic people permit classification into specific types.

Methods for Answering the Question

If the assumptions —treatment for aphasia is efficacious, various types of treatment exist, and aphasic people can be classified into various types— are correct, then it is necessary to formulate a design or designs that will answer the question, "Are various methods of therapy efficacious for various types of aphasic patients?". Certainly, not all designs are appropriate for answering the question. And, the treatment research conducted to date does not answer the question, because the designs employed are not appropriate for answering the question.

Purpose of This Paper

The purpose of this paper is twofold: review the literature that has compared various methods of therapy for various types of aphasia and discuss designs that will answer the question, "Are various methods of treatment efficacious for various types of aphasic patients?".

THE EVIDENCE

Methods for determining the efficacy of language treatment for aphasia are, generally, two —group treatment designs and single-subject designs. Within each category are a variety of approaches. For example, group treatment designs can be single treatment group, comparison of treatments, treatment vs. no-treatment, and comparison of treatments with a no-treatment control. Single-subject designs include the A-B-A, withdrawl design; multiple baseline design; alternating treatments design; cross-over design; and changing criterion design. Representatives of both types of designs have their strengths and weaknesses, depending on the research question that is asked. The following discussion focuses on group designs.

Approximately 30 aphasia group treatment studies have been conducted. The authors of most of these efforts imply their results are "positive"; treatment for aphasia is efficacious. Few investigations, however, have met the scientific requirements necessary to demonstrate treatment results in significantly more improvement than no treatment. None has demonstrated that one type of treatment is efficacious for one type of aphasia but not for another type of aphasia. The following designs do not permit answering that question.

Single Group Designs

This design selects a group of patients, administers a pretreatment outcome measure, provides treatment, repeats the pretreatment outcome measure post-treatment, and compares pre- and post-treatment performance. Improved performance post-treatment is interpreted as demonstrating the efficacy of treatment. The weakness of this design, of course, is that it does not indicate whether improvement results from the treatment or time. Investigations by Frazier and Ingham (1920); Butfield and Zangwill (1946); Wepman (1951); Marks, Taylor and Rusk (1957); Sands, Sarno and Shankweiler (1969); and Lesser et al. (1986) employed a single group design. All reported improvement in the patients treated, but none indicated what kind of treatment is efficacious for which kinds of aphasic patients or whether treatment is efficacious.

Poeck, Huber and Willmes (1989) also employed a single group design, however they "corrected" for spontaneous recovery, "as determined by a previous investigation." If one accepts their "correction," one can accept that their treatment was efficacious. However, the design does not permit determining whether various methods of aphasia therapy are efficacious for various types of aphasic patients.

The single treatment group design will not answer the question, "Are various methods of aphasia therapy efficacious for various types of aphasic patients?". Different types of aphasic patients are combined in the single treatment group, and the treatment is either one type or multiple types that cannot be separated in the results to determine the efficacy of any specific type. Moreover, the single group design will not answer the question about the efficacy of treatment.

Comparison of Treatments Designs

The comparison of treatments design assigns patients randomly to one treatment or another, collects pretreatment performance on an outcome measure, administers the specified treatments, repeats the pretreatment outcome measure post-treatment, and compares improvement between the treatment groups to determine whether one treatment resulted in more improvement than the other. Several comparison of treatments investigations have been conducted.

Wertz et al. (1981) compared traditional, stimulus-response individual treatment with group treatment in which there was no direct manipulation of language deficits. Significant differences in improvement between the two groups were few. Those that did exist favored individual treatment. Two investigations in the United Kingdom (David, Enderby and Bainton, 1982; Meikle et al., 1979) compared treatment administered by speech pathologists with treatment administered by trained volunteers. In both studies, both treatment groups improved, but there were no significant differences between speech pathologist treated patients and volunteer treated patients. Hartman and

Landau (1987) compared conventional aphasia treatment with patient and family counseling and found no significant differences in improvement between the two groups.

The comparison of treatments design is limited to demonstrating one treatment is the same as, better than, or worse than the other. It will not demonstrate whether either treatment is efficacious. Wertz et al. (1981) attempted to address the question of efficacy by examining improvement in their individual and group treated groups after six months postonset; after significant spontaneous recovery is believed to have stopped. Significant improvement in both groups between six and 12 months postonset suggested that both treatments were efficacious.

Like the single treatment group design, the comparison of treatments design will not indicate which type of treatment is efficacious for which types of aphasic patients, and it will not demonstrate whether either treatment is efficacious. Thus, it is inappropriate for answering the question, "Are various methods of therapy efficacious for various types of aphasic patients?".

Treatment vs No-Treatment Design

The treatment vs no-treatment design constitutes the classical clinical trial designed to determine treatment's efficacy. The classical clinical trial requires random assignment of patients who meet selection criteria to treatment and no-treatment groups. Only one investigation (Lincoln et al.,1984) has met this requirement. Other investigations (Vignolo, 1964; Basso, Capitani and Vignolo, 1979) have used a self-selected no-treatment group or patients assigned to a waiting list (Hagen, 1973) as a no-treatment control group.

Lincoln et al. (1984) assigned patients randomly to treatment and no-treatment groups. Comparison of groups after 24 weeks indicated no significant differences in improvement between the two groups. Vignolo (1964) and Basso, Capitani and Vignolo (1979) compared performance between treated patients and a self-selected no-treatment group --patients who could not participate in treatment because of geographical location or who elected not to participate for other reasons. Vignolo (1964) observed that treatment had a positive influence on improvement if the treatment was initiated between two and six months postonset and if it continued for more than six months. Basso, Capitani and Vignolo (1979) reported treated patients made significantly more improvement than untreated patients. Hagen (1973) compared performance between treated patients and a no- treatment group that was placed on a waiting list for treatment. After 12 months, he observed the treated group made significantly more improvement than the no-treatment patients.

While the treatment vs no-treatment design provides evidence about the efficacy of treatment, it does not indicate which patients respond best to which types of treatment. Thus, this design is not appropriate for answering the question, "Are various methods of therapy efficacious for various types of aphasic patients?".

Comparison of Treatments With a No-Treatment Control Group

This design compares two or more types of treatment with each other and compares the treatment groups with a no-treatment control group. Typically, patients who meet selection criteria are assigned to the different treatment groups or the no-treatment group; pretreatment performance on an outcome measure is collected; the treatment groups are treated with the specified treatment and the no-treatment group is followed; at the end of the treatment trial, the outcome measure is readministered; and improvement is compared among groups. Four investigations have employed the comparison of treatments with a no-treatment control group design.

Shewan and Kertesz (1984) assigned patients randomly to one of three treatments —language oriented therapy (LOT), stimulation-facilitation therapy (ST), or stimulation therapy provided by nonspeech-language pathologists (nurses) (UNST). In addition, patients who met selection criteria but did not want treatment or who were unable to receive treatment were followed in a self-selected, no-treatment group. The LOT and ST treated patients made significantly more improvement than the no-treatment patients; LOT, ST, and UNST groups did not differ significantly; and UNST did not differ significantly from no-treatment.

Wertz et al. (1986) assigned patients randomly to clinic treatment by speech-language pathologists, home treatment by trained volunteers, and no-treatment. Comparison among groups after 12 weeks of treatment indicated the clinic group made significantly more improvement than the no-treatment group, clinic and home treatment groups did not differ significantly, and home treatment did not differ significantly from no-treatment.

Prins, Schoonen and Vermeulen (1989) assigned patients randomly to either a systematic treatment program for auditory language comprehension (STAC) or conventional stimulation therapy (STIM). A third group served as a self-selected no-treatment control--patients had reached their limits of recovery according to their speech therapists. Comparison of groups post-treatment indicated no significant differences in improvement among groups.

Katz and Wertz (1990) assigned patients randomly to three groups—computerized reading treatment, computer stimulation (nonlinguistic cognitive tasks and computer games), and no-treatment. Comparison of improvement among groups at the end of a six-month treatment trial indicated the computer reading treatment group made significantly more improvement than both the computer stimulation and the no-treatment groups. There was no significant difference in improvement between the computer stimulation group and the no-treatment group.

The comparison of treatments with a no-treatment control group design will indicate whether one treatment is superior to another and whether either or both treatments are superior to no-treatment. Thus, the design will answer the question of efficacy and the question about various treatments. It will not, however, indicate whether various methods of treatment are efficacious for various types of aphasia.

DESIGNING RESEARCH TO ANSWER THE QUESTION

None of the reported investigations on the treatment of aphasia provides an answer to the question, "Are various methods of therapy efficacious for various types of aphasia?". One might cull the data of individual patients contained in the group research in search of an answer. For example, one might divide, retrospectively, a group of treated patients into aphasic types in an attempt to determine whether one type improved more than the others when provided the specified treatment. However, there are reasons aphasic patients may or may not improve —size and site of lesion, severity of aphasia, time postonset, etc.— other than type of aphasia and type of treatment administered. Moreover, retrospective speculation differs from prospectively gathered empirical evidence in a design appropriate for answering the question. Thus, an answer to the question, "Are various methods of aphasia therapy efficacious for various types of aphasic patients?", requires an appropriate research design that will indicate one or more methods are or are not efficacious for one type of aphasia and not another. Three designs are appropriate, a single-subject alternating treatments design with replications; a group design that tests a specific treatment, blocks on aphasic types, and includes no-treatment control groups; and a group design that compares two or more treatments with a single type of aphasia and includes a no-treatment control. The former is necessary to develop and test the treatment method with various types of aphasic patients, and the latter are essential to demonstrate efficacy.

Single-Subject Alternating Treatments Design

The single-subject alternating treatments design, shown in Figure 1, will indicate which of two treatments appears most effective with a single aphasic patient. It will not indicate whether either treatment is efficacious because there is no control for time, i.e., whether improvement would have occurred without treatment. One could combine an alternating treatments design with a multiple baseline design, shown in Figure 2, to obtain some information about efficacy. For example, the alternating treatments would be applied to one set of stimuli or one behavior and a second set of stimuli or behavior would be followed in an untreated (baseline) condition. Improvement on the treated stimuli or behavior with one treatment or the other and no improvement or less improvement on the untreated stimuli or behavior could imply that the treatment was efficacious. However, the implication about efficacy is limited to the specific patient being treated and cannot be generalized to imply the treatment would be efficacious with other aphasic patients, even those with the same type of aphasia. Thus, the

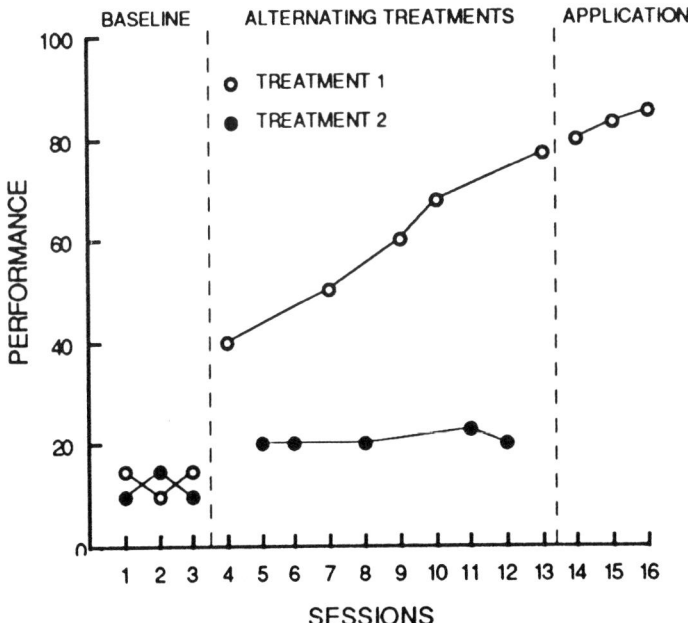

FIGURE 1. Single-subject alternating treatments design, including baseline (no-treatment), treatment (comparison of treatment 1 and 2), and application (continuation of the effective treatment) phases.

single-subject alternating treatments design is best for developing and refining a treatment; obtaining information about for whom the treatment is appropriate, through replications with other patients; and providing pilot data to justify a group study designed to test the efficacy of the treatment.

As explained by Rosenbek, LaPointe, and Wertz (1989), the alternating treatment design requires determining performance in a number of pre- treatment, baseline sessions to ensure it is stable. Treatment is initiated by assigning two treatments randomly over a number of treatment sessions. The patient receives the assigned treatments across sessions until it is apparent one treatment is more effective than the other. When this is established, an application phase— continuation of the effective treatment—is instituted to demonstrate the effective treatment results in continued or improved performance.

The alternating treatments design will provide preliminary information about which treatments are most effective for which patients. For example, Melodic Intonation Therapy (MIT) (Sparks, Helm and Albert, 1974) is reported to be an effective treatment for improving speech in Broca's aphasia. Similarly, verbal repetition is also reported to be effective in improving speech in some patients with Broca's aphasia (Wertz, LaPointe and Rosenbek, 1984). Testing which of the two methods is most effective for Broca's aphasia can be

68   Efficacy of various methods

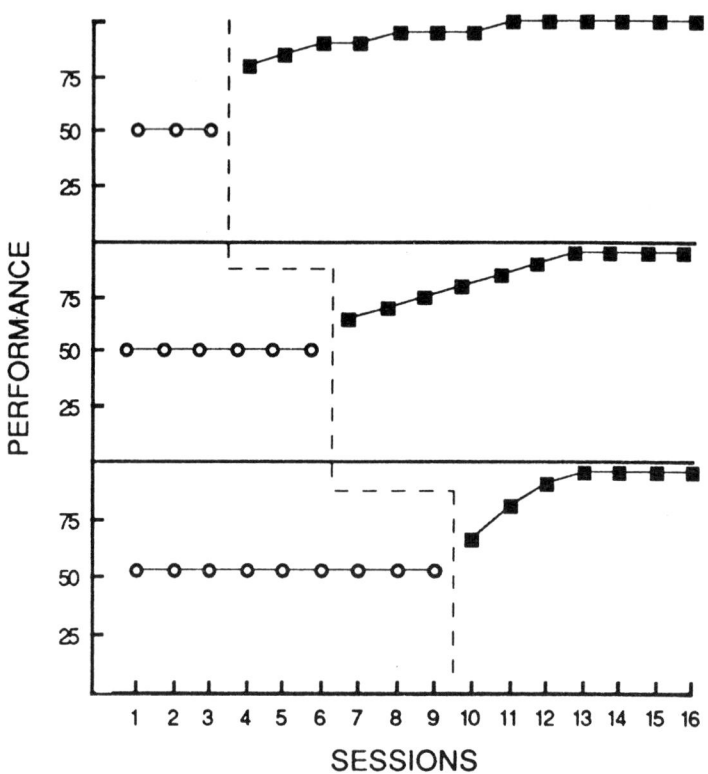

FIGURE 2. Single-subject multiple baseline design showing baseline (pretreatment) performance on all behaviors and subsequent intervention with treatment on behavior 1 in session 4, behavior 2 in session 7, and behavior 3 in session 10.

accomplished with an alternating treatments design. Initially, one selects a patient who displays unequivocal Broca's aphasia to participate in the study. After a stable baseline is established pretreatment, the two treatments, MIT and repetition, are assigned randomly across treat- ment sessions, and treatment is initiated. During the treatment phase, the patient's performance will indicate that one treatment is more effective than the other, both treatments are equally effective, or neither treatment is effective. If one treatment is more effective than the other, the less effective treatment is discontinued, and the patient continues treatment in the applica- tion phase with the effective treatment.

This type of single-subject alternating treatments design should indicate whether MIT or repetition or both is/are effective treatments for Broca's aphasia. If an untreated set of stimuli or behavior was followed in an untreated baseline, some inferences about efficacy with the treated patient can be made. Moreover, the results provide a comparison for replications. For example, if one or both methods are effective or not effective, the design can be replicated with another patient who displays Broca's aphasia. Consistent results in subsequent replications with additional Broca's aphasia patients should indicate whether MIT and/or verbal repetition are or are not effective treatment for Broca's aphasia. Similarly, replication of the design with patients displaying other types of aphasia will test MIT's and/or verbal repetition's effectiveness with aphasias other than Broca's. Eventually, single-subject alternating treatments design replications should provide sufficient evidence to justify testing a hypothesis in a group treatment design.

Comparing a Treatment With Types of Aphasic Patients and No-Treatment Groups

Group designs permit generalization to the population from which the subjects are selected on a statistical basis. Thus, with adequate sample sizes to provide acceptable power, a treatment can be compared between or among types of aphasic patients and no-treatment groups to answer the question, "Are various methods of therapy efficacious for various types of aphasia?". For example, if replications of the single-subject alternating treatments design suggest MIT is effective for Broca's aphasia but not Wernicke's aphasia, the efficacy of MIT for Broca's and Wernicke's aphasia can be tested in a group design. The null hypothesis would be, "There is no difference in response to MIT by Broca's and Wernicke's aphasia patients. Appropriate sample sizes would be estimated; selection criteria established; patients selected for the appropriate groups-- Broca's, Wernicke's, no-treatment; pretreatment outcome measures administered; the treatment trial conducted; pretreatment measures readministered post-treatment; and improvement among groups compared. This straight-forward approach should indicate whether MIT is efficacious for Broca's or Wernicke's aphasia patients.

Comparison of Treatments for a Type of Aphasia With a No-Treatment Control

A second group design that would answer the question is to select two treatments, one demonstrated by single-subject alternating treatment designs to be effective with a specific type of aphasia and one demonstrated to be ineffective; compare them with two groups of aphasic patients that represent the same type of aphasia; and include a no-treatment control group. For example, if replications of the single-subject alternating treatments design suggest MIT is effective for Broca's aphasia patients and general stimulation-facilitation therapy is not effective or less effective, the efficacy of MIT for Broca's aphasia can be tested in a group design. The null hypothesis would be, "There is no difference in response to MIT and stimulation-facilitation therapy by Broca's aphasia patients. Appropriate sample sizes would be estimated; selection criteria established; patients selected for the appropriate groups —Broca's patients who receive MIT, Broca's patients who receive stimulation-facilitation therapy, and

Broca's patients who receive no-treatment; pretreatment outcome measures administered; the treatment trial conducted; pretreatment measures read- ministered post- treatment; and improvement among groups compared. This straightforward approach should indicate whether MIT is a more appropriate treatment for Broca's aphasia than stimulation-facilitation therapy and whether MIT or stimulation-facilitation therapy are efficacious treatments for Broca's aphasia.

Requirements In Single-Subject and Group Designs

Unfortunately, what is called a "straight-forward approach" often travels a path littered with obstacles. Appropriate designs require utilization of appropriate methods. When these methods are ignored, results are confounded. The following must be employed to obtain a valid answer to the question, "Are various methods of therapy efficacious for various types of aphasic patients?".
SELECTION CRITERIA. There is ample evidence to indicate numerous variables may influence an aphasic patient's response to treatment (Darley, 1972, 1975; Wertz, 1978, 1983, 1985, 1987). These may include, but are not limited to, age, time postonset, etiology of aphasia, site and size of the lesion, neurologic history, medical and psychosocial status, sensory and motor status, education, severity of aphasia, handedness, etc. Acceptable treatment studies control these variables by establishing selection criteria. If one is interested in the effects of various methods of therapy on various types of aphasia, one wants to control for the contaminating influences of age, time postonset, etiology, etc. Thus, aphasic patients who are compared in replications of alternating treatment designs and patients who are assigned randomly to groups in group designs must be as similar as possible. Mixing etiology, multiple episodes, and hemispheric involvement (Lincoln et al., 1984), for example, is unacceptable in singlesubject replications or treatment groups. Heterogeneity must be reduced, and it is reduced by rigid selection criteria and analysis of covariance of results.
SPECIFYING VARIOUS METHODS. If one is interested in whether various methods of therapy are efficacious for various types of aphasic patients, one must ensure that the methods are, indeed, various. Similarities exist in seemingly dissimilar methods. For example, MIT appears different from some of the intersystemic treatments described by Wertz, LaPointe and Rosenbek (1984). However, on closer inspection both utilize gesture, rate, and stress, and both fall within the theoretical framework of intersystemic reorganization developed by Luria (1970). Results of single-subject alternating treatments designs will indicate how "various" treatment methods are and whether aphasic patients respond to different methods differently. This certainly suggests the necessity of exploring different treatment methods in single-subject designs prior to investing the time, expense, and effort required by a group design.
SPECIFYING TYPES OF APHASIC PATIENTS. If one is interested in whether various methods of therapy are efficacious for various types of aphasic patients, one must ensure that aphasic patients are, indeed, different in type. As indicated earlier, controversy exists regarding the existence of aphasic types. Further,

systems for typing aphasic patients show poor agreement (Wertz, Deal and Robinson, 1984). And, type of aphasia has been demonstrated to change over time (Kertesz and McCabe, 1977). Thus, there is a need to demonstrate the types of patients entered in a treatment study are classified validly and reliably. Again, the results of replicating single-subject alternating treatments designs should indicate the types of aphasia and how different types of aphasia must be for profitable investigation in a group design.

RANDOM ASSIGNMENT AND NO-TREATMENT GROUP. Assignment of treatments in the single-subject alternating treatments design and assignment of patients who meet selection criteria to groups in group designs must be random. This controls experimenter and clinician bias, and it assists in equating groups on a number of variables that may influence response to treatment. Some (Shewan and Kertesz, 1984) balk at random assignment of patients to a no-treatment group, considering it unethical to withhold treatment. Nevertheless, a no-treatment group or groups is/are essential to demonstrate the efficacy of various methods of therapy for various types of aphasia. Self-selected no-treatment groups are not acceptable, because there are numerous reasons why a patient rejects treatment or does not seek it elsewhere if restricted from treatment by geographical location. Further, self-selected no-treatment groups typically differ from treatment groups on variables, e.g., severity, time postonset, that may influence response to treatment. One solution to ethical objections to a randomly assigned no-treatment group is use of a deferred treatment group. Patients who meet selection criteria are assigned randomly to a treatment group or groups and a deferred treatment group who receives no-treatment during their early participation in the study and are treated later. This provides a randomly assigned no-treatment group for comparison with patients who receive treatment during their early participation in the study. Wertz et al. (1986) utilized this approach by assigning patients to a deferred treatment group that was not treated for the first 12 weeks after entry and then received treatment for the next 12 weeks.

COMPLIANCE. Study patients must receive the prescribed amount of treatment in the intensity and duration specified. If patients do not receive the prescribed amount of treatment, they are dropped from the study. A "treatment" group in which only 26 percent of the patients receives close to the prescribed amount of treatment (Lincoln et al., 1984) is unacceptable. In addition, study patients must receive only the amount and type of treatment prescribed. Concurrent treatment during participation in a treatment study is forbidden.

SAMPLE SIZE. The most frequent problem in group treatment studies is failure to obtain the required sample size. The number of patients required must be estimated before initiating the study. Similarly, effect size and power should be stated. For example, the investigator makes an a priori statement such as, "In order to detect a difference of 15 points in the outcome measure with a Type I error rate of .05 and with a power of .90, a sample size of 35 in each group is necessary." This provides the number of patients necessary; the effect size, 15 points in the outcome measure; acceptable statistical significance, $p < .05$; and the power, .90. The latter is especially important in treatment studies, because power reduces the risk of a Type II error, failure to reject the null hypothesis when the null hypothesis is false.

SUMMARY AND CONCLUSIONS

Are there various methods of therapy that are efficacious for various types of aphasia? The evidence collected to date does not answer the question. Previous efforts have been directed at demonstrating whether aphasia treatment, generally, is efficacious or whether one treatment differs from another. An answer to the question will come from employing single-subject alternating treatments designs with replications to test specific treatments with specific types of aphasic patients and provide evidence to justify well-controlled group studies that will determine whether specific treatments are efficacious for specific types of aphasic patients. Appropriate group designs are either comparing a treatment with types of aphasic patients and no-treatment groups or comparison of treatments with a type of aphasia with a no-treatment control. Single-subject alternating treatments designs with replications can be conducted in most treatment centers. Group designs will require multicenter treatment trials to meet the requirements of rigid selection criteria, specification of various methods, specification of types of aphasic patients, random assignment and no-treatment groups, compliance with the study protocol, and acquisition of required sample size.

References

Basso, A., Capitani, E. and Vignolo, L. 1979. Influence of rehabilitation of language skills in aphasic patients: A controlled study. *Archives of Neurology*, **36**: 190-196.
Butfield, E. and Zangwill, O. 1946. Re-education in aphasia: A review of 70 cases. *Journal of Neurology, Neurosurgery, and Psychiatry*, **9**, 75-79.
Darley, F.L. 1972. The efficacy of language rehabilitation in aphasia. *Journal of Speech and Hearing Disorders*, **37**: 3-21.
Darley, F.L. 1975. Treatment of acquired aphasia. In W.J. Friedlander (ed.), *Advances in Neurology, Vol. 7. Current reviews of higher nervous system dysfunction* (pp. 111-145). New York: Raven Press.
Darley, F.L. 1982. *Aphasia*. Philadelphia: W. B. Saunders.
David, R.M., Enderby, P. and Bainton, D. 1982. Treatment of acquired aphasia: Speech therapists and volunteers compared. *Journal of Neurology, Neurosurgery, and Psychiatry*, **45**: 957-961.
Davis, G.A. and Wilcox, M.J. 1985. *Adult aphasia rehabilitation: Applied pragmatics*. San Diego: College-Hill Press.
Frazier, C.H. and Ingham, S.D. 1920. A review of the effects of gunshot wounds of the head: Based on the observation of two hundred cases at U. S. General Hospital No. 11, Cape May, N.J. *Archives of Neurology*, **3**: 17-40.

Goodglass, H. and Kaplan, E. 1983. *Boston diagnostic aphasia examination.* Philadelphia: Lea and Febiger.
Hagen, C. 1973. Communication abilities in hemiplegia: Effect of speech therapy. *Archives of Physical Medicine and Rehabilitation,* **54**: 454-463.
Hartman, J. and Landau, W.M. 1987. Comparison of formal language therapy with supportive counseling for aphasia due to acute vascular accident. *Archives of Neurology,* **24**: 646-649.
Helm-Estabrooks, N., Fitzpatrick, P.M. and Barresi, B. 1982. Visual action therapy for global aphasia. *Journal of Speech and Hearing Disorders,* **47**: 385-389.
Holland, A. 1977. Some practical considerations in aphasia rehabilitation. *In M. Sullivan and M. Kommers (Eds.), Rationale for adult aphasia therapy* Pp. 167-180. Omaha: University of Nebraska Medical Center.
Katz, R.C. 1986. *Aphasia treatment and microcomputers.* San Diego: College-Hill Press.
Katz, R.C. 1987. Efficacy of aphasia treatment using microcomputers. *Aphasiology,* **1**: 141-149.
Katz, R.C. and Wertz, R.T. 1990. Computerized hierarchical reading treatment in aphasia. Paper presented to the Fourth International Aphasia Rehabilitation Congress, Edinburgh, Scotland, September.
Kertesz, A. 1979. *Aphasia and associated disorders: Taxonomy, localization, and recovery.* New York: Grune and Stratton.
Kertesz, A. and McCabe, P. 1977. Recovery patterns and prognosis in aphasia. *Brain,* **100**: 1-18.
Lesser, R., Bryan, K., Anderson, J. and Hilton, R. 1986. Involving relatives in aphasia therapy: An application of language enrichment therapy. *International Journal of Rehabilitation Research,* **9**: 259-267.
Lincoln, N.B., McGuirk, E., Mulley, G.P., Lendrem, W., Jones, A.C. and Mitchell, J.R.A. 1984. Effectiveness of speech therapy for aphasic stroke patients: A randomised controlled trial. *Lancet,* **1**: 1197-1200.
Luria, A.R. 1970. *Traumatic aphasia: Its syndromes, psychology, and treatment.* The Hague: Mouton.
Marks, M., Taylor, M., and Rusk, H.A. 1957. Rehabilitation of the aphasic patient: A summary of three years' experience in a rehabilitation setting. *Archives of Physical Medicine and Rehabilitation,* **38**: 219-226.
Meikle, M., Wechsler, E., Tupper, A., Benenson, M., Butler, J., Mulhall, D. and Stern, G. 1979. Comparative trial of volunteer and professional treatments of dysphasia after stroke. *British Medical Journal,* **2**: 87-89.
Poeck, K., Huber, W. and Willmes, K. 1989. Outcome of intensive language treatment in aphasia. *Journal of Speech and Hearing Disorders,* **54**: 471-479.
Prins, R.S., Schoonen, R. and Vermeulen, J. 1989. Efficacy of two different types of speech therapy for aphasic stroke patients. *Applied Psychologinguisitics,* **10**: 85-123.

Rosenbek, J.C., LaPointe, L.L. and Wertz, R.T. 1989. *Aphasia: A clinical approach.* Boston: A College-Hill Publication, Little, Brown and Company.
Sands, E., Sarno, M.T. and Shankweiler, D. 1969. Long-term assessment of language function in aphasia due to stroke. *Archives of Physical Medicine and Rehabilitation,* 50: 202-206.
Schuell, H., Jenkins, J.J. and Jimenez-Pabon, E. 1964. *Aphasia in adults; Diagnosis, prognosis, and treatment.* New York: Hoeber Medical Division, Harper and Row Publishers.
Shewan, C.M. and Kertesz, A. 1984. Effects of speech and language treatment on recovery from aphasia. *Brain and Language,* 23: 272-299.
Sparks, R., Helm, N.A. and Albert, M. 1974. Aphasia rehabilitation resulting from melodic intonation therapy. *Cortex,* 10: 303-316.
Vignolo, L. 1964. Evolution of aphasia and language rehabilitation: A retrospective exploratory study. *Cortex,* 1: 344-367.
Wepman, J.M. 1951. *Recovery from aphasia.* New York: Ronald Press.
Wertz, R.T. 1978. Neuropathologies of speech and language: An introduction to patient management. In D.F. Johns (ed.), *Clinical management of neurogenic communicative disorders* Pp. 1-102. Boston: Little, Brown and Company.
Wertz, R.T. 1983. Language intervention context and setting for the aphasic adult: When? In J. Miller, D.E. Yoder and R. Schiefelbusch (eds.), *Contemporary issues in language intervention, ASHA Reports 12* Pp. 196-220. Rockville, MD: American Speech-Language- Hearing Association.
Wertz, R.T. 1985. Neuropathologies of speech and language: An introduction to patient management. In D.F. Johns (ed.), *Clinical management of neurogenic communicative disorders,* 2nd ed. Pp. 1-96. Boston: Little, Brown and Company.
Wertz, R.T. 1987. Language treatment for aphasia is efficacious, but for whom? *Topics in language disorders,* 8: 1-10.
Wertz, R.T., Collins, M.J., Weiss, D., Kurtzke, J.F., Friden, T., Brookshire, R.H., Pierce, J., Holtzapple, P., Hubbard, D.J., Porch, B.E., West, J.A., Davis, L., Matovitch, V., Morley, G.K. and Resurreccion, E. 1981. Veterans Administration cooperative study on aphasia: A comparison of individual and group treatment. *Journal of Speech and Hearing Research,* 24: 580-594.
Wertz, R.T., Deal, J.L. and Robinson, A.J. 1984. Classifying the aphasias: A comparison of the Boston Diagnostic Aphasia Examination and the Western Aphasia Battery. In R.H. Brookshire (Ed.), *Proceedings of the Conference on Clinical Aphasiology* Pp. 40-47. Minneapolis: BRK Publishers.
Wertz, R.T., LaPointe, L.L. and Rosenbek, J.C. 1984. *Apraxia of speech in adults: The disorder and its management.* New York: Grune and Stratton.

Wertz, R.T., Weiss, D.G., Aten, J.L., Brookshire, R.H., Garcia-Bunuel, L., Holland, A.L., Kurtzke, J.F., LaPointe, L.L., Milianti, F.J., Brannegan, R., Greenbaum, H., Marshall, R.C., Vogel, D., Carter, J., Barnes, N.S. and Goodman, R. 1986. Comparison of clinic, home, and deferred language treatment for aphasia: A Veterans Administration cooperative study. *Archives of Neurology*, **43**: 653-658.

3 Efficacy of language systematic learning approaches to treatment

Luise Springer and Klaus Willmes

INTRODUCTION

In numerous therapy studies it has been shown that intensive and professional treatment of aphasic impairments is efficacious (Basso et al. 1979; Poeck et al. 1989; Shewan and Kertesz, 1984; Wertz et al. 1981; 1986). Some large scale studies in the 80's demonstrating no superiority of professional treatment compared to trained volunteers or no effect of aphasia therapy at all. (David et al. 1982, Meikle et al. 1979, Lincoln et al. 1984) have been critized for several reasons (Huber et al. 1983). Intensity of treatment was very limited and methods and contents of therapy were not specifically aimed at the particular patterns of the patients' aphasic disorders.

A seemingly positive aspect of these large-scale studies was the randomized assignment of patients to the different treatment methods compared in combination with blind assessment of outcome. This argument, however, must be qualified as well. Given the heterogeneity of aphasic symptoms and their combination it is hard to conceive that a uniform treatment approach administered to a large sample will be effective for the whole sample (Howard, 1986).

Consequently, modern aphasia therapy research has taken a different route. In various single case or small group studies the efficacy of specific intervention techniques and of specific materials for the treatment of particular aphasic symptoms was compared (Basso et al. 1979; Helm-Estabrooks et al., 1981; Howard et al., 1985; Huber et al., 1978; Jones, 1986; Seron et al., 1979; Shewan and Kertesz, 1984; Weniger et al., 1987; Wiegel-Crump and Koenigsknecht, 1973). Methodologically, such studies call for the intra-subject comparison of the different approaches in so called cross-over designs. These are either available for a single patient (Coltheart, 1989) or for (two) groups of patients for which the order of treatment methods has been varied systematically (Brunner and Neumann, 1987; Grizzle, 1965; 1974). More recently, the cognitive neuropsychological approach has also been introduced to locate the patient's impairments within some hypothesized model of normal language processing and subsequently to derive appropriate therapeutic methods for remediation or compensation of that "functional lesion" (Byng and Coltheart, 1986; Coltheart, 1989; Hatfield and Shewell, 1989; Howard and Patterson, 1987; Lesser, 1987; Seron and Deloche, 1989). The derivation of therapeutic methods from these models is not straightforward. Rather, the models help to identify the nature of the disorders and the processing strategies used by the patient. The therapist then has to choose among the various treatment methods available or even to design specific materials and new procedures, which may be confined to the individual patient under study.

In the following, two experimental studies carried out at the Neurology Department and the School of Logopedics in Aachen will be described. In each study, two treatment methods were compared within aphasic patients. The presentation will concentrate on methodological aspects concerning study design and statistical evaluation as well as on the different schools of aphasia therapy compared.

In study 1, dialogue discourse was trained via the use of Wh-questions and prepositions in related answers. A linguistically structured (learning) approach was compared to a stimulation approach in a two-period cross-over design for two groups of 6 aphasic patients each. Statistical methods for the evaluation of individual patient's improvement were used as well.

In study 2, the traditional, pragmatically oriented PACE approach (Davis and Wilcox, 1980; 1981; 1985; 1986) was compared to a modified PACE approach, in which linguistically structured material was used for the treatment of lexical semantic impairments. Four patients were studied individually with a double cross-over design for treatment methods, two for each of the sequences A-B-A-B or B-A-B-A, respectively.

EXPERIMENTAL RESULTS

Study 1 Training the use of Wh-questions and prepositions in dialogues.

One of the reasons why aphasic patients in general have difficulties with actively participating in conversations is that they have problems with understanding and asking questions. In this study Wh-questions and appropriate elliptic answers with a prepositional phrase were embedded in conversational dyads in combination with a situational context (frame). One example from the 12 scenarios used as therapy material is the following:

Frame: It is a weekend. Mrs. Peters wants to stay in bed a little bit longer. Mr. Peters asks her.
Dialogue: (*When*) do you want to have breakfast?
Frame: Mrs. Peters is yawning. She answers:
Dialogue: (*In*) half an hour.

Each frame is 5 sentences long. It begins with a short description of the situational context. The dialogue always starts with a temporal interrogative particle (when (German: Wann), how long (Wie lange)). After a short additional situational context the subsequent elliptic answer starts with a temporal preposition depending on the context and the question itself.

The training of each discourse always proceeded in five steps as depicted in Figure 1.

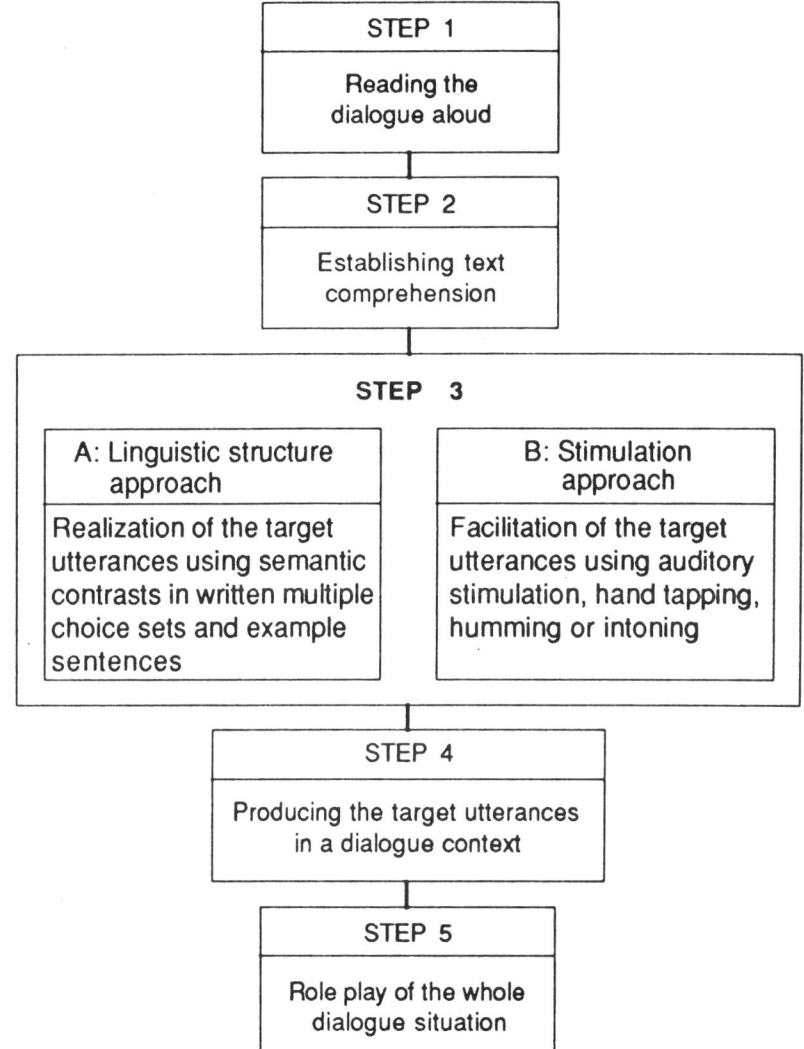

FIGURE 1. Steps of the two treatment approaches compared in study 1: linguistically oriented learning approach (linguistic structure approach, method A) vs. stimulation approach (method B)

Both approaches differ in the central step 3. The therapist either demonstrated the linguistic characteristics of target and distractor (method A: linguistically structured learning approach) or provided stepwise auditory and prosodic stimulation (method B) ranging from tapping and intoning to speaking according to the prosodic pattern of the target phrase as known from Melodic Intonation Therapy (Albert et al., 1973; Sparks and Holland, 1976; Schuell et al., 1964). Steps 1 and 2 were identical for both approaches. They only served to establish sufficient text comprehension of the situational context. Steps 4 and 5, again identical, were introduced to practice the target utterances - which were delt with in isolation in step 3 - in a dialogue context first (step 4) and then in a role play (step 5). In order to illustrate the crucial differences between both approaches to treatment, procedural details for step 3 of the two methods are described.

In the linguistically structured learning approach the patient is given two multiple choice sets in written form, one for question words and one for prepositions. The items were selected to demonstrate semantic contrasts. The interrogative set consisted of the two targets (how long, when) and five distractors (with what (*womit*), who (*wer*), how much (*wieviel*), how (*wie*), why (*warum*)). In the preposition set there were seven targets to be selected depending on the preceding question word used: until (*bis*), since (*seit*), before (*vor*), after (*nach*), at (*am*), in (*im*), at (*um*). First, the patient is requested to read all the interrogative words aloud. Then he has to look at the context as well and to decide by himself which of the particles fits into the first blank. After the patient has chosen one item by pointing and then reading it, he has to read out the whole dialogue . Again, phonological or morphological errors are not concentrated on, but corrective feedback is provided. In case of difficulties or incorrect choices, the different meanings of the interrogatives and prepositions are demonstrated using verbal explanations and exemplar applications as well as schematic drawings and visual symbols, which illustrate the semantic function of a given element. Verbal explanations concentrate on the different uses of the temporal Wh-particles, e.g., by telling that 'when' (wann) is used to ask for a particular point in time as opposed to 'how long' (wie lange), which asks for a stretch of time. As a visual memory aid a big black point is introduced compared to a black bar with a start- and an endmarker. For the other Wh-words there are appropriate verbal and visual aids as well. Exemplary applications are provided by the therapist using the same Wh-particles in a different context, e.g., 'When (*wann*) will your spouse come to visit you?' as opposed to 'How long (*wie lange*) have you been in the hospital?'. The wordings of the examples are not totally fixed, but the examples center around the actual situation of the patients.

In the stimulation approach the melodic and rhythmic contour of the target utterance is trained first using hand tapping, humming or intoning (Sprechgesang) according to the prosodic pattern of the phrase. The prosodic aids are then gradually reduced. Errors are not pointed out explicitly but auditory stimulation is repeated instead. These stimulation techniques are well known from Melodic Intonation Therapy (Albert et al., 1973). A more precise description of the other steps has been given elsewhere (Springer et al., 1988).

The control tests included the same units and regularities as the practiced material, but were embedded in different linguistic contexts, in order to assess so called trivial learning effects. Furthermore, spatial expressions were introduced (where, on, under, to, etc.), in order to examine non-trivial learning effects. The patient had to use these expressions in closure tasks. There were 20 items for Wh-questions and prepositions each. The whole set was doubled because both oral completion and selection from multiple choice sets of written words were required.

The scoring of each response was on a 0 – 2 scale. A score of 1 was given for a partially correct solution. For oral tasks this was the case if the correct Wh-particle was used but if phonological or morphological errors were present in the remaining utterances. For written tasks, partial credit was given if patients exhibited problems with reading aloud or with choosing the adequate alternatives from the MC-sets. During testing the responses of the patients were noted on a scoring sheet. The final score for each item was based on a consensus judgement of both therapists involved in the study. The design of the therapy experiment is shown in Table 1.

		Method A		Method B	
GROUP 1	Pre-test	Linguistically oriented learning (L)	Post-test 1	Stimulation approach (S)	Post-test 2
GROUP 2	Pre-test	Method B Stimulation approach (S)	Post-test 1	Method A Linguistically oriented learning (L)	Post-test 2
DURATION	1–2 days	max. 2 weeks (6 sessions)	1–2 days	max. 2 weeks (6 sessions)	1–2 days

TABLE 1. Two-period cross-over design of therapy study 1

In the two-period cross-over design both sequences of the linguistically structured learning approach and the stimulation therapy are present. With such a design it is possible to separately assess two types of treatment effects. The direct treatment effect expresses the impact of one treatment period measured as the difference in performance in the control test before and after treatment. This direct effect of one approach is assumed to be present irrespective of whether it is the first or the second one in the order of treatments. In addition, one assumes an after-effect for the second period, which expresses the possible impact of the first therapy period on the second one. It may well be that the order in which both therapy methods are administered can influence the final outcome across the whole study period, irrespective of the efficacy of either treatment method in isolation. Thus, besides choosing the most adequate treatment approach per se, determination of the optimal sequence of both treatment methods was of interest as well.

Twelve patients having either moderate Broca's or Wernicke's aphasia according to an examination with the Aachen Aphasia Test (AAT, Huber et al.,

1984) prior to the therapy experiment were included in the study. In order to ensure basic lexical semantic, phonological and graphemic abilities, patients were excluded who had a severe impairment in the Written Language and Comprehension subtests of the AAT and whose spontaneous speech was characterized by very severe semantic and phonemic problems. In all patients etiology was vascular, all were right handed; there were 4 female and 8 male patients. Duration of aphasia was at least 3 months (median 11 months, range 3 – 69); median age was 49.5 (range 27 – 66).
Assignment to one sequence of therapy methods was such that groups were approximately balanced with respect to type and severity of aphasia as well as age and time post onset.

RESULTS:

Group data

First, both treatment approaches can be compared across the two groups of patients. Mean performances in the control tests are shown in Figure 2. Results are presented separately for temporal and non-trained local Wh-particles and prepositions. Global inspection of the graphs gives the following impression. For Wh-particles in the written modality (Figure 2a) line segments connecting pre- and post-treatment performances are almost parallel when the same treatment method is used. The steepness of the lines for the linguistically structured learning approach is much higher than for the stimulation approach, indicating superiority of the former method. On the other hand, in Figure 2b for the written modality parallel lines for the first treatment period in combination with highly different gradients for the second period indicate differential after-effects. Patients receiving linguistically structured therapy after stimulation improved still further, whereas the other group remained stable.

For the non-trained local Wh-particles (Figure 2c) there is a pattern of effects for the oral modality similar to the temporal Wh-particles but no strong effect whatsoever in the written modality. For the non-trained local prepositions (Figure 2d) there are no sizeable effects at all.

For an inferential statistical examination of both types of effects, nonparametric rank test procedures for a two-period cross-over design were used (Brunner and Neumann 1987). They allow for a separate assessment of direct and after-effects of both therapy methods. If there are significant differences in the amount of after-effects for the two types of therapy, a comparison of the direct therapy effects has to be restricted to the first treatment period only. Otherwise, both treatment periods can be considered together. As the dependent variable for this cross-over analysis, the difference score between pre- and post-test 1 and post-test 1 and 2 respectively is computed for each of the eight control tasks. The results are summarized in Table 2. The left part of the table gives the mean difference score (change in performance) for each of the two treatment periods and each patient group. From these an estimate of the difference between the after-

Systematic learning approaches 83

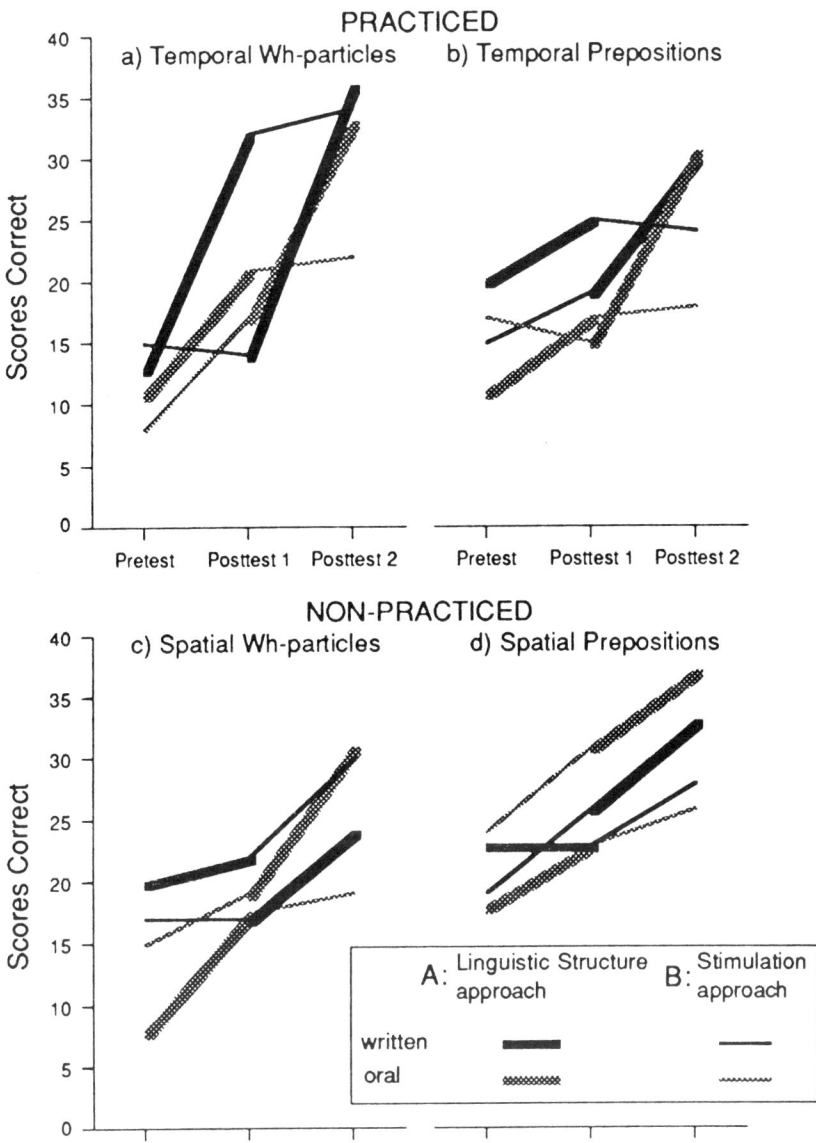

FIGURE 2. Mean performances for patients in the three control examinations for each of the two sequences of treatments (n=6 each), A-B vs B-A, A: linguistically oriented learning approach, B: stimulaton approach separately for
(a) trained temporal Wh-particles
(b) trained temporal prepositions
(c) non-trained spatial (local) Wh-particles
(d) spatial (local) prepositions.
Solid black lines indicate written performances, unfilled lines oral performances; thick lines are for method A, thin lines for method B.

effects of the stimulation (S) on the linguistically structured learning approach (L) and the reverse after-effect is computed by means of the following difference in mean change scores (Table 2):

(L period 2 - L period 1) - (S period 2 - S period 1)

These estimates are shown in the right half of the table. Furthermore, the table gives the respective estimates for the differences between direct treatment effects both for the first treatment period alone and for the two periods taken together. The resulting p-values for the nonparametric tests of whether these differences are significantly different from zero are given in brackets. Only if there are no differential after-effects at a liberal type-I error level of 0.10, is the test for differences concerning direct effects for both treatment periods carried out.

Generally, the direct effect of the linguistically structured learning approach is numerically larger than that of the stimulation method, i.e. the differences in the rightmost column of Table 2 are positive. In contrast, the reverse is true for the comparison of the after-effects, i. e., the sequence stimulation approach preceding linguistically structured therapy is more advantageous than the other way round. But only some of these effect differences are significantly different from zero. Most of the significant effects are present for the trained temporal function words. Patients profit more from the linguistically structured learning approach for Wh-particles in the written modality and for temporal prepositions in the oral modality. The stimulation approach has, however, a significantly larger after-effect for all temporal tasks except for the temporal Wh-particles in the written modality. Looking at the non-trained local function words, there is only one significant effect: the linguistically structured learning approach is superior to the stimulation approach for local Wh-particles in the oral modality.

For an overall comparison of trained (temporal) and non-trained (local) tasks, scores for both types of function words and modalities were summed up and the resulting differences between control test applications and their respective estimates of effect differences were divided by four for a better comparison with the other entries in Table 2. Now there were only significant direct and after-effects for temporal items, such that the linguistically structured learning approach was superior to the stimulation approach. But the after-effects of the stimulation approach were larger than those of the linguistically structured learning approach. Furthermore, there was significant overall improvement in each condition when performance of the second post-test was compared to that of the pretest by means of a permutation test for the dependent two-sample problem (Edgington, 1987). This was also true for the unpracticed local expressions, even though the amount of overall improvement was again significantly larger for the practiced temporal expressions again tested with the above mentioned permutation test.

THERAPY SEQUENCE (number of tasks, max. 8) Patient	THERAPY PERIOD 1						THERAPY PERIOD 2						THERAPY PERIODS 1&2				
	mastery pretest	++	+	(+)	∅	–	mastery postt.1	++	+	(+)	∅	–	++	+	(+)	∅	–
L/S																	
1		3		1	1		2			2	6		4	3		1	
2	1	3^1	1		4^1		1		1	1	7^2	1	5^1		1^1	2	
3	1	2	1		4^1		1				6^1		3^1			5^1	
4		4^1	1		3^1		1	1^1			7^1		2^1	2		4	
5	1		1	1^1	6		2	2^2	2^1	3^1	1^1		4^3	1^1	1	2^1	
6		2	1		4	1			2	3	3		2^1	2		4	
Total	3	14^2	8	3^1	22^3	1	6	3^3	4^1	9^1	30^5	2	20^7	8^1	2^1	18^2	0
S/L																	
7				4	4	1	1	4^3	2^1	1			6^5	1	1		
8		3^1		7^1			1	7^6					7^6	1			
9		1	2	3	2			7^6	1^1				8^7				
10		1	3	2	2		1	2^2	2^1		4^1		6^3	2^2			
11		2^1	2	1	1				1		5	1	1	3		3	1
12		1	1	1	6			7^2		1			8^2				
Total	0	5^1	5^1	7	29^1	2	3	29^{19}	6^3	1	11^2	1	36^{23}	7^2	1	3	1

* According to exact McNemar test (one-tailed) for each of the 8 control tasks; summarized from tables A1 & A2 of the appendix: ++, =, (+), significant improvement at p≤.01, .05, .10; – sign. deterioration

TABLE 2 (Preceding page). Analysis for the two-period cross-over comparison of the two therapy methods: descriptive statistics includes estimates of differences between after-effects and between direct effects for both therapy methods together with p-values for the nonparametric rank tests (Brunner and Neumann 1987); in case of significant differences for after-effects at p <.10 only tests for differences between direct effects for period 1 are valid.

Analysis of individual patients

Besides the group results reported so far it is of interest to study treatment effects separately for each individual patient. For each of the eight control tasks, performances before and after a therapy period are compared itemwise. By treating the 20 items of a set of tasks with a score of 0-2 as 40 dichotomously scored items, a 2 x 2-table with 40 entries is obtained. The exact one-tailed McNemar test (Siegel, 1956) was used to compare the number of improvements with the number of deteriorations (see also Willmes, 1990). Besides looking for (significant) improvement one can also investigate whether a certain level of mastery (Popham, 1978; Willmes, 1990) has been reached. Using the binomial model, one can calculate the minimum number of correct scores that has to be attained for a test of 40 items in order to reach a specified mastery level of 90%. From tables of the binomial distribution one can infer that 32 out of 40 are sufficient for the 95%-confidence interval around 32 to cover a probability of 90% for a correct solution. The results for each individual patient summarized across all 8 tasks (Wh-particles vs. prepositions, temporal vs. local, written vs. oral) are reported in Table 3. It is obvious that most of the significant improvements were obtained for the language systematic therapy method, both for written and oral tasks. For each patient the number of control tasks showing different degrees of significant changes in performance is given separately for each of the two treatment methods as well as for the whole treatment period. For the majority of patients there was more improvement after the linguistically structured learning treatment period, i.e., there was a larger number of tasks showing significant changes at the 1%- and 5%- level in both patient groups. Only patient 5 was very much in contrast to this pattern of performance. This patient suffered from additional speech apraxia. Thus, it is conceivable that he profited much more from the stimulation techniques. To a lesser degree the same holds true for patient 6. Furthermore, the righthand part of Table 3 again illustrates that the sequence stimulation before linguistically structured training had a larger effect. There were considerably more tasks for which significant improvement as well as mastery at the second post test could be observed.

Three conclusions can be drawn from this study. First, specific training irrespective of the method applied led to overall improvement of linguistic performance. Second, trivial learning effects were larger and more widespread than non-trivial ones. Third, the linguistic structure approach was more effective than the stimulation approach. It was most favorable when preceded by a period of stimulation. It appears that stimulation techniques prepare the patients optimally for structured linguistic learning even in the more chronic stages of the impairment.

		MEAN CHANGE IN PERFORMANCE				EFFECT ESTIMATES (p-values*)			
		Sequence L/S (n=6)		Sequence S/L (n=6)		Differences in after effects	Differences in period 1 only	Direct effects period 1 & 2	
TYPE OF TASK		Period 1 L§	Period 2 S§	Period 1 S	Period 2 L	S-L (p-value)	L-S (p-value)	L-S (p-value)	
WH- PARTICLES	temporal: written	14.3	0.8	-0.7	19.5	3.7 (.269)	15.0	16.8	(<.0001+)
	oral	9.7	2.3	7.2	13.5	8.7 (.025+)	2.5 (.274)	6.8	
	local: written	2.0	5.7	2.5	7.0	1.8 (.472)	0.5	0.4	(.458)
	oral	5.8	3.3	2.7	8.2	1.7 (.397)	3.2	4.0	(.028+)
PRE- POSITIONS	temporal: written	6.0	-1.2	5.0	11.7	11.8 (<.0001+)	1.0 (.414)	6.9	
	oral	6.8	2.8	1.8	12.7	4.8 (.083+)	5.0 (.022+)	7.4	
	local: written	1.8	3.0	6.5	6.5	8.2 (.129)	-4.7	-0.6	(.225)
	oral	5.3	2.2	4.3	5.3	2.2 (.361)	1.0	2.1	(.394)
ALL FUNCTION WORDS**	temporal	9.2	1.2	3.4	14.4	7.3 (.016+)	5.9 (.002+)	9.5	
	local	3.8	3.5	4.0	6.8	3.5 (.132)	-0.3	1.5	(.310)

* Nonparametric rank test results (p-values two-tailed)
** Values divided by 4 for better comparability with the results for single types of tasks
\+ Significant difference
§ L: Linguistically oriented learning approach; S: Stimulation approach

TABLE 3 (Preceding page). Summary of individual changes in performance for each of the 8 control tasks for each patient in each therapy period and across the whole study period. Mastery - expressed by the superscript digits - is assumed if the individual performance probability is compatible with a criterion probability of 0.90 according to the binomial model using a 95%-confidence interval.

Study 2   Comparison of traditional and modified symptom specific PACE therapy

Pragmatically oriented speech pathologists agree that therapy should focus on the use of language in communicative contexts rather than setting the goal of therapy only in terms of linguistic entities that have been lost or are available only inconsistently.

PACE therapy, as originally proposed by Davis and Wilcox (1980; 1981; 1985; 1986) fits into this general communicative approach. Since the patient is encouraged to convey a message by using any verbal or nonverbal means available, language systematic demands such as specific lexical selection and construction or precise phonological realization only play a minor role in the traditional PACE approach. The therapist does not try to engage the patient in relearning structures of the language but to optimize via indirect techniques those communicative capacities, that are only mildly affected in the patient. Although Davis and Wilcox (1985) as well as Edelman (1987) suggest to use therapy material appropriate for the specifity of a patient's symptoms, linguistic parameters are only loosely controlled and no metalinguistic demands are imposed on the patient. However, such an emphasis on 'natural' communication may be too limited for (re)learning specific verbal skills. Therefore, we integrated language systematic considerations into the treatment approach (Springer et al., 1991). As an example we chose the treatment of lexical semantic impairments. In the traditional PACE approach (method A) patients had to identify individual line drawings of objects which were presented in random order. The target name was unknown to the therapist. Patients had free choice of the naming modality. For example, they could say the name, could try to write the name, could choose to describe the object, could use gestures or could even try to draw the object. The therapist would never correct the patient directly, but would only tell the patient whether the message was understood or not.

The modified PACE approach (method B) used semantic classification tasks. Instead of only one picture, the patient was confronted with an array of 22 pictures, a subset of which the patient had to sort out as belonging to a semantic class such as tools, toys, fruits, etc. The superordinate category was written on a card. This card together with the random array of 22 pictures was given to both the patient and the therapist, who were separated by a screen, as suggested by Clerebaut et al. (1984). The screen makes sure that subjects are required to convey new information to their partners, which is the essential feature of PACE therapy. Each decision on class membership had to be conveyed to the partner. The therapist would give feedback on the accuracy of classification in addition to the 'natural' feedback of the traditional approach. For the training, a total of 220 items taken from 10 different classes (vehicles, tools, fruits, vegetables, furniture, wild animals, office supplies, toys, musical instruments, clothing) was used. For the 12 members of each class there were also 10 distractors. High, medium and low prototypicality of target items was distinguished. Associated objects as well as members of another, loosely related class were selected as distractors. An example is given in Table 4. Testing was done in the traditional PACE setting. Assessment material was selected according to the same lexical criteria as the training material. It consisted of 60 practiced and 60 unpracticed

TARGETS Prototypicality of class membership		
high	medium	low
ball pen	stamp	calendar
pencil	sharpener	wastepaper basket
ruler	ink	pocket calculator
rubber	glue	dividers
DISTRACTORS		
associated object		other class
watch		shopping net
comb		file case
(pair of) glasses		box
newspaper		suitcase
cigarette		basket

TABLE 4. Linguistic structure of the training material used in study 2; example: semantic class 'office utensils'

items, but practice turned out to have no significant influence on the degree of improvement. The patients' attempts to identify the line drawings of objects were scored on the 6-point PACE communication scale and on a 5-point language systematic naming scale, which reflects semantic accuracy of the naming response. Training and testing were carried out in a double cross-over design as depicted in Figure 3. Each treatment phase comprised 5 one-hour sessions over five days.

Three patients with marked lexical-semantic difficulties were treated. Two had many semantic paraphasias and circumlocutions both in spontaneous speech and in the confrontation naming subtest of the AAT. In the comprehension subtest they failed on several items because they confused word and meanings.

All three patients D.Z. (female, 62 yrs.), W.B. (female, 47 yrs.) and U.Z. (male, 48 yrs.) were in the chronic stage at least 4 years after stroke. The first two patients were classified on the basis of the AAT performance as being mildly global, the third was assigned a moderate Wernicke's aphasia by means of a nonparametric discriminant analysis procedure (ALLOC) for the AAT-scores.

Patient No.	Time Course				
	T_1	T_2	T_3	T_4	T_5
1		A	B	A	B
2		B	A	B	A
3		A	B	A	B
4		B	A	B	A
etc.					
A, B: different treatment methods $T_1 \ldots T_5$: control tests					

FIGURE 3. Design of study 2

In addition, a fourth patient H.M. (male, 55 yrs) with mild chronic Broca's aphasia (16 months post onset) but with severe speech apraxia was studied. As opposed to the first three patients he showed no noticeable semantic disorders but moderate word retrieval problems, his comprehension performance was almost normal (percentile rank 91, AAT subtest comprehension). In part, his 'word-finding' difficulties may be attributed to articulatory groping. For this patient, semantic classification tasks incorporated in method B were expected to have no additional beneficial impact on his verbal and non-verbal abilities.

RESULTS

The control test performances of the first three patients with semantic disorders are depicted in Fig. 4. Overall, significant improvement was almost exclusively present after a treatment period with the modified PACE approach. This was true irrespective of the order of treatment methods. Surprisingly, both assessment scales revealed that method B was more efficacious.

For patient H.M. with speech apraxia the results are shown in Fig 5. The pattern of performances is quite different from the other three patients.

The language systematic AAT score remained low throughout. There was even some significant deterioration after the initial PACE-therapy period, which was only compensated for in the final modified PACE treatment period. In contrast, the communication score went up from one testing to the next, irrespective of type of treatment.

Statistical evaluation of changes in performances had to be carried out in a way different from the first study, because item responses were given graded scores. Basically, there are two statistical approaches. In the first one, item scores are summed up within semantic classes, such that there are 10 class totals per patient and time of testing. The ten classes are taken as replications in order

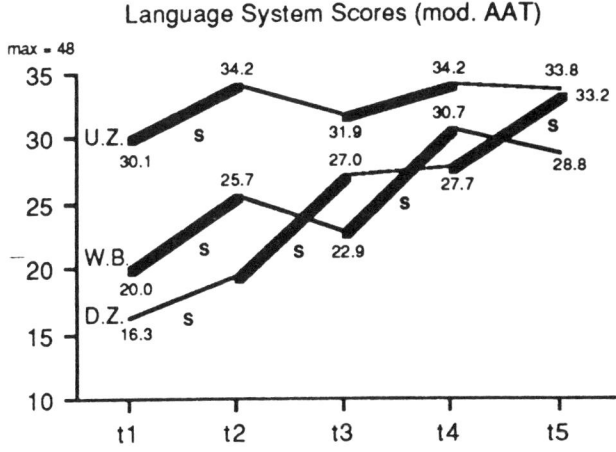

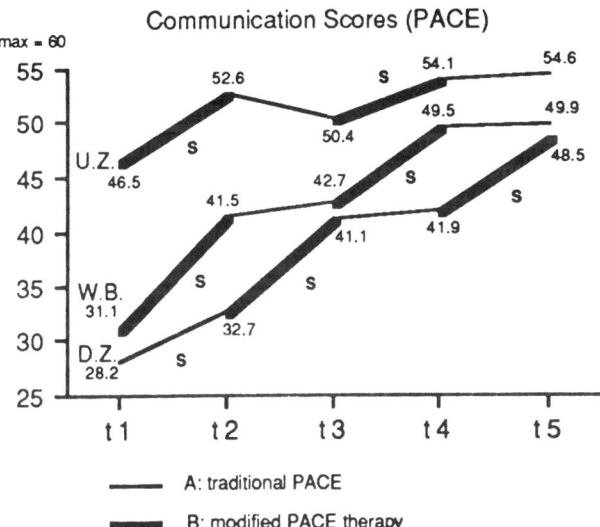

FIGURE 4. Control test performances of the three patients with lexical-semantic difficulties, s: significant differences (one-tailed randomisation test)

to test for therapy effects per patient by means of randomization tests (Edgington, 1987). Structurally, one has the equivalent of a pre-post design usually evaluated with a dependent samples t-test or the Wilcoxon matched pairs signed ranks test. With the randomization test the semantic class-totals can be evaluated themselves—not only ranks—without assuming normality. Computations can efficiently be carried out via the PC package StatXact (version

2, 1991) or a routine by Dallal (PC-Pitman, 1985) which was also used in the statistical evaluation of study 1. Exact p-values for the comparison of performances before and after each therapy period are reported in Table 5. The second mode of evaluation is based on performance per item irrespective of semantic class membership. Computation of class totals might mask intraclass-variations in performance.

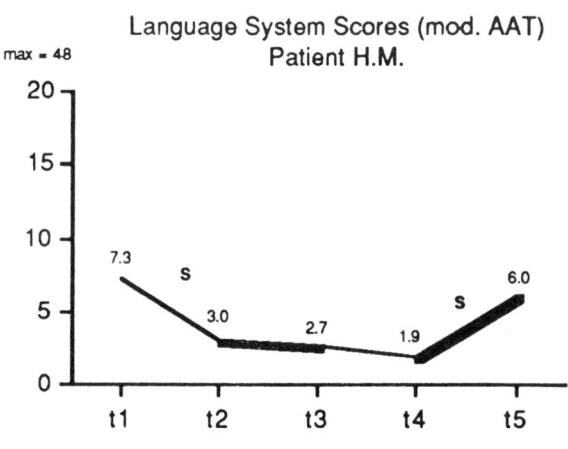

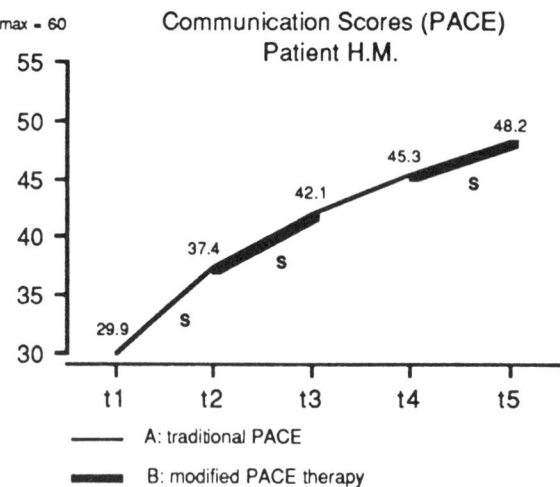

FIGURE 5. Control test performances of the patient with speech apraxia, s: significant (see Fig.4).

For each therapy period a c x c- contingency table can be formed containing the 120 pre-post response combinations for each patient separately for both AAT- and PACE-scale. One appropriate statistical test procedure for changes in performance is a generalization of the McNemar test for correct/false-scoring to

Patient	Statistical test*	THERAPY PERIOD							
		1		2		3		4	
		AAT score	PACE score	AAT score	PACE score	AAT score	PACE score	AAT score	PACE score
		A		B		A		B	
D.Z.	Random test	.0518	.0489	.0078	.0069	.3633	.3408	.0117	.0010
	Marg. hom.	.0559	.0113	.0001	.0001	.3496	.3381	.0036	.0002
H.M.	Random test	.0039	.0058	.3828	.0196	.1485	.0469	.0039	.0215
	Marg. hom.	.0014	.0003	.4206	.0084	.2427	.0210	.0031	.0166
		B		A		B		A	
U.Z.	Random test	.0030	.0010	.1289	.0508	.1309	.0058	.4424	.3594
	Marg. hom.	.0001	.0001	.0618	.0024	.0760	.0017	.4130	.3478
W.B.	Random test	.0030	.0010	.0742	.2852	.0010	.0010	.1455	.4061
	Marg. hom.	.0001	.0001	.0659	.2606	.0001	.0001	.0985	.3996

* all p-values one-tailed
TABLE 5. Exact p-values for the comparison of performances before and after each therapy period for method A resp. method B both using a randomization test for the comparison of semantic class total scores or a test for marginal homogeneity in two-way tables with ordered categories

several ordered response categories (0-4, resp. 0-5). Such an exact test for "marginal homogeneity" is available in the StatXact program package as well. For a comparison with the results for semantic class totals the exact one-tailed p-values of these generalized McNemar tests are also given in Table 5. For all 4 patients the results are very similar to the randomisation test evaluation.

In case of lexical-semantic difficulties the modified PACE approach was clearly more efficacious than the traditional one in enabling patients to convey specific new information by either identifying or circumscribing (line drawings of) objects in interaction with the therapist. These effects were visible both for rating of communicative but also of linguistic skills. The result can easily be explained. Improvement of lexical-semantic abilities leads to or is even a prerequisite for better communicative skills. Compensation by nonverbal means may be supportive, but there was no evidence for the view that with linguistically unspecific communicative stimulation, as provided in the traditional PACE approach, improvement of verbal or even nonverbal communicative skills can be achieved in patients with predominant lexical semantic impairments. When there are only mild semantic impairments but e.g. severe motor speech planning deficits the effects may be completely different.

The fourth patient improved continuously under both treatment methods but only with respect to nonverbal communicative skills. These could be stipulated by the PACE setting common for both methods because the semantic difficulties of that patient were in fact mild. There was no reason for the expressive verbal abilities to improve since neither method was adequate for their treatment.

CONCLUSIONS

The results of the two studies can be summed up as follows: The linguistic structure approach when aimed at specific impairment of linguistic knowledge is more favorable in chronic aphasic patients than a purely stimulating or communicatively activating approach. This general statement must however be qualified.

First, the setting for practicing linguistic structure and for pre- and post-training control was "pragmatic" in both studies in the sense that either a discourse pattern (first study) or communicative interactions (second study) were chosen for the general treatment and testing setting. It is conceivable that this is a prerequisite for the linguistic structure approach to be effective. Learning of grammatical structure alone might not be successful. More important, pragmatic settings are more likely to induce transfer to everyday communication. This is not to say that any treatment on any linguistic level requires pragmatic embedding. Problems like sublexical phonological disorders, speech apraxia or word retrieval difficulties call for stepwise training procedures dealing with the impairments in isolation (Hatfield et al., 1977).

Second, the linguistic structure approach should always be symptom specific. Most promising effects can be expected for patients in whom linguistic knowledge is affected as opposed to difficulties of access/retrieval of lower level processing disorders such as speech apraxia. It should be understood that some residual linguistic knowledge must be preserved, which precludes the linguistic structure approach in severe chronic global aphasia.

Third, for more complex linguistic demands such as and text production, which require integration of information from several levels of processing, the structure approach may fail completely if the processing capacities of the patients are too limited. In such a situation, the training of strategies is preferable together with attempts to improve disorders of working memory.

Fourth, the proper sequencing of different methods for the treatment of a patient should be given more attention using appropriate cross-over designs and related statistical evaluation techniques. It may well be that specific, linguistically structured treatment methods are most efficacious if they are preceded by a phase of stimulation even in case of chronic aphasia. This stimulation is likely to reactivate residual or inconsistently available capacities.

Fifth, transfer of improved linguistic performance from the therapy setting into everyday spontaneous language and communicative interactions still lacks demonstration in our therapy studies as in most others. To do so is a methodologically thorny thing. Ultimately, success or failure of patients and therapists to achieve transfer to everyday language demands will decide whether the symptom specific and language structured approaches will remain as central paradigms of aphasia therapy.

Acknowledgement

We would like to thank Ellen Haag, Walter Huber and Ralf Glindemann together with whom the two experimental studies were carried out.

References

Albert, M.L., Sparks, R. and Helm, N.A. 1973. Melodic intonation therapy for aphasia. *Archives of Neurology*, **29**: 130-131.
Basso, A., Faglioni, P. and Vignolo, L. 1979. Influence of rehabilitation on language skills in aphasic patients: A controlled study. *Archives of Neurology*, **36**: 190-196.
Brunner, A. and Neumann, N. 1987. Non-parametric methods for the 2-period-cross-over design under weak model assumptions. *Biometrical Journal*, **29**: 907-920.
Byng, S. and Coltheart, M. 1986. Aphasia therapy research: metholodogical requirements and illustrative results. In E. Hjelmquist and L.B. Nilsson (eds.) *Communication and handicap*. (Elsevier, Amsterdam).
Clerebaut, M., Coyette, F., Feyereisen, P. and Seron, X. 1984. Une méthode de rééduction fonctionelle des aphasiques: la P.A.C.E., *Rééeduction Orthophonique*, **22**: 329-345.
Coltheart, M. 1989. Aphasia Therapy Research: A Single-Case Study Approach. In C. Code and D.J. Müller (eds.) *Aphasia Therapy*, 2nd London: Whurr.
Dallal, G.E. 1985. PC-PITMAN: Randomization tests, version 2.0. (USDA Human Nutrition Research Center on Aging, Tufts University, Boston).
David, R., Enderby, P. and Bainton, D. (1982) Treatment of acquired aphasia: Speech therapists and volunteers compared. *Journal of Neurology, Neurosurgery and Psychiatry*, **45**: 957-961.
Davis, G.A. and Wilcox, M.J. 1980. A critical look at PACE-therapy. In R.H. Brookshire (ed.) *Clinical Aphasiology Conference Proceedings*. Minneapolis, Minn.: BRK Publishers.
Davis, G.A. and Wilcox, M.J. 1981. Incorporating parameters of natural conversation in aphasia treatment. In R. Chapey (ed.), *Language Intervention Strategies in Adult Aphasia*, 1st ed. Baltimore: Williams and Wilkins.
Davis, G.A. and Wilcox, M.J. 1985. Adult Aphasia Rehabilitation: *Applied Pragmatics*. San Diego, Cal.: College-Hill Press.
Davis, G.A. 1986. Pragmatics and treatment. In R. Chapey (Ed.) *Language intervention strategies in adult aphasia*. 2nd ed., Baltimore: Williams and Wilkins.
Edelman, G. 1987. PACE: *Promoting Aphasics' Communicative Effectiveness*. Bicester, Oxon: Winslow Press.

Edgington, E.S. 1987. *Randomization Tests*. 2nd ed. New York: Dekker.
Grizzle, J.E. 1965, 1974. The two-period change-over design and its use in clinical trials. *Biometrics*, **21**: 467-480 and corrections in *Biometrics*, **30**: 727.
Huber, W. 1992. Therapy of Aphasia. Comparison of various approaches. In N. v. Steinbüchel, D.Y. von Cramon, E. Pöppel (eds.) *Neuropsychological Rehabilitation*. Heidelberg: Springer.
Huber, W., Mayer, I. and Kerschensteiner, M. 1978. Untersuchungen zur Methode und zum Verlauf der Therapie von phonematischem Jargon bei Wernicke-Aphasie. *Folia phoniatrica*, **30**: 119-135.
Huber, W., Poeck, K., Springer, L. and Willmes, K. 1983. Comments on 'Treatment of acquired aphasia: speech therapists and volunteers compared'. *Journal of Neurology, Neurosurgery and Psychiatry*, **46**: 691-692.
Huber, W., Poeck, K. and Willmes, K. 1984. The Aachen Aphasia Test. In F.C. Rose (ed.) *Progress in Aphasiology*. New York: Raven Press.
Hatfield, F.M., Howard, D., Barber, J. and Jones, C. 1977. Object naming in aphasics – the lack of effect of context for realism. *Neuropsychologia*, **15**: 117-127.
Hatfield, F.M. and Shewell, C. 1989. Some applications of linguistics to aphasia therapy. In C. Code and D.J. Müller (eds.) *Aphasia Therapy*, 2nd ed. London: Whurr.
Helm-Estabrooks, N.A., Fitzpatrick, P.M. and Barresi, B. 1981. Response of an agrammatic patient to a syntax stimulation program for aphasia. *Journal of Speech and Hearing Disorders*, **46**: 422-427.
Howard, D. 1986. Beyond randomised controlled trials: the case for effective case studies of the effects of treatment in aphasia. *British Journal of Disorders of Communication*, **21**: 89-102.
Howard, D., Patterson, K.D., Franklin, S., Orchard-Lisle, V. and Morton, J. 1985. The treatment of word retrieval deficits in aphasia: a comparison of two therapy methods. *Brain*, **108**: 817-829.
Jones, E.V. 1986. Building the foundations of sentence production in a non-fluent aphasic. *British Journal of Disorders of Communication*, **21**: 63-82.
Lesser, R. 1987. Cognitive neuropsychological influences on aphasia therapy. *Aphasiology*, **1**: 189-200.
Lincoln, N.B., Mulley, G.P., Jones, A.C., McGuirk, E., Lendrem, W. and Mitchell, J.R.A. 1984. Effectiveness of speech therapy of aphasic stroke patients. *Lancet*, **1**: 1197-1200.
Meikle, M., Wechsler, E., Tupper, A., Benenson, M., Butler, J., Mulhall, D. and Stern, G. 1979. Comparative trial of volunteer and professional treatments of dysphasia after stroke. *British Medical Journal*, **2**: 87-89.
Poeck, K., Huber, W. and Willmes, K. 1989. Outcome of intensive language treatment in aphasia. *Journal of Speech and Hearing Disorders*, **54**: 471-479.

Popham, W.J. 1978. *Criterion-referenced Measurement.* Englewood Cliffs, N.J.: Harper and Row.
Schuell, H.M., Jenkins, J.J. and Jimenez-Pabon, E. 1964. *Aphasia in Adults: Diagnosis, Prognosis and Treatment.* New York: Harper and Row.
Seron, X., Deloche, G., Bastard, V., Chassin, G. and Hermand, N. 1979. Word-finding difficulties in learning transfer in aphasic patients, *Cortex*, 15: 149-155.
Seron, X. and Deloche, G. 1989. *Cognitive Approaches in Neuropsychological Rehalitation.* London: Lawrence Erlbaum.
Siegel, S. 1956. *Nonparametric Statistics for the Behavioral Sciences.* New York: McGraw-Hill.
Shewan, C.M. and Kertesz, A. 1984. Effects of speech and language treatment on recovery of aphasia. *Brain and Language*, 23: 272-299.
Sparks, R.W. and Holland, A.L. 1976. Method, melodic intonation therapy for aphasia. *Journal of Speech and Hearing Disorders*, 41: 287-297.
Springer, L., Glindemann, R., Huber, W. and Willmes K. 1991. How efficacious is PACE-therapy when 'Language Systematic Training' is incorporated? *Aphasiology*, 5: 391-401.
Springer, L., Willmes, K. and Haag, E. 1988. Training the use of Wh-questions and prepositions in dialogues: a comparison of two different approaches in aphasia therapy. Paper presented at the Third International Aphasia Rehabilitation Congress, Florence.
StatXact 1989. Statistical Software for Exact Nonparametric Inference. Cambridge, Mass: Cytel Software Corporation.
Weniger, D., Springer, L. and Poeck, K. 1987. The efficacy of deficit-specific therapy materials, *Aphasiology*, 3: 215-223.
Wertz, R.T., Collins, M.J., Weiss, D.G., Kurtzke, J.F., Friden, T., Brookshire, R.H., Pierce, J., Holtzapple, P., Hubbard, D.J., Porch, B.E., West, J.A., Davis, L., Matovitch, V., Morley, G.K. and Resurreccion, D. 1981. Veterans Administration Cooperative Study of Aphasia: A comparison of individual and group treatment. *Journal of Speech and Hearing Research*, 24: 580-594.
Wertz, R.T., Weiss, D.G., Aten, J.L., Brookshire, R.H., Garcia-Bunuel, L., Holland, A.L., Kurtzke, J.F., LaPointe, L.L., Milianti, F.J., Brannegan, R., Greenbaum, H., Marshall, R.C., Vogel, D., Carter, J., Barnes, N.S. and Goodman, R. 1986. Comparison of clinic, home and deferred language treatment for aphasia. Veterans Administration Cooperative Study. *Archives of Neurology*, 43: 653-658.
Wiegel-Crump, C. and Koenigsknecht, R.A. 1973. Tapping the lexical store of the adult aphasic: Analysis of the improvement made in word retrieval skills. *Cortex*, 9: 410-417.
Willmes, K. 1990. Statistical methods for a single case study approach to aphasia therapy research. *Aphasiology* 4: 415-436.

Part II Linguistic foundations of aphasia rehabilitation methods

Implicitly at least, models of language rehabilitation depend on the authors' views about the nature of language, the nature of language processing, and the nature of aphasia. One's conception of the grammar and of the way it is represented in the brain influences the manner in which rehabilitation is approached. Papers in this section examine methods which explicitly rely on specific linguistic theories, e.g., structuralism, transformational generative grammar, or government and binding to elaborate language therapy models. This section also explores the role of linguistics, as seen by various authors. It asks such questions as whether linguistics is useful in the diagnosis of language disorders and/or as an aid in the development of explicitly principled therapeutic strategies.

For instance, in a behaviourist account, language tends to be equated with vocabulary, structures are assumed to be stored in memory and later retrieved, and language is assumed to be learned by listening and repetition. In a modular account, language is assumed to be represented independently of cognitive or emotional systems, independently of memory. Whether language structures are considered to be linear, "like beads on a string" (Jones, 1986), or hierarchically organized, in which case the mastery of one structure is assumed to presuppose that of an other (Hatfield and Shewell, 1983), will have an impact on the way treatment is planned. Likewise, the nature of aphasia will shape the kind of intervention that will be implemented, whether aphasia is regarded as a general defect of intelligence manifested most obviously in language (Goldstein, 1948), or a deficit specific to language, and if so, whether it is viewed as the disruption of the general process of language (Schuell and Jenkins, 1959), or of specific components; whether it is considered to be a performance deficit (Weigl and Bierwisch, 1970) or a competence deficit (Whitaker, 1971); or again, whether it is taken as the disruption of various levels of the microgenetic processing of the evolutionarily determined thought- language continuum (Brown, 1979).

Chapter 4 identifies the various contributions of linguistics to language rehabilitation methodology. If aphasia is the breakdown of the linguistic code, linguistics—the formal study of the language code—should provide a means by which to determine what specific parts of the grammar are unavailable or dysfunctional. At the very least, a linguistic theory provides a hierarchy of complexity of linguistic structure that can be used in the preparation of materials for diagnostic purposes, as examined in Chapter 5. As it turns out, the "linguistic" principles mentioned by various authors are not so much the principles of theoretical linguistics (e.g., transformational, stratificational, or relational grammars; tagmemics, generative semantics or government and binding theory), as those of applied linguistics (i.e., pedagogical grammars) and they refer mainly to surface structures. Indeed, there is a close parallel between language therapy and foreign language teaching, as each has been practised. That is, both therapy and foreign language teaching have evolved over the years from the teaching of metalinguistic knowledge to the practice of discourse in communicative situations.

One strategy used to compensate for a lack of grammatical competence has been greater reliance on pragmatic cues. This is examined in Chapter 6. There is indeed evidence that paralinguistic skills, such as gestures and pantomime, enhance communicative effectiveness (Helm-Estabrooks, Fitzpatrick and Barresi, 1982). For example, patients exposed to carefully graded filmed language instruction did not demonstrate significant gains in syntax, lexicon or semantics but did improve general communication by learning how to capitalize on, and make use of, situational cues and gestures (Di Carlo, 1980).

References

Brown, J.W. 1979. *Mind, brain, and consciousness.* New York: Academic Press.
Di Carlo, L.M. 1980. Language recovery in aphasia: Effect of systematic filmed programmed instruction. *Archives of Physical Medicine and Rehabilitation*, 61: 41-4.
Goldstein, K. 1948. *Language and language disorders.* New York: Grune and Stratton.
Hatfield, F.M. and Shewell, C. 1983. Some applications of linguistics to aphasia therapy. In C. Code and D.J. Muller (Eds.), *Aphasia Therapy.* London: Edward Arnold. Pp. 61-75.
Helm-Estabrooks, N., Fitzpatrick, P.M. and Barresi, B. 1982. Visual action therapy for global aphasia. *Journal of Speech and Hearing Disorders*, 47: 385-389.
Jones, E.V. 1986. Building the foundations for sentence production in a non-fluent aphasic. *British Journal of Disorders of Communication*, 21: 63-82.
Schuell, H.M. and Jenkins, J.J. 1959. The nature of language deficit in aphasia. *Psychological Review*, 66: 45-67.
Weigl, E. and Bierwisch, M. 1970. Neuropsychology amd linguistics: topics of common research. *Foundations of Language*, 6: 1-18.
Whitaker, H.A. 1971. *On the representation of language in the human brain.* Edmonton: Linguistic Research.

4 Linguistic foundations of rehabilitation methods

Joanne Wilk and Michel Paradis

INTRODUCTION

Within aphasia therapy programmes, treatment necessarily reflects an underlying theory of language, whether implicit or explicit. In contrast to "classical" therapy which is based on tradition or intuition, Hatfield and Shewell (1983) maintain that therapy must be "rational" in that it should be based on a "well-thought-out theory". However, even classical therapeutic programmes are based on some theory of language (although not necessarily "well-thought out" ones); it is implicit in the design. In examining this relationship between theory and therapy, Behrmann and Lieberthal (1989) state that theory guides intervention and therapy contributes to the evolution of theory. According to Byng (1988), theoretical development must also account for results from previous studies based on earlier versions of a theory.

In this chapter, the existing theories of language, both explicit and implicit, underlying aphasia therapy will be examined. First, an overview of theories applied in aphasia therapy that view language in terms of context-independent grammar will be presented. Within this paradigm, detailed aspects of language—the levels of phonetics, phonology, morphology, syntax, semantics, and the process of naming—have been applied to rehabilitation treatments and will also be reviewed. In response to the presumed inadequacy or unrealistic nature of therapy based on purely linguistic aspects of language, a trend in aphasia therapy based on a theory of the communicator's pragmatic competence has emerged. This theory will also be examined.

THEORIES OF LANGUAGE I: CONTEXT-INDEPENDENT GRAMMAR

Behaviourism

In the aphasia therapy literature, several trends exist with respect to assumptions about language. The first trend to be examined follows a behaviourist perspective, although not exclusively. According to this view, aphasia is perceived as the disruption of the general process of language (Culton and Ferguson, 1979; Davies and Grunwell, 1975; Davis, 1973; Kinsey, 1986; Kushner and Winitz, 1977; Montgomery, 1971; Naeser, 1975; Podraza and

Darley, 1977; Sarno, Silverman and Sands, 1970; Sefer, 1978; Sefer and Shaw, 1972). Language, within this paradigm, refers only to a surface level. Complexity of language is generally equated with length of output such that, within therapy programmes, the number of words in a determines its level of difficulty. These suppositions are assumed under the "classical" approach to aphasia therapy. More specifically, in some rehabilitation methods, language is equated with vocabulary, where "regaining vocabulary" is the main goal (Montgomery, 1971; Sefer, 1978) and is assessed only in terms of production (versus comprehension). Production is understood by some authors to be a reflection of comprehension (Culton et al., 1979; Kushner et al., 1977). It is also noted by these authors that comprehension precedes production in the process of re-learning language. As part of this emphasis on production, Sefer et al. (1972) note that the general language process includes all language (performance) modalities including articulation. While the process of comprehension is not defined, "grammar" is acknowledged in terms of linguistic structures which generally refer to s (Culton et al., 1979; Davies et al., 1975; Davis, 1973; Kushner et al., 1977; Naeser, 1975; Sparks and Holland, 1976). These structures are then "trained" in rehabilitation. Kushner et al. (1977) suggest that once these linguistic structures are understood, they are stored in memory, then retrieved later for use in production. It is also suggested that grammar becomes internalized through problem-solving (Culton et al., 1979; Kushner et al., 1977).

In terms of language (re-)acquisition, there is a consensus that language skills are learned (Culton et al., 1979; Davies et al., 1975; Davis, 1973; Naeser, 1975; Sparks et al., 1976). For instance, aphasia therapy has been based on training material from the first lesson of a program used for teaching "language comprehension" (Kushner et al., 1977). Aphasia therapy programmes following this paradigm have also been based on the revision of a TESL program (Naeser, 1975), on a language instruction method for deaf children (Davis, 1973), and on the explicit teaching of the rules of grammar, pronunciation and spelling (Sefer, 1978; Sefer et al., 1972). Theoretical differences arise over the status of the language learner in the learning context. While Kushner et al. (1977) hold that the language learner is an active participant in the language learning process, Sefer (1978) claims that language is learned passively by listening and through repetition. In fact, this author claims that tasks which draw attention away from simple listening and repeating are destructive to memory; why these tasks are deleterious is not explicated.

Theoretical Linguistics

Another approach to language in aphasia therapy identifies language according to principles of theoretical linguistics. Hatfield (Hatfield, 1972; Hatfield et al., 1983), in a review of linguistic theory and its application to aphasia therapy, suggests that therapeutic programmes be based on principles of structuralist linguistics. Accordingly, it is hypothesized that each language is composed of a

Linguistic foundations 103

closely structured autonomous system, or network, capable of being described without reference to the cognitive or emotional state of the individual speakers, or to such factors as fatigue, memory or attention (Hatfield, 1972, p. 64). An attempt is made by these authors to combine structuralism with the concept of generative transformational grammar by proposing that language is not only rule governed but also hierarchically organized. Language is considered hierarchical in that there exists a linguistic hierarchy such that, in aphasia rehabilitation, mastery of one structure may presuppose that of another (Hatfield et al., 1983), and it is rule governed in that the language "code" operates by a set of rules. However, a distinction made by Hatfield and her colleagues between "deep structure" and "surface structure" does not follow the theory of generative grammar that these authors are trying to establish. While referring to deep structure as beyond the surface phenomena, these authors are actually referring to the level of semantics. This is evident by their use of the term "the level of meaning relations" (p. 69) as equivalent to deep structure.

These authors (Hatfield, 1972; Hatfield et al., 1983) as well as others (Miller, 1989; Muma, Hamre and McNeil, 1986; Prins, Shoonen and Vermeulen, 1989; Ross, 1989), decompose language into distinct levels. The levels distinguished include phonetics (Miller, 1989; Ross, 1989); phonology (and in particular, segmental phonology [Hatfield et al., 1983] and prosody [Hatfield et al., 1983; Miller, 1989]); morphology; morphosyntax (Prins et al., 1989); syntax; and lexical semantics (Hatfield et al., 1983; Prins et al., 1989). Unfortunately, many of these levels are not well defined in their application to therapeutic programmes. Miller (1989) suggests that, despite the discreteness of these levels or "formal features" of language, the notions of connectivity, contrast and dynamism are crucial in language. Miller also emphasizes "functional features" of language—interactions between optimal speech, movement and language—since language functions within motor constraints. Muma et al. (1986) provide an outline of historical trends within each level. Specifically, Muma and his colleagues state that within semantic theory there has been a shift from generative semantics, with emphasis on semantic functions and relations, to interpretive semantics which stresses presuppositions and implicatures and finally to an extension to theory-of-the-world. In syntactic theory, the shift has been from structuralist models of language to generative transformational theory to present-day generative syntax. Finally, the author outlines the trend in phonological theory which has seen a shift in emphasis from the phoneme to syllable structure, stress and other prosodic domains.

While Muma asserts that these current trends in linguistic theory are invaluable for furthering our knowledge of aphasia, these trends have not been reflected in the aphasia therapy literature. In fact, despite consideration by many authors of the importance of linguistic theory in aphasia therapy (Crystal, 1972; Hatfield, 1972; Hatfield et al., 1983; Lesser, 1989; Miller, 1989; Muma, 1986; Ross, 1989), there has been no systematic application of a specific linguistic theory to a rehabilitation programme for aphasic patients (Lesser, 1989). Linguistic theory has only provided for assessment and analysis of linguistic deficits in aphasia (Hatfield, 1972; Lesser, 1989; Miller, 1989) There is, however, agreement about what linguistic theory can provide for aphasia therapy: a hierarchy of complexity of linguistic structures (Lesser, 1989; Ross, 1989).

There is also some consensus about the limitations of linguistics applied to aphasia therapy. First, linguistics cannot provide learning strategies that may be specific to the demands of the aphasic disorder (Hatfield, 1972). Second, while linguistic theory is important in the prognosis of the specific linguistic deficit, it must be combined with psycholinguistics and cognitive science to provide an overall theory of language behaviour in everyday situations (Hatfield, 1972; Ross, 1989).

Language Units

Some aphasia therapy treatment programmes have been based on theories about the specific units of linguistic structure. These hypotheses will be reviewed according to the categories of phonetics and phonology, morphology, syntax, semantics and lexicon. Discussions of naming will also be examined.

PHONETICS AND PHONOLOGY

Principles of phonetics and phonology have been applied to language treatment concerned with the level of sounds. To begin with, the principles of articulatory phonetics have been applied to phonologically-based intervention therapy. Miller (1989) states that articulatory phonetics supplies the information necessary to understand how and where the articulators are placed in the production of speech sounds. In therapy for speech disorders, including apraxia, this information is used to explain to patients how to physically form speech sounds (Dabul and Bollier, 1978; Miller, 1989). Distinctive feature theory is also presupposed in many phonologically based therapies. In rehabilitation programmes, this theory is used to define and assess the patient's phonological system. The resulting information may then be applied in devising a patient-specific rehabilitation programme (Hatfield, 1972; MacMahon, 1972; Miller, 1989). The principle of phoneme discrimination derived from distinctive feature theory—that the greater the difference between the distinctive features of neighbouring segments, the simpler the discrimination will be between them—has also been applied to specific therapies. In one programme, Sentence Level Auditory Comprehension (SLAC), phoneme discrimination is trained through the use of minimal pairs (Naeser, Haas, Mazurski and Laughlin, 1986). In another, (Systematic Therapy program for Auditory Comprehension Disorders in Aphasic Patients (STACDAP), phoneme discrimination identification tasks and auditory discrimination (same/different) tasks are used for the rehabilitation of patients with phoneme discrimination deficits (Prins et al., 1989).

Language rehabilitation programmes based on phonological theory focus on the level of the phoneme and the syllable. Typically at the phonemic level, patients are trained first to master individual phones and then (re)learn how to combine them. This method is used both for patients with a speech apraxia where the main goal is the production of speech sounds (Dabul et al., 1978) and for patients with conduction aphasia where the objective is controlled phonemic production (Cubelli, Foresti and Consolini, 1988,). Miller (1989) proposes a similar technique for patients with speech disorders called phonetic derivation, whereby one sound is derived via progressive approximation from another.

Attention in phonologically-based intervention has also focused on the syllable. According to some authors, the syllable is posited as a minimal unit of analysis at the level of phonological processing. In a programme for the rehabilitation of neologistic jargon aphasia, patients were given the number of syllables in a stimulus word as a cue in a picture-naming task in order to restore control of auditory feedback mechanisms (which are defined as language repetition and reading aloud) (Kotten, 1982). Similarly, Cubelli et al. (1988) devised a task for the rehabilitation of conduction aphasia where the patient must combine given syllables to form a stimulus word. In a second task, the patient must choose the correct syllables from an array. These two tasks were designed specifically to enable the patient to visualize the phonemic structure of words since the authors hypothesize that one of the causes of conduction aphasia may be the result of a deficit of the phonological processing system. In particular, phonemic paraphasias typical of conduction aphasia are hypothesized as errors at the level of phonological planning. These authors also utilize individual phonemes in therapy in a similar manner. In terms of phonological processing, monosyllabic words are held to be the most manageable for some aphasic patients. In a therapy programme designed for aphasics with writing impairments, Seron, Deloche, Moulard and Rousselle (1980) propose a three-level word-battery beginning with mono- and bisyllabic words (with neither consonant clusters nor graphic complexities), progressing to words with up to four syllables. In a different approach based on a cognitive neuropsychological account of naming, Lesser (1987) suggests that, in the rehabilitation of phonemic paraphasias, the patient may benefit from decomposing polysyllabic words into manageable single syllables. This method is proposed as a "classical technique", but is justified by positing that these simpler forms may facilitate association with the semantic lexicon. In terms of perception, it is implied by Laughlin, Naeser and Gordon (1979) that the syllable is the most basic unit of perception for words and phrases when these authors hypothesize that increased syllable duration within the MIT procedure will aid nonfluent aphasics with word and phrase production.

Phonological complexity within phonological intervention therapy has generally been defined in terms of consonant clusters. Cubelli et al. (1988), in a task for conduction aphasics, intentionally control phonological complexity of stimuli, defined as "sequences of consonants" (p. 243). Similarly, in the first level of a three-level word-battery for the remediation of aphasics with writing disorders, Seron et al. (1980) deliberately omitted consonant clusters, while the second more difficult level included longer words with consonant clusters. Rosenbek, Lemme, Ahern, Harris and Wertz (1973), on the other hand, explicitly hypothesize that there is a hierarchical arrangement of speech sounds and clusters based on empirical findings from apraxic adults. This principle is used to justify commencing therapy with "easy" phonemes, then systematically progressing through the hierarchy to more difficult ones. These authors also suggest that the initial phoneme of a word has special phonological status when they state that "careful selection of stimuli with regard to initial sound will increase response adequacy" (p. 463). The notion of prosody within phonological processing has also been noted in the aphasia therapy literature. An abstract phonological level is explicitly proposed by Davies et al. (1975) in a linguistic evaluation of their patient, which is held to include the stress pattern, number of syllables, vowels and consonantal features of a word. While this hypothesis was noted in the assessment, it was not incorporated within the structured treatment programme. Similarly, Kotten (1982), in an account of the disturbed phonology of neologistic utterances includes prosody in the definition of phonology and states that the combination of phonemes and intonation are rule governed. Here also, these hypotheses are not applied to the therapy programme proposed for aphasics with neologistic jargon aphasia. Together, Davies et al. (1975) and Kotten (1982) contend (in the manner previously mentioned) that there are language-specific phonological systems and/or rules which native speakers will never violate. Kotten further claims that there are language-specific intonation rules.

Miller (1989) also notes the importance of prosody, especially grammatical stress and intonation, in word meanings. Based on the premise that control over pitch and loudness will allow patients with speech disorders to express grammatical and affective concepts, Miller proposes that contrastive stress drills will be effective for rehabilitation. Contrastive stress drills include producing different meanings by altering the stress/intonation pattern with the same basic syntactic structure and/or using a carrier word or phrase with contrastive stress to give different meanings with different filler words.

Implications of the organization of the phonological system based on theories of reading and writing have also been found in the aphasia therapy literature. In order to enable conduction aphasic patients to relearn phonological coding abilities, Cubelli et al. (1988) gave patients written syllables which were to be combined to form words which were then to be read aloud. In a second task, individual letters, which were also to be combined to form words, were given to the patient. Justification for these tasks was given on the basis of specific therapeutic principles, including having the patient focus on the formal structure of the phonological word form through metalinguistic judgments. These therapeutic methods imply that, minimally, there must be written letter representations within the phonological system such that writing and reading aloud proceeds through combining single written letter representations and/or written syllable representations to form words. Alternatively, Seron et al. (1980) implicitly presuppose that written letters are, at some point, represented in auditory form. This is evident by the criteria set by these authors for the word battery of a programme for the treatment of aphasics with writing impairments. Deliberately excluded from the first level of the word battery are written words containing graphic complexities, i.e., instances in which one phonemic value may be represented by two or more different spellings. Lesser (1987) combines both of these notions and presents a cognitive neuropsychological transcoding model of reading aloud and writing to dictation. This model includes (1) an orthographic-to- phonological conversion route for reading which is dependent on regularity of spelling; (2) a lexical route for reading aloud common words with irregular spellings without accessing meaning; and, (3) a main route consisting of direct reading through meaning, via the cognitive system. Lesser also supports the separation of a lexicon for written words ("visual input lexicon") and spoken words ("auditory input lexicon"). This model is the basis for the diagnosis of such reading disorders as surface and deep dyslexia. Similar models have been proposed for writing to dictation, repeating spoken words and copying written words. Lesser maintains that there are two ways in which therapy may proceed based on these models. One tactic is to reactivate the impaired function and the other is to reorganize the remaining pathway by finding an alternate route. Aphasia therapy has been found to emphasize the reorganizational approach. An example of this approach is offered by Hatfield (1983) who devised a task to aid patients with deep dysgraphia to re-learn to spell function words by associating them with content words of similar spellings (Lesser, 1987). De Partz (1986), on the other hand, proposed a method based on a reactivation approach. The programme was designed to reactivate the grapheme-phoneme conversion route of a patient with deep dyslexia. The first part of this method included teaching the patient to associate letter names with familiar words such as family names then learning to associate the letter name with its phonic sound by prolonging the pronunciation of the initial phoneme of the associated word (Lesser, 1987).

Thus, theories applied in aphasia therapy at the phonological level include principles of articulatory phonetics, the notions of a basic unit and phonological complexity, prosody and the principle of a universal phonological system and theories of the phonological system involved in reading and writing.

MORPHOLOGY

Emphasis on morphological theory in aphasia therapy is not extensive. However, some therapy programmes have included tasks for the rehabilitation of the morphological system. For example, Burton, Burton and Lucas (1988) designed 14 general language tasks suitable for all types of aphasia, one of which was relevant to morphology, in order to examine the role of the microcomputer in aphasia therapy. The pertinent task tested compound words where a word is matched to another word so as to make a third, compound word. "Car - pet" was offered as an example (p.484). It can be inferred from this example that the authors presuppose that words at some level are derived from syllables. In a different account, Sefer (1978) states that monosyllabic words which have simple spellings will be easier for aphasic patients to manage in therapy. This therapeutic strategy suggests that the author presupposed that words at some point are processed on the basis of spelling. While these studies do not explicitly attempt to rehabilitate disordered morphological processing, a study by MacMahon (1972) does examine morphological breakdown in aphasia. In a review of linguistic research into aphasia, the author presents one study of morphological breakdown (Goodglass and Hunt, 1958) which examined the parallel between the acquisition and breakdown of morphological inflection. The results of this study suggested that morphemes are hierarchically ordered in terms of difficulty or abstractness. For example, the possessive marker is more difficult than the plural marker.

Other researchers have suggested an intimate relationship between morphology and syntax. In STACDAP (Prins et al., 1989), a multi-level programme for the rehabilitation of receptive language deficits, a "morphosyntactic level" was designed for patients having problems with the comprehension of various morphological aspects and syntactic constructions. These tasks involve the training of verb tense and aspect, gender and case of personal and possessive pronouns, and word order in active and passive s. Jones (1986) contends that s are not made up of words like beads on a string but are complicated patterns of interactions mapping the relationship between semantics, syntax, phonology and pragmatics. Thus, he further states, it is important in therapy procedures for non-fluent aphasics to be aware that certain omissions or substitutions of phonological units may come about from the units' syntactic role. Sefer et al. (1972) reiterate this notion and conclude that therapy should strive to increase length of retention span and familiarity with different forms of s which will help the aphasic person overcome these difficulties.

More specifically, Pierce (1982) states that aphasic patients may be unable to use morphological surface structure markers in the comprehension of syntactic patterns that deviate from the simple subject-verb-object present tense form. As a result, Pierce recommends adding additional surface structure markers as a means to compensate the patient's inability to use the syntactic information conveyed by morphemes in a . Davis (1973) further asserts that some morphological processes, such as inflections, are classified as minor transformations to s in accordance with a modified version of transformational grammar. Based on this hypothesis, Davis developed the Sentence Construction Board which allows for the orderly presentation of syntactic patterns increasing gradually in complexity.

In conclusion, where aphasia therapy programmes have included morphological rehabilitation, they have focused both on individual morphological processes, and the interdependent relationship between morphology and syntax.

SYNTAX

Presuppositions about structure in aphasia therapy have followed both linear and non-linear accounts. A common view of syntax, which tends to follow the paradigm of aphasia as the disruption of a non-hierarchical language process,

Linguistic foundations 109

interprets s linearly. In this "classical approach" to the treatment of non-fluent aphasia, it is assumed that construction begins at the level of words, where they are combined to create phrases which are eventually combined to produce s (Hatfield et al., 1983; Sarno et al., 1970; Sefer et al., 1972). This is reflected in therapy which begins by "training" word retrieval, then phrases and finally s. It has also been implicitly hypothesized that new s are created by filling in the "slots" of a basic pattern with different lexical items (Culton et al., 1979; Davies, 1975; MacMahon, 1972; Naeser, 1975; Shewan, 1976; Sparks and Holland, 1976). In therapy, the basic pattern is drilled with the intention of generalizing to different lexical items. Naeser (1975) further suggests that slots may be predefined. For instance, in one type of pattern, "that" is bound to the first slot of a specific type. It has also been suggested that there is an order of difficulty for s (e.g., present is simpler than past [Culton et al., 1979]), and types (Hatfield et al., 1983), but justification for these claims was not provided. In these cases, the therapy programme provides a fixed progression of s/sentence types that are to be trained. In contrast, Helm-Estabrooks and her colleagues (Helm-Estabrooks, Fitzpatrick and Barresi, 1981; Helm-Estabrooks and Ramsberger, 1986) propose the use of a syntactic hierarchy for agrammatic patients founded upon the observation of the order of appearance of syntactic structures in agrammatic patients in general. In this case, the syntactic structures are elicited through a story-completion exercise.

Some linear accounts of syntax claim to follow the principles of transformational grammar. However, this theory has been found to be misinterpreted in its application. For example, in one therapy programme explicitly based on Chomsky's theory of transformational grammar, basic notions of this theory are present, but the premise of the approach is inherently linear. In one rehabilitation programme, aphasic patients are required to fill in the slots for three different patterns (Naeser, 1975). Each type has three slots that must be filled in with a specified constituent. However, the progression from Type I to Type III involves a change in the type of constituent (i.e., transitive versus intransitive verb), not a change in the actual structure. Similarly, the Sentence Construction Board (Davis, 1973), an electronic device with lights that represent constituents and their transformations, was also devised using the model of transformational grammar. In this case, as was stated earlier, transformational grammar theory was modified to include morphological processes as "minor transformations" to basic sentence types. Hatfield et al. (1983) also maintain that their assumptions about language are grounded in transformational theory. However, as previously mentioned, this assumption is predicated upon the assumption that syntax is decomposed into a deep structure—the "level of meaning relations"—and a surface structure—the level of "basic sentence types". These authors are posing a relationship between sentence structure and meaning and do not actually apply the model to a rehabilitation regimen.

Models of sentence processing according to this linear approach are also found in some aphasia therapy programmes. Following the notion of sentences as fill-in-the-slot structures, it was proposed that, when individuals process two-word phrases, the subject slot may have to be filled in mentally when the subject is not given (Culton and Ferguson, 1979). The formulation of this hypothesis was based on the results of a "comprehension training" program in which the aphasic subjects found two-word sentence constructions more difficult than the longer sentences. This was in contrast to the authors' predicted results of a syntactic progression of difficulty where sentence complexity was defined in terms of length (i.e., the number of words). Another interpretation of these results proposed by these authors suggests that since longer sentences contain more "language material", redundancy may play a role in sentence processing; but they do not elaborate upon this concept further.

Davis and Tan (1987) propose "reverse chaining" as a more natural way of forming sentences. According to this account, in order to construct a sentence, the last word is found first, then the last two words, and so on, until the complete sentence is formed. This technique was directly applied to a language stimulation programme by these authors as a means for an aphasic patient to produce complete sentences. In contrast, in a study comparing behavioural and stimulation approaches (Thompson and McReynolds, 1986), "forward chaining" (the opposite of reverse chaining) is suggested as a means of sentence processing (within the behavioural approach) where it is used in therapy as a form of modelling.

While the linear approach to the rehabilitation of syntax in aphasia offers models of sentence structure and processing, the possibility of a different "unspecified" theory of sentence formulation is acknowledged by Sparks et al. (1976) within Melodic Intonation Therapy (MIT).

In contrast to the linear explanation, more recent, non-linear accounts of syntax have been hypothesized and applied to language therapy programmes. In general, these assumptions follow principles of generative grammar.

For example, the theory was applied in the development of the Sentence Construction Board (Davis, 1973). As previously noted, this electronic device represents syntactic constituents and their transformations by means of branching lights. In another explicit application of transformational grammar, it was stated that the training of the basic underlying "simple active affirmative declarative", which can take many different surface forms, would generalize to untrained exemplars with the same underlying structure (Shewan, 1976).

In accordance with a generative account of syntax, an abstract underlying syntactic structure has been proposed. In one instance, a therapeutic programme for non-fluent aphasia was based on the hypothesis that normal sentence processing occurs in two levels, namely, the functional level and the positional level. (Jones, 1986, p.65). The functional level is said to include the selection and insertion of lexical items into a predicate-argument structure, whereas at the positional level, surface structure markers are implemented. The resulting therapy goal is to rehabilitate the client's ability to map meaning relations between semantics and syntax. This assumption is fairly common in sentence rehabilitation programmes and will later be reviewed in detail.

Linguistic foundations 111

In a second case, the notion of generative response class has been applied to syntax rehabilitation programmes (Kearns and Salmon, 1984). According to this hypothesis, despite possible developmental and grammatical differences, there may exist an underlying functional relationship between similar surface forms in the same "response class." Thus, there will be generalization from a trained form to an untrained form within the same class. This generalization effect is consistent with generative grammar such that reactivating/reacquiring one structure will generate structures that have the same underlying syntactic structure. Based on this theory, it has been predicted that, given their common surface forms, training auxiliary "is" or copula "is" verbs may result in generalized responding to the untrained reciprocal (Kearns and Salmon, 1984), and that the training of one Wh-interrogative will generalize to other Wh-interrogatives (Thompson and McReynolds, 1986; Wambaugh and Thompson, 1989), and finally, that training one locative will generalize to the use of other locatives (Thompson, McReynolds and Vance, 1982). In clinical trials it was generally found that generalization occurred only within a linguistic form and not across forms. These authors concluded that the linguistic forms of inquiry therefore did not belong to the same response class despite their similarities in surface structure.

A different hypothesis concerned with deep structure and sentence processing has also been applied to a syntactic comprehension rehabilitation programme. This theory proposes that during sentence processing, the listener uses a canonical word order strategy that states the following: Assign a Subject-Verb-Object deep structure relationship to a Noun-Verb-Noun surface structure unless otherwise marked (Pierce, 1982). Deviating surface structure features, for example the passive verb form, signal the irregularities to the normal listener so that the sentence may be appropriately comprehended. Based on this theory, the author proposes that, by increasing the syntactic constraints and adding redundancy to a sentence through the addition of surface structure markers, auditory comprehension of sentences will improve (Pierce, 1982). It can be assumed that this notion was also presupposed in STACDAP (Prins et al., 1989) where, at the level of morphosyntax, the progression of difficulty depends on the number and type of surface structure markers.

The interactions between syntax and other linguistic elements, including phonology, morphology and semantics, have also been considered in aphasia therapy programmes. The interaction between morphology and syntax is apparent in STACDAP (Prins et al., 1989), a multi-level auditory comprehension programme for patients with a wide range of language deficits. At the level of morphosyntax, morphological aspects of grammar, like inflection, are combined with syntactic constructions, like word order and active and passive forms, to form a hierarchy of difficulty at the sentence level. At this level, patients with aphasia of any type (with the exception of global aphasia) may be trained through this progression of difficulty to improve their comprehension of various morphological aspects and syntactic constructions.

In other cases, an explicit distinction has been made between phonological and syntactic as well as semantic functions. In "classical" therapy, the difficulty that agrammatic patients have with certain words has been attributed to the syntactic function of those words. Thus sentences are trained in therapy. However, Jones (1986) points out that words in a sentence have more than just a syntactic function. They also carry a phonological component which may have an effect on how the agrammatic patient understands and produces those words. As a concrete example, Jones (1986) states that prepositions such as "in", "on", or "up" may be classified according to their syntactic function—transitive or intransitive, or they may also be classified according to whether or not they receive stress. In this case, intransitive prepositions take stress but the transitive ones do not. Hence, the difficulties that prepositions pose for the agrammatic patient may not be based purely on syntax. Therefore in therapy, the clinician should be aware of the influence of phonology on syntax. Lesser (1987) is in general agreement with this distinction between syntax and phonology, and, like Jones (1986), stresses that the actual disorder in agrammatism occurs at a lower level than the phonological production of an utterance. These authors further propose that the first stage of sentence planning begins at the semantic level. As previously noted, Jones (1986) follows the proposal that sentence formation proceeds by the speaker first choosing the appropriate lexical items, then creating a predicate-argument structure, and finally assigning the lexical items to the roles within that structure. Lesser (1987) presents a similar model of sentence formation based on psycholinguistic research that asserts that mapping from the semantic lexicon to the phonological lexicon occurs simultaneously with the organization of a syntactic planning frame which arranges the lexical items and assigns the frame elements appropriately. The sentence is then ready for phonological assembly.

Byng (1988) also bases a therapy for "sentence processing deficits" on an assumption about mapping in sentence processing. In this case, the author follows the hypothesis of Schwartz, Linebarger and Saffran (1985) who suggest that agrammatic patients are unable to co-ordinate descriptions of sentence form and descriptions of sentence meaning (Byng, 1988, p. 632). Based on this "mapping hypothesis", Byng proposes that sentence comprehension is composed of three parts. First, a syntactic parsing mechanism determines the grammatical relations within the target sentence, then the thematic relations attached to the verb are found, (i.e., the arguments bound to the verb are specified), and finally, this information is combined through some as yet unspecified mapping procedure and the sentence is comprehended. While this account is only intended to explain sentence comprehension, the author states that in production there is mapping in reverse.

In rehabilitation programmes based on the intimate relationship between syntax and semantics, it is agreed that therapy for patients with syntactic deficits should begin at the level of meaning relations. Therapy according to this view has been found to focus on training specific relations, i.e., element and category, (Kotten, 1982); on (re-)learning the predicate-argument structure (Jones, 1986);

on the differentiation of various thematic roles (Lesser, 1987); and on the comprehension and production of thematic relations (Byng, 1988).

According to the mapping hypothesis, it is the verb's argument structure that dictates the additional elements that must be included in the sentence (Lesser, 1987). This theory—that the verb plays a crucial role in the specification of content words—is implicitly supported by those authors who claim that sentence completion tasks aid in naming (Chin Li, Kitselman, Dusatko and Spinelli, 1988; Chin Li and Williams, 1990; Helm-Estabrooks, Emery and Albert, 1987; Kotten, 1982; Podraza et al., 1977; Seron, Deloche, Bastard, Chassin and Hermand, 1979; Wiegel-Crump and Koenigsknecht, 1973). As a therapeutic technique, naming is facilitated with the use of an open-ended sentence: The verb's argument structure constrains the type of word that can be used to complete a sentence and thus the search for the label is limited to only those words which successfully meet the verb's criteria.

A theory about performance at the sentence level has also been applied by Kohn and Smith (1990) in therapy designed to improve performance errors through sentence repetition. According to these authors, in comprehension, a temporary phonological trace is stored in working memory. Performance will then be affected by many factors, including length (the number of words and syllables), and richness of semantic content (which is manipulated by functors and light verbs which carry little semantic meaning in contrast with substantives). Since a hierarchy of difficulty is not presented, it is not possible to discern whether "rich" semantics helps or hinders performance.

In summary, according to a non-linear framework for syntax, many of the authors reviewed proposed features of syntactic structure, and the relationship of syntax to phonology, morphology and semantics. A mapping hypothesis has also been proposed which offers an explanation of sentence processing. Finally, a hypothesis about sentence performance was presented.

SEMANTICS AND LEXICON

Many language rehabilitation programmes are founded on principles of semantics and the lexicon. Most semantic treatment programmes either explicitly or implicitly presuppose some kind of organization of the lexicon. These presuppositions of meaning structure and classification will be surveyed.

Meaning is assumed to be systematically and hierarchically organized in the lexicon. It is systematic in that meaning has been hypothesized to be organized in discrete categories. It is hierarchical in that a top-down model of semantic representation has been proposed based on Rosch's (1975) category levels—superordinate, basic and subordinate (Behrmann and Lieberthal, 1989). Behrmann et al. (1989) hypothesize that if a category-specific, hierarchical organization occurs, then it follows that treatment should represent this organization. Treatment developed according to this view includes the retraining of category-specific items with the prediction of generalization to untreated lexical items within the same category and across categories (Behrmann et al., 1989; Hatfield et al., 1983; Seron et al., 1979; Wiegel-Crump et al., 1973). Generalization to untreated lexical items within the same category has been found. Generalization across categories has generally been found to be minimal. These predictions are consistent with the hypothesis that word-retrieval difficulties are the result of a loss of access to the lexicon, and not a loss of the lexical store (Seron et al., 1979; Wiegel-Crump et al., 1973). Thus, it is concluded that treatment of categories is more efficacious than the "classical" approach of training a large number of random lexical items.

This postulation of a hierarchical organization of the semantic system is one of the common presuppositions applied to semantic treatment. On the basis of neuropsychological observation, it has subsequently been posited that meaning is accessed from the general superordinate knowledge to more specific, subordinate details. The application of this hypothesis to therapy consists of first teaching the superordinate features of a given category and then familiarizing the patient with the details of particular items within the category. The use of categories in therapy to rehabilitate word-finding deficits has also involved the use of category-related lexical items at the superordinate or basic level as a cue for the target word (Chin Li et al., 1990; Howard, Patterson, Franklin, Orchard-Lisle and Morton, 1985a, 1985b; Podraza et al., 1977; Ross, 1989; Seron et al., 1979; Wiegel-Crump et al., 1973). While the results of many of these studies confirm the usefulness of these cues in facilitating word finding, Podraza et al. (1977) found that the presentation of semantically related words proved to be counterproductive for naming, and hypothesized that while de-blocking of the category may have occurred, it may have also confused the patient. The confusion may have resulted from a difficulty in selecting the proper label from the alternatives.

Other therapeutic techniques have included matching a basic category item to its superordinate category and matching one basic category item to two other items within the same category (Burton et al., 1988), and answering a yes/no question requiring the patient to access the meaning of the name (e.g., "Is a cat an animal?" [Howard et al., 1985b]). Auditory word-to-picture matching has also been used as a facilitator of word retrieval. This involves having the patient point to a picture of a spoken target from an array including semantically related foils (Howard et al., 1985a, 1985b; Prins et al., 1989). Pointing to a semantically related picture of a spoken target word is a second method, but it should be mentioned that this technique was also developed on the notion that auditory stimulation of one lexical item would serve to de-block the category, thus facilitating the retrieval of all other members of that category (Howard et al., 1985a).

In terms of the internal structure of the categories themselves, Behrmann and Lieberthal (1989) note that double dissociations have been found between words with concrete and abstract features as well as animate and inanimate features which suggest the possible arrangement. Behrmann et al. (1989) point out that, despite neuropsychological evidence to support the dissociation of animate and inanimate features, it is possible that this dichotomy may be due to the distinction between items on the basis of functional versus physical/sensory attributes. Thus, it is insinuated that the feature-defined characterizations that describe the categories should be taken into account in the formation of semantic rehabilitation therapy. Similarly, Hatfield and Shewell (1983) propose that the categories are defined by these distinctive features such that lexical items are identified and contrasted by presence or absence of certain attributes such as "animate" versus "inanimate". It is further stated by these authors that in any semantic dichotomy, the more common attribute will be understood as unmarked while the other less common feature will be considered marked. It is advised that in therapy, the clinician should draw attention to meaning-differentiating or meaning-integrating features.

In addition to the categorial representation of the semantic system, another prevalent assumption considers the semantic system as a network of semantic associations. In many semantic treatment programmes, different types of cues are used to facilitate word-retrieval. These cues implicitly suggest lexical organization since they imply that the label of a lexical item must be associated with the cued information in some way. The nature of this organization has not yet been specified.

Specific relations revealed by the postulation of cues as facilitators in the process of word retrieval include antonymy (Hatfield et al., 1983) and synonymy (Burton et al., 1988; Hatfield et al., 1983; Wiegel-Crump et al., 1973). It has also been proposed that lexical items that are used more frequently in speech are more accessible than infrequently used words and thus should be included as targets in therapy (Hatfield et al., 1983; Helm-Estabrooks et al., 1987; Kotten, 1982; Montgomery, 1971; Ross, 1989; Sefer, 1978; Sefer et al., 1972). According to this view, a bottom-up model of word-retrieval is presupposed by one researcher, such that cueing with an infrequently occurring word will facilitate retrieval of a more common and frequently occurring one (Montgomery, 1971). De-blocking is an alternative explanation for this effect.

In some cases, semantic cues expose underlying relationships between lexical items and their non-linguistic attributes. For instance, lexical items must, in some way, include in their description associated pragmatic features such as representational gestures (Helm-Estabrooks et al., 1987; Ross, 1989; Seron et al., 1979; Wiegel et al., 1973) as well as the description of the action carried out by the word (Chin Li et al., 1990), the situational context in which the target occurs (Chin Li et al., 1990; Kotten, 1982; Ross, 1989), and descriptive "encyclopedic" information such as functional descriptions (Ross, 1989), the target's colour or its composition (Burton et al., 1988) and tactile information (Helm-Estabrooks et al., 1987). Reference is also made to other characterizations of the meaning of words such as its concreteness or "imageability" (Chin Li and Williams, 1990; Hatfield et al., 1983; Helm-Estabrooks et al., 1987; Ross, 1989) and "affective aspects" (Helm-Estabrooks et al., 1987; Ross, 1989; Sefer et al., 1972). In fact, Sefer et al. (1972) postulate that the "personal significance" of a word has more influence on its accessibility than abstractness or length.

Lesser (1987) considers the distinction made by some linguists between semantics and knowledge of the world. "Semantics" refers to the paradigmatic relationships among word meanings which may be conceived of as hierarchically organized, as patterned in associative networks, or as organized in semantic fields. This is in contrast with the "residual" knowledge we have about words that is not part of a word's hierarchical or associative structure. In addition, it is hypothesized that both denotative and connotative aspects of word meaning are included within the lexicon, and that they may not be coextensive with conceptual knowledge or knowledge of the world. This conjecture is said to be applicable to therapy since the connotative aspects of meaning have been found to be spared in aphasic patients.

All of these aspects of meaning have been employed in the facilitation of word-finding deficits and reveal the presupposed network of associations within the lexicon.

An important relationship and distinction between the phonological form of a word and its semantic representation has been suggested by many authors concerned with the rehabilitation of naming in aphasia. This segregation is implied by the use of phonological cues to facilitate word-finding. These cues include the use of the first phoneme (Helm-Estabrooks et al., 1987; Podraza et al., 1977) and/or syllable in a word (Wiegel-Crump et al., 1973).

In support of the use of phonological cues to assist in word-finding, it has been hypothesized that, at the phonological level, certain abstract phonological features other than the initial phoneme may be available to the patient with word-finding difficulties. These features include the number of syllables, primary stress, and phonetically similar words. Thus in word-retrieval, the phonemic cue provides the first phonetic element in the available abstract form which will assist the patient in locating the remainder of the phonetic pattern of the target word (Podraza et al., 1977).

Rather than having the clinician supply a phonemic cue, another therapeutic technique entails having the patient self-generate these cues. In one study, it was hypothesized that in order to generate a phonemic cue one must (1) indicate the initial letter of the word which cannot be retrieved, (2) convert this letter into its corresponding phoneme, and (3) use the phonemic cues as an aid to word retrieval (Bruce et al., 1987). Since it was found by these authors that patients with Broca's aphasia had difficulty with this process of transformation, a computer was employed to supply the missing information in therapy. Similarly, Ross (1989) suggests that word-finding difficulties may be facilitated through self-generated cues by either finding the first letter on an alphabet board or searching aloud or silently to find the first sound or syllable to access the lexical item via the phonological route. The use of the first letter as a cue for improving written naming in neologistic jargon aphasia has also been proposed (Kotten, 1982). However, a theoretical basis for this approach was not provided.

The separation between the phonological and semantic systems has also been proposed by some researchers who claim that phonological processing occurs at a later stage than does semantic processing. Based on this notion, it has been hypothesized that, depending on the nature of the word-finding difficulty, semantic cues may be more facilitative than phonemic cues or vice versa. However, in clinical trials Howard et al. (1985b) found a relatively small advantage for semantic cues over phonological cues in naming objects, whereas Chin Li et al. (1990) found that phonemic cues were beneficial in increasing word naming ability in nouns, and semantic cues proved advantageous for verbs. The differential effects of cueing have also been explained by a mapping hypothesis which states that items from the semantic lexicon are mapped onto their phonological shapes in the phonological lexicon, after which their phonetic realization is prepared and sent for motor output. Thus, depending on where the breakdown occurs, different cues will be more facilitative than others (Lesser, 1987).

Similarly, in a programme for the rehabilitation of conduction aphasia, an implicit assumption that the phonological form is dissociated from the semantic, and that phonological processing occurs at a different stage is made. In this reeducation program for facilitating repetition abilities, it is maintained that patients with conduction aphasia tend to look at the meaning but not the form of language and therefore, exercises must prevent the use of compensation strategies employing the spared lexico-semantic system.

General features of the lexicon itself have also been proposed and applied to aphasia therapy. In the rehabilitation of agrammatism, Hatfield and Shewell (1983) expressly presuppose a primitive grammatical system consisting of a small lexicon composed of nouns, verbs or verbal forms (probably -ing forms), adjectives and adverbs. Accordingly, in therapy, the patient practices expressing elementary propositions in response to thematic picture stimuli, the structures of which comply to this primitive system.

More recently, Behrmann and Lieberthal (1989) suggest that there are at least two functionally independent semantic systems—one visual and one verbal. Consequently, remediation may be necessary for both pictures and/or objects, and for words. A further distinction between concepts or knowledge of the world and the verbal semantic lexicon has also been made (Lesser, 1987). Denotative aspects of meaning within the verbal semantic lexicon are also distinguished from connotative aspects to form a model of the cognitive/semantic system. Based on this model, the type of therapy depends on the stage of breakdown.

While most therapeutic programmes that centre on the rehabilitation of naming focus their attention on nouns, the relationship between the semantic structure of verbs and their relation to syntax has also been considered. As noted earlier, verbs have been hypothesized to carry argument structure and as a result, it can be inferred that these thematic relations are associated with the verb in the lexicon (Byng, 1988; Jones, 1986; Lesser, 1987).

NAMING

Since semantic treatment programmes usually involve the remediation of naming deficits, different theories of naming have been hypothesized. Naming models hypothesized in a cognitive neuropsychological framework are common in their application to aphasia therapy. As previously mentioned, a neuropsychological model of naming has been proposed by Lesser (1987). According to this model, first, through the connection from the semantic lexicon to the phonological lexicon, the phonological form is found. Next, the phonemic pattern is selected and arranged in order. Finally the form is phonetically realized, planned and executed. Therapy is planned according the stage at which a breakdown has occurred. In a second account, Chin Li et al. (1990) suggest a neuropsychological information processing model of picture naming based on Howard and Orchard-Lisle's (1984) model. According to this account, following initial picture recognition, the appropriate semantic representation is retrieved. This is used to activate the corresponding entry in the phonological output lexicon. This phonological representation is then converted into a motor articulatory program, which finally results in the spoken word (p.56).

It is thus speculated that the phonemic cue operates at the articulatory level such that facilitation with this type of cue suggests an intact semantic system. Conversely, if a semantic cue is beneficial, a complication may exist within the semantic system itself. Thus, the use of a phonemic cue would be redundant (assuming the phonological system is intact).

In another study, Howard et al. (1985a) account for picture naming with prompts (i.e., phonemic cues, target word prestimulation) through the combination of models, including the logogen model of Morton (1979) and Morton and Patterson (1980) which states that, on the basis of activation of an input logogen unit (spoken or written word), a set of semantic features corresponding to the stimulus word is activated in the cognitive system. In output, these semantic features from the cognitive system are used to address a phonological output logogen, which contains an abstract phonological specification of the articulatory word form (p. 75).

A model of semantic representation that distinguishes two levels of representation, one visual, the other verbal, is also included to account for the facilitative effect of non-lexical information (a picture) on naming. It is concluded however, that despite the beneficial effect of prompts on naming, they may not be effective in facilitating word-finding abilities over time. These authors further argue that techniques that require patients to access the semantic representation corresponding to the picture name (word-to-picture matching with either auditory or visual presentation, and semantic judgements) may have longer lasting effects than some prompts, despite the lack of a verbal response from the patient. It is further cautioned that even this technique may not be effective in therapy since positive effects have only been documented to last 24 hours.

In a different report of naming, Hanlon, Brown and Gerstman (1990) base their hypothesis on a microgenetic model of language. According to this model, the development of a name will go through a series of stages in a manner which recapitulates the evolution of the brain. Consequently, it is surmised that pointing with the right hemiplegic limb to the object to be named will assist the nonfluent aphasic in a confrontation naming task. It was concluded by these authors that functional activation of the proximal motor system might tap into early levels in speech production and thereby possibly facilitate verb expression.

Various other explanations of naming have also been proposed. One type of therapy uses a memorization technique to stimulate the acquisition of meaning. Generally, a target word is repeated frequently as stimulation for the patient to learn (memorize) the word (Helmick and Wipplinger, 1975; Kushner et al., 1977). In this auditory comprehension training approach, it is postulated that linguistic information should be firmly established in memory storage prior to training in retrieval (Kushner et al., 1977).

Edwards (1987), on the other hand, emphasizes the importance of pragmatics in naming. This author notes that in the process of reference identification in normal communicative contexts, pragmatic features such as prosody, gesture and context are used. Thus attention in therapy should focus on context as well as the linguistic unit to be trained.

In naming tasks, aphasic patients tend to perseverate—use the same response across tasks—thus interfering with their ability to successfully name. In Helm-Estabrooks and Albert's (1987) Treatment of Aphasic Perseveration (TAP), it is hypothesized that the mechanism causing perseveration can be raised to a conscious level and actively inhibited.

Many aphasic treatment programs are based on theories of semantics and the lexicon. These hypotheses cover the organization of meaning in the lexicon as well as assumptions about the structure of the lexicon itself. Furthermore, theories of naming have also been proposed, including, among others, those that follow a neuropsychological paradigm.

In summary, many aphasia therapy programmes are based on theories of language in terms of "linguistic structures". Some of these accounts presuppose a theory based on structuralist principles, while others are founded on principles of generative grammar. When focusing on the rehabilitation of specific types of linguistic structures, aphasic treatment programmes have been based on theories of phonetics and phonology, morphology and syntax—linear and non-linear accounts—and on theories of the organization of meaning and the structure of the lexicon. Specific hypotheses about the process of naming have also been applied to rehabilitation programmes.

THEORIES OF LANGUAGE II: COMMUNICATIVE COMPETENCE

Contrary to the view of language as context-independent grammar, many authors regard language in terms of natural communication. Language is said to have evolved as an interactive tool (Gildston and Gildston, 1986). Thus both grammatical and pragmatic competence are considered necessary for the interpretation and use of language in context (Edwards, 1987; Muma et al., 1986). Lesser (1989) refers to this approach as a sociolinguistic one, but this term suggests that only social circumstances affect language. However, the communication paradigm refers to more than just sociolinguistic competence, and therefore the general term "pragmatics" is more suitable. Pragmatic competence has been theorized in the aphasia therapy literature to include the use of discursive context, extralinguistic context, paralinguistic context, conventions of natural conversation, world knowledge, and humour for the interpretation of meaning in natural (context-dependent) language. In therapy, these factors are included within the physical setting and the structure of therapy itself.

Since language is viewed as more than just linguistic constituents (Green, 1984), or speech comprehensibility (Miller, 1989), emphasis in aphasia therapy in this framework is placed on successful communication (Potter and Goodman, 1983; Weniger, Springer and Poeck, 1987). In other words, emphasis is placed on the patient getting the message across (Aten, 1986; Davis and Wilcox, 1981). A "whole person, whole context" perspective is taken (Miller, 1989, p.296). As a result, the ultimate goal of therapy is maximum communication potential versus production of specific linguistic structures as presented earlier.

General theories about language following the communication paradigm have been suggested. Edwards (1987) hypothesizes a top-down model of language that begins at discourse factors and works down to the molecular or linguistic units of language. The significance of the functions of language is also noted as important in the creation of therapeutic goals. These functions include establishing interpersonal relationships, regulating the behaviour of others, satisfying material needs or desires, exploring and organizing environments and exchanging messages and information (Green, 1984: 36; see also Wambaugh and Thompson, 1989).

Pragmatics

The foundation of communication therapy rests on assumptions about pragmatics. These presuppositions, to reiterate, include the use of context (discursive, paralinguistic and extralinguistic), rules of natural conversation, world knowledge, and humour which aid not only in the interpretation of linguistic utterances, but also in the ability to communicate.

DISCURSIVE CONTEXT

The nature of language in a discursive context rather than isolated in a vacuum is emphasized by many authors concerned with the rehabilitation of communicative abilities. Discursive factors presupposed in aphasia therapy include elements that affect discourse at both the microstructural and the macrostructural levels of discourse.

Mechanisms necessary at the microstructural level of discourse—the level of individual sentences and their relations—have been proposed and applied to therapy. Weniger and her colleagues (1987) state that in therapy it is more efficient to treat linguistic forms associated with discourse within a speech context rather than in isolation. For example, in therapy, these authors suggest the training of modal verbs. In German, modals function to express the speaker's intention which helps to define an utterance's illocutionary force which is crucial communicatively in discourse. It is also suggested that pronominal and deictic forms are viable treatment targets because, in discourse, they are used to refer back to information conveyed in previous sentences, or to designate persons, objects, events and locations in the listener's field of reference.

Features at the global level of discourse, the macrostructural level, have also been identified and implemented in aphasia therapy. First, the importance of cohesion in the overall structure of discourse for comprehension has been addressed. It has been suggested that the notion of cohesion is important as a strategy in communicating with a person with aphasia. More specifically, it is important that those persons in contact with an aphasic patient keep related topics together in their discourse in order to better ensure listener comprehension (Lubinski, 1981). Similarly, it is recommended that linguistic materials within Melodic Intonation Therapy (MIT) should include cohesive, meaningful discourse to further promote comprehension (Sparks and Deck, 1986).

Secondly, the significance of a single theme or topic in conversation is also emphasized as an important guideline in forming any programme to increase communication ability (Edwards, 1987; Ross, 1989). Aten (1986) contends that a unitary topic offers structure and redundancy and endorses the use of personally relevant, frequently occurring social topics in therapy. Waller and Darley (1978) state that prior knowledge of the topic of conversation is also beneficial, if not essential, to the deciphering of discourse. Thus, topic prestimulation may be incorporated into therapy at the discourse level. Green (1984) also suggests the integration of natural discourse into therapy. More precisely, therapy may include the conversational notion of anticipating a partner's next action. This concept can be accounted for by positing some form of discourse schemata that is shared among speakers in a community.

Thus, the use of linguistic cues that link sentences and the use of global features of discourse have been suggested as features to be emphasized in aphasia therapy.

PARALINGUISTIC CONTEXT

The notion of paralinguistic context as an intricate part of the language system to be incorporated into aphasia therapy has also been found in the literature. This type of context includes the use of gesture and prosody in language comprehension and production.

Gestures and pantomime have been recommended as a therapeutic technique to increase the effectiveness of communication. In describing treatment methods within Functional Communication Treatment (FCT), Aten (1986) advises the clinician to investigate and reinforce any mode of communication, including gestures, that may better enable the patient to communicate effectively. Promoting Aphasic's Communicative Effectiveness (PACE) therapy (Davis and Wilcox, 1985), while also encouraging the use of gestures for communication, suggests that gestures should only be used supplementarily and not as replacements for verbal expression. The use of gesture has also been applied to a client-specific programme for a patient with Wernicke's aphasia. The goal of this treatment plan was the production of expressive speech. This was initially achieved through the use of gestures to convey a message, followed by the production of short linguistic utterances, enhanced with gestures when necessary (Edwards, 1987). Gesture has also been implemented as a strategy for the conversational partner of an aphasic as a means to assist comprehension (Green, 1984; Lubinski, 1981).

Alternative communication systems based on gesture and pantomime have also been endorsed in aphasia rehabilitation. These programs are generally designed for the globally aphasic individual who does not respond to more conventional methods of rehabilitation. A sign language program based on the Amerind gestural code system is one example of such a program developed after the failure of augmentative communication devices. Unfortunately, while the patient in this study was able to learn the gestures in therapy, there was no spontaneous generalization (Coelho, 1987).

Similarly, Visual Action Therapy (VAT) teaches the patient with global aphasia to produce representational gestures for visually absent stimuli through the manipulation of real objects (Helm-Estabrooks, Fitzpatrick and Barresi, 1982). Results with VAT were more promising with patients showing improved ability to produce symbolic gestures as well as some improvement in auditory and reading comprehension. VAT was developed on the hypotheses that pantomime-based communication is independent of verbal communication, that hand gestures require less refined motor control than motor movements required for speech, and that these movements are under unilateral control of an assumed intact hemisphere, and finally, that such movements can be observed and monitored by the patient. It is also assumed that the conceptual system remains intact because it is presumed to be independent from the linguistic system.

In another study, a group therapy role-playing activity was implemented in order to help aphasic patients cope with their morbid level of language functioning in familiar "stress" and "nonstress" situations (Schlanger and Schlanger, 1970). Pantomime role playing was advised for the more severely impaired patients for whom verbal communication was not possible.

Facial expression has also been included as a paralinguistic factor affecting meaning in communication. Many researchers regard the meaning of facial expression as an important component of any rehabilitation programme designed to facilitate communication in natural contexts (Green, 1984; Weniger et al., 1987).

Prosody—features superimposed on a linguistic utterance such as suprasegmental features and intonation—is also an intricate part of the language system, affecting both comprehension and production. Both non-grammatical and grammatical prosody have been incorporated in aphasia remediation. When speaking with an individual with aphasia, Lubinski (1981) advises the conversational partner to be aware of non-grammatical prosodic features, including rate of communication, articulation, volume and pauses because of its effect on the comprehension of spoken language.

A programme concerned with grammatical prosody is Melodic Intonation Therapy (MIT)(Sparks and Holland, 1976; Sparks and Deck, 1986). MIT is based on the premise that grammatical prosody and, in particular, sentential intonation, rhythm, and stress, are bound to the linguistic system. This intervention programme was designed to facilitate verbal language in nonfluent aphasia through the exaggeration of melody, tempo and rhythm of utterances. It was also found to be beneficial for phonological intervention in verbal apraxia. It was further hypothesized by these authors that, because of its similarity to music, prosody is processed in the right hemisphere, and thus MIT involves interhemispheric reorganization of language.

EXTRALINGUISTIC CONTEXT

It has been suggested that extralinguistic context is the fundamental concern of pragmatic theory and thereby pragmatic therapy (Muma, 1986). Investigators' presuppositions about the extralinguistic contexts necessary for communication in aphasia therapy refer to internal variables and external contexts.

It has been stated that spoken language establishes the patient in a social context through the style and accent in which he or she speaks and in the content of what is said, including social small talk (Lesser, 1989). The therapist must therefore be responsive to this aspect of language, as well as to the deficits in the linguistic system itself, when assessing the needs of the patient.

Specific internal variables included within remediation refer to personal attributes that the individual brings to the conversational interaction which will affect the person's ability to communicate. In particular, Lubinski (1981) refers to the communicator's own physical, intellectual and emotional characteristics, as well as to his/her attitudes, roles and experiences as examples of internal variables affecting communication that must be taken into account in therapy.

External contexts noted include situational and sociolinguistic factors. As a general guideline to pragmatic treatment, it is suggested that contexts encouraging spontaneous communication should be offered in therapy (Ross, 1989). More specifically, Green (1984) submits that treatment tasks should highlight common daily communicative activities rather than the structured linguistic tasks typical in the clinical setting. The goal, according to this approach, is to emphasize the function of language, which is communication. In other words, in therapy it is important to remember that a statement, spoken in real life, is never detached from the situation in which it has been uttered and therefore contextual therapy tasks need to be chosen (Green, 1984).

Green (1984) also stresses the importance of situational variables for natural language comprehension. While it is held that the ideal therapy session should occur in the client's own environment, alternative suggestions are offered to simulate a natural atmosphere when these conditions are not available. Green maintains that since background noise, poor acoustics, and cross-conversations occur in real life situations, they should be simulated in the clinical setting. Following the goal of natural communication, the author states that under such situations redundancy and misunderstanding may be accepted as normal, and transactions may be accomplished with little speech. Lubinski (1981), on the other hand, proposes strategies to enhance successful communication that include controlling external factors such as the transmission and reception of the message, the physical distance between the communicants, and the amount of interference and distortion from the physical environment.

The participants in the communicative situation are also noted as having an effect on the interaction. All participants bring with them both novel and shared knowledge as well as personal characteristics and experience which may affect many aspects of communication including, among others, the choice of lexical items and syntactic structure. Since the formulation of utterances in conversation depends on the information shared by the communicants, PACE therapy includes this notion of naturalness (Davis and Wilcox, 1981). A problem occurs in PACE and other intervention strategies that recommend natural communication since the concept of new and shared information between the participants becomes altered. That is, in the therapy situation the clinician is the one to select the message to be transmitted and thus is able to anticipate the new information that the patient may be attempting to communicate. Green (1984) also advocates the inclusion of the concept of shared information in therapy. It is also noted that the clinician should be sensitive to the diversity of characteristics and experiences that both the therapist and patient bring to the rehabilitation process (Lubinski, 1981).

Lubinski (1981) also refers to the socio-cultural environment as an essential aspect of communication. This includes the standards set by society concerning the behaviour of individuals in different roles and situations. The ways in which we learn these expectations are also emphasized as important in defining the relationship of individuals to their total environment. Learning is said to occur through both social interaction and informal interaction within specific groups such as the family, clubs and small work units. By remaining aware of these factors, the clinician will better be able to understand the aphasic patient and help to enhance the communication atmosphere of the patient's daily life.

NATURAL CONVERSATION

The concept of rules of natural conversation has also been incorporated within pragmatic therapy in aphasia since it is held to be a necessary part of a communicator's pragmatic competence. A fundamental premise of communication-based rehabilitation is that therapy sessions, regardless of the setting, follow principles of natural communication which predominantly consists of conversation.

Davis and Wilcox (1981; 1985) provide a theoretical rationale of context in action (i.e., conversation) on which their therapy, Promoting Aphasic's Communicative Effectiveness (PACE), is based. The principles of PACE follow three presuppositions of natural conversation. The first principle states that an exchange of new information occurs (Davis and Wilcox, 1981; 1984; Holland, 1983). This principle is based on Clark and Haviland's (1977) "Given-New Strategy" which states that in any conversation, there is new information exchanged and there is given information that is shared between the participants. The new information is then incorporated into the given information. This strategy is presumed by many authors as the basis of conversational exchanges that should be incorporated into the therapy situation (Chapey, 1981; Davis and Wilcox, 1981; 1984; Pierce and Wagner, 1985; Waller and Darley, 1978). The second principle asserts that the participants have a free choice as to which communicative channels they may use (Davis and Wilcox, 1981). This proposition follows the first in that new information is exchanged through the interaction of both verbal and nonverbal signs (Chapey, 1981). Thirdly, Davis and Wilcox (1981) emphasize that the conversational dyad has the minimum requirement of two participants, the speaker and the hearer. A reciprocity rule further states that when one participant holds the role of speaker, the other synchronously becomes the listener. Nonverbal cues are believed to signal the desire to exchange roles. This notion of turn-taking is also assumed to be essential to promoting the naturalness of language intervention (Edwards, 1987; Schlanger and Schlanger, 1970).

The importance of natural feedback in communication is also considered in communication treatment. Natural feedback refers to the situation in which the speaker is aware that the hearer does not understand the message and thus the speaker may be required to repeat it in a more precise manner. Furthermore, successful transmission of the message is rewarded with the listener's comprehension of it. This type of feedback is preferred in communication therapy over the behaviouristic type of feedback familiar to the structured therapy situation—e.g., positive reinforcement (Aten, 1986; Davis and Wilcox, 1981; 1985; Schlanger and Schlanger, 1970).

In addition, Lubinski (1981) includes the rules of opening a conversation as a component of a communicator's pragmatic competence. These rules allow the participant to know that a communicative interaction is desired. Thus, it is suggested that in order to improve communicative effectiveness in an individual with aphasia, a cue signalling the beginning of a conversation should be supplied.

Cooperative principles of conversation are also included within the rules of natural conversation. Following Grice's conversational maxims (1975), Lubinski (1981) and Davis (1986) acknowledge that, in conversation, the speaker's contribution will be informative, and the context will be relevant, unambiguous and will avoid obscurity. It is advised that these maxims be mirrored in therapy.

WORLD KNOWLEDGE

World knowledge contained within the linguistic code is also identified as a factor influencing the comprehension of sentences. Pierce and Wagner (1985) derive a therapy for syntactic decoding based on Deloche and Seron's (1981) hypothesis that states that within the single-sentence boundary, world knowledge, in the form of semantic plausibility, affects the comprehension of reversible active sentences. In addition, on the basis that both linguistic and extralinguistic contextual information is available for the interpretation of specific syntactic structures in normal discourse, these authors hypothesize that prior semantically supportive context aids in the comprehension of even difficult syntactic structures like reversible passive sentences. Thus prior linguistic context can be used as a strategy to assist the aphasic individual in interpreting passive structures. It is concluded that in discourse processing, the contextual sentence established a core of knowledge which restricted, on semantic grounds, what could occur next.

Similarly, Waller and Darley (1978) hypothesize that integration and contextual prerequisites are necessary for processing linguistic material. Thus, world knowledge contained inside and outside the sentence boundary is said to affect comprehension. Hence, the authors encourage the use of discourse in therapy so as to utilize this natural process of comprehension.

HUMOUR

Finally, humour is proposed as a therapy facilitator based on the premise that the positive effect of humour on humans is widely accepted as an a priori condition for successful communication (Potter and Goodman, 1983). A neuropsychological rationale is given by these authors that states that humour is subserved by the right hemisphere and thus humour (specifically, laughter) may serve to heighten right-hemispheric activity in the language relearning process and may possibly activate a new pathway to the left hemisphere. Consequently, it is concluded that humour, laughter, and amusement should be incorporated into therapy for theoretical reasons, as well as to promote relaxation in both the patient and clinician. A specific therapeutic technique is offered whereby a tape of laughter is provided for positive reinforcement.

In summary, communication-based aphasia therapy has developed out of a theory of pragmatics based on the contribution of discursive, paralinguistic, and extralinguistic contexts, rules of conversation, world knowledge and humour in the interpretation and use of context-dependent grammar.

THEORIES OF LANGUAGE III: OTHER ACCOUNTS

In the last section of this chapter, theories of language presupposed in aphasia therapy based on factors other than grammar and pragmatics will be reviewed. These include, among others, neuropsychological and neurological accounts.

Neuropsychological Models

The underlying premise of psycholinguistic models of language within the neuropsychological paradigm states that mental processes are organized in modules which are psychologically and neuropsychologically real (Lesser, 1989). Hence, in its application, various information-processing models are applied to identify a precise linguistic processing deficit and the route(s) through which re-activation or re-organization might occur (Ross, 1989).

One example of this approach is a programme designed to teach individuals with conduction aphasia to control production in repetition tasks (Cubelli et al., 1988). Three types of conduction aphasia are identified by these authors based on the point of breakdown within the cognitive system. The first type is caused by a deficit of the phonological processing system with the result of a profusion of phonemic paraphasias in both spontaneous speech and repetition. The second type results from a disorder at the articulatory output control mechanism. The resultant behaviour includes many articulation substitutions. A deficit in auditory-verbal short-term memory has also been found to create a syndrome consistent with conduction aphasia, and is characterized by normal spontaneous speech with circumlocutions and word-finding difficulties. A set of therapeutic principles and procedures concerning the type of task and the stimuli to be used is offered only for the first type of conduction aphasia since it was found to be the most common and is linguistic in nature.

A second transcoding model put forth by Lesser (1987) specifies the interaction of a number of modules necessary in the process of reading or writing a word. In its simplified form, this model states that for reading a written word or for writing a spoken word, a visual input lexicon or an auditory input lexicon respectively, interact with a cognitive system where meaning is derived. If the goal is speaking, then the information is sent to a phonological output lexicon or to an orthographic output lexicon if the requirement is to write a word to dictation. This model, in its entirety, has been used to describe such syndromes as surface and deep dyslexia. The inference that can be made about the module of malfunction can assist the therapist in choosing the most efficient goals of intervention.

Neuropsychological research has also found that symbolic behaviour may exist partially independent from the linguistic system. While it was previously held that global aphasia was the disruption of the entire language system and therefore was untreatable, research suggests that, while there is a deficiency in the linguistic system, the cognitive skills necessary for language are retained in global aphasia (Helm-Estabrooks, 1988; Helm-Estabrooks et al., 1982). Consequently, these authors propose an alternative communication system, Visual Action Therapy (VAT), consisting of symbolic gestures. VAT has been found to be efficacious for improving not only communication ability, but also auditory and reading comprehension. One explanation of the unexpected enhancement of language comprehension skills has been hypothesized as the result of the reintegration of some of the conceptual systems necessary for linguistic performance.

As has been alluded to, the right hemisphere has been implicated in the language process and the presumption of its linguistic role has been used as the basis for aphasia rehabilitation. Melodic Intonation Therapy (MIT) (Sparks et al., 1976; Sparks et al., 1986), as described earlier, subscribes to the rationale that the right hemisphere is involved in the processing of propositional language except for the final integrative process which takes place in the temporal lobe of the left hemisphere (Sparks et al., 1986). In addition, the processing of prosody is identified as the domain of the right hemisphere such that it will remain intact in aphasia resulting from a deficit in the left hemisphere. The tasks of MIT focus on the prosody of sentences with the goal being the rehabilitation of propositional language. The presupposition here is that activation of the right hemisphere will stimulate reorganization of the interhemispheric processing of language and will increase language processing within the intact right hemisphere (Sparks et al., 1986). In order for MIT to be successful, it is important that the commissural fibres of the hemispheres be intact. While this argument about the proposed efficacy of MIT seems quite enticing, Berlin (1976) warns that alternative physiological routes may also account for the results of MIT.

As previously stated, the processing of humour has also been cited as being within the domain of the right hemisphere. Correspondingly, it has been hypothesized that humour/laughter stimuli may serve to heighten right-hemispheric activity in the language relearning process. As in the theoretical function of Melodic Intonation Therapy, the right hemisphere may be performing a mediational task in language encoding by activating a new pathway to the left hemisphere (Potter et al., 1983).

A microgenetic model of language has also been applied to a rehabilitation programme for aphasia. This model equates language with cognition. The microgenetic model presented by Hanlon et al. (1990) states that the cognitive process (language) unfolds in a unidirectional (bottom-up) progression traversing distributed levels and cognitive stages in a manner that recapitulates the course of forebrain evolution. In particular, for language processing, this model suggests that the aphasic patient's use of the contralateral shoulder to point to a picture to be named will facilitate naming since this type of gesture will activate inhibited levels of language in the left hemisphere. In other words, the gesture will stimulate these levels.

CONCLUSION

In conclusion, there is no single view of the nature of language in aphasia therapy. Models of language range from the purely linguistic to those based on physical reductionism.

Two trends in the nature of aphasia rehabilitation have been observed. First, therapy has moved from focusing specifically on linguistic structures as observed in "classical" and "programmed" treatments, to programmes that concentrate on the rehabilitation of the patient's communicative abilities. Recently, Chin Li, Kitselman, Dusatko and Spinelli (1988) compared a traditional stimulation technique to a procedure following the principles of PACE in the facilitation of naming. Improvement was only found in the PACE condition, suggesting the importance of communicative strategies in the interpretation of language. (It is not the purpose of this report to review the efficacy issue. For an overview of this topic, see Wertz, this volume, and Shewan, 1986). A second trend towards aphasia therapy based on neuropsychological information processing models is also evident.

While these current approaches imply an abandonment of a purely linguistic approach, theoretical linguistics and psycholinguistics remain valuable resources for the assessment of aphasic disorders.

Clearly a reciprocal relationship exists between theory and therapy: Any research that reveals the nature of language will benefit the individual with aphasia, and any therapeutic technique that benefits the individual with aphasia will reveal specific characteristics of the language system. Clearly, both the therapist and the researcher, while realizing different goals, can benefit from one another's endeavours.

References

Albert, M.L. 1989. Experimental approaches to aphasia therapy. *Journal of Neurolinguistics*, 4: 427-34.
Albert, M.L., Bachman, D., Morgan, A. and Helm-Estabrooks, N. 1987. Pharmacotherapy for aphasia. *Neurology*, 37 (Supplement 1): 175.
Aten, J.L. 1986. Functional Communication Treatment. In R. Chapey (ed.), *Language Intervention Strategies in Adult Aphasia*. Baltimore: Williams & Wilkins. Pp. 266-76.
Behrmann, M. and Lieberthal, T. 1989. Category-specific treatment of a lexical-semantic deficit: A single case study of global aphasia. *British Journal of Disorders of Communication*, 24: 281-99.
Berlin, C.ii. 1976.. On: Melodic Intonation Therapy for aphasia by R.W. Sparks and A.L. Holland. *Journal of Speech and Hearing Disorders*, 41: 298-300.
Bruce, C. and Howard, D. 1987. Computer-generated phonemic cues: An effective aid for naming in aphasia. *British Journal of Disorders of Communication*, 22: 191-201.
Burton, E., Burton, A. and Lucas, D. 1988. The use of microcomputers with aphasic patients. *Aphasiology*, 2: 479-91.
Byng, S. 1988. Sentence processing deficits: Theory and therapy. *Cognitive Neuropsychology*, 5: 629-76.
Chapey, R. 1981. An introduction to language intervention strategies in adult aphasia. In R. Chapey (ed.), *Language Intervention Strategies in Adult Aphasia*. Baltimore: Williams and Wilkins. Pp. 3-14.

Chin Li, E., Kitselman, K., Dusatko, D. and Spinelli, C. 1988. The efficacy of PACE in the remediation of naming deficits. *Journal of Communication Disorders*, 21: 491-503.

Chin Li, E. and Williams, S.E. 1990. The effects of grammatic class and cue type on cueing responsiveness in aphasia. *Brain and Language*, 38: 48-60.

Clark, H.H. and Haviland, S.E. 1977. Compreheniosn and the given-new contract. In R.O. Freedle (ed.), *Discourse Production and Comprehension*. Norwood, N.J.: Ablex.

Coelho, C.A. 1987. Sign acquisition and use following traumaticbrain injury: Case report. *Archives of Physical Medicine and Rehabilitation*, 68: 229-31.

Crystal, D. 1972. The case of linguistics: A prognosis. *British Journal of Disorders of Communication*, 21: 239-50.

Cubelli, R., Foresti, A. and Consolini, T. 1988. Reeducation strategies in conduction aphasia. *Journal of Communication Disorders*, 21: 239-49.

Culton, G.L. and Ferguson, P.A. 1979. Comprehension training with aphasic subjects: The development of five automated language programs. *Journal of Communication Disorders*, 12: 69-82.

Dabul, B. and Bollier, B. 1978. Therapeutic approaches to apraxia. Journal of Speech and Hearing Disorders, 41: 268-76.

Davies, C.L. and Grunwell, P. 1975. A new approach to the treatment of severe dysphasia: A case study. *British Journal of Disorders of Communication*, 10: 142-48.

Davis, G.A. 1973. Linguistics and language therapy: The sentence construction board. *Journal of Speech and Hearing Disorders*, 38: 204-14.

Davis, G.A. 1986. Pragmatics and treatment. In R. Chapey (ed.), *Language Intervention Strategies in Adult Aphasia*. Baltimore: Williams and Wilkins. Pp. 250-65.

Davis, G.A. and Tan, L.L. 1987. Stimulation of sentence production in a case with agrammatism. *Journal of Communication Disorders*, 20: 447-57.

Davis, G.A. and Wilcox, M.J. 1981. Incorporating parameters of natural conversation in aphasia treatment. In R. Chapey (ed.), *Language Intervention Strategies in Adult Aphasia*. Baltimore: Williams and Wilkins. Pp. 169-93.

Davis, G.A. and Wilcox, M.F. 1985. *Adult Aphasia Rehabilitation: Applied Pragmatics*. San Diego, CA: College-Hill Press.

Deloche, G. and Seron, X. 1981. Sentence understanding and knowledge of the world: Evidence from a sentence-picture matching task performed by aphasic patients. *Brain and Language*, 14: 57-69.

Edwards, S. 1987. Assessment and therapeutic intervention in a case of Wernicke's aphasia. *Aphasiology*, 1: 271-76.

Gildston, H. and Gildston, P. 1986. Hypnotherapy in adult aphasia rehabilitation. *Folia Phoniatrica*, 38: 301.

Goodglass, H. and Hunt, J. 1958. Grammatical complexity and aphasic speech. *Word*, **14**: 197–207.
Green, G. 1984. Communication in aphasia therapy: Some of the procedures and issues involved. *British Journal of Disorders of Communication*, **19**: 35-46.
Grice, H. 1975. Logic and conversation. In P. Cole and J. Morgan (eds.), *Syntax and Semantics: Speech Acts*. New York: Academic Press.
Hanlon, R.E., Brown, J.W. and Gerstman, L.J. 1990. Enhancement of naming in nonfluent aphasia through gesture. *Brain and Language*, **38**: 298-314.
Hatfield, F.M. 1972. Looking for help from linguistics. *British Journal of Disorders of Communication*, **7**: 64-81.
Hatfield, F.M. and Shewell, C. 1983. Some applications of linguistics to aphasia therapy. In C.Code and D.J. Muller (eds.), *Aphasia Therapy*. London: Edward Arnold. Pp. 61-75.
Helm-Estabrooks, N. 1988. The application of neurobehavioral research to aphasia rehabilitation. *Aphasiology*, **2**: 303-08.
Helm-Estabrooks, N., Emery, P. and Albert, M.L. 1987. Treatment of Aphasic Perseveration (TAP) program: A new approach to aphasia therapy. *Archives of Neurology*, **44**: 1253-255.
Helm-Estabrooks, N., Fitzpatrick, P.M. and Barresi, B. 1981. Response of an agrammatic patient to a syntax stimulation program for aphasia. *Journal of Speech and Hearing Disorders*, **46**: 422-27.
Helm-Estabrooks, N., Fitzpatrick, P. and Barresi, B. 1982. Visual Action Therapy for global aphasia. *Journal of Speech and Hearing Disorders*, **47**: 385-89.
Helm-Estabrooks, N. and Ramsberger, G. 1986. Treatment of agrammatism in long-term Broca's aphasia. *British Journal of Disorders of Communication*, **21**: 39-45.
Helmick, J.W. and Wipplinger, M. 1975. Effects of stimulusrepetition on the naming behavior of an aphasic adult: A clinical report. *Journal of Communication Disorders*, **8**: 23-39.
Holland, A.L. 1983. Language intervention in adults: What is it? In J. Miller, D.E. Yoder and R. Schiefelbusch (eds), *Contemporary Issues in Language Intervention*. Rockville, Maryland: ASHA. Pp. 3-14.
Howard, D. and Orchard-Lisle, V. 1984. On the origin of semantic errors in naming: Evidence form the case of a global aphasic. *Cognitive Neuropsychology*, **1**: 163-90.
Howard, D., Patterson, K., Franklin, S., Orchard-Lisle, V. and Morton, J. 1985a. The facilitation of picture naming in aphasia. *Cognitive Neuropsychology*, **2**: 49-80.
Howard, D., Patterson, K., Franklin, S., Orchard-Lisle, V. and Morton, J. 1985b. Treatment of word retrieval deficits in aphasia: A comparison of two therapy methods. *Brain*, **108**: 817-29.

Jones, E.V. 1986. Building the foundations for sentence production in a nonfluent aphasic. *British Journal of Disorders of Communication*, **21**: 63-82.
Kearns, K.P. and Salmon, S.J. 1984. An experimental analysis of auxiliary and copula verb generalization in aphasia. *Journal of Speech and Hearing Disorders*, **49**: 152-63.
Kinsey, C. 1986. Micro computer speech therapy for dysphasic adults: A comparison with two conventionally administered tasks. *British Journal of Disorders of Communication*, **21**: 125-34.
Kohn, S.E. and Smith, K.L. 1990. The remediation of conduction aphasia via sentence repetition: A case study. *British Journal of Disorders of Communication*, **25**: 45-60.
Kotten, A. 1982. Therapy of neologistic jargon aphasia: A case report. *British Journal of Disorders of Communication*, **17**: 61-73.
Kushner, D. and Winitz, H. 1977. Extended comprehension practice applied to an aphasic patient. *Journal of Speech and Hearing Disorders*, **42**: 296-305.
Laughlin, S.A., Naeser, M.A. and Gordon, W.P. 1979. Effects of three syllable durations using the Melodic Intonation Therapy technique. *Journal of Speech and Hearing Research*, **22**: 311-20.
Lesser, R. 1987. Cognitive neuropsychological influences on aphasia therapy. *Aphasiology*, **1**: 189-200.
Lesser, R. 1989. Aphasia: Theory-based Intervention. In M.M.Leahy (ed.), *Disorders of Communication: The Science of Intervention*. New York: Taylor and Francis. Pp. 189-205.
Linn, L. 1967. Sodium amytal in treatment of aphasia. *Archives of Neurology and Psychiatry*, **58**: 357-58.
Lubinski, R. 1981. Environmental language intervention. In R. Chapey (ed.), *Language Intervention Strategies in Adult Aphasia*. Baltimore: Williams and Wilkins. Pp. 223-45.
MacMahon, M.K.C. 1972. Modern linguistics and aphasia. *British Journal of Disorders of Communication*, **7**: 54-63.
Miller, N. 1989. Acquired speech disorders: Applying linguistics to treatment. In K. Grundy (ed.), *Linguistics in Clinical Practice*. New York: Taylor and Francis. Pp. 281-300.
Montgomery, J. 1971. The importance of seeing red: Self-teaching techniques for adult aphasia. *Journal of Speech and Hearing Disorders*, **36**: 250-51.
Morton, J. 1979. Some experiments on facilitation in word and picture recogntion and their relevance for the evolution of a theoretical position. In P. Kolers, M. Wrolstad and H. Bouma (eds.), *Processing of Visible Language*. New York: Plenum.
Morton, J. and Patterson, K.E. 1980. A new attempt at an interpretation, or an attempt at a new interpretation. In M. Coltheart, K.E. Patterson and J.C. Marshall (eds.), *Deep Dyslexia*. London: Routledge and Kegan Paul.

Muma, J.R., Hamre, C.E. and McNeil, M.R. 1986. Theoretical models applicable to intervention in adult aphasia. In R. Chapey (ed.), *Language Intervention Strategies in Adult Aphasia*. Baltimore: Williams and Wilkins. Pp. 277-83.

Naeser, M.A. 1975. A structured approach to teaching aphasics basic sentence types. *British Journal of Disorders of Communication*, **10**: 70-6.

Naeser, M.A., Haas, G., Mazurski, P. and Laughlin, S. 1986. Sentence Level Auditory Comprehension treatment program for aphasic adults. *Archives of Physical Medicine and Rehabilitation*, **67**: 393-99.

Pierce, R.S. 1982. Facilitating the comprehension of syntax in aphasia. *Journal of Speech and Hearing Research*, **25**: 408-13.

Pierce, R.S. and Wagner, C.M. 1985. The role of context in facilitating syntactic decoding in aphasia. *Journal of Communication Disorders*, **18**: 203-13.

Podraza, B.L. and Darley, F.L. 1977. Effect of auditory prestimulation on naming in aphasia. *Journal of Speech and Hearing Research*, **10**: 669-83.

Potter, R.E. and Goodman, N.J. 1983. The implementation of laughter as a therapy facilitator with adult aphasics. *Journal of Communication Disorders*, **16**: 41-8.

Prins, R.S., Shoonen, R. and Vermeulen, J. 1989. Efficacy of two different types of speech therapy for aphasic stroke patients. *Applied Psycholinguistics*, **10**: 85-123.

Rosch, E. 1975. Cognitive representations of semantic categories. *Journal of Experimental Psychology: General*, **104**: 192-233.

Rosenbek, J.C., Lemme, M.L., Ahern, M.B., Harris, E.H. and Wertz, R.T. 1973. A treatment for apraxia of speech in adults. *Journal of Speech and Hearing Disorders*, **38**: 462-72.

Ross, A. 1989. Applying linguistics to aphasia. In K. Grundy (ed.), *Linguistics in Clinical Practice*. New York: Taylor and Francis. Pp. 205-23.

Sarno, M.T., Silverman, M. and Sands, E. 1970. Speech therapy and language recovery in severe aphasia. *Journal of Speech and Hearing Research*, **13**: 607-23.

Schlanger, P.H. and Schlanger, B.B. 1970. Adapting role-playing activities with aphasic patients. *Journal of Speech and Hearing Disorders*, **35**: 229-35.

Schwartz, M.F., Linebarger, M.C. and Saffran, E.M. 1985. The status of syntactic theory of agrammatism. In M.L. Kean (ed.), *Agrammatism*. New York: Academic Press.

Sefer, J.W. 1978. Principles of aphasia therapy. In Y. Lebrun and R. Hoops (eds.), *The Management of Aphasia*. Amsterdam and Lisse: Swets and Zeitlinger B.V. Pp. 66-75.

Sefer, J.W. and Shaw, R. 1972. The use of psycholinguistic principles in the treatment of aphasia. *British Journal of Disorders of Communication*, 7: 87-9.

Seron, X., Deloche, G., Bastard, V., Chassin, G. and Hermand, N. 1979. Word-finding difficulties and learning transfer in aphasic patients. *Cortex*, 15: 149-55.

Seron, X., Deloche, G., Moulard, G. and Rousselle, M. 1980. A computer-based therapy for the treatment of aphasic subjects with writing disorders. *Journal of Speech and Hearing Disorders*, 45: 45-58.

Shewan, C.M. 1976. Facilitating sentence formulation: A case study. *Journal of Communication Disorders*, 9: 191-97.

Shewan, C.M. 1986. The history and efficacy of aphasia treatment. In R. Chapey (ed.), *Language Intervention Strategies in Adult Aphasia*. Baltimore: Williams and Wilkins. Pp. 28-43.

Sparks, R.W. and Deck, J.W. 1986. Melodic Intonation Therapy. In R. Chapey (ed.), *Language Intervention Strategies in Adult Aphasia*. Baltimore: Williams and Wilkins. Pp. 320-32.

Sparks, R.W. and Holland, A.L. 1976. Method: Melodic Intonation Therapy for aphasia. *Journal of Speech and Hearing Disorders*, 41: 287-97.

Thompson, C.K. and McReynolds, L.V. 1986. Wh-interrogative production in agrammatic aphasia: An experimental analysis of auditory-visual stimulation and direct-production treatment. *Journal of Speech and Hearing Research*, 29: 193-206.

Thompson, C.K., McReynolds, L.V. and Vance, C.R. 1982. Generative use of locatives in multiword utterances in agrammatism: A matrix-training approach. In R.H. Brookshire (ed.), *Clinical Aphasiology Conference Proceedings*. Minneapolis: BRK. Pp. 289-97.

Vendrell, P. and Junque, E. 1989. Principles of aphasia therapy in Luria' theory: Basic introducing remarks. *Journal of Neurolinguistics*, 4: 155-60.

Waller, M.R. and Darley, F.L. 1978. The influence of context on the auditory comprehension of paragraphs by aphasic subjects. *Journal of Speech and Hearing Research*, 21: 432-45.

Wambaugh, J.L. and Thompson, C.K. 1989. Training and generalization of agrammatic aphasic adults' wh-interrogative productions. *Journal of Speech and Hearing Disorders*, 54: 509-25.

Warrington, E.K. and McCarthy, R. 1987. Categories of knowledge. *Brain*, 110: 1243-296.

Weniger, D., Springer, L. and Poeck, K. 1987. The efficacy of deficit-specific therapy materials. *Aphasiology*, 1: 215-22.

Wiegel-Crump, C. and Koenigsknecht, R.A. 1973. Tapping the lexical store of the adult aphasic: Analysis of the improvement made in word retrieval skills. *Cortex*, 9: 410-18.

APPENDIX

Behrmann, M. and Lieberthal, T. 1989. Category-specific treatment of a lexical-semantic deficit: A single case study of global aphasia. *British Journal of Disorders of Communication*, 24: 281-299.

Subject Area: Global aphasia with lexical-semantic comprehension deficit.
Therapy Area: Semantics; lexicon.
Therapy: Apply cognitive neuropsychological model, in this case a model of semantic processing, to treatment programme based on specific deficit of patient.
Purpose: To describe semantic rehabilitation programme based on the assumption that meaning is organized in a category-specific, hierarchical fashion. Aim was to improve patient's comprehension of single lexical items through category-specific rehabilitation and to assess whether improvement obtained on item-specific retraining could be generalized both within the same category and whether carry-over to other categories would occur.
Method: *Subjects*: One subject with global aphasia; semantic deficit at single word level.
Design:: Therapy plan divided into selection stage, treatment stage and post-therapy evaluation. Within 3 categories to be treated, 2 groups of words: treated (group X) and untreated control (group Y).
Procedure: 15 1-hour therapy sessions over approximately 6 weeks.
Stage 1: Aimed at teaching meaning at a general level of description (superordinate features of each category);
Stage 2: Aimed at teaching specific details of the group X items, leading to precise identification of these items.
Semantic features distinctive to each category were explained first and reiterated until the patient understood them. Individual items were presented in both printed and spoken form. Verbal name matched to a pictorial representation of the referent. Post-therapy measure taken one week following completion of the 3 treated categories.
Results: Significant improvement in post-therapy scores observed compared to pre-therapy scores. Superior performance on treated items in treated categories evident; carry-over to untreated items within same categories demonstrated. Minimal generalization to items in untreated categories.
Discussion: The patient acquired superordinate information relatively easily and demonstrated generalization at this level, but he was unable to utilize basic level object information unless it had been specifically addressed in treatment. (based on Rosch's (1975) hierarchy of semantic representation). Generalization to related items observed; reveals that patient was able to acquire new linguistic skills and to carry-over his newly acquired knowledge to other untreated items. Minimal carry-over across categories lends support to hypothesis that the semantic system is organized in a categorical fashion.

Bruce, C. and Howard, D. 1987. Computer-generated phonemic cues: An effective aid for naming in aphasia. *British Journal of Disorders of Communication*, 22: 191-201.

Subject Area: Broca's aphasia; microcomputers.
Therapy Area: Naming.
Therapy: Use of computer generated phonemic cues to aid in word retrieval leading to ability to self-generate these cues for naming.
Hypothesis: There are three stages in the process of generating phonemic cues in naming: 1) indicate initial letter of the word which cannot be retrieved; 2) convert this letter into its corresponding phonemic cue; 3) use the phonemic cue as an aid to word retrieval. Since patients with Broca's aphasia have not been found to possess all three abilities, supplying a missing stage, specifically, converting a letter into its corresponding sound, will aid naming.
Method: *Subjects*: 5 Broca's aphasic patients; word-finding difficulties; ability to benefit from phonemic cues; at least 6 months post-onset.
Procedure: 100 pictures presented—50 during training and again in test, 50 previously unseen for post-therapy assessment. Two experimental conditions: Aid: computer aid available: 9 consonants on 9 keys which produced the corresponding phoneme+schwa; Control: computer aid not available.
Discussion: 4/5 showed significantly better naming in the aid condition compared to control condition. Patients able to generalize their use of the aid to previously unseen pictures. Repeated phonemic cueing and repetition of the name over a number of days had some effect on a word's accessibility; a single use of a phonemic cue has only very short-lasting effects.
Conclusions: Computer generated phonemic cue provided extra activation to entries in a lexicon of phonological word forms. Computer aid has specific, positive effects in treatment and as a prosthetic device.

Burton, E., Burton, A. and Lucas, D. 1988. The use of microcomputers with aphasic patients. *Aphasiology*, 2: 479-491.

Subject Area: Microcomputer—as a tool.
Purpose: To explore the potential for microcomputers in aphasia therapy.
Method: *Subjects*: 99 patients from various speech therapy clinics in England chosen at therapist's discretion. No restrictions.
Procedure: Set of 14 general language stimulation programs; Format: one item on left-hand side of screen with one related/correct item to be chosen from a list on the right-hand side. Tasks ranged from pre-reading—matching shapes—to higher level semantic tasks.
Results: 73% of therapists reported "some improvement"; 27% not sure. Age a major predictor: the older the patient, the more help they needed, and the more apprehensive about using a computer. Patients who could use their preferred hand to operate computer appeared to find programs more enjoyable. Some problem with ambiguity of answers. In conventional therapy, patients are

encouraged to discuss the possibilities and assess the most likely answer. (Not possible, at present, with a computer.)
Conclusions: concentrate on using the computer for presentation and practice of straightforward, unambiguous tasks.

Byng, S. 1988. Sentence processing deficits: Theory and therapy. *Cognitive Neuropsychology*, 5: 629-76.

Subject Area: Difficulty comprehending reversible simple active declarative, passive, or locative sentences.
Therapy Area: Syntax; mapping hypothesis; thematic roles.
Therapy: Individualized remediation therapy based on model of sentence processing based on mapping hypothesis.
Purpose: Attempt to meet criteria for cognitive neuropsychological remediation based on model of sentence comprehension and production through mapping hypothesis.
Background: *Criteria for a neuropsychological treatment study:* Assess disorder within framework of a multicomponent information-processing system; treat disordered component and observe effects on other components of system; unexpected or impossible results based on the model would be reason to re-examine the validity of the model.
Mapping (Deficit) Hypothesis (Schwartz et al., 1985): Agrammatic patients are unable to co-ordinate descriptions of sentence form and descriptions of sentence meaning.
Sentence Processing Model (based on Mapping assumption): Components of sentence comprehension: 1) establish grammatical relations in the sentence, yielding "sentence form"; 2) parsing/syntactic analysis; produces a hierarchical description of syntactic structure of sentence. May also include establishing thematic relations of the particular verb; 3) integration of the information yielded by the first two procedures: mapping of grammatical relations/syntactic functions on to thematic roles. (Authors suggest possible methods by which mapping mechanism may work, but do not commit to any one method.)
For sentence production, suggested that there must be mapping in reverse.
Hypothesis: Possible to individually test the components of comprehension process to locate site (or sites) of individual patient's deficit.
The mechanism that maps thematic roles and grammatical relations may be a central mechanism common to all modalities of input and output. Only when a comprehension deficit is due to impaired mapping will there be parallel deficits in comprehension and production.
Method: *Case 1:* Moderate/severe expressive and mild receptive dysphasia; observed impairment in inability to map thematic relations onto other representations.
Treatment Program: Comprehension of locative sentences (written sentences only). For 4 propositions selected, 5 reversible sentences devised for each. Each sentence accompanied by 2 pictures, correct and their reverse-role picture.

Patient's task was to select the correct picture for the sentence. Clues provided: Each proposition had an associated meaning card which diagrammatically explained the meaning; practice cards indicating correct answer and corresponding test cards given for home use.
Results: Improvement generalized across modalities. Evidence suggesting that the patient had learned a principle, not just a strategy about mapping thematic roles onto grammatical relations. A second treatment programme was developed to improve comprehension of abstract words through picture-word matching therapy and dictionary therapy—find synonyms to word through dictionary. Results of this treatment programme suggest that the treatment has restored some lexical entries that had been lost rather than restore the whole lexical system in some way.
Case 2: Broca's aphasic patient.
First Treatment Program: Remediation of Mapping Deficit II: similar to Case 1; unsuccessful results.
Second Treatment Program: Aims: 1) to improve sentence comprehension of reversible active declarative sentences by encouraging perception of the thematic roles of agent and theme related to their position in the sentence; and 2) on the basis of the hypothesis about the relationship between a mapping deficit in production and comprehension, to improve production of simple sentences as a result of the procedure carried out in 1.
Results: Implications for theories of sentence processing: Case 1: improved comprehension of thematic relations in all modalities of input, but also across different types of thematic relations and in output tasks where mapping can be observed. Improvement in tasks not involving mapping was not found. Outcome lends strong support to mapping deficit hypothesis as an interpretation of some of this patient's comprehension and production problems; Case 2: performance on comprehension of sentences involving the thematic relations used in therapy improved in both modalities; structured production of sentences also improved.
Conclusions: Studies suggest that remediation can be effective even for very persistent forms of language impairment. Also suggest that the outcome of therapy studies has much to offer the cognitive neuropsychologist interested in purely theoretical questions, so that a truly co-operative approach between clinicians and theorists could have benefits for researchers and patients alike.

Chin Li, E., Kitselman, K., Dusatko, D. and Spinelli, C. 1988. The efficacy of PACE in the remediation of naming deficits. *Journal of Communication Disorders*, **21**: 491-503.

Subject Area: Severe fluent aphasia.
Therapy Area: Naming.

Therapy: Use of PACE to facilitate naming.
Purpose: To compare the relative effectiveness of PACE and Schuell's traditional stimulation in an aphasic patient with naming deficits.
Method: *Subjects:* One subject with severe fluent aphasia.
Design: ABCBC time-series design; treatment phases included: traditional stimulation therapy (B) and PACE therapy (C).
Traditional stimulation therapy: Both auditory and visual stimulation used to elicit naming responses. In addition to picture stimuli, auditory-verbal models were supplied which included use of carrier phrases, sentence completion, and associated words. If patient did not succeed in naming the specific target words, corrective feedback and cueing techniques were employed. Most facilitating cues were presentation of initial phoneme and use of printed word.
PACE: natural interaction sequences in which the client and clinician alternate turns communicating the identity of an object depicted on a card. Only the sender of the message knows the identity of the object. During the patient's "turn" as sender, client encouraged to use any effective means of communication including gestures and verbal circumlocutions. During clinician's turn as sender, s/he models the use of effective channels and types of communicative behaviours. Both object and action pictures are used during PACE therapy.
Results: Observable improvement during the PACE phases of therapy, a trend absent during the traditional phases. Improvement associated with PACE was evidenced on confrontation naming and picture description tasks. With PACE, patient able to use communicative strategies such as effective circumlocutions and effective multiple responses (paired gestures with verbal circumlocutions). Patient's mood and responsiveness to communication also improved during PACE therapy.
Conclusions: Study provides evidence for the effectiveness of PACE therapy when compared to traditional stimulation therapy.

Chin Li, E. and Williams, S.E. 1990. The effects of grammatic class and cue type on cueing responsiveness in aphasia. *Brain and Language*, 38: 48-60.

Subject Area: Different aphasic types: Broca's, Wernicke's, conduction and anomic.
Therapy Area: Naming; grammatic class (noun and verb).
Therapy: Comparison of effects of semantic and phonemic cueing for confrontation naming.
Purpose: To compare cueing responsiveness in verb versus noun target words; focus on phonemic and semantic cues.
Hypothesis: If the retrieval mechanism and lexical organization differ in verbs versus nouns, it is likely that responsiveness to cues would also vary with grammatical class.
Method: *Subjects:* 36 aphasic patients: 10 Broca's, 10 Wernicke's, 8 conduction, 8 anomic.

Test Stimuli: Line drawings to elicit confrontation naming responses; words represented a wide range with regard to frequency of occurrence; word length varied from one to three syllables.
Confrontation naming tests: 1) noun targets associated with semantic cues; 2) noun targets associated with phonemic cues; 3) verb targets associated with semantic cues; 4) verb targets associated with phonemic cues.
Phonemic cue: initial sound of word, followed by a 2-word carrier phrase.
Semantic cue: description of target followed by carrier phrase. 3 categories: 1) superordinate: named the class of which the target was a member (used for noun target words); 2) function associate: used a verb to designate the action carried out by or on the target (used for noun target words); 3) functional context: designated the situation in which the target occurred (used for noun and verb targets). Note: all categories considered equal (no significant differences found previously).
Procedure: Subjects completed 4 confrontation naming tests (42 items each); Noun subtests: "Tell me the name of this"; Verb subtests: "Tell me what the person is doing".
Results: On noun targets words, phonemic cueing found to be more beneficial than semantic cueing.
Conclusions: Nouns and verbs are represented differently in the lexicon and these differences are accentuated by the aphasic pathology. If this is true, a lapse or breakdown in naming would involve different mechanisms in nouns as compared to verbs; hence response to cueing differs for these 2 grammatic classes. (Did not hold true for anomic aphasic patients). A neuropsychological information-processing model of naming and/or imageability may present a plausible explanation for differences in the lexical access of nouns versus verbs. It is likely that aphasic patients retrieve semantic information with greater ease in nouns as compared to verbs. Thus, a semantic cue is more useful in the verb-naming situations in which the patient lacks the necessary semantic components. During noun naming, the semantic cue is likely to be redundant. Most of the patients derived greater benefit from phonemic cueing which is likely to operate at a later stage of processing—the articulatory level. In the case of anomic aphasic patients, the particular nature of their semantic deficit renders the semantic cue potent in both naming situations; but more so during verb naming.

Coelho, C.A. 1987. Sign acquisition and use following traumatic brain injury: Case report. *Archives of Physical Medicine and Rehabilitation*, **68**: 229-31.

Subject Area: Severe aphasia.
Therapy Area: Symbolic gestures.

Therapy: Manual signing taught in structured environment.
Purpose: Report of case of a traumatically brain-injured patient who remained a nonfunctional oral communicator eight months after injury. After repeated failures at establishing a functional, nonverbal means of communication, a basic sign vocabulary was introduced and after several weeks began to be acquired.
Method: *Subjects*: 21-year-old female; severe aphasia and oral-verbal apraxia.
Procedure: Manual signing; vocabulary selection: 20 basic signs believed to be useful for communicating patient's needs and wants within the hospital. Training procedure: 1) imitation; 2) recognition; 3) expression. Criterion for sign acquisition: when patient produced target response at expression stage.
Results: Patient had acquired 47 signs, of which only 16 were noted to be used by the patient in a semiappropriate fashion to express needs and wants, and only then in very structured interactions when specific cues were provided. Some sign combinations were also presented for training but were never acquired.
Conclusions: Simply acquiring a vocabulary of signs does not guarantee their communicative use. However, aphasic patients can acquire a variety of signs when presented within a structured training program. The total number of signs which can be acquired is related to the overall severity of the aphasia. There is a paucity of evidence that these acquired signs can be used for functional communication. When selecting a corpus of signs for training, the iconicity of sign features and motoric complexity of sign production may contribute to a specific sign's ease of acquisition.

Crystal, D. 1972. The case of linguistics: A prognosis. *British Journal of Disorders of Communication*, **221**: 239-50.

Subject Area: Linguistics applied to speech therapy; mostly concerned with child language and acquisition. The main ideas can be applied to aphasia therapy.
Purpose: To speculate about the ideal world of 'therapeutic linguistics' and to lay down practicable guidelines for the future.
Proposal: The following items should be completed by linguistics for use by therapists: 1) manual describing features of normal language (in particular, developmental aspects), both spoken and written; 2) manual of fully developed language; 3) full description of linguistic characteristics of the various categories of disorder, including a means of assessing different kinds and degrees of divergence from the language norms; 4) set of techniques capable of describing all significant linguistic features in the study of a particular case; 5) scheme for evaluating language patterns in terms of relative complexity, and thus a set of recommendations concerning the order in which linguistic forms and structures could be presented in the treatment of a disorder; 6) set of explanatory principles able to account for the specific acquisition and breakdown of language in relation to anatomical, physiological, neurological, psychological and other states; 7) an introductory exposition to the conceptual and terminological apparatus used.

Conclusions: Linguistics can offer for the therapist tested analytic skills. Linguistics does not have all the answers; it can provide a more explicitly principled therapy. In conclusion, linguists and speech therapists should work together.

Cubelli, R., Foresti, A. and Consolini, T. 1988. Reeducation strategies in conduction aphasia. *Journal of Communication Disorders*, **21**: 239-249.

Subject Area: Conduction aphasia.
Therapy Area: Phonemic errors in oral and written production; repetition.
Therapy: Reeducation program for reproductive difficulties (use of words and sentences).
Purpose: To present a reeducation program following therapeutic principles based on neuropsychological findings from, and assumptions made about, conduction aphasic patients.
Therapeutic principles: 1) aim is to teach patients to control phonemic production and to prevent paraphasic errors by means of orientating their attention to formal structure of their expression; 2) exercises consist of metalinguistic judgements of phonological and syntactical structure of words and sentences; 3) visual stimuli used; 4) exercises must prevent compensation effects of spared lexical-semantic system.
Method: *Subjects*: 3 conduction aphasic patients. Aphasia resulted from a deficit of the phonological processing system; disorder of mechanisms of phonemic selection and seriation.
Exercises: 1) Patient given object (or picture) and must choose the corresponding written word among visually similar alternatives. distractors created by means of single letter substitution or consonant or syllable metathesis of target. Length and complexity—sequences of consonants—controlled; 2) patient must match a scene to the written sentence describing it. Alternative sentences include: correct one, preposition substitution, morphological transformation; 3) patient given a picture; syllables of corresponding word on three sheets of paper. Patient must compose the correct sequence and read it. In a second step, the syllables must be chosen from distractors; 4) patient given a picture, then shown a letter, and must decide if the letter belongs to the target. After presentation of single letters, patient must compose word sequence with those he has chosen; 5) patient given anagram sentence task. Complexity of syntactical structures of sentence increases.
Results: All patients improved their performance in linguistic skills. Reeducation/language stimulation program successful. Language stimulation per se cannot totally explain the linguistic improvement of the patients.

Culton, G.L. and Ferguson, P.A. 1979. Comprehension training with aphasic subjects: The development of five automated language programs. *Journal of Communication Disorders*, **12**: 69-82.

Subject Area: Aphasia—general disorder.
Therapy Area: Sentences; vocabulary and linguistic structures.
Therapy: Use of automated comprehension training program for improved comprehension of sentence structures.
Purpose: To investigate the effectiveness of an automated comprehension training program with adult aphasic subjects. Focus on the training of linguistic structures through comprehension rather than production based on hypothesis that comprehension precedes productions. Emphasis on valid and reliable program development, presentation and assessment, including response generalization from one stimulus situation to another, rather than repetition of stimuli.
Method: *Subjects*: 7 left-hemisphere damaged adults.
Materials: Core vocabulary compiled from three lists of most frequently occurring words and phonetic contexts. Length of linguistic structures ranged from 2 to 6 words.
Format: Frame 1: Auditory stimulus presented (blank frame); Frame 2: Auditory stimulus repeated with visual stimulus presentation; Frame 3: Auditory stimulus repeated again, with multivisual stimulus presentation (4 visual stimuli, one of sentence). Alternative retention-generalization frames were interspersed throughout the programs in order to assess retention of vocabulary and understanding of linguistic structures that had been previously problem-solved; these frames were constructed by changing key words in previously-trained structures.
Results: No pattern of increasing difficulty found in scores as length of structures increased; often longer structures easier to comprehend. Explanation: 2-word phrases are processed by the aphasic patient in a different manner from complete sentences. For example, the aphasic individual had to mentally supply the noun subject to process the picture accurately, therefore information not presented auditorily had to be retrieved from the patient's own repertoire.
Conclusions: Programmed instruction is attractive because it allows consistent implementation of learning theory principles, while automated programming is most attractive because it has labour-saving potential. Automated comprehension training successfully assisted aphasic patients in processing certain linguistic structures. Paradigm included specific content training, but also provided opportunity for generalization; clients who do not generalize language learning from one stimulus situation to another will not be functionally rehabilitated.

Dabul, B. and Bollier, B. 1978. Therapeutic approaches to apraxia. *Journal of Speech and Hearing Disorders*, **41**: 268-76.

Subject Area: Apraxia.
Therapy Area: Sequencing of speech sounds.

Therapy: (A) Sequencing of speech sounds; (B) facilitation techniques in acquisition of correct articulatory postures; (C) facilitation of speech sound sequencing through graphic cues.
Purpose: To present 3 therapeutic techniques for rehabilitation of apraxic adults; 2 case histories are offered in support of speech sound sequencing techniques.
Method: *Subjects*: 2 apraxics with aphasia; some previous aphasia therapy; at least 9 years post-onset.
Procedure: Stages to improve sequencing of speech sounds: 1) mastery of individual consonant phones; 2) rapid repetition of consonant+/a/; 3) build-up of phones into syllables; 4) word-attack by phones and syllables. For the acquisition of correct articulatory postures, the clinician assumes the role of auditory model and utilizes visual and placement techniques—oral directions and physical manipulation. Graphic material used in the breakdown of words into syllables.
Results: Improvement of sequencing errors after speech sound sequencing therapy where no such improvement found with prior aphasia therapy.

Davies, C.L. and Grunwell, P. 1975. A new approach to the treatment of severe dysphasia: A case study. *British Journal of Disorders of Communication*, **10**: 142-48.

Subject Area: Severe dysphasia.
Therapy Area: Syntax (sentences); verbal and written output.
Therapy: Structured treatment programme; to increase verbal output and improve performance on written tasks.
Purpose: Attempted the application of linguistic principles in the treatment of a case of long-standing severe dysphasia who had previously failed to make any progress with expressive speech ability despite intensive therapy.
Method: *Subjects*: 1 patient with moderate to severe auditory comprehension problem and dyspraxia.
Procedure: Part I: substitution drills beginning with SVO sentences. Each structure presented in the following modes: written texts; copying sentences; repeating with clinician; reading alone; producing sentences orally from visual (picture) stimulus alone. Supplementary technique of rearranging sentences' word order. Concentration on use of content words in complete sentences.
Part II: increase verb vocabulary, incorporate various verb tense forms, adjectives, pronouns, comparatives, prepositions and question words, in appropriate sentence structures, in that order. The basic sentence pattern was provided in each case and the fillers for the content slots changed appropriately. Choice of structures was determined by informal criterion of difficulty in relation to sentence length, concept encoded and previous therapeutic experience.
Results: No real spontaneous production of language after therapy. Strict adherence to learned pattern necessary, any deviations caused chaos. Expressive language used post-therapy was derived exactly from the treatment programme.

Conclusions: Structured approach used is most suitable for severely dysphasic. It might be worthwhile to plan treatment for less severe dysphasia on these principles, but with a modified, more flexible approach.

Davis, G.A. 1973. Linguistics and language therapy: The sentence construction board. *Journal of Speech and Hearing Disorders*, **38**: 205-14.

Subject Area: Dysphasia—general disorder.
Therapy Area: Syntax (sentences).
Therapy: Sentence Construction Board as apparatus to facilitate therapy involving sentence structures.
Hypothesis: Grammatical patterns are used as frameworks for the expression of ideas, as frameworks within which words are inserted. At prescribed syntactic levels, patients should use various words within the given pattern. Once a particular syntactic pattern is mastered, the pattern should be modified in gradual steps so that the patient may rely on previous experience for success at levels of greater complexity. (Derived largely from a language instruction method for deaf children.)
Method: *Subjects*: No test subjects; suggested that adults with aphasia will be most helped if they have some abilities to comprehend simple instructions, to recognize and distinguish colours, to read (and to read aloud), to retrieve some basic vocabulary, and to maintain at least 2-3 items in short-term memory.
Materials: Sentence construction board: lights serve as cues for grammatical classes on one level, i.e., Adj, N, V. Bottom row of branching lights symbolizes further differentiation within each category, for example, within an NP.
Procedure: Patient given picture and asked to describe it with a sentence. The patient's response is then shown on the sentence construction board. Each word is placed in appropriate place, depending on sentence type being learned. In other words, show the patient where his words are placed in surface structure.
Conclusions: The sentence construction board is useful for patients with a wide range of syntactic deficiencies.

Davis, G.A. and Tan, L.L. 1987. Stimulation of sentence production in a case with agrammatism. *Journal of Communication Disorders*, **20**: 447-57.

Subject Area: Agrammatism of Broca's aphasia.
Therapy Area: Sentence production; verbal expression.
Therapy: "Schuellean" stimulation approach for progress in verbal expression; improving sentence production in persons with agrammatism of Broca's aphasia.
Purpose: To analyze the spread of effect (generalization) of three treatment phases of equal length which follow the stimulation approach. Also, to emphasize that the clinician may adjust the program while still meeting some of the original objectives.

Method: *Subject:* One 41-year-old female with left-hemisphere CVA; 6 months post-onset; moderate-to-severe Broca's aphasia.
Materials: 3 sets of pictures containing 10 drawings of familiar activities.
Procedure: Stimulation procedure consisting of steps for eliciting the best possible description of a picture. As a form of "loose training", linguistic cues and patient's verbal responses could be variable as long as they were appropriate with respect to each picture.
Step 1: Repetition of phrases as each picture presented. Length of stimulus increases as repetition ability improved. Step 2: Involved retrieving single words that would represent any element of the pictures. Designed for practice of propositional utterance without having to attempt entire sentence. Step 3: Sentence completion began with eliciting the last word of a sentence, then the last two words and so on, until a complete sentence was produced. (Designed to elicit more spontaneous and complete sentence production, starting with reverse chaining.) These steps were flexibly applied in order to elicit the most spontaneously complete response. Progress measured by length of utterance.
Results: Effect restricted to items being trained.
Conclusions: Studies have shown aphasia treatment produces results, but are not informative about spread of effect. A client can be improving in a treatment activity while showing no gains with communicative use of language, especially in other settings.

Edwards, S. 1987. Assessment and therapeutic intervention in a case of Wernicke's aphasia. *Aphasiology*, 1: 271-76.

Subject Area: Wernicke's aphasia.
Therapy Area: Comprehension.
Therapy: "Holistic" approach includes the use of PACE therapy, which creates the conditions under which structured linguistic aspects may also be facilitated; psychological support for spouse/family as well as education about what type of input will facilitate communicative interactions and what can be expected from the patient.
Hypothesis: The social needs of the patient and the family are as much the focus of therapy as the consideration of the linguistic deficit. Issue of the important relationship between assessment and remediation.
Method: *Subjects:* 1 male client diagnosed with Wernicke's aphasia.
Procedure: General advice given to wife about her input to her aphasic husband: restrict input to short but grammatical utterances; slow speech but maintain normal prosody; repeat and stress key phrases; gesture where appropriate; supplement speech by writing key words; listen for pronominalization; use context to help in reference identification; participate in conversation turn-taking; give unequivocal feedback using both verbal and non-verbal means.
Specific Aphasia Therapy: therapeutic approach aimed to encompass both context-oriented strategies where he worked from discourse factors down and strategies to strengthen molecular units of language. PACE therapy used (use

non-linguistic strategies to compensate for damaged linguistic system). Exercises devised to improve control of phoneme production, starting with control of gross movements of the articulators. Facilitation of comprehension without contextual clues: selection tasks with an increasing number in the array; various parts of speech, nouns, verbs, adjectives, adverbs, pronouns and prepositions were used as target words; phrases and clauses containing contrastive morphemes and tense were also used; tasks were given to stimulate the patient's use of his semantic system: categorization and word association tasks.
Results: Auditory and written comprehension showed improvement on BDAE. Improved social life reported.
Conclusions: For the fluent patient, the ability to participate in structured psycholinguistic tasks might be severely limited by comprehension loss. Overriding need is to enhance communicative competence which involves considerations such as effective topics and context of interaction, including the conversational partner. Attention to these factors creates the conditions under which structured linguistic aspects may also be facilitated, namely those aspects which, while dominant in the linguistic assessment, could not be addressed directly in the remedial procedures. For the subject of this study, the two aspects of therapy, promoting communication and attention to specific linguistic factors, changed during the course of therapy, but both remained prime concerns.

Green, G. 1984. Communication in aphasia therapy: Some of the procedures and issues involved. *British Journal of Disorders of Communication*, **19**: 35-46.

Subject Area: Adult aphasia (all types).
Therapy Area: Naturalistic communication; discourse.
Therapy: Communication therapy.
Purpose: To address some of the pertinent issues relating to communication therapy with adult aphasic patients and to provide some indications of possible management procedures and future research directions in the area. There is a need to place greater emphasis in management programmes on aphasic patients using language in everyday contexts.
Issues: Traditional structured treatment programs ignore fact that language evolved primarily as an interactive tool for various purposes. Aphasia therapy needs to focus on such functions as establishing interpersonal relationships, regulating the behaviour of others, satisfying material needs or desires, exploring and organizing environments and exchanging messages and information. In order to do this it is necessary to take into account contextual and situational variables within which the language elements may be used, as both the utterance and situation are required for the comprehension and usage of words. Other features to be included involve non-linguistic activities of the communicative interaction such as anticipating possible subsequent moves by the other speaker, determining what information is shared and what is new, and using and

interpreting gesture, nonsubstantive utterances ("oh", "hmm", etc.) and facial expressions.
Possible Knowledge Bases: Studies of child language development and disorders; discourse and pragmatic models based on normal populations; discourse and pragmatic analyses of impaired communicators. Applicability of the material to aphasic population must be considered. Important to clarify whether normal or effective communication is focus of therapy.
Assessment: Traditional aphasia tests (BDAE, etc.) are not informative about how the aphasic patient communicates or uses residual language skills; alternatives are available or being developed (e.g., CADL, FCP, CET). More information is necessary in order to devise tests that will also provide sufficient information for detailed management planning. For appropriate treatment techniques to be selected within a communicative model, it is necessary to utilize detailed client specific assessments, which determine how an aphasic patient communicates with his/her partners. In addition, attention must be paid to the varying nature of the aphasic's skills as they fluctuate according to the people, context and situations involved.
Training of Communicative Partners: If communication is an interactive process, then the people with whom the aphasic patient interacts should be actively encouraged to learn how to maximize the aphasic patient's communication potential.
Treatment Methods with the aphasic patient: For naturalistic therapy, the same standards that apply for normal communication can also be followed in therapy. Tasks need to focus upon context typical of daily interactions. The constrained highly structured settings common in therapist-client didactic procedures need to be minimized so that language is used for the different purposes for which it evolved. Note that conversational discourse structure differs from the discourse structure of tests and remediation programmes. Variables or stimulations that occur in everyday conversation, such as background noise, poor acoustics, and cross-conversations, should be incorporated into treatment.
Conclusions: It is suggested that naturalistic communication therapy for the adult aphasic patient will involve language work situated within a framework of the total communicative process and will encourage active language usage as an interactive medium within different situational and communicative contexts. Communication therapy entails specific contextual activities where language is used actively for different reasons by the aphasic patient. Communication therapy aims primarily to enable the aphasic patient to function in the community, regardless of the residual aphasia.
Summary: In order for aphasic patients to benefit maximally from aphasia therapy, management programmes should place a greater emphasis upon the usage of language within naturalistic contexts.

Hanlon, R.E., Brown, J.W. and Gerstman, L.J. 1990. Enhancement of naming in nonfluent aphasia through gesture. *Brain and Language*, **38**: 298-314.

Subject Area: Nonfluent aphasias.
Therapy Area: Naming; gesture.
Therapy: Combination of pointing gesture with verbal naming to enhance naming abilities.
Purpose: To determine if unilateral gestural movements simultaneous with attempts at oral-verbal expression (confrontation naming) would facilitate expressive capacity in severe aphasic patients.
Hypothesis: Communicative pointing gestures employing the proximal (i.e., shoulder) musculature of the right hemiplegic limb of nonfluent aphasic patients will facilitate confrontation naming; effect probably limited to aphasic patients with anterior damage, nonfluent speech, misarticulation, and hemiplegia.
Background: Hypothesis based on a microgenetic model which holds that action formation traverses distributed levels in brain and cognition in a process recapitulating the course of forebrain evolution. Thus, functional activation of the proximal motor system of the right hemiplegic arm of anterior nonfluent aphasic patients might tap into early levels in speech production and thereby possibly facilitate verb expression.
Method: *Subjects*: 24 aphasic patients with moderate to severe naming impairment; 8 Broca's, 6 Wernicke's, 4 global; 4 transcortical motor; and 2 anomics.
Procedure: Shown pictures of objects which they were to try to name. Instructed to produce the following movements immediately upon seeing a pictures, while trying to name the pictured object: Condition 1: "Point to the pictured objects with your right arm"; Condition 2: "Make a fist with your right hand"; Condition 3: "Point to the pictured objects with your left arm".
Results: the production of deictic or pointing gestures through the activation of the proximal motor system of the hemiplegic side appears to have a facilitatory effect on naming performance among the nonfluent aphasic subtypes. Moreover, the engagement of the right proximal system facilitated naming more than right hand engagement or a comparable gesture with the motorically intact left arm.
Conclusions: One explanation of the proximal facilitation of naming involves the engagement of older proximal motor systems of the right arm (left hemisphere) in the production of an expressive movement (pointing gesture). This may enable nonfluent aphasic patients to tap into or access subsurface levels of language processing, resulting in increased naming performance.

Hatfield, F.M. 1972. Looking for help from linguistics. *British Journal of Disorders of Communication*, 7: 64-81.

Therapy Area: Linguistics: levels of phonology, morphology, syntax, lexicon.
Therapy: Apply linguistics to description, assessment, and remediation procedures.

Hypothesis: Linguistics will devise a better investigative methodology for speech pathology.
Background: A "linguistic" approach treats each language as a closely structured autonomous system, or network, one capable of being described without reference to the cognitive or emotional state of the individual speakers, or to such factors as fatigue, memory, or attention. In applying linguistics to aphasic phenomena, one is ultimately concerned with the degree to which the patient's knowledge of Saussure's (1969) langue is preserved (versus parole). Or, somewhat similarly, Chomsky's competence-performance dichotomy (1965): Competence: speaker's own knowledge of the code of the language-community (langue is the code).
Contemporary linguistics uses linguistic levels: 1) phonology; 2) morphology; 3) syntax; 4) lexicon. At each level, there are questions to be answered by a linguistic approach: 1) What are rules for generating code system of speaker's community? 2) Is patient obeying these rules? 3) Is he operating any set of rules, in which case, which ones and how do they relate to target set of rules? 4) What modalities are intact/affected (spoken/written)?
Linguistic approach applied to speech-language pathology: 1) A linguistic approach should be applied firstly to a descriptive analysis of the aphasic patients speech; 2) attempt theoretical classification of aphasic speech disorders or of aphasic speakers; 3) suggest remedial measures to be undertaken. For example: Phonological disorder: Articulatory phonetics applied to description and analysis; phonological level: Linguistics can, for the speech pathologist, provide tests to analyze data, provide "good" rules and most common rules, and answer questions regarding speech irregularities; morphological and syntactic Levels: Go beyond surface phenomena; underlying unifying theory would help therapists devise most efficient therapy; linguistics can be helpful to investigate each patient's linguistic deviations separately on specific levels (disorders not necessarily on one level only).
Conclusions: If a linguistic explanation can be found for the majority of patients' deviations, and if their errors in most cases can be shown to be part of a coherent system, our understanding of the problem will gain greatly. However, not realistic for every problem of verbal behaviour. Therefore do not confine approach to purely linguistic.

Hatfield, F.M. and Shewell, C. 1983. Some applications of linguistics to aphasia therapy. In C. Code and D.J. Muller (eds.), *Aphasia Therapy*. London: Edward Arnold. Pp. 61-75.

Purpose: Examines need to apply some of the insights of modern linguistic science to the remediation of aphasia.
Summary: *Structural Linguistics*: branch of synchronic linguistics which looks at each language as a set of interrelated systems of elements—sounds, words, etc.— which have no validity independent of the relations between them; language is hierarchically organized and rule-governed; most linguistic-aphasic

papers published in the past 20 years have focused on analysis rather than therapy.
Application of Linguistics to the Analysis of Aphasia Before Therapy: General principles underlying the aims of assessment and therapy: therapy should be throughout *rational* and also *specific* to the form of aphasia. *Rational:* Therapy is based on a well-though-out theory; in contrast with therapy based on tradition or on intuition unsupported by explicit interpretation of past experience. *Specific:* Therapy is carefully chosen for each particular form of the disorder; reject unitary theory; accept concept of qualitatively distinct patterns of language deficit, with clear implications of the need for different types of remediation.
Linguistic Refinements in Assessment: Four levels of linguistic analysis: 1) segmental phonology and prosody; 2) morphology and syntax; 3) the lexical semantic; 4) functional aspects of communication for the individual in society. Breakdown may occur at one level only, or across several levels.
Aphasia Therapy with a Linguistic Basis: Therapy aims to help a patient to move from a given state of functioning at one level of language to a higher, more elaborated state.
Selection of Linguistic Structures: Should be communicatively useful to the patient and presented in a linguistic hierarchy based on the premise that mastery of one structure may presuppose that of another; order of presentation suggests an inherent order of difficulty of linguistic structures; hierarchy not necessarily devised by following the order of child language development or EFL procedures.
Application of Some Theories: Transformational-generative grammar: An embedded sentence will be more easily understood by Broca's aphasic patients if it is broken down into its constituent propositions. Structuralist approach: Items owe their identity to meaning-differentiating features. Therefore, in therapy, use of phonemic contrasts; same principle may be applied to syntactic or lexical units, which contrast with each other in the same "slot" in the sentence/utterance (e.g., "is" contrasts with "was").
Other examples are provided concerning therapy for word-retrieval and semantic reinforcement, remediation of severe agrammatism, and therapy for patients with residual syntactic skills.
Conclusions: Linguistics fosters the specification and planning of appropriate and effective language remediation. It could be misleading to rely on the findings of linguistics alone, without concurrently examining for a multitude of possible disturbances—of perception, attention, orientation, cognition or memory. Moreover, although the discipline of linguistics can identify aspects and degree of language breakdown and give guidelines as to the direction that therapy should take and the order in which items and structures should be practised, it cannot specify learning strategies. Linguistics, in collaboration with psycholinguistics and cognitive psychology, can contribute a great deal to the understanding of what to train and how learning or re-learning can best be achieved.

Helm-Estabrooks, N. 1988. The application of neurobehavioral research to aphasia rehabilitation. *Aphasiology*, 2: 303-08.

Subject Area: Global aphasia, severe non-fluency (anarthria), and auditory comprehension and grammatical disorders.
Therapy: Global aphasia: visual symbol systems, VAT (Visual Action Therapy); severe non-fluency: MIT (Melodic Intonation Therapy); auditory comprehension and grammatical disorders: HELPPS (Helm Elicited Language Program for Syntax Stimulation).
Purpose: To describe the efforts that have been made to apply neurobehavioral research findings to clinical practice of aphasia rehabilitation.
Treatments: *Global aphasia:* Utilize visual/gestural inputs and outputs rather than auditory/verbal. It has been found that globally aphasic patients retain a rich conceptual system and that their language impairment exceeds their cognitive impairment. Further evidence that at least some of the cognitive operations necessary for natural language are retained in global aphasia and that global patients can communicate, at least on a basic level, by using an alternative symbol system.
Therapy: VAT: Findings from VAT supports the notion that global patients do have the capacity for symbolic behaviour and that non-verbal training can deblock or re-establish linguistic skills. Therefore, global aphasia is no longer regarded as a total loss of language or symbolic capacity.
Severe Non-fluency: Neurobehavioral research suggests that singing may be retained in aphasia because it is mediated by the undamaged right hemisphere.
Therapy: MIT: Utilizes this principle and employs high-probability phrases and sentences presented in a performance hierarchy according to techniques of programmed instruction.
Auditory Comprehension and Grammatical Disorders: For rehabilitation of sentence level comprehension, there is a hierarchy of difficulty emerging from neurolinguistic studies.
Therapy: HELPSS: a hierarchy and story completion approach adopted for a syntax stimulation programme.
Conclusions: Neurobehavioral findings have been applied directly to aphasia rehabilitation programmes. However, our understanding is incomplete. Researchers must formulate a hypothesis about the neuropsychological mechanisms underlying aphasia symptoms to improve chances of arriving at more effective treatment approaches.

Helm-Estabrooks, N., Emery, P. and Albert, M.L. 1987. Treatment of Aphasic Perseveration (TAP). A new approach to aphasia therapy. *Archives of Neurology*, **44**: 1253-255.

Subject Area: Transcortical sensory, nonfluent, conduction aphasia; moderate to severe perseveration.
Therapy Area: Naming.

Therapy: Treatment of Aphasic Perseveration (TAP): manipulate those factors known to induce perseveration.
Hypothesis: Perseveration may be an integral component of aphasic symptoms. Consequently, treatment of perseveration itself may improve language performance in aphasic patients.
Purpose: To test this hypothesis, devised a technique to reduce and/or eliminate perseveration called Treatment of Aphasic Perseveration (TAP).
Method: *Subjects*: transcortical sensory, nonfluent, and conduction aphasia; moderate to severe perseveration.
Design: Single-subject experimental design; alternate TAP with other standard treatment program (Case 1: Sentence Level Auditory Comprehension Treatment (SLAC); Case 2: Voluntary Control of Involuntary Utterances (VCIU); Case 3: approach designed to increase appropriate verbal output).
TAP Program: General strategies used to aid patient in confrontation naming. Explain to patient concept of perseveration; encourage patient to give no response or ask for help rather than perseverate on previous response. If patient consistently perseverates on particular word, sensitize by writing incorrect perseverative response on paper, show it to patient, tear it and leave it in his field of vision as reminder; every time he begins to say the word again, point to "word pieces" quickly to help him inhibit this response.
Some Specific Strategies: Provide descriptive sentence for stimuli; sentence completion; provide initial phoneme of target word; provide pantomime associated with object (gestural cue); sing a song or recite a rhyme that might elicit target response; tactile cue; unison speech; melodic intonation, etc.
Results: *Case 1*: transcortical sensory aphasic patient; perseveration moderate. Patient improved with TAP (e.g., his naming raw scores increased while his perseveration decreased); *Case 2*: nonfluent aphasic patient; perseveration severe. Confrontation naming improved better with standard treatment, but TAP was more effective in reducing perseveration. *Case 3*: conduction aphasia; perseveration severe. Comparable improvement in confrontation naming with both methods, but TAP was significantly more effective in decreasing perseveration.
Conclusions: In all 3 patients, the TAP program was effective in reducing perseveration and improving language performance. Program is based, in part, on the hypothesis that perseveration can be raised to a conscious level and actively inhibited by the patient, allowing for a correct, non-perseverative response. TAP technique was consistently better than standard therapy in reducing perseveration.

Helm-Estabrooks, N., Fitzpatrick, P.M. and Barresi, B. 1981. Response of an agrammatic patient to a syntax stimulation program for aphasia. *Journal of Speech and Hearing Disorders*, **46**: 422-27.

Subject Area: Agrammatism.
Therapy Area: Syntax.

Therapy: Syntax Stimulation Program (SSP): Teach sentences to agrammatic patients to improve communicative effectiveness; expand their phrase length and syntactical repertoire.
Hypothesis: Since it is an encoding deficit which appears to diminish communicative effectiveness in the agrammatic patient, therapeutic efforts should be directed toward expansion of a patient's phrase length and syntactic repertoire.
Method: *Subject*: One 35-year-old, right-handed male with classical agrammatism.
Procedure: Utilizes sentence constructions that are presented based on a hierarchy of difficulty for Broca's aphasia patients—a hierarchy based on performance capacities of agrammatics.
Story Completion Format: Level A: Patient required to produce a delayed repetition of the target response; Level B: Patient required to complete story with a self-retrieval target response. All stories accompanied by simple action figure line drawing.
Results: Improved expressive performance. Limited ability to generalize. Once a sentence type had been trained, patient produced new exemplars of that type in spontaneous speech.
Conclusions: It appears that even presumably stable agrammatic patients can respond positively to a specific treatment approach such as SSP.

Helm-Estabrooks, N. and Ramsberger, G. 1986. Treatment of agrammatism in long-term Broca's aphasia. *British Journal of Disorders of Communication*, 21: 39-45.

Subject Area: Agrammatism; expressive disorder; Broca's aphasia.
Therapy Area: Syntax.
Therapy: Helm Elicited Language Program for Syntax Stimulation (HELPSS) —designed to improve expressive skills/verbal output skills to a level more in keeping with receptive ability.
Purpose: To examine the response of agrammatic patients to the HELPSS treatment programme. Treatment programme based on syntactic hierarchy of difficulty shown by agrammatic patients.
Method: *Subjects*: 6 nonfluent, agrammatic (Broca's) aphasic patients.
Procedure: Story Completion Format: Used to elicit 11 sentence types; 2 task levels and multiple exemplars for each syntactic construction. Level A: Clinician relates short story that ends with target sentence; elicit target sentence with question. Level B: Target sentence omitted from story. Once performance criterion reached for Level B, go to Level A of next sentence type in hierarchy until have completed all 11 constructions.
Results: All patients able to complete HELPSS programme although the number of sessions varied. Significant improvement in expressive skills; positive changes in receptive skills but improvement not significant.

Conclusions: Treatment programs based on linguistic assets and deficits of specific aphasia syndrome may prove effective with groups of patients who demonstrate these behavioral features.

Helmick, J.W. and Wipplinger, M. 1975. Effects of stimulus repetition on the naming behaviour of an aphasic adult: A clinical report. *Journal of Communication Disorders*, **8**: 23-9.

Subject Area: Mild aphasia—general; naming disorder.
Therapy Area: Naming; vocabulary.
Therapy: Emphasis on stimulation repetition of vocabulary and not response repetition.
Purpose: To specify a program for increasing the naming ability of an aphasic adult, and to examine the effectiveness of stimulation repetition.
Method: *Subjects*: 54-year-old male; left-cerebral damage secondary to CVA; mild aphasia with primary deficit in naming skills.
Materials: Stimulus vocabulary consisted of 45 words, either nouns or verbs
Procedure: Experimental Conditions: Max, Min, Non-T (nontreatment); Max and Min identical except for amount of stimulus repetition provided: Program Min: 4 program steps presented once for a total of 6 stimulus repetitions. Program Max: 24 stimulus repetitions; each of 4 steps 4 times.
Program: 1) identification: Therapist utters name of picture; 2) contextual cue: Therapist uses stimulus word in short simple sentence; 3) differentiation: Therapist points to and names stimulus picture out of 4 pictures; 4) tracing-copying: Therapist prints stimulus word and then reads the printed word aloud; patient traces printed word while therapist names word; patient copy-writes word while therapist reads word aloud. Each word is presented verbally as part of a linguistic unit—for example, in combination with an article for 1, 3, 4, and in sentence form for 2—and accompanied by a pictorial depiction of the word.
Pretests were done to measure the effect of treatment: patient asked to name, without stimulation, the pictures that had been presented during the previous day's therapy session and at end of treatment, name all 45 words.
Results: Patient identified more stimulus pictures under each of 2 therapy conditions than under Non-T.
Conclusions: The application of a planned treatment program enhanced naming skills. However, no difference between the two conditions of stimulus repetition. Therefore, use reduced amount of stimulus repetition; it is a more economical use of therapy time, thereby allowing for incorporation of additional variables into the program, if necessary. Gains noted in Non-T condition most likely reflect effects of generalization from therapy conditions (and spontaneous recovery).

Howard, D., Patterson, K., Franklin, S., Orchard-Lisle, V. and Morton, J. 1985a. The facilitation of picture naming in aphasia. *Cognitive Neuropsychology*, 2: 49-80.

Subject Area: Aphasia, all types.
Therapy Area: Naming; lexicon; logogen model of naming.
Therapy: Facilitation of naming through accessing the semantic representation corresponding to the picture name using word-to-picture matching with either auditory or visual presentation, and semantic judgments.
Purpose: A series of 4 experiments to investigate the effects of a number of treatments on the ability of aphasic patients to retrieve picture names, at some time after the treatment is applied. Experiments in this paper all use paradigm of facilitation—one application of a single technique with a view to assessing its specific effects at some later time.
EXPERIMENT 1
Purpose: Question of whether auditory word-to-picture matching facilitates picture naming by aphasic patient, and whether the effects are stable over a time interval of twenty minutes.
Method: *Subjects*: 15 aphasic patients; 9 men, 6 women with left CVA.
Procedure: 16 naming failures assigned to 4 experimental treatments conditions: 1) pointing and naming: items treated by pointing on auditory command to one of 4 pictures and then, after 6 intervening events, target pictures presented for intermediate naming; presented again for naming in post-test; 2) pointing and naming control: intermediate naming without prior treatment by auditory word-to-picture matching; tested again in post-test; 3) pointing only: auditory word-to-picture matching as treatment; no intermediate naming practice; presented for naming in post-test; 4) pointing only control: no experimental treatment; presented for final naming in post-test.
Pre-test: single pictures presented for naming until 16 correct (fillers) and 16 failed. Treatment: four blocks, each comprised of auditory word-to-picture matching for 5 pictures (2 target, 3 fillers), followed by intermediate naming test for 5 pictures. Post-test: all experimental items (pre-test successes and 4 groups of failures) presented for naming immediately after final treatment block.
Results: For pictures that could not be named in the pre-test, found that patients had greater success in naming in post-test, if, in the interval, they had pointed to the picture on auditory command rather than having had a second opportunity to try to name it. Effect numerically greatest if intermediate opportunity to name picture followed shortly after word-to-picture matching; this facilitation was not significantly greater than in condition without intermediate naming.
Conclusions: Plausible explanation: Picture pointing requires the subject to retrieve, on the basis of the heard word, a semantic representation of a lexical entry and then to search for a picture whose semantic representation matches it.
EXPERIMENT 2

Purpose: To differentiate whether facilitation resulting from auditory word-to-picture matching was due to an effect on the process of picture recognition or due to priming at a semantic level.
Method: *Subjects:* 20 adult aphasic patients; 7 classified as anomic, 7 as Broca's and 6 as conduction aphasic patients.
Procedure: Experimental Conditions: 1) Same: distractors were semantically related to the target; 2) associate: point to picture of a different object from the same semantic category as the target; 3) control: no pointing; controlled for effects of repeated opportunities to retrieve a picture name.
Pre-test; treatment (same as Experiment 1); post-test 1: followed immediately after treatment section; post-test 2: identical post-test re-presented 24 hours later.
Conclusions: Confirms results found in Experiment 1 and concludes that auditory word-to-picture matching has an effect that is stable for at least 24 hours. Facilitation is a result of priming of a representation or process at a semantic level.

EXPERIMENT 3

Purpose: Investigated whether facilitation depends crucially upon presentation of either the picture or its spoken name.
Method: *Subjects:* 9 aphasic patients, all of whom of either Broca or anomic type.
Procedure: Experimental Conditions: 1) Word: treated by matching the written word to one of a choice of four pictures; 2) semantic judgment: subjects did not see the pictures; they simply had to answer a yes/no question which required access to the semantics of the name of a picture failed in the naming pre-test; 3)Control: no treatment. All experimental items were presented for intermediate naming six items after treatment, and again in a post-test immediately afterwards, which was repeated 2 weeks later. Pre-test; treatment; Post-test 1: immediately after treatment section; Post-test 2: same post-test repeated 2 weeks later.
Results: Both semantic judgments and written word-to-picture matching resulted in effective facilitation of naming which persisted for at least 20 minutes.
Conclusions: Assumption: all of these effects reflect priming at the same level. Conclude that facilitation is not due to priming of any process in picture recognition. Common factor between all of the tasks which, in these studies, resulted in substantial facilitation lasting for at least 20 minutes: they were all comprehension tasks requiring the subjects to access a semantic representation corresponding to the picture name.

EXPERIMENT 4

Purpose: Investigated the effects of 3 techniques that provide information about the phonological form of picture names: repetition, rhyme cues and rhyme judgments.
Method: *Subjects:* 8 aphasic patients; left CVA.
Procedure: Experimental Conditions: 1)Repetition: repeat the name of a picture that could not be named in the pre-test; 2)Rhyme judgments: required patient to

judge whether the picture name rhymed with another word; 3)Rhyme cue: aphasic patient presented with picture and told a rhyme; the patient then had five seconds to attempt to retrieve the picture name.
Pre-test; Treatment; Post-test 1: immediately after end of treatment; Post-test 2: same post-test repeated after an interval of 30 minutes.
Conclusions: Cueing techniques that only provide phonological information about difficult-to-retrieve names may have a brief beneficial effect on aphasic's naming performance; but this effect dissipates rapidly.
General Discussion: Durable facilitation of aphasic word retrieval is a consequence of treatment techniques that require the patients to access the semantic representation corresponding to the picture name, and this is contrasted with the short-term effects of techniques that provide patients with information about the phonological shape of the name.
Implications for Aphasia Therapy: Techniques that require patients to access the semantic representation corresponding to the picture name—word-to-picture matching with either auditory or visual presentation, and semantic judgments—result in dramatically improved accessibility of the name that lasts for at least 24 hours, even though such techniques do not involve the patient in articulating the picture name. This does not, of course, imply that using these techniques repeatedly will result in long-term therapeutic improvement.

Howard, D., Patterson, K., Franklin, S., Orchard-Lisle, V. and Morton, J. 1985b. Treatment of word retrieval deficits in aphasia. A comparison of two therapy methods. *Brain*, **108**: 817-829.

Subject Area: Aphasia, all types with word-finding problems.
Therapy Area: Naming; lexicon.
Therapy: Semantic and phonological techniques used to facilitate naming.
Purpose: Contrast phonological and semantic techniques in a therapy paradigm to determine (1) whether facilitation effects are cumulative over sessions, and (2) whether such cumulative effects, if found, will result in genuinely lasting effects on the accessibility of picture names.
Method: *Subjects*: 12 aphasic patients: 6 Broca's, 4 mild conduction, 2 anomic patients.
Procedure: Pretherapy: from naming task of 300 pictures, chose 80 failures to assign at random to experimental conditions; Therapy: each target item treated 3 times, once with each of 3 techniques included in the type of therapy (semantic or phonological) being applied.
Experimental Conditions and Techniques: Semantic condition: 1) pointing to the picture out of a set of 4 semantically related pictures on spoken request; 2) matching the written word to the appropriate one in this same set of 4 pictures; iii)answering a yes/no question requiring the patient to access the meaning of the name. Phonological condition: 1) repeating the picture name; 2) attempting to produce the name with the aid of a phonemic cue; 3) judging whether the name rhymed with another word. Naming Control: Pictures presented for naming

during therapy with the same frequency of opportunities for naming as the treated items. Baseline Control: Pictures presented for naming only in post-therapy tests and so were not seen or named at all during the course of therapy. Post-therapy: Presentation of all experimental and control pictures interspersed with some pretest successes, for naming; administered 1 week and 6 weeks after the end of each therapy period.
Results: Evaluation of therapeutic techniques for aphasic naming impairments has produced evidence of reliable benefit from intensive therapy lasting over short periods. Results suggested a small degree of generalization to untreated items. Effects are not permanent.
Conclusions: Suggest 2-stage model for aphasia therapy: 1)use facilitation techniques to access a word; 2)use other techniques to consolidate this access, for example, include those techniques that emphasize the use of the word in a communicative context. Found only a small advantage for semantic as compared with phonological therapy.

Jones, E.V. 1986. Building the foundations for sentence production in a nonfluent aphasic patient. *British Journal of Disorders of Communication*, 21: 63-82.

Subject Area: Non-fluent aphasia; Broca's aphasia; agrammatism.
Therapy Area: Semantics and syntax.
Therapy: To improve spoken output; to achieve accessibility to mapping meaning relations between semantics and syntax.
Purpose: To devise a therapeutic program based on "theoretically argued" hypothesis of the underlying structure of sentences. This is in opposition to the classical approach that sentences are strings of words.
Method: *Subjects:* 41-year-old male; Broca's type aphasia; 6 years post-lesion; showed no evidence of meaning relations between words.
Procedure: Initial programme concentrated on verb information retrieval. Strategies included acting out the verb using real people and toy figures (this proved to be a failure). Therefore, decided to base therapy on the patient making judgements about sentences and not on him making the sentences.
Step 1: present written sentence and patient requested to "block off"—to recognize words which "go together"; once identification of verb established, patient asked to label the units by writing 'verb' underneath; Step 2: Once verb recognized consistently, concept of actor introduced by explaining that it marked the answer to the question 'who' or 'what' undertook the activity; required to first identify verb, then label subject with the correct question word; Step 3: Theme (or object) argument introduced using verbs which require such an argument obligatorily; question words, "who", "what", used; Step 4: Verbs introduced which demanded question "where". These were verbs which take an obligatory argument in a prepositional phrase; Step 5: Gradually other question words were added answering questions about the activity relating to 'when', 'why', 'how'; Step 6: Other tasks introduced to reinforce what had hopefully been grasped to

date. For example, patient given a written sentence in which one obligatory argument within a PP had been inserted in the wrong place and asked to judge the suitability of the sentence (patient did not generalize what he had learned, but could now grasp the concepts); Step 7: Gradual introduction of more complex structures (previously only active sentences); introduced passives. Initially introduced by using irreversible sentences in which patient could rely on pragmatic knowledge. Embedded clauses and subordinate clauses introduced with emphasis on identification of all verbs first and initial use of the same actor for both verbs using a pronoun substitute.
At each step introducing more complex mapping, the processing load was initially decreased by manipulation of lexical items to allow use of pragmatic knowledge and then such a "crutch" was removed.
Results: Evidence of sentence structure emerging in output. Evidence that patient grasped concept of meaning relation and their mapping.
Improvement in use of prepositions, auxiliaries and determiners even though not directly worked on.
Conclusions: Improvement arose because recognition given to patient's underlying problem; by working on verbs and the roles they assign to the other constituents in the sentence, his problems had been confronted directly. Of equal importance to programme was strategy of emphasis on input rather that output. It is important to recognize that Broca's type patients are not a homogeneous group and researchers must identify which level of processing the problems come from.

Kearns, K.P. and Salmon, S.J. 1984. An experimental analysis of auxiliary and copula verb generalization in aphasia. *Journal of Speech and Hearing Disorders*, **49**: 152-163.

Subject Area: Broca's aphasia.
Therapy Area: Syntax.
Therapy: Train production of auxiliary verb which should generalize improvements to production of copula verbs due to functional relationship (same response class).
Purpose: to explore the relationship between auxiliary and copula verbs for two patients with aphasia. Specifically, the primary purpose was to determine whether aphasic subjects who were trained to produce '"is" in auxiliary sentences would generalize their production of "is" to copula contexts. In addition, generalization from trained to untrained exemplars of auxiliary "is" was also investigated.
Hypothesis: Auxiliary and copula "is", which are structurally identical verb forms, will be functionally related as members of a single response class despite their developmental and grammatical differences.
Response Class Generalization: Response generalization occurs when untreated behaviours, which are topographically similar to treated behaviours, are produced within the training setting.

Method: *Subjects*: 2 Broca's aphasic patients.
Procedure: Single-subject reversal design (ABAB): 1) Baseline (A): subjects' verbal production of the present tense auxiliary and copula '"is" in sentence contexts was measured; 2) Auxiliary "is" Training (B): trained subjects to produce the third person present tense verbal auxiliary 'is' in sentence contexts; Phases: (i) Imitation phase: imitate verbal model of target phrase; (ii) Spontaneous phase: subject asked to describe depicted action; 3) Reversal Training (A): subjects trained to produce their baseline, telegraphic sentence forms in which auxiliary or copula verbs were deleted; 4) Auxiliary "is" Retraining (B): subjects were again trained to produce the third person present tense auxiliary "is" in sentence contexts. Imitation and spontaneous training procedures used.
Results: Considerable amount of generalization to untrained auxiliary "is" items was obtained after the subjects received treatment on a restricted number of auxiliary "is" items.
Conclusions: Data partially support hypothesis. Auxiliary "is" treatment facilitated copula "is" + predicate adjective responding for both subjects. This finding supports the notion that auxiliary and copula verbs are related through response class membership. However, the analyses of copula "is" + predicate nominative and locative forms were not consistent with this notion. Despite the mixed findings reported for the response class analysis, the subjects consistently generalized "is" production from auxiliary to copula predicate adjective contexts despite the different grammatical functions of these verbs. Improvement not carried over to spontaneous speech production. Support conclusion that generalization is not automatic and should be actively planned for rather than expected.

Kinsey, C. 1986. Microcomputer speech therapy for dysphasic adults: A comparison with two conventionally administered tasks. *British Journal of Disorders of Communication*, **21**: 125-133.

Therapy Area: Microcomputer as a tool; comparison of computer versus conventional aids.
Hypothesis: In order to determine if the computer can be an effective learning tool for dysphasics, must establish if performance levels using computer equals the performance levels in a conventional situation—for example, the use of pictures and/or words on cards.
Advantages of microcomputers: perform repetitive tasks with endless patience; patients do not make errors in front of clinician, which presumably lessens sense of failure; combination of audio and visual presentation of tasks may improve length of attention.
Method: *Subjects*: 12 dysphasic patients with unilateral dominant hemisphere lesions.
Procedure: 4 sets of multiple choice tasks: 2 comprise Test 1; other 2 comprise Test 2. Within each test were non-linguistic—visual discrimination and

matching of simple rectangular designs, and linguistic items—reading comprehension of simple nouns. Test 1: spotting odd one out from 3 words (and shapes [non-linguistic]); Test 2: rearrangement of upper-case letters to make one of 3 lower-case words provided. Test 1 and 2 were given by both computer and conventional means.
Results: Once patients became familiar with computer, time recorded to complete task was comparable with time recorded in conventional situation. Regardless of different levels of capability, the individuals needed no encouragement to use the computer following initial interaction with it. Patients enjoyed using computer; attention levels high.
Conclusions: Microcomputers may provide an effective learning tool.

Kohn, S.E., Smith, K.L. and Arsenault, J.K. 1990. The remediation of conduction aphasia via sentence repetition: A case study. *British Journal of Disorders of Communication*, **25**: 45-60.

Subject Area: Conduction aphasia.
Therapy Area: Repetition; speech fluency.
Therapy: Sentence repetition.
Hypothesis: Treatment programme based on sentence repetition (in conduction aphasia, repetition is usually considered to be a target of treatment, as opposed to an approach to treatment.) Hoped to reinforce the prosodic experience of producing complete sentences. Expected that improvement would carry over to spontaneous speech because of an increased motivation to continue speaking without correcting errors.
Method: *Subject:* One 74-year-old right-handed male with left CVA; conduction aphasia.
Procedure: Sentence Repetition Treatment Programme.
Pre-/Post-test of sentence repetition: composed of 30 sentences that varied along variables known to affect performance in aphasia: 1) number of words; 2) number of syllables per word; 3) richness of semantic content, manipulated by: i) functor sentences—contain functors and light verbs (i.e., main verbs that are similar to functors in that they carry little semantic meaning—do, take, give); ii) substantive sentences—contain substantives (i.e., content words) for major lexical categories and few functors.
Treatment Materials: consisted of sentences for repetition. First treatment session introduced sentences that were within and just beyond patient's initial repetition ability; sentence difficulty was increased gradually over successive sessions along the same variables as were included in the pre-/post-test.
Pre-test and Cookie Theft subtest administered prior to first session of sentence repetition treatment. During treatment sessions, the therapist read aloud each sentence and asked the patient to repeat it after her. After about 2 months of treatment, patient reassessed with post-test and Cookie Theft subtest of the BDAE.

Results: Speech recovery had plateaued prior to treatment. After treatment, patient displayed significant improvement in his sentence repetition, both in terms of the number of content words and the portion of sentences he produced correctly. Improvements generalized to other speech contexts.
Conclusions: Improvement seems attributable to the repetition treatment programme. The patient's improvement supports original assessment that his difficulty in repeating sentences does not reflect a deficit that is specific to repetition, but a general deficit in speech programming that affects all modalities.

Kotten, A. 1982. Therapy of neologistic jargon aphasia: A case report. *British Journal of Disorders of Communication*, **17**: 61-73.

Subject Area: (Neologistic) jargon aphasia.
Therapy Area: Comprehension; semantic relations.
Therapy: Improve understanding of spoken and written language for minimal communication following specific guidelines (as given below).
Hypothesis: Therapy for those afflicted with jargon aphasia should concentrate equally on the following essential aspects: 1) Improving understanding of spoken and written language by reconstituting semantic relations; 2) Controlled use of "yes" and "no"; 3) Improving written word usage with goal of establishing writing as a minimal communication technique; 4) Establishing a minimal self control in language repetition and reading aloud (auditory feedback).
Method: *Subject*: One 71-year-old female; left cerebral haemorrhage; 6 months post-onset; spontaneous speech; oral naming and repetition consisted of complete neologistic jargon.
Procedure: Followed the aforementioned essential aspects: 1) In course of therapy, effort was made to develop understanding for both auditory and graphic material at the same time. Major effort concentrated on rebuilding the main semantic relations (syntax largely neglected in comprehension exercises): i) element + category; ii) elements + their properties; iii) element + material; iv) part-whole; v)action-object. Simple sentences with highly frequent words were read to the patient, who had to decide whether the statement made sense or not. Answers could be given verbally or by pointing to yes/no cards. Later, multiple choice sentence completion tasks were used. In all sentences, direct object had to be filled in; 2) Use of "yes" and "no" were first trained extra-communicatively, e.g., in highly controlled tasks without any resemblance to normal communicative use of language; 3) To improve written naming, client given the first letter as a clue. Also oral repetition and a dictation exercise; no hints or other aids given. Finally, a written completion task followed by written naming again; 4) Training of repetition using minimal pairs. Basic programme for practising vowels corresponded to the basic vowels as represented in the vowel triangle.

Results: Learning took place, especially in written material (but patient did not use it as a means of communication). Comprehension improved (better comprehension together with correct use of "yes/no" serves for simple communicative device). Auditory feedback mechanisms could be established only with often-trained material (therefore, the basic disturbances in the auditory analyzing mechanism could not be improved).

Kushner, D. and H. Winitz 1977. Extended comprehension practice applied to an aphasic patient. *Journal of Speech and Hearing Disorders*, **42**: 296-305.

Subject Area: Aphasia—as general disorder.
Therapy Area: Lexical items.
Therapy: "Comprehension Training": verbal responses elicited through problem solving.
Purpose: To examine the effect of comprehension training on language production by an aphasic patient.
Background: therapy based on Schuell's "Controlled auditory stimulation"--input without problem solving—with additional assumptions: 1) Language is acquired through comprehension. Comprehension is defined as the pairwise relationship between environmental events and linguistic units. (Production is the outcome of comprehension); 2) grammar becomes internalized through practice in problem solving. (Problem solving is a conceptual operation essential for language comprehension). Verbal responses should be elicited through problem solving but not forced.
Method: *Subjects*: left temporal lobectomy; deficiencies in verbal production and in comprehension; German.
Procedure: Training material was first lesson of program for teaching comprehension of German. Forced-choice response using pictorial events. Patient not encouraged to produce words, just listen; experimenter showed picture, pointed to it and then word was spoken. 19 different common nouns represented and repeated frequently in 106 frames.
Conclusions: Comprehension training contributed importantly to patient's improvement. As the patient regains language, alter stimulus setting so patient has opportunity to indicate understanding by responding verbally, making use of acquired linguistic knowledge. Auditory comprehension is the primary avenue through which linguistic structures are facilitated or relearned. Success in production should reflect achievement in comprehension.

Laughlin, S.A., Naeser, M.A. and Gordon, W.P. 1979. Effects of three syllable durations using the Melodic Intonation Therapy technique. *Journal of Speech and Hearing Research*, **22**: 311-20.

Subject Area: Nonfluent aphasia syndromes: Broca's, Mixed, and Global aphasia.

Therapy Area: Syllable duration in MIT.
Therapy: Increase duration of syllable in MIT to facilitate goal of MIT.
Purpose: To manipulate syllable duration and measure its effect on phrase production ability in a small sample of nonfluent aphasic subjects; to determine if a prolonged syllable duration would improve subjects' verbal performance with MIT technique.
Method: *Subjects:* 4 patients with left hemispheric cerebrovascular lesions and 1 with bilateral lesions; nonfluent: 2 global, 2 mixed and 1 Broca's.
MIT Steps: 1) Unison humming; 2) unison intoned phrase production; 3) clinician fades as patient attempts intoned production alone; 4) patient repeats intoned production alone; 5) patient intones appropriate response to clinician's intoned question.
Procedure: Subjects produced words and phrases at 3 syllable durations: regular speech duration, and 2 expanded MIT intoned durations (1.5 seconds and 2.0 seconds). 60 stimulus phrases and single word items were administered by live voice to each patient. (MIT does not use pre-recorded stimulus material.) The stimulus phrases and words were not balanced for phonologic or syntactic complexity. All durations were approximate.
Results: All patients, whether with moderate or severe aphasia, had significantly more correct phrase productions with prolonged intoned conditions than with regular speech condition. As MIT steps become more difficult, the percentage of correct phrase production decreased across all syllable durations.
Conclusions: Further study is recommended to determine whether the same effect can be obtained when increased syllable duration of 2.0 seconds is used only on the normally stressed syllables.

Lesser, R. 1987. Cognitive neuropsychological influences of aphasia therapy. *Aphasiology*, 1: 189-200.

Therapy Area: Reading disorders in aphasia—surface dyslexia.
Purpose: Three psycholinguistic models are reviewed—exical transcoding, lexical retrieval for naming, sentence production—and therapies derived from each.
Background: Psycholinguistic approach to aphasia therapy makes specific predictions about what might be the location of problem in an individual patient and offers a rationale for treatment.
Models: 1)*Transcoding Model:* Two separate lexicons proposed (orthographic and phonological) which both interact with the "cognitive system" to derive meaning. Surface dyslexics are thought to rely on an orthographic-to-phonological conversion route for reading most words. Deep dyslexics show impaired direct lexical route from visual input lexicon to phonological output lexicon and appear to read mainly by means of the cognitive system. In order to plan rehabilitation for reading difficulties in aphasia, adopt one of 2 strategies: either aim to reactivate the impaired function, or help the patient to find an alternative means of achieving the same performance by a roundabout route.

Review of Therapy: For deep dyslexia: helped patients learn to spell function words again by associating them with content words of similar spelling. For surface dyslexia: reorganization method—program designed to re-teach basic spelling rules; also taught mnemonics to associate sets of words spelled in similar ways. Also for deep dyslexia: strategy to attempt to re-instate the grapheme-phoneme route so that this could be used to reconstruct the deficient lexicon. This therapy exhibited a double re-organization—firstly through the use of strategies to restore a malfunctioning route, then the employment of this route as an alternative strategy for reading to reinforce the impaired cognitive system.
2)*Stages in Retrieving Names:* Distinction made between the semantic lexicon and the phonological lexicon. Stages in naming: 1) Items from the semantic lexicon mapped on to their phonological shapes in the phonological lexicon; 2) select and arrange in the correct order the phonemic pattern needed for its outputting; 3) plan for the phonetic realization of the patterned word and finally, transmission of the planned motor movements to achieve the appropriate articulations.
Review of Therapy: For therapy, given model, find impaired stage, then plan therapy with goal of either reactivation or re-organization. Reactivation therapy for a semantic disorder: improve naming by semantic techniques which only require patient to make semantic decisions about word meaning, and matching the word with its referent. For other stages of naming impairment, for example patients with difficulty in mapping the semantic lexicon onto the phonological, use of phonemic cues to reactivate mapping. For phonemic paraphasia: teach patients to split the target word into manageable single syllables.
3)*Sentence Production:* Based on model: 1) Message level representation; 2) functional level representation; 3) positional level representation; 4) phonetic level representation; 5) articulatory level representation. For aphasia therapy, interested in how a speaker gets from message level (the idea which the speaker wishes to communicate) to functional level (multi-phrasal planning level in which the identity and role of lexical items is at issue) and from functional to positional level. This involves mapping from semantic lexicon to phonological lexicon which occurs simultaneously with the organization of a syntactic planning frame, which arranges the lexical items in position and assigns the frame elements appropriately. The sentence is then ready for phonological assembly. Agrammatism, seen as a deficit in the retrieval of predicative words, is a problem that may arise from disruption in this system.
Review of Therapy: colour-coding to mark verb, subject and object. Use Wh words to mark the key roles of who, what, where, and when, and progress patient through a hierarchy of sentences.
Conclusions: Therapy proposed needs to be complemented by approaches which draw on the pragmatics and function uses of communication, and which extend language work beyond the synthetic clinical tasks of naming, spelling and generating sentences to describe pictures, taking into account the sociolinguistic background of the patient.

Lesser, R. 1989. Aphasia: Theory-based Intervention. In M.M. Leahy (ed.), *Disorders of Communication: The Science of Intervention.* New York: Taylor & Francis. Pp. 189-205.

Topic: Relationship between theory and practice in aphasia therapy.
Theoretical Bases of Intervention: *Neurological Theory:* has specified syndromes on which therapy programmes have been devised, e.g., HELPSS for agrammatism of Broca's aphasia, SLAC for comprehension difficulties in Wernicke's aphasia. Notion of right hemisphere stimulation to assist in language recovery by the left hemisphere or to develop latent language in the right hemisphere, e.g., MIT.
Unitary Theory: one aphasia; dominant influence: Schuell; not a powerfully explanatory or testable theory.
Linguistic Theory: For aphasiology, has primarily provided frameworks for description and analysis. For clinical practice, it has provided theory-based notions of rank order of complexity in language, for example distinctive features in phonology, phrase structure trees in syntax and sense relations in semantics. Attempts to apply specific linguistic theories to aphasia have not yet been made.
Sociolinguistic Description: Application: the therapist should be concerned in restoring, not some abstract norm of language, but the colloquial spoken language of individual patients in their particular speech communities and social networks. Sociolinguistics: language establishes the individual in a social context through the style and accent in which he/she speaks and in the content of what is said, including social small talk; the aphasia therapist should be sensitive to this. Interest in application from the study of conversational exchanges, turn-taking, cohesive devices for linking sections of discourse.
Psycholinguistic Models: Cognitive neuropsychological approach. Notion: mental processes are organized in modules: modules are psychologically and neuropsychologically real, in that brain damage seems to show that the language system can fractionate in such a way as to disturb modules selectively. Best models concern processing of single words—reading and naming—but also include sentence processing. To apply neuropsychological model to the remediation of reading or spelling difficulties in a patient, it is necessary first to be able to identify which hypothetical modules or processes may be malfunctioning. Effectiveness of Intervention has not really been possible to ascertain because some theories (e.g., linguistic, sociolinguistic, cognitive neuropsychological) have not been applied often enough to compare, if at all.

MacMahon, M.K.C. 1972. Modern linguistics and aphasia. *British Journal of Disorders of Communication,* 7: 54-63.

Therapy Area: linguistics applied to aphasiology; justification of the use of linguistic theory to describe aphasia.

Purpose: To give some idea of the type of linguistic research that has been undertaken into aphasia.
Main Arguments: Following an analysis of a language disorder (aphasia) by a linguist, classifications, neurological theories of localization, and treatment methods can be suggested. An arrangement of single levels of language with its features described in one or more types of aphasia is given below:
Phonology: The use of sounds to differentiate the meanings of words. Tendencies in phonological breakdown across languages: 1) a disturbance of the phonological structure of the word. (No aphasic has ever tried to use a non-existent structure in normal language); 2) Changes in the actual phoneme: a phonological unit is still pronounced but its phonetic characteristics have changed. The phonological system in aphasia is subject to systematic disordering, the breakdown is not haphazard. A clear distinction should be drawn between an articulatory (neuromuscular) and a phonological disturbance.
Semantics: Very little written at time of this publication. A few attempts have explored aspects of aphasia using concepts such as semantic fields and sense-relations.
Grammar: Linguistic levels of morphology and syntax. A sentence is constructed by two processes: selection of items and combination of items in sequence; in aphasia, one or the other can be disturbed. Therefore, there are two types of disorders: a selection disorder and a combination disorder. Research shows aphasic patients do not lose the parts of speech, rather there is a severe reduction in their use of most of them but this is directly connected with the frequency with which they occur in normal language.
Morphology: Morphemes are hierarchically ordered in a normal grammar, for example, in English, possessive has a more complex structure than the plural.
Conclusions: In order to clear the way for the development of vigorous linguistic investigations of aphasia which are going to be of practical use to the speech therapist, the time has come to begin thinking about the methodology of applied clinical linguistics.

Miller, N. 1989. Acquired speech disorders: Applying linguistics to treatment. In K. Grundy (ed.), *Linguistics in Clinical Practice*. New York: Taylor & Francis. Pp. 281-300.

Subject Area: Dysarthria; dyspraxia.
Therapy Area: Articulation; speech.
Therapy: Apply notions of acoustic and articulatory phonetics to assess breakdown and devise an appropriate therapy programme.

Summary: *Linking Assessment and Therapy—Formal Features:* The function of speech is to convey meaning. Assumed that extracting meaning firstly rests on the ability to identify structures, word boundaries and sounds within the words. While assessment might deal in discrete units and static postures, therapy must aim for connectivity, contrast and dynamism.
Linking Assessment and Therapy—Functional Features: Assesses the interaction between speech, movement and language. Aim of therapy is to provide optimal language functioning within the given motor constraints. Through their interaction and interdependence, motor and linguistic strengths can be manipulated at one point to compensate for weaknesses at another. Lexicon, syntax, phonology and phonetics can be accommodated to restricted movement.
Targeting for Therapy: After detailing all factors contributing to the problem, a rank order list of crucial variables should be drawn up. From this, decide what is causing the greatest deficit for the listener and what, in turn, is causing that.
The Context of Linguistic Contribution: Maximize communication, not just speech comprehensibility.
Treatment: Articulatory phonetics supplies the knowledge of how and where to place the articulators in teaching reacquisition of sounds.
Contrastive Stress and Tonal Variation: By manipulating stress and intonation different implications of a statement can be conveyed. Incorporate minimal pair work. Segmental accuracy may be influenced by stress, position of word in a syntactic structure, or word class
Conclusions: Linguistics provides a principled framework and descriptive method for analyzing and measuring communicative performance. It provides the background theory and knowledge for: (a) what is normal speech, (b) what variables interact to produce normal intelligible speech and therefore in turn, (c) what features might be targets for therapy and be manipulated to regain effective communication. The detailed analyses possible within a linguistic framework enables the clinician to target precise loci of breakdown, thereby making therapy at once directly relevant and parsimonious. Linguistics also makes possible the careful charting of progress (or deterioration) and breaking down of therapy into graded sub-goals. It is noted that a broad linguistic approach incorporating macro- and micro-analysis fits into and stresses a whole person, whole context perspective.

Montgomery, J. 1971. The importance of seeing red: Self- teaching techniques for adult aphasia. *Journal of Speech and Hearing Disorders*, 36: 250-251.

Subject Area: Aphasia.
Therapy Area: Vocabulary and colour.
Therapy: Use of colour to increase attention span necessary for vocabulary exercises.

Hypothesis: Colour might be the key to maintaining and prolonging attention span in task of vocabulary recovery. Attention span, although it has its limitations, should respond to a variety of colour stimulation.
Method: *Subject*: author suffered from aphasia and conducted and designed experiments through personal experience.
Procedure: Words block-printed with crayon in a variety of colours to create a new interest. Concept of "instant definition": as lengthy definitions of words written down, for example "proclivity" and "intrepid", it was more enjoyable and effective to attempt a more "instant definition"—a word could be recalled in a shorter time in a more pleasant way if personalized definition made. For example, for the word "pertinence", subject defined it as "That's the point." Also, the author purposely "over-trained" in increasing vocabulary with words that might be considered superfluous because as each word and definition was written down, it was found to facilitate recall of other "standard" words.
Conclusions: These experiences with colour therapy—"instant definition" and "over- training"—are presented with the thought that there cannot be too many alternatives in speech therapy for the adult aphasic; where one method does not succeed, another may.

Muma, J.R., Hamre, C.E. and McNeil, M.R. 1986. Theoretical models applicable to intervention in adult aphasia. In R. Chapey (ed.), *Language Intervention Strategies in Adult Aphasia*. Baltimore: Williams & Wilkins. Pp. 277-83.

Topic: Theoretical assumptions underlying the clinical management of aphasia.
Summary: *Five criteria for construct validity:* 1) Relativity: behaviours are related to each other; 2) conditionality: contexts of verbal behaviour; importance of contextual influences; 3) complexity: pertains to intact systems and process (cognition, language and aphasia are complex); 4) dynamism: behaviour changes with use or lack of use; clinical assessment is a continuous process; 5) ecology: the inner-relationship between individuals and their environment.
Theoretical Models: There is no "direct" treatment possible, only treatment based on models of the presumed nature of the disorder. "Catch-22" in models: clinical management approaches are explicitly or implicitly based on assumptions (models) concerning the essence of language dysfunction in aphasia. Catch-22 Solution: become familiar with assumptions basic to many models.
Statistical Model of Language: Rooted in probability. Various word frequency measures have been applied to aphasia, e.g., word frequency lists. Short-lived as linguistic theory (finite state). Rather than a focus on word frequency measure, the clinician may want to incorporate recent evidence concerning the richness and complexity of individual lexicons.
Behavioral Model: Stimulus-response and mediation principles used for explaining language.
Cybernetic Model: Speech model with emphasis on continuous sensory influences. Refers to systems which automatically regulate themselves by

comparing feedback to signals being sent out. Speech is similarly guided by feedback obtained from tactile and proprioceptive receptors in the vocal tract, and by auditory feedback.
Neurological Model: Lack of integration of formal language description and physiological processes
Cognitive Model: Cognitive bases of language learning and use may be considered from 3 perspectives: 1) Indirect: causality, alternative means, symbolic play and delayed imitation (Piaget); 2) direct: categorization and symbolization; also, intention, felicity conditions or cooperative principles or social contracts, proposition (semantic functions and relations), presupposition, and reference; 3) mental operations: construction and utilization processes for decoding messages and planning and execution processes for encoding messages.
Linguistic Model : Linguistic theory has 3 main components (all have potential for advancing the understanding of aphasia): 1)Semantic theory: shift from generative semantics (semantic functions and relations) to interpretative semantics (presupposition and implicature) and more recently the extension to theory-of-the-world; 2)Syntactic theory: structuralistic models → generative transformational theory → generative syntax; 3)Phonological theory: emphasis has shifted from phoneme → syllable structure, stress and other prosodic dimensions.
Pragmatic Model: Major shifts in pragmatic or communication theory has occurred in recent years, some of which revive syntax. Central issues: role of context in social commerce and the function of utterances against a backdrop of relevant presuppositions within social contracts to cooperate.

Naeser, M.A. 1975. A structured approach to teaching aphasics basic sentence types. *British Journal of Disorders of Communication,* **10**: 70-6.

Subject Area: Expressive disorder.
Therapy Area: Syntax; verbal output.
Therapy: Structured approach with three levels corresponding to three basic sentence types.
Purpose: To investigate a structured approach teaching 3 basic sentence types to adult aphasic patients on the basis that rehabilitation of verbal language performance on the syntactic level will be the result of improvement of competence.
Method: *Subjects:* 4 male aphasic patients; good verbal output and auditory input; 1 control received conventional therapy.
Procedure: Language drills as used in TESL are modified for rehabilitation of aphasics. Teach three basic sentence types in adult aphasics:

TYPE	POSITION		
	1	2	3
I to be	NP	be	Pred.(NP)
Example:	That	is	a house.

II	Transitive V Example:	NP The woman	V opens	NP the door.
III	Intransitive V Example:	NP Soldiers	V march.	ø

Pictures represent 3 basic sentence types. Questions asked to elicit target sentence: Type I: "What is that?"; Type II and III: "What happens here?" Examiner looking for: 1) Production of the basic sentence type: is each position or slot filled properly? 2) Production of pattern carryover: is the same verb form and tense that was used with practice picture used with test picture? 3) Production of subject-verb agreement: does person marker for verb agree with subject?
Results: All subjects improved in their production of all basic sentence types.
Conclusions: Results favourable towards a structured approach to teaching basic sentence types to aphasic patients who are capable of certain skills. Patients with expressive type of aphasia most likely to benefit.

Naeser, M.A., Haas, G., Mazurski, P. and Laughlin, S. 1986. Sentence Level Auditory Comprehension treatment program for aphasic adults. *Archives of Physical Medicine and Rehabilitation*, **67**: 393-399.

Subject Area: Comprehension deficits.
Therapy Area: Syntax; phonology; naming.
Therapy: Sentence Level Auditory Comprehension (SLAC): a three-level approach to improving comprehension of sentence level material.
Purpose: 1) To present the SLAC treatment program and investigate whether it could improve language comprehension test scores in chronic aphasic persons with comprehension deficits; 2) to investigate whether there exists a relationship between CT scan lesion sites and good response to the SLAC treatment program; and 3) to investigate whether there exists a relationship between pre-SLAC language comprehension test scores and good response to the SLAC treatment program.
SLAC Treatment Program: 3-level approach to improving comprehension of sentence level material. Major purpose to train the auditory modality to focus on verbal information without visual cues (i.e., never written). All levels involve use of a Language Master tape recorder device and sets of cards with prerecorded stimuli.
Level I: trains the patient to discriminate (same/different) CVC word pairs which differ by only one phoneme; Level II: trains patient to identify a target word which has been prerecorded in the stimulus "frame" sentence; Level III: trains the patient to identify the target word within more complex sentences which have been prerecorded.

Some patients perform better with Level III sentences than with Level II sentences. Level III provides more redundancy with the longer sentences. When all levels have been completed at the 80% correct response level for a given word list, another CVC word list can be used. The SLAC program includes 47 CVC word lists where the initial consonant varies, and 54 lists where the final consonant varies. The word lists where the initial consonant varies were considered easier because they rhyme.
Method: *Subjects*: 16 aphasic patients; left CVA
Experimental Conditions: 1) Group 1, chronic cases, SLAC only: 7 aphasic patients; 2) group 2, chronic cases, SLAC plus other treatment programs: 5 aphasic patients; 3) group 3, acute cases, SLAC plus other treatment programs: 4 aphasic patients. (Chronic = 4 months postonset or more; acute = less than 2 months postonset). Note that "other treatment programs" not defined.
Results: Chronic aphasic patients can improve with treatment initiated later than 6 months postonset. SLAC treatment program has the advantage that it can be used with mild or more severe aphasic patients with comprehension deficits and can be adjusted rapidly to the patient's own level; even a severely impaired aphasic patient can learn to do the same/different judgments. Patients can practice with Language Master by themselves, as often and as long as they want.
Conclusions: SLAC may be effective alone or in combination with other treatment programs in patients having mixed, global or Wernicke's aphasia, either acute or chronic. SLAC appears to increase auditory attention span, although no formal testing in this area was carried out.

Pierce, R.S. 1982. Facilitating the comprehension of syntax in aphasia. *Journal of Speech and Hearing Research*, **25**: 408-413.

Subject Area: Auditory comprehension skills, but impaired syntactic comprehension (fluent and nonfluent).
Therapy Area: Syntactic comprehension; word-order and tense distinctions.
Therapy: Additional surface markers are used to improve comprehension of tense and word-order distinctions.
Hypothesis: To answer the following questions: 1) Can an aphasic listener's comprehension of past and future tense be facilitated further by addition of a different set of tense markers? 2) Will the presence of an additional marker facilitate the comprehension of word order distinctions? 3) Does the presence of a relative pronoun facilitate comprehension of sentences containing an object-relative clause?
Background: Most aphasic subjects are restricted in the number of strategies which they use to decode word-order and tense distinctions. They rely on the following strategy: assume sentence is canonical form S-V-O present tense unless otherwise marked. However, when a sentence deviates, functor words and morphemes are not used by aphasic subjects, therefore poor interpretation results.

Method: *Subjects*: 18 aphasic patients with left CVA; divided into high and low comprehension groups (fluency not controlled).
Procedure: Each subject required to point to 1 of 3 photographs to indicate interpretation of a stimulus sentence played via tape recording. 6 syntactic distinctions studied: 3 tense distinctions (irregular past, regular past, future); 2 word-order distinctions (S/O order in reversible passive, direct/indirect object in preposition of destination); and 1 relativization distinction (object-relative clause). For each distinction, 12 sentences were presented, 6 of which contained an additional surface structure marker particular to that distinction, the other 6 did not.
Results: Additional surface structure markers were facilitatory for the comprehension of tense and word-order distinctions.
Conclusions: Aphasic subjects tend to over-apply a general strategy based on canonical word order and present tense structure. Therefore, they may be able to overcome this problem to some extent when sentences contain ample syntactic restriction and redundancy so that the listeners are able to decode the syntactic relationships in the sentence without having to process the basic syntactic morphemes or markers.

Pierce, R.S. and Wagner, C.M. 1985. The role of context in facilitating syntactic decoding in aphasia. *Journal of Communication Disorders*, **18**: 203-213.

Subject Area: Aphasia—fluent and nonfluent.
Therapy Area: Context and syntax; semantic constraints and passive comprehension.
Therapy: Use of context to aid in comprehension of different syntactic structures.
Hypothesis: Comprehension of reversible sentences should be better when preceded by a supportive contextual sentence than when they are either preceded by a nonsupportive contextual sentence or presented in isolation.
Background: Within a single sentence boundary, several factors influence comprehension, e.g., world knowledge. Typically in discourse, additional linguistic and extralinguistic information is available to the listener. Listeners use the Given-New Strategy: given information relates to previously stored information and new information needs to be bridged or related to the existing information. Therefore when given information readily exists in the listener's knowledge or in prior context of narrative, comprehension is facilitated.
Method: *Subjects*: 14 aphasic; 38-91 years of age; 10 nonfluent, 4 fluent.
Procedure: 2 sentence constructions: 1) reversible active sentences; 2) reversible passive sentences. Each sentence construction is presented in 3 different contextual settings: (i) semantically supportive context; (ii) semantically nonsupportive context; (iii) no context (in isolation). Each contextual sentence was followed by the question: "Who/Which was verb-ed?"—forced-choice

paradigm. Each target sentence altered to generate 2 additional sentences that contained the same verb but different nouns.
Conclusions: Aphasic patients can use prior linguistic context to assist in the comprehension of passives. The aphasic patients' use of the semantic constraints established by context to determine an agent-object relationship replaces the canonical strategy used by aphasic patients when context is lacking. Results reinforce the notion that aphasic comprehension deficits are comprised of difficulty in: 1) understanding deep structure relationships depicted by the sentence; and 2) recognizing the underlying relationships as they are reflected by surface structure markers. Semantically predictive context helps aphasics overcome 2, but not 1.

Podraza, B.L. and Darley, F.L. 1977. Effect of auditory prestimulation on naming in aphasia. *Journal of Speech and Hearing Research*, 20: 669-83.

Subject Area: Aphasia—general; naming/word finding disorder.
Therapy Area: Naming.
Therapy: Study compares the effects of presenting 4 types of auditory cues — phonetic cue, open-ended sentence, set of 3 words containing the target word, set of 3 semantically related words, and no cue—on naming performance of aphasic patients.
Hypothesis: The use of auditory cues should facilitate naming in aphasics. (Word recognition and word recall are different processes with word recognition being easier because it can be accomplished on the basis of partial information.)
Method: *Subjects:* 5 aphasic patients; 1 with dysarthria; 2 with apraxia.
Procedure: Experimental Conditions: A) phonetic prestimulation: presented initial phoneme of pictured word with neutral vowel; B) open-ended sentence prestimulation: open-ended sentence designed to elicit the target word; C) target-word prestimulation: 2 words unrelated to picture semantically or phonologically, third was target word (order randomized); D) prestimulation via semantically related words: 3 related words given aloud; E) control: absence of any cue. All 5 conditions presented to every patient during each of 5 sessions.
Conclusions: No hierarchy of effectiveness of cues found, therefore must base therapy on patients' individual needs. Prestimulation with words semantically related does not appear to be as effective a technique for facilitating naming as prestimulation with the target word and may in fact be counterproductive; it may confuse them. The goal of using pictures in language therapy is not to teach the patient the name of a set of pictures, but to use the pictures to stimulate the language processes of word understanding or recognition and word recall. It is hoped that such stimulation will improve the patients' basic language functioning abilities so that they will better understand the language around them and be able to find the words they need to communicate their thoughts and needs to others.

Prins, R.S., Schoonen, R. and Vermeulen, J. 1989. Efficacy of two different types of speech therapy for aphasic stroke patients. *Applied Psycholinguistics*, 10: 85-123.

Subject Area: Aphasia, all types.
Therapy Area: phonology, lexicon, morphosyntax.
Therapy: Systematic Therapy program for Auditory Comprehension Disorders in Aphasic Patients (STACDAP).
Hypothesis: Attempted to answer following questions: 1) Does language therapy in general have a positive effect on recovery from aphasia, i.e., do patients who receive treatment improve more than untreated controls? 2) Is treatment with a systematic therapy program that is directed at specific language disabilities more effective than traditional stimulation therapy? 3) If systematic treatment of auditory language comprehension disorders is indeed effective, what aspects of language comprehension recover most, and to what extent will treatment effects generalize from practised to nonpractised language skills?
Purpose: Issues addressed by comparing effects of STACDAP with those of stimulation (STIM) therapy (currently most widely used).
Method: *Subjects*: 32 aphasic patients; disorder of auditory language comprehension; those in control group (CONT= no treatment) had treatment discontinued because patients had reached their limits of recovery according to their former speech therapists.
Procedure: Pre- and Post-tests: Part I: auditory comprehension tests comprising 3 linguistic levels: phonology, lexical-semantics, and morphosyntax; Part II: subtests for linguistic skills of auditory comprehension, reading comprehension and oral expression. Part I was used as practice material in treatment to STACDAP group, but not in STIM group.
Methods of Rehabilitation: STIM and STACDAP.
STACDAP: Training of auditory language comprehension; all tasks involve the matching of spoken stimulus with corresponding picture, unless otherwise indicated.
Level 0: Nonverbal: For patients with a serious (global) aphasia and various cognitive-perceptual problems. Exercises intended to train basic functions like recognizing, matching and pointing to different pictures, paying attention to the instruction, concentrating on a task, etc. Exercises: 1) picture matching; 2) recognizing meaningful non–linguistic sounds.
Level I: Phonology: For patients who have problems with distinguishing between acoustically similar phonemes. Exercises: 1) phoneme discrimination (identify); 2) auditory discrimination (same/different task).
Level II: Lexicon: For patient having problems with the understanding of single words and/or word series. Exercises: 1) word recognition: single words; 2) word recognition: series (match 2 or 3 spoken stimulus words with their pictorial equivalents; all alternatives belong to the same semantic category); 3) word recognition: synonyms.

Level III: Morphosyntax: For patients having problems with the comprehension of various morphological aspects and syntactic constructions.; all tasks involve the matching of a spoken stimulus sentence or phrase with its corresponding picture. Exercises: 1) word order N+V, (N+V)+(N+V); 2) conjugation: verb tense and aspect; 3) inflection: number of NPs; 4) inflection: personal and possessive pronouns; 5) Adj+N, (Adj+N)+(Adj+N), and comparatives; 6) word order: prepositions of place and direction; 7) word order: active and passive sentences.
Exercises trained successively in order of increasing complexity. Combination of specific therapeutic strategies varies from patient to patient and depends on the nature and complexity of the STACDAP task being trained.
Results: aphasic groups (STACDAP, STIM, and CONT) did not differ significantly on any outcome measure as a result of the therapy given.
Conclusions: Main findings of study seem to indicate that 1) treatment with a systematic therapy program for auditory comprehension disorders yields no better results than the conventional stimulation therapy; 2) both kinds of therapy investigated do not have any clear effect on recovery from aphasia. Since there is no effect of treatment type, question about generalization effects on different linguistic levels and in various modalities becomes irrelevant. Possible points of criticism: methodological aspects and the authors' interpretation of the results. Some evidence that speech therapy may lead to an improvement of functional language abilities, even years after stroke.

Rosenbek, J.C., Lemme, M.L., Ahern, M.B., Harris, E.H. and Wertz, R.T. 1973. A treatment for apraxia of speech in adults. *Journal of Speech and Hearing Disorders*, 38: 462-72.

Subject Area: Apraxia of speech and mild-to-moderate aphasia.
Therapy/Therapy Area: Restore volitional-purposive communication.
Hypothesis: Therapy for brain-damaged patients—apraxic and aphasic— should follow essential principles: 1) All therapy activities should be organized according to task continua so that the patient does not struggle and so that they work at a high level of success during each session; 2) brain injury in most instances will require that the patient take longer and work harder at regaining or learning lost (articulatory) skill, so clinician should plan for intensive and extensive drills; 3) because the patient is an adult with a history of normal speech and language comprehension and use, mastery of meaningful and useful verbal communication should be emphasized as early as possible; 4) efficient self-correction must be encouraged; 5) the brain injury may have imposed physiological limits on the patient's speech production mechanism. If so, clinician should teach compensatory movements.
For apraxia of speech, therapy should 1) concentrate on the disordered articulation and hence be different from language stimulation and auditory and visual processing therapies appropriate to the aphasias; 2) emphasize the regaining of adequate points of articulation and the sequencing or articulatory gestures; 3)

provide conditions such that the patient can advance from limited, automatic-reactive speech to appropriate, volitional-purposive communication.
Method: *Subjects:* 3 severely apraxic adults with mild-to-moderate aphasia. *Procedure:* Patients were taught volitional control of 5 utterances using 8-step continuum based on given principles. Utterances varied in length from 1-7 words. No word was greater than 3 syllables.
Results: Case 1 and 2 achieved volitional control of target utterances with a minimum of struggle and frustration.
Conclusions: 8-steps, with exception of step 4, represented a legitimate task continuum.

Ross, A. 1989. Applying linguistics to aphasia. In K. Grundy (ed.), *Linguistics in Clinical Practice*. New York: Taylor & Francis. Pp. 205-23.

Summary: *Theoretical Background of Aphasia:* Typology of aphasia based on localizationist model.
Aphasic Symptoms and Levels of Language: Phonetics and Phonology: Problems in language expression: apraxia of speech; phonemic paraphasia; neologisms. Problems of language reception: word comprehension may fail at phonological level, resulting in impaired recognition or misinterpretation as a word with a similar phonological structure. Central problem of phonological processing (combination of above symptoms).
Semantics: Problems in language expression: word-retrieval errors including circumlocution, semantic paraphasia, verbal paraphasia, perseverative paraphasia, neologism, anomia. Problems in language reception: lexical semantic comprehension; relational semantic comprehension.
Syntax: Problems in language expression: agrammatism, paragrammatism. Problems of language reception and expression may be due to a central problem of syntactic processing.
Pragmatics: Non-verbal signals preserved to assist language expression. Contextual information, prosody preserved and, including personally relevant topics, aid language reception. Pragmatic deficits: failure to recognize situations, feigning understanding, ignoring turn-taking signals, and indulging in a press of speech; also, failure to initiate conversation, change topics or employ a full range of illocutionary acts. Pragmatic deficits may be secondary to the overall linguistic deficit.
Other Characteristics of Language Expression: Jargon; automatic responses; recurrent utterances. Other Factors in Language Reception: Deficits in short term memory; auditory deficits.
Treatment Procedures: Overall aim for patient to gain greatest possible independence in communication. Two approaches: 1) therapy programmes which concentrate on specific aspects of language; 2) guide patient and their personal contacts towards a greater understanding of the linguistic disorder and a

knowledge of how linguistic performance can be maximized in everyday contexts.
Treatment Directed at Specific Aspects of Language: Involves application of techniques which will achieve learning: by re-teaching language rules (competence); or by re-activating and re-organizing impaired modalities (performance) to bring about an improved access to preserved knowledge of language; combined approach of teaching and improving performance skills
Linguistic treatment programmes involving a systematically designed remediation plan containing a hierarchy of levels of complexity can be devised for all aspects of language: where speech does not seem possible at a functional level, reorganization of word, phrase and sentence production via preserved prosody—MIT; also, HELPSS: programme based on hierarchy of difficulty observed in patients with agrammatism. Therapy programmes incorporating a hierarchy of levels of difficulty are also relevant to semantic treatment: apply various cues and find which are most effective in accessing target words. e.g., initial phoneme, semantic association, description, situational context, orally spelled word, functional description, superordinate, representational gesture. Cues which guide the patient to the target can then be adapted for use in learning programme and will either serve as self-generated cues later or restore semantic skills.
Cognitive Neuropsychology: Carefully specified, goal-directed treatment based on single case studies. Various information-processing models are applied to identify a precise linguistic processing deficit and the route(s) by which re-activation or re-organization might be brought about.
Microcomputers: Can be advantageous method of teaching specific aspects of language in a structured manner
Pragmatics: Although in PACE therapy linguistic accuracy is not required, it offers a context in which linguistic skills can be used spontaneously whilst supported by other aspects of communication.
Conclusions: Linguistic treatment is a core need for most individuals with aphasia but may be of limited effect without the involvement of others and a concern that the skills encountered are applied in everyday situations.

Sarno, M.T., Silverman, M. and Sands, E. 1970. Speech therapy and language recovery in severe aphasia. *Journal of Speech and Hearing Research*, **13**: 607-623.

Subject Area: Severe expressive-receptive aphasia.
Therapy Area: Verbal behavior.
Therapy: Programmed instruction.
Hypothesis: To investigate whether programmed speech therapy enhances language recovery in severe aphasia and whether or not the method of speech

therapy used affects recovery if it occurs. Study concerned with 2 hypotheses about the treatment of severe aphasia: 1) severe aphasic patients can learn; 2) programmed instruction is an effective method for teaching them. (Results did not support either hypothesis.)
Method: *Subjects:* 31 subjects with severe aphasia; virtually no speech function and little understanding of speech; alert.
Procedure: Experimental Conditions: 1) programmed instruction: designed on basis of a presumed hierarchy of skills, thought to be representative of the order of acquisition of skills in the recovery process: imitation of body movements → visual recognition → writing and auditory comprehension → oral production. Vocabulary used: 2 nouns, 2 numbers, 2 adjectives. Sequence of presentation: single words with nouns first, followed by numbers, then colours; combination of numbers and nouns; combination of numbers, colour and noun into 3-word phrases; 2) nonprogrammed speech therapy: clinicians given vocabulary to be taught in each modality; to be taught in any way or order, or in as many modalities as felt necessary; 3) control: no treatment received.
Results: Significant differences in learning between treatment groups for only 2/10 behaviors: visual recognition tasks requiring matching words and pictures —nonprogrammed improved more than control. Programmed instruction was not more effective than either of the other groups. Patients did not appear to have capacity necessary to transfer or generalize information.
Conclusions: Gains small for all 3 groups, therefore patients with severe aphasia do not benefit from speech therapy. Ineffectiveness of speech therapy related to severity of aphasia.

Sefer, J.W. 1978. Principles of aphasia therapy. In Y. Lebrun and R. Hoops (eds.), *The Management of Aphasia.* Swets & Zeitlinger B.V.: Amsterdam and Lisse. Pp.66-75.

Subject Area: Aphasia therapy—principles.
Purpose: To outline specific principles that aphasia therapy should follow.
Principles: Aphasia is defined as a reduction of available vocabulary and of verbal retention span in all language modalities: reading, writing, auditory understanding and speech; "available vocabulary" because the patient has not lost his vocabulary, it is not available to him.
Aphasia therapy should follow the preceding principles: 1) Strong controlled auditory stimulation using natural units of language at the patient's level should take precedence over any artificial type of therapy such as attempting to teach the patient rules of grammar or rules of pronunciation or spelling; 2) material used for stimulation should be meaningful to the patient; 3) tasks such as rhyming, which draw attention away from simple listening and repeating are destructive to memory; the more natural the stimulation and response, the better for recall; 4) stimulation must be meaningful to the patient, not only in word commonality, but in length of unit of language the patient is able to process; 5) rehearsal, in the form of many stimulations and responses to the same word or phrase, is

important. Because the aphasic is not able to hold a stimulus very long, it must be repeated over and over so stimuli will "get in".
Remind the patients to listen. It can be explained that this is how they learned language the first time. Words used should be of high frequency, or common words; monosyllabic words are easier because of ease of spelling and they are often common/frequent words. Systematic repetition of sentences is used; aphasic patients operate under restrictions of length of unit to be processed; there is a relationship between the length of unit and syntactic complextiy. As language, specifically vocabulary, increases, aphasic patients are able to use the rules of their language.
We do not teach aphasic patients language, we stimulate their own language processes to work better. For patients who have severe difficulty speaking, just stimulate language processes and do not worry about articulation; when the patient "gets the word in" his articulation will improve.
Conclusions: Start treatment at level at which the patient breaks down. Treatment should include language stimulation in all language modalities with primary emphasis on auditory stimulation. Material should be controlled for length and complexity.

Sefer, J.W. and Shaw, R. 1972. The use of psycholinguistic principles in the treatment of aphasia. *British Journal of Disorders of Communication*, 7: 87-89.

Subject Area: Therapy—psycholinguistic principles.
Therapy: Stimulation therapy; stimulation of disrupted processes to function maximally, using intensive auditory stimulation, but not necessarily stimulation through auditory channels alone.
Hypothesis: Aphasia is an interference with language processes resulting from brain damage. The interference disrupts both analysis and integration of verbal messages. This disruption causes a reduction of available vocabulary and verbal retention span in all language modalities. The language storage system is at least relatively intact. Therefore, restimulate the language processes.
Background: Chomsky (1965): both competence and performance are involved in language use; the aphasic patient has the comptence but the performance is impaired through language disruption; aphasics follow the same rules of language (i.e., have the competence) as non-aphasics; the underlying structure of language is preserved.
Therapy: Determine first where the level of disruption occurs, e.g., at the single word, phrase, or sentence level. Treatment should begin at the individual patient's level, or just above it. Examples of therapy: patient can be asked to point, repeat, read aloud together with clinician, spell, tell about and write units at each level of difficulty. Manipulate at any level to increase difficulty by removing one mode of stimulus presentation, increasing length of stimulus, or changing the response mode. Words used should be very high frequency or common words; monosyllabic words are easier because of ease of spelling and

because there are common and varied associations connected with high frequency words.
Main function of treatment is to stimulate the language that is familiar to each patient. The patient will know the difference once it "gets in" to the system. Errors made in generating sentences often indicate a failure to discriminate between singular and plural, between tenses of verbs, and between nominative and objective forms of pronouns. Therapy aimed at increasing length of retention span and familiarity with different forms of sentences should help the aphasic overcome these difficulties.
As the patient recovers, association errors tend to increase. These errors indicate the patient is getting more language, but that retrieval is defective. The aphasic needs a larger vocabulary to make association errors. At the same time, use of connected speech occurs: commonly-used phrases appear earliest. Short sentences follow, first with errors in plural forms, gender of pronouns and errors in verb forms. As the patient's language increases, these errors begin to be self corrected spontaneously. In other words, as language increases, aphasic ipatients are able to use the rules of their language.
Another indication of improvement is the patients' use of their own pronunciation; as patient begin to make the vocabulary their own, and it becomes usable, they begin to pronounce words in their own way. This phenonemon extends to syntax as well when patients begin to paraphrase the sentence given them.
Conclusions: We do not teach aphasic patients language, we stimulate their own language processes to work better. Therapy should be used which concentrates on performance factors, using stimulation for better auditory discrimination, easier retrieval of lexical items and to increase verbal retention span.

Seron, X., Deloche, G., Bastard, V., Chassin, G. and Hermand, N. 1979. Word-finding difficulties and learning transfer in aphasic patients. *Cortex*, **15**: 149-155.

Subject Area: Word-finding difficulties.
Therapy Area: Naming; generalization.
Therapy: Word-finding difficulties therapy based on restoration of access processes to lexicon rather than on lexical items themselves.
Purpose: To study learning transfer, the effect of word-finding difficulties and their evolution on the semantic structure of the lexicon.
Method: *Subjects*: all demonstrated word-finding difficulties.
Tests: Naming test: picture categories: animals, food, house, clothes, tools, action verbs. Semantic classification test: written words; examiner presented and spoke superordinate word. Below this word, there were two columns: in the left one, the patient had to place items belonging to the semantic field thus defined; in the right one, the other items. Items could be divided into 3 categories: 1) words belonging to the semantic field; 2) words not belonging to the semantic

field but having some features in common with it; and 3) words unrelated to the semantic field. Two types of errors: expansion when a word not belonging to a semantic field was included, and narrowing in the opposite case.
Procedure: Experimental Conditions: two groups of aphasic patients were considered: four patients were trained according to a classical extensive method, i.e., using a great number of lexical items (control group); four other patients were trained according to an intensive method, i.e., based on access strategies to lexicon and relating to few lexical items (experimental group).
Pre-test: naming and semantic classification test; then 20 therapy sessions (traditional or experimental); post-test.
Training method intensively used all facilitating processes capable of eliciting the emission of the right word (oral and written association or related word cues, gestures, lead-in phrase, etc.), thus changing according to the patient's pathology.
Results: Experimental therapy proved to be more efficient than the traditional therapy, since there were significantly fewer errors in post-test as compared to pre-test. All patients succeeded better in the semantic classification test than in the naming test.
Learning Transfer in Experimental Group: Naming test: except in one patient, there was a very significant decrease in the percentage of errors between the pre-test and the post-test, both for drilled and non-drilled superordinate categories. It seemed that, in these 3 cases, there was a learning transfer. Semantic Classification test: In 2 cases, global improvement observed, thus there seemed to exist a slight learning transfer to non-drilled categories.
Conclusions: Results of this study demonstrate that in certain cases at least, word-finding difficulties therapy based more on restoration of access processes to lexicon than on lexical items themselves proves more efficient than the traditional method. Supports the theories which consider word-finding difficulties as an impairment in item retrieval skills rather than a loss of words. Problem of the relations between word-finding difficulties and semantic structures of the lexicon still remains.

Seron, X., Deloche, G., Moulard, G. and Rousselle, M. 1980. A computer-based therapy for the treatment of aphasic subjects with writing disorders. *Journal of Speech and Hearing Disorders*, **45**: 45-58.

Subject Area: Aphasic subjects with writing impairment.
Therapy Area: Writing words.
Therapy: Computer-aided dictation.
Purpose: To test a rehabilitation method that allows for controlling the subject's writing activity in a dictation situation and is based on avoiding visual reinforcement of wrongly chosen letters; with the traditional method, it is difficult to prevent and/or correct occurrence of errors.
Method: *Subjects:* 5 aphasic patients with writing impairments.

Procedure: A typewriter keyboard and visual display connected to a computer was programmed to signal errors immediatly and to prevent their visualization. A word list dictated by the clinician was used. The computer checked spelling and determined if the letter should be displayed on the screen.
1) Procedure using small frames (SF): number of frames=number of letters in word; If incorrect letter: a)and letter not in word, nothing displayed and buzzer sounded; and b)letter belonged, buzzer sounded and letter displayed below appropriate frame. If patient made one or more errors, same word was to be typed again. 2) Procedure using a large frame (LF): length rectangle proportional to number of letters in word. Therefore, the patient is given fewer cues. 3) Procedure without frame (NF): no frames; patient had to type "blank" when completed typing in word.
Three Word Batteries: B1: mono- or bisyllabic words with neither consonant clusters nor graphic complexities; words selected for their high frequency in written French; B2: longer words -- mono-, bi- and trisyllabic; included graphic complexities; high and low frequency words used; B3: used words varying in length from 2-4 syllables, some with irregularities of spelling.
Pretest: write list of 50 words from dictation. Treatment: began with B1 using SF, then progressing through LF, NF. B2 and B3 followed with same progression (SF, LF, NF). Posttest: 2: (i) given upon completion of treatment program; (ii) given 6 weeks post-therapy.
Results: Transfer of learning observed at first posttest where words were handwritten rather than typed. Improvement decreased at second posttest.
Conclusions: The differential correction system, the framing of the words, and appropriate letter positioning seem to have a positive role in the control of writing activity.

Shewan, C.M. 1976. Facilitating sentence formulation: A case study. *Journal of Communication Disorders*, 9: 191-197.

Subject Area: Broca's aphasia; aphasia—general.
Therapy Area: Sentence formation.
Therapy: Sentence construction.
Purpose: To determine effects of training on sentence formulation abilities of Broca's aphasia patients; to compare sentence formulation skills and use in 2 conditions: 1) written cue; 2) written and auditory cue, for declarative sentences describing a picture ("sentence training"); assessment of appropriateness of a sentence's syntactic form and its semantic content.
Method: *Subject:* 64-year-old male; 21 months post-CVA; moderate aphasia of Broca's type.
Procedure: Sentence Type: simple active affirmative declarative which could have surface forms of S-V, S-V-O, or S-V-PP. (Chosen since relatively easy linguistically). 30 pictures assigned randomly to 3 groups: Group 1: picture presented with 6 cards with linguistic constituent written on them: 2 S (NP), 2 V (VP), and 2 O (NP) or 2 (PP). From these, the patient selected the 3

appropriate for the picture stimulus; sequence them to make sentence, then orally read sentence; Group 2: provided both auditory and written form of verb (VP) as cues. Patient was to make up sentence about picture using cue; Group 3: control; no training.
Correct response: concerned with semantic and syntactic aspects: looked at number of correct linguistic constituents and appropriate lexical items.
Results: Training did result in an increased number of grammatically correct sentences, although the increases occurred on trained stimuli only.
Conclusions: Possible explanations for the lack of demonstrated syntactic generalization: may have abstract strategy, but was not able to apply it to unfamiliar material; may have needed more training before firmly established a strategy or rule; patient could not abstract a strategy at all.
Significant increase in proportion of appropriate lexical items in the responses on both trained and untrained items suggests that training influence generalized at least to some extent. Nature of cue provided not as important for this patient as the presence of some cue. Results consistent with clinical findings that auditory and viusual stimulation can facilitate word retrieval.

Shewan, C.M. 1986. The history and efficacy of aphasia treatment. In R. Chapey (ed.), *Language Intervention Strategies in Adult Aphasia*. Baltimore: Williams & Wilkins. Pp. 28-43.

Summary: *Aphasia Treatment Prior to World War II:* very little emphasis on treatment; in the few studies reported, the seeds from which many later issues about the efficacy, content, and process of language treatment are found.
Post WWII: increased interest and enthusiasm for speech therapy. Patients, primarily of traumatic etiology, showed substantial gains following treatment. 1950s: extended programs to stroke patients, with general success reported. Efficacy of language treatment highly disputed.
Challenge of the Early 1970s: acknowledgment that more research including controlled studies were necessary if the efficacy question were to be answered. Questioned therapy versus spontaneous recovery.
Approaches to Studying Efficacy: problem of control groups. Problem of enormous time commitment and numerous methodological problems, e.g., ethical problems. In general, studies with and without control groups support notion that language therapy is efficacious.
Factors Potentially Influencing the Effects of Treatment: Influence of the therapist: speech-language pathologist's assessment provides the basis for decisions about whether treatment is recommended and what the content of the treatment plan should be. Due to shortage, personnel other than speech-language pathologists have been taught to provide treatment, but it is not known with confidence whether they provide as effective treatment; Amount of Treatment: wide discrepancies in the intensity and duration of treatment; Interaction of Time and Treatment: providing treatment early results in greater language gains than providing treatment late, although significant gains are not restricted to the early

group. Even when control for spontaneous recovery, language treatment effects are reported as generally significant and positive; Influence of Type of Treatment: evidence supports positive benefits of treatment. Little evidence comparing types of treatment; what there is allows us to conclude little (programmed vs. non-programmed). Cannot make any statements about whether one type of treatment is more effective than another.

Patient Variables: Age: age does not seem to be a reliable predictor of language outcome; Type of Aphasia: all types of aphasic patients are capable of improving with treatment and no type guarantees it. Global aphasic patients have consistently been reported as having a poor prognosis and make the least gains if any at all, and their gains often do not become functional. Initial Severity of Aphasia: in general, patients with severe aphasia make the least gains, but some do make significant recovery and regain functional communication skills; Etiology of Aphasia: consistent reporting that aphasic patients with traumatic etiology recover to a greater extent than those with cerebrovascular disorder.

All of these patient variables can be influential and all interact. Little is known about these factors in isolation and their interactions.

Conclusions: History is instructive in several senses: 1) It provides a context from which to view an issue; 2) it may explain why and how we got where we are; 3) it prevents us from going the same route again; 4) it may provide clues for future directions.

Problem of comparison: Efficacy studies reviewed differ in how they approached the question, how they gathered and analyzed their data, and how convincing their conclusions were about the efficacy of treatment. Despite these variations, aphasic patients benefit significantly from language treatment even after the period of spontaneous recovery.

Sparks, R.W. and Holland, A.L. 1976. Method: Melodic Intonation Therapy for aphasia. *Journal of Speech and Hearing Disorders*, **41**: 287-297.

Subject Area: Severely aphasic; global aphasia.
Therapy Area: Emphasis on recovery of fomulation of propositional language rather than motor aspects of speech production; also, reducing the frequency of phonemic errors in some aphasic patients.
Therapy: Melodic Intonation Therapy (MIT).
Purpose: To describe principles of MIT and MIT stages.
Background: Physiological model presented that would account for MIT success in terms of right cerebral hemisphere dominance for music and speech prosody. Crucial to differentiate between intoning propositional sentences/phrases and singing songs. MIT uses a limited range of musical notes.

Basic Principle: intoned pattern is based on one of several speech prosody patterns which are reasonable choices for a given sentence, depending on inference intended. It is composed of 3 elements: melodic line, rhythm, points of stress. Intonation patterns which resemble those of well-known songs are always avoided, since regression to the words of the songs occurs if such melodies are used. Variation of higher or lower pitch dictated by speech prosody patterns of the sentence are more important than accuracy of pitch.

Principle of Gradual Progression underlies form of MIT: the patient is guided through a sequence of steps which increase the length of the units, diminish dependency on clinician and diminish reliance on intonation. At the end of the program, the patient should be capable of using spoken prosody for uttering the sentences embedded in the program structure.

MIT: 4 Levels of Difficulty:
1) No linguistic component; establishes process of intoning melody patterns and accurately handtapping rhythm and stress of each pattern; 2) move from requiring subject to tap the rhythm and stress of clinician's intoned utterance to responding to a request from the clinician for a repetition of the target sentence; 3) program's difficulty increases by fading participation of the clinician, by introducing enforced delay of responses so that some element of retrieval is introduced, and finally, by requiring the subject to give appropriate intonted responses to intoned question from the clinician regarding elements of information in the sentences. In addition, opportunity for correction of errors is introduced by incoporation of a backup procedure; 4) primarily concerned with return to normal speech prosody. Latency between stimulus and response is increased to produce decay of repetition skill and to increase efficiency of retrieval. Training sentences are longer and more complex; transition back to speech prosody is facilitated by a technique called Sprechgesang (sung speech)—melodic line remains the same as the intoned sentence of the preceding step, except that the constant pitch of the intoned words is replaced by the variable pitch of speech.

Language units: at each level, 12-20 short sentences or phrases; should be content-related and as relevant as possible to the patient's needs and background. Sentences and phrases at Level 2 do not exceed 4 words; Levels 3 and 4 are longer and more complex.

Currently, emphasis is on task progression or form rather than on linguistic content of the units. Some attention must now be granted to grammatical structure of language units, perhaps in terms of linguistic spontaneity, by using a changing lexicon in appropriate slots of controlled sentence types, or perhaps in some other as yet unspecified way.

Post-MIT: transition to a program of continuing language therapy might include modification of the last level, which would increase the number of questions prompted by the subject matter of the sentence being used.

Conclusions: MIT is best for aphasic patients whose auditory comprehension exceeds their verbal expression. Procedures of MIT should be modified to meet the needs of individual patients.

Thompson, C.K. and McReynolds, L.V. 1986. Wh-interrogative production in agrammatic aphasia: An experimental analysis of auditory-visual stimulation and direct-production treatment. *Journal of Speech and Hearing Research*, 29: 193-106.

Subject Area: Agrammatism.
Therapy Area: Syntax; wh-interrogatives; generalization.
Therapy: Training the production of specific syntactic patterns, in this case Wh-interrogatives, results in rapid acquisition of target structures and generalization within form.
Purpose: To compare the effects of 2 treatments on agrammatic subjects' use of Wh-interrogatives in complete sentences. Of primary interest was examining the acquisition and generalization of Wh responses and the effectiveness of each for maintaining these behaviors. Response generalization to untrained exemplars within and across interrogatives was assessed. In addition, stimulus generalization to elicited language samples was examined and maintenance of trained responses was measured.
Method: *Subjects*: 4 agrammatic aphasic individuals; comprehension of Wh questions.
Design: Alternating-treatments design in combination with a multiple-baseline design across behaviors and a multiple-baseline design across subjects. Involved 3 phases: 1) baseline; (2) application of either direct-production treatment or auditory-visual stimulation treatment; and 3) application of both direct production and auditory-visual stimulation treatment.
Procedure: Subjects asked to "Ask a question about the picture." Four Wh-interrogatives (what, where, who and why) chosen for analysis. Interrogatives trained using one or both treatments.
Auditory-visual Stimulation: Derived from principles associated with a stimulation approach; designed to provide multi-modal (auditory and visual/orthographic) stimulation of target responses; multi-modal input, repetitive auditory stimulation and provision of further stimulation for incorrect responses; correct production never forced.
Direct-production Treatment Incorporated principles associated with behavioral approach; designed to provide subjects with guided practice in producing targeted responses. One stimulus provided prior to each response; Modeling (entire sentence modeled), forward chaining (examiner sequentially modeled the first 2 words of the target response and then the remaining words of the target response, instructing the subject to repeat each portion as it was modeled) and response-contingent verbal feedback incorporated.
Results: Direct-production treatment consistently more effective than auditory-visual stimulation treatment in facilitating production of Wh- interrogatives in the agrammatic aphasic subjects under study.
Conclusions: Training the production of specific syntactic patterns results in rapid acquisition of target structures at a high level. Response generalization within interrogative forms occurred parallel to acquisition of trained responses,

regardless of treatment approach. Finding suggests that once a pattern for question production is learned, it can be successfully used in untrained contexts. For response generalization across interrogative contexts, training effect did not generalize to untrained interrogative forms. These data suggested that Wh-interrogatives may not be members of the same response class, even though they belong to the same linguistic or structural class, for a functional relationship was not observed between them. Data suggest that aberrant interrogative production may result from a "loss" of rules or information about how to use them appropriately to generate grammatically complete and accurate utterances, and that, because rules for production of different Wh forms are different, it is necessary to train each rule individually.

Thompson, C.K., McReynolds, L.V. and Vance, C.E. 1982. Generative use of locatives in multiword utterances in agrammatism: A matrix-training approach. In R. H. Brookshire (ed.), *Clinical Aphasiology Conference Proceedings*. Minneapolis: BRK. Pp. 289-297.

Subject Area: Agrammatism.
Therapy Area: Syntax; locatives; generalization.
Therapy: Train specific linguistic structures systematically via matrix training procedure.
Hypothesis: To investigate the effects of a matrix-training procedure on the acquisition and generalization of locatives in multiword utterances by agrammatic aphasic subjects. Specifically, to investigate the effects of treatment on: 1) the acquisition of the locatives "behind" and "beside" within the context of NP + is + PP utterances; 2) generalization of the locatives "behind" and "beside" within the context of NP + is + PP utterances to items within trained and untrained matrices; and 3) generalization of the constructions of interest to spontaneous speech.
Method: *Subjects*: 2 subjects with Broca's aphasia.
Materials: 36 drawings depicting locative relations of interest divided into 4 groups of 9 items each and used to elicit NP + is + PP responses. Each group of 9 items made up a 3 x 3 language matrix. Each matrix contained 3 nouns and 3 PPs. Ex. *Row*: bike, tree, rake; *Column*: behind tent, behind house, behind fence (9 sentences in all).
Design: Multiple baseline design across behaviors.
Procedure: One sentence per session was trained using modeling, forward chaining and response contingent verbal feedback. Sentences trained in a diagonal sequence. Generalization probes administered daily following treatment. Follow-up probes administered at 4 and 6 month intervals.
Results: 1) Treatment was effective in facilitating verbal production of target locatives within the context of NP + is + PP utterances for both subjects; 2) generalization to untrained items within trained and untrained language matrices occurred when a limited number of exemplars were trained; 3) production of locative responses within NP + is + PP contexts was maintained for several

month following treatment; 4) an increase in the use of NP + is + PP utterances occurred during spontaneous speech for one patient, but not for the other.
Data show the acquisition, generalization and maintenance effects of matrix-training procedures on production of targeted locatives within the context of NP + is + PP constructions in agrammatic aphasic subjects.
Conclusions: Data from this study suggest that locatives may not constitute a response class even though they belong to the same linguistic class, because changes in production of untrained locatives did not occur after application of treatment to other locatives. Therefore, these findings suggest that generalization cannot be expected to occur even to topographically similar responses. It is suggested, then, that treatment be systematically applied to specific linguistic structures at the level of the agrammatic aphasic patient's residual language ability and that a mechanism for assessing the generalization effects of treatment, such as a matrix training procedure, be incorporated into treatment.

Wambaugh, J.L. and Thompson, C.K. 1989. Training and generalization of agrammatic aphasic adults' Wh-interrogative productions. *Journal of Speech and Hearing Disorders*, **54**: 509-25.

Subject Area: Broca's aphasics; agrammatism.
Therapy Area: Wh-interrogative productions; syntax; generalization.
Therapy: Modeling, forward chaining, response contingent verbal feedback used to train interrogative production; story completion items used to elicit responses.
Purpose: To examine the response and stimulus generalization effects of training wh-interrogative production in Broca's aphasic subjects with agrammatism.
Experimental Questions: 1. Will training of one wh-interrogative form facilitate production of a different wh form? 2. Will training of ('what'/'where' + copula [is] + nominative) interrogative constructions facilitate production of grammatically different "what"/"where" interrogative constructions? (e.g., "what'/'where' + transposed NP); 3. Will training of Wh-interrogative constructions result in generalization across settings and person? 4. Will sequential modification facilitate generalization to conditions in which stimulus generalization failed to occur?
Response generalization occurs when untrained behaviors emerge as a result of training. Stimulus generalization occurs when trained responses are used in conditions different from the original training condition (e.g., across settings, persons, objects, or situation).
Method: *Subjects:* 4 aphasic adults diagnosed with Broca's aphasia.
Stimuli: 4 story completion stimulus items used to elicit "what" and "where" questions. Stimulus items designed to elicit: 1) what + copula (is) + nominative questions (e.g., What is the answer?); 2) where + copula (is) +

nominative questions (e.g., Where is the dog?); 3) what + transposed NP (TNP) questions (e.g., What is he eating?); 4) where + TNP (e.g., Where is he hiding?).
Design: Multiple baseline design across behaviors, subject, and settings used to assess treatment effects.
Procedure: 5/10 (what + copula [is] + nominative) and (where + copula [is] + nominative) items randomly selected and used for training. Remaining 5 from each set and all TNP and misc question were untrained and used for probing response generalizations. Modeling, forward chaining, and response contingent verbal feedback used to train interrogative production of the Wh-questions used in training. Story completion items used to elicit responses.
Generalization Training: Sequential Modification: treatment was initiated to facilitate generalization across stimulus conditions by extending story completion items to situations involved in novel social interactions.
Results: Question production treatment facilitated production of trained (what/where + copula [is] + nominative) constructions. All subjects' responses during treatment exceeded those observed during baseline. Treatment applied to 'what' interrogatives did not facilitate production of 'where' interrogatives, nor did treatment applied to "where" interrogatives facilitate production of "what" interrogatives. Supports observation that Wh-interrogatives do not operate as a single functional response class even though they belong to the same linguistic class. Negligible generalization to grammatically different Wh-interrogatives. Sequential modification found to facilitate generalized responding across conditions for some subjects.
Conclusions: Aphasic patients may not maximally benefit from training focused upon production of specific constructions, but that training function (e.g., requests for information) as opposed to structure might result in greater generalization.

Weniger, D., Springer, L. and Poeck, K. 1987. The efficacy of deficit-specific therapy materials. *Aphasiology*, 3: 215-222.

Subject Area: Aphasia (mild speech and comprehension deficits).
Therapy Area: Communication; discourse; situational context.
Therapy: When attempting to improve a patient's conversational abilities, treat linguistic forms associated with discourse within a speech context rather than in isolation.
Purpose: In order to investigate the efficacy of speech contexts in language therapy, speech contexts were constructed in which certain discourse features, in particular pronominal forms and modal verbs, were to be practised. (Decision to use modal verbs was motivated by the fact that in German, modulations in terms of permission, obligation and volition are expressed quite systematically by modal verbs.) Also, hypothesis that therapy materials can influence improvement, irrespective of the therapeutic techniques used. Therapy materials

which are based on the structural principles of the impaired linguistic component to be treated lead to significantly larger generalization effects.

Background: Communicative interactions are characterized by a variety of discourse features including non-verbals. Features of discourse involve specific linguistic capacities: the use of pronominal and deictic forms referring to persons, objects, events and locations in the listener's field of reference. Effective communication depends greatly on the availability of the linguistic forms which make reference to antecedents.

A communicatively crucial aspect of an utterance is its illocutionary force. The propositional meaning of an utterance always has some interpretation tied to it, i.e., it must be understood in terms of some intent of the speaker and, consequently, some expected response on the part of the addressee. The illocutionary force of an utterance is frequently expressed by means of modal verbs or verbs of "mood".

Method: Subjects: 8 aphasic patients; spontaneous speech and language comprehension moderately to mildly impaired.

Materials: 10 situational contexts in which dialogues were embedded. Each of the these "scenarios" or "scripts" was 9-12 sentences long. Scenarios began with a brief description of the situational context. This description was followed by a dialogue taking place between 2 individuals. The utterances of the 2 individuals were of varying syntactic structure, as required by the course of the conversation. However, the utterance of each conversant always contained a modal verb. Modal verbs, translated, were: "can", "may", "must" and "want to".

Procedure: For each of 10 scenarios, there were 4 practice runs. First run: scenario given to patient in written form; patient read it aloud with therapist. In the dialogue utterances of each scenario, the modal verbs were missing and had to be filled in by the patient. Patient given 4 word cards, on each of which a modal verb was printed. Second run: patient had to supply the modal verb without the support of such word cards. Third and Fourth runs: dialogue utterances were missing in the scenario.

Control test: administered to determine improvement in conversational abilities: 40 brief descriptions of an everyday life situation; each description followed by a question about some element of the situation worded in such a way that the response had to contain a modal verb. Checked for memory effects.

Results: Patient with a short duration of illness as well as patients with a chronic aphasia achieved significant improvements after therapy. Contrary to expectations, improvements in the use of the linguistic forms which were specifically treated were not observed. (Nature of methodology may offer explanation.)

Conclusions: Results suggest that the more interrelated the linguistic structures to be treated are, the less specific treatment effects will be.

Wiegel-Crump, C. and Koenigsknecht, R.A. 1973. Tapping the lexical store of the adult aphasic: Analysis of theimprovement made in word retrieval skills. *Cortex,* **9**: 410-18.

Subject Area: Amnesic aphasia (word retrieval deficit).
Therapy Area: Word-retrieval; naming; generalizability.
Therapy: Therapy program of concentrated language drill leads to significant improvement in accuracy of response in word retrieval skills for picture-naming by adult aphasics with predominant amnesic deficits.
Experimental Questions: Question of whether word-retrieval reflects an underlying loss of efficiency in retrieving words from the lexical store or a reduction in the lexical store itself. Related question is whether improvement in word retrieval skills generalizes during the course of therapy from words drilled on in therapy to additional non-drilled words. No generalization expected if problem is reduction in lexical store.
Method: *Subjects*: 4 adult aphasics with amnesic aphasia (word retrieval deficit).
Procedure: Pretreatment and Testing Methods: 1) Picture naming task: picturable words controlled within 5 superordinate categories of household items, clothing, food, living things and actions words; 2) picture-referent matching: match one picture with another picture from the same superordinate category.
Treatment: Drilled and Non-drilled items: selected randomly from patients failures. Equally divided among 4 or 5 categories represented in the pretest. No drill words selected from category "Foods", in order to determine if improvement made on drilled items in therapy would generalize both to non-drilled items within the same superordinate categories as the drilled items and also to items within a superordinate category not presented in therapy.
Each drilled word was presented 10 times as a single word, followed by 5 simple sentences incorporating the drill word, followed by 10 more repetitions of the word in isolation. Both auditory and visual stimulation used to elicit language. Language acquisition enhanced by the use of repetitive stimulation.
Clinician enabled patient to respond successfully at each level of recovery; clinician supplied the patient with a host of auditory-verbal clues in addition to presenting the picture stimulus. These auxiliary stimuli included gestures, associated words, synonyms, carrier phrases, and the initial phonemes or syllable of the desired response. Stimulating and eliciting language, not forcing the patient to respond.
Results: Referent matching abilities superior to picture-naming skills. Improvement in picture-naming in all 5 superordinate categories. Therefore, generalization from categories drilled on in therapy to other categories.
Conclusions: Based on results, disturbance of lexical retrieval. Pretreatment ability of referent-matching gives evidence to indicate that the aphasic's lexical concepts are intact and that only the verbal label is lacking.

5 Linguistic principles underlying two aphasia tests

Margaret Meth, Loraine Obler, and Patricia Walsh

The purpose of the present paper is to examine two standardized tests of aphasia, *The Minnesota Test for Differential Diagnosis of Aphasia* (MTDDA) by Hildred Schuell (1973) and *The Boston Diagnostic Aphasia Examination* (BDAE) by Goodglass and Kaplan (1983). By examining the tests and related materials we will determine the linguistic presuppositions and principles upon which they and their subtests have been based. This exercise is of interest because authors of aphasia tests are rarely explicit about the linguistic theories on which their tests are based, yet, we maintain, it is impossible to devise an aphasia test without some motivating assumptions about language structure per se.

Language breakdown in aphasia has traditionally been studied by professionals in varied disciplines in order to hypothesize about the way language is stored in the brain. Early research was conducted by neurologists who appear to have devised individualized test materials to explore the problems of the patients they reported on. Standardized testing of such breakdown became desirable in the late 1940's with the influx of World War II veterans who had suffered head trauma and resulting aphasia. One of the widely used early assessment tools to perform such testing was the MTDDA (Schuell, Jenkins and Jimenez, 1964: 1). More recently, the *Porch Index of Ccommunicative Ability* (PICA) (1967) and the BDAE (1972) have become most widely used.

Because these tests were being developed in the United States during the period when the discipline of linguistics was developing, we are justified in asking the extent to which the authors of these aphasia batteries incorporated linguistic principles into the tests. Schuell and her colleagues developed materials for the MTDDA in the 1950's at a time when structural linguistics was already well established. Experimental versions of the test were dated from 1955-1958; first publication of the test was in 1965. Chomsky introduced transformational generative linguistics with his 1957 book "Syntactic Structures", and elaborated on it in his 1965 book, "Aspects of The Theory of Syntax", as his ideas were gaining greater influence in linguistics. Neither aphasia battery to be discussed in this paper itself explicitly refers to linguistic theory, however, elements of the linguistic notions of the eras in which they were conceived can be abstracted from the test materials and related writings.

In order to determine unarticulated linguistic principles that the authors subscribed to, we took the following factors into account: (1) the test philosophy of each set of authors; (2) overall test structure; (3) structure and content of the test materials; (4) diagnostic categories. Several issues appeared of interest as we focused on linguistic principles underlying the tests: (1) the role of linguistic structure and/or rules; (2)the relative importance of linguistic levels and hierarchies (e.g. phonology, syntax); (3) the relative importance of language abilities and modalities (e.g. naming, comprehension, reading, speaking).

In the next section, the structure of each of the two tests focused on in this paper is described in the light of these issues. The following section identified linguistic principles underlying each test and the final section compares the two tests.

DESCRIPTION OF TESTS

Minnesota Test For Differential Diagnosis of Aphasia

Schuell and her colleagues developed Form 1 of the MTDDA in the late 1940's and made a research version available in 1955. In 1965 the test was first published; a second edition was published in 1973. The final version is an aphasia battery which is composed of five sections for assessment of auditory comprehension, visual and reading disturbances, speech and language disturbances, visuomotor and writing disturbances, and disturbances of numerical relations and arithmetic processes (See Appendix A for a list of tests on the other 5 subsections.)

The goal of assessment is to provide a basis for therapeutic intervention (Spreen and Risser, 1981: 101). Aphasics are classified into seven levels of disorder, rated from mild to severe (see Table 1). The test design was such that the earlier tasks required little or no comprehension. They progressed from imitating nonsense syllables to mono- and then bisyllabic words, to phrases and then sentences, to the generation self-formulated utterances. This type of hierarchy is followed, in somewhat varied fashion, throughout each section on the assessment battery (see Appendix A).

Within the speech and language disturbances subsection, for example, the subtests were ordered as follows: imitating gross movements of the speech musculature, rapid alternating movements, repeating monosyllables, repeating phrases, counting to 20, naming days of the week, completing sentences, answering simple questions, giving biographical information, expressing ideas, producing sentences, describing a picture, naming pictures, defining words, retelling a paragraph.

THE ROLE OF LINGUISTIC STRUCTURE OR RULES

With respect to linguistic structure, the MTDDA considers phonological structure relevant, but not syntactic structure. Schuell et al. (1962: 365) explicitly state that the battery does not contain any tests "requiring specific syntactic usages or transformations". Within the section which assesses speech and language disturbances, the Repeating Monosyllables subtest begins with

CLASSIFICATION	LANGUAGE TREATMENT	SENSORY MOTOR DEFICITS	RECOVERY
Simple Aphasia	Mild reduction of available language in all modalities	None	Excellent
Aphasia with visual involvement	Mild reduction of available language in all modalities	Impaired visual discrmination, recognition and recall	Excellent for language, but reading and writing recover slowly
Mild aphasia with persisting dysfluency	Mild aphasia with associated verbal dysfluency	Proprioceptive disturbance	Excellent for language, but conscious control over speech execution remains necessary
Aphasia with scattered findings compatible with generalized brain damage	Moderate aphasia with additional problems (i.e. lability)	Various visual and motor speech disorders	Limited
Aphasia with sensory motor involvement	Severe reduction of language in all modalities	Impaired discrimination, perception and production of phonemes	Limited
Aphasia with intermittent auditory imperception	Almost complete loss of functional language skills in all modalities	Severe auditory impairment of perception	Limited but functional
Irreversible aphasia syndrome	Severe language impairment	Can be numerous	Recovery of functional language is poor

TABLE 1. Schuell's Major Aphasia Syndromes

monosyllabic CV words. Items progress deliberately from simple plosives to more complex consonant clusters deliberately. On tests involving sentence length material (e.g. yes/no), it is not syntactic structure that determines the item difficulty but rather sentence length. Granted, modifying clauses extend sentence length, but no further syntactic manipulations are employed.

THE RELATIVE IMPORTANCE OF LINGUISTIC LEVELS AND HIERARCHIES

Of the strict descriptivist linguistic levels, (phonology, morphology and syntax) the MTDDA treats only phonology separately. Some linguists treat lexicon and/or semantics as a separate level and the MTDDA does indeed consider "vocabulary" independently (e.g. in asking for naming or word definitions).

Hierarchy is reflected in the MTDDA in that all subtests progress from easy to hard. The hierarchical principles, however, are not strictly linguistic; rather they are based primarily on memory load (i.e. many subtests progress from shorter to longer items). Schuell and Jenkins (1959, P.65) state that the MTDDA was not designed to furnish specific information on the relation of the nature of language tasks to specific levels in the language hierarchy, and does so only in the most general manner (p.65).

THE RELATIVE IMPORTANCE OF LANGUAGE ABILITIES[1] AND MODALITIES[2]

Given assertions of Schuell et al. (1964) that language is unitary, one might expect to find no prioritizing of language modalities or abilities. Nevertheless, there is substantial evidence that comprehension is the basic modality according to Schuell (1973). The authors of the MTDDA state "probably all aphasic patients show some impairment of auditory processes, because language, learned by ear, remains dependent upon discrimination, recognition, and recall of learned auditory patterns, and upon auditory feedback processes" (Schuell, 1973: 5). The primacy of comprehension is reflected in the fact that it is the first major subsection of the test.

The authors present speaking as the next most important modality and one closely linked to comprehension. Schuell, Jenkins and Carroll (1962) state that the speech mechanism is a functionally educated structure whose output is related to integration of language as well as to organization of movement patterns (p. 358). The speaking tests (in the third subsection) include 15 subsections as compared to the 9 for Auditory Disturbances, 9 for Visual and Reading Disturbances, 4 for Visuomotor and Writing Disturbances, and 4 for Numerical Relations and Arithmetic Processes.

---

[1] By language abilities we mean the different ways language can be used (discourse, repetition, naming, etc.).

[2] By modalities, we mean the four classic ones: reading, writing, comprehension and speaking.

The second subsection is the Visual and Reading disturbances subsection, but this does not necessarily mean that Schuell and colleagues see reading and writing as second in importance to comprehension. Rather, we suspect, they place Visual and Reading Disturbances here to obtain evidence about neurological deficits such as hemionopsia that may influence subjects' performance on later components of the test. Writing is the final task tested, suggesting, the oral modalities are primary as compared to the written ones.

Boston Diagnostic Aphasia Examination

The BDAE is an aphasia battery whose general aim is to diagnose the existence and type of aphasia by measuring a wide range of performance initially. The result of the assessment suggests cerebral localization for site of lesion as well as language strengths and weaknesses as a guide for therapy. Five areas of language are assessed: conversational and expository speech, auditory comprehension, oral expression, understanding written language, and writing (see Appendix B). In addition, a set of supplementary tests provide a more psycholinguistically based assessment of comprehension and production.

The oral expression section is the section most relevant to this paper. Oral Expression subtests include: Oral Agility, Automatized Sequences, Recitation, Singing and Rhythm, Repetition of Words, Repeating Phrases, Word Reading, Responsive Naming, Visual Confrontation Naming, Animal Naming (Fluency), and Oral Sentence Reading. With the exception of the Nonverbal Agility portion of the Oral Agility subtest, all subtests are scored based on articulation ability and use of paraphasias.

THE IMPORTANCE OF LINGUISTIC STRUCTURE OR RULES

As compared to the MTDDA, no phonological structuring or hierarchy seems to be built into the BDAE, except insofar as "Repetition of Words" ends with a tongue twister. Syntactic structures are tested in the supplementary materials, in particular active vs passive construction. Several of the BDAE subtests are designed to explore semantic fields. Clusters of semantic categories such as colors, shapes, numbers, objects and actions are presented for auditory comprehension and confrontation naming. The express inclusion of both objects and actions may be seen as an appreciation of the semantic/syntactic categories of nouns and verbs.

THE RELATIVE IMPORTANCE OF LINGUISTIC LEVELS AND HIERARCHIES

The tests in the BDAE are set up according to a difficulty hierarchy progressing from words to phrases to sentences. As in the MTDDA, a hierarchy of length is incorporated into the repetition and paragraph comprehension subtests.

Different linguistic levels starting with words are each given equal importance in error analysis. For example, both the semantic and phonological levels are considered when paraphasic errors are qualitatively analyzed. No phonology hierarchy appears to be built into any subtest. Linguistic hierarchies are not in evidence in other areas of the test. In the word comprehension test, for example, words are organized into semantic categories only. The use of high and low probability words and sentences in the repetition, reading and writing tasks appears to reflect a belief by Goodglass and Kaplan that frequency of use determines availability of the linguistic unit to the system. The more frequently a sequence of words is called upon to be used in output, the less likely it will be impaired and the more likely it will return following breakdown of the system.

A hierarchy of grammatical complexity is considered in the supplementary tests. In the test proper, the authors of the BDAE apparently find no need to separate syntactic and lexical aspects of sentence comprehension.

THE IMPORTANCE OF LANGUAGE ABILITIES AND MODALITIES

The BDAE reflects more modern neurolinguistic theory than the MTDDA in that it attempts to assess the interaction of varied modalities while at the same time it is concerned with localizing the area of the brain where the breakdown has occurred.

Goodglass and Kaplan intended their test to reflect equivalent weight for all language abilities and modalities. "The subtests of the batteries in most cases represent alternative windows that enable us to infer the status of our underlying capacity" (1983: 4). However, they do give priority to speaking by opening with the Conversational and Expository Speech section, and they include Comprehension Subtests before returning to Oral Expression. Reading and Writing are last, suggesting, it would appear, a somewhat secondary status.

LINGUISTIC PRINCIPLES REFLECTED IN THE TESTS

The analysis of the linguistic principles incorporated into the MTDDA and the BDAE focus on 5 questions that arose from comparing the tests and the linguistic premises that appeared to motivate them: (1) Is language unitary or modular[3] in structure? (2) Is language hierarchical? (3) Is auditory language

---

[3] We recognize that several of the terms we employ in our analyses, phonological, *metalinguistics* and *modularity*, are perhaps employed anachronistically. That is, the tests' authors are not likely to have read the terms prior to the publication of the tests. We have employed these terms because the concepts these terms express today were reflected in one or both tests.

Although the term metalinguistics was in use as early as the 1940's it seems to have had sociolinguistic as well as semantic implications. In recent decades, (e.g. Hakes, 1982) metalinguistic awareness is defined as the ability to reflect upon and manipulate the structural components of spoken language. This is the sense with which we employ the term in this paper.

basic among the modalities? (4) Is there a basic linguistic level? (5) Is metalinguistic ability a component of language behavior?

Linguistic principles underlying the MTDDA

Principles: (1) Language is unitary; (2) Language abilities are hierarchically organized; (3 Auditory language is basic among language modalities; (4) Phonology is basic among linguistic levels; (5) Metalinguistic abilities are part of language behavior.

LANGUAGE IS UNITARY

The authors of the MTDDA view language as a unitary system, within which aphasics range from more to less impaired. "We consider aphasia primarily a language deficit upon which various perceptual and sensorimotor deficits concomitant with brain damage may or may not be superimposed. The observed language deficit crosses all language modalities and is characterized by impairment of retrieval of the learned code" (Schuell, Jenkins and Jimenez-Pabon, 1964: 104). Schuell and Jenkins (1959) further state, that there is "a single dimension of deficit present in all aphasia" (p. 66.).

Classifications of severity are based on amount of deficit rather than type of deficit (mild, moderate, or severe with or without sensory motor components) (see Table 1). They specify: "It seems meaningful to define aphasia as a reduction of available language that crosses all language modalities and may or may not be complicated by perceptual or sensorimotor involvement, by various forms of dysarthria, or by other sequelae of brain damage." (Schuell, 1973: 4). The MTDDA treats perception and sensory motor components as processes separate from the language capacity.

Although the MTDDA uses length as an important parameter, lengthy sentences are currently examined for elements other than number of units or "memories" in the sentence. Repeating Digits (A8), versus common object word retrieval in Serial Item Identification (A4), and Sentence Repetition (A9) are all structured according to length. Syntactic phenomena that reflect structural depth, such as embeddedness, theta role placement, and missing constituents are not considered within this test (as it is in other analyses, e.g., Hildebrandt and Caplan, 1989; Shapiro, 1990).

LANGUAGE ABILITIES ARE HIERARCHICALLY ORGANIZED

Hierarchy is built into the MTDDA, however, it is not the linguists' hierarchy so much as a hierarchy of difficulty, complexity or severity. Schuell's diagnostic categories suggest hierarchical levels (see Table I). "The reduction of language in aphasia may be mild or severe or somewhere in between" (Schuell, 1973: 4).

Hierarchy is reflected in the treatment procedure. The speech language pathologist is instructed to start with "the level" where the patient still has some

success, and work from that point (Schuell, Jenkins and Jimenez-Pabon, 1964: 335). "Improvement, when it occurs is orderly and predictable" (Schuell et al., 1964: 114).
 A simple to complex hierarchical structure can be seen in all of the subtests of the MTDDA. Oral peripheral movement assessment precedes assessment of automatic speech which precedes assessment of idea expression via self generated and self formulated responses on the speech and language portion of the test. In terms of lexicon, more familiar words are presented first, followed by those that are less familiar; monosyllabic before multisyllabic. Word modifiers are used to add complexity to syntax. Propositional speech is viewed as more complex than a sentence fill in task and therefore, is presented later.

AUDITORY LANGUAGE IS BASIC AMONG LANGUAGE MODALITIES

Auditory comprehension appears basic in the MTDDA. "There is also always demonstrable impairment of auditory processes in aphasia" (Schuell et al., 1964: 115). Schuell's treatment approach, the Auditory Stimulation Approach, according to Duffy (1986: 189) employs strong, controlled, and intensive auditory stimulation of the impaired symbol system. The arrangement of the subtests on the MTDDA reflects Schuell's belief that auditory comprehension is primary to language and reading; the first modality to be tested is the auditory mode. This approach is the primary tool to facilitate and maximize the patient's reorganization and recovery of language.

IF ANY LINGUISTIC LEVEL IS PRIMARY, IT IS PHONOLOGICAL.

No linguistic level is seen as primary, however, in the speech and language subtests phonological features appear to be incorporated. There are a number of subtests where the hierarchical order is based upon the order of phoneme acquisition and the ease of phoneme production. For example, the subtests requiring repetition of monosyllables begins with bilabials, and ends with consonant clusters. Within the fricatives, /f/ and /v/ precede /s/, /z/, /š/ and /θ/. Affricates follow. Only in the phrase repetition task is /tš/ employed. These phonological parameters are additionally built into both comprehension and production subtests, however, they could have been included in the reading subtest, and were not.
 With respect to other linguistic levels, factors other than strictly linguistic ones seem to motivate the test items. Memory load underlies the increasing sentence length on the Repeating Sentences task; the Yes-No questions and Oral Reading subtest become progressively longer, but not in a syntactically principled fashion. Rather, increasing sentence length seems to have been the authors' goal. Retelling a paragraph is scored in large part on the number of propositions ("memories") included. In sum, no linguistic level can be said to predominate.

METALINGUISTIC ABILITIES ARE PART OF LANGUAGE BEHAVIOR.

Although Schuell and her colleagues did not use the term "metalinguistics", several of the subtests in the MTDDA may be interpreted as metalinguistic tasks. Defining Words and Producing Sentences (the task involves creating a sentence, in oral or in written modality, in response to a single word stimulus) are metalinguistic tasks. The fact that these tasks occur late in the Speech and Language Disturbance portion of the MTDDA (11th and 14th out of 15 subtests) suggests that Schuell and her colleagues appreciate that metalinguistic tasks require complex processing.

Linguistic principles underlying the BDAE

Principles: (1) Language is modular in structure; (2) Language abilities are hierarchically organized; (3) No language modality is considered basic; (4) Production is primary; (5) Metalinguistics is not really language.

LANGUAGE IS MODULAR

The most restricted sense of the term "modular" is that of Fodor (1983). Basic to this understanding of modularity is the notion of encapsulation whereby subsystems of the human language system (Fodor restricts his theorizing to comprehension of language) are autonomous and isolated from context, belief, meaning, etc.

Fodor does not apply this notion to brain organization per se, but the parallels are obvious between the Fodor notion of modularity and a strict localizationist approach to aphasiology. While modules as Fodor postulated them do not divide along strictly linguistic lines, but rather reflect automatic processing devices working in the service of the language faculty, Swinney et al. (1989), point out that differentiation among the modules can give insight into the neurological structuring of cognitive architecture. In this way, the BDAE may be construed as modular in that it assures there are not only differences in severity of aphasia, but also important differences in aphasia type related to specific areas of the brain which are lesioned.

A weaker form of the notion of modularity asserts the relative independence of specific language processing systems without insisting on their encapsulation. By this more popular usage, the BDAE is clearly modular in its treating each language task as potentially spared or impaired unrelated to other language tasks, as Goodglass et al. 1966 indicate. For example, on lexical tests they use a set of words from several different semantic fields throughout a number of subtests because loss of naming in one semantic field (e.g. color naming) can be independent of loss in other semantic fields (e.g. numbers). Moreover, the same lexical items are employed across a number of subtests (e.g. word discrimination, visual confrontation naming) because one or more "input system(s)" can be impaired while others may be spared.

One further indication that Goodglass and Kaplan subscribe to a relatively modular approach lies in the summary test profiles. The way they are structured indicates that these authors expect that performance on one linguistic task can be unrelated to performance on another.

LANGUAGE ABILITIES ARE HIERARCHICALLY ORGANIZED

In the strictly linguistic sense, language is not treated particularly hierarchically in the BDAE. That is, phonology, morphology, syntax, and semantics are not tested independently and in an ordered fashion.

Deep and surface structures enter into two of the supplementary subtests; subject of verb complement and passive subject-object discrimination. In the latter, patients are asked to identify the agent and experiencer of passive sentences, thus requiring them to recognize the deep structure underlying the passive.

Complexity hierarchies structure the test throughout. Simple, shorter items begin each subtest, and longer more complex ones occur as it progresses.

ORAL MODALITIES ARE BASIC

The first section of the BDAE is the Conversational and Expository speech task, suggesting the primacy of oral discourse. Reading and writing modalities, by contrast, are last, and have fewer items. Moreover, for two of the three summary scales, testers score only the oral modalities. The Aphasia Severity Rating Scale is a "scale of capacity for oral communication" (Goodglass and Kaplan, 1983: 30); The Rating Scale Profile of Speech Characteristics includes measures of speech production and one of auditory comprehension. (The remaining scale is for repetition which involves both production and, potentially, comprehension).

NO LINGUISTIC LEVEL IS PRIMARY

The emphasis on discourse in the BDAE, points to a view of language wherein syntax, lexicon, and semantics, are interactive and none of them is given primacy. Moreover, among linguistic levels syntax is treated with most sophistication in the BDAE. Phonological constructs do not enter into the structuring of subtests that might have included them as they did in the repetition test of the MTDDA.

METALINGUISTICS IS NOT LANGUAGE BEHAVIOR

Only one task that has traditionally been construed as a metalinguistic task is included in the BDAE: the Animal Naming Task in which patients are asked to generate a list of animal names. Animal names are generated as timed objects of thought rather than generated in the course of natural conversation.

Some people consider any language test to include metalinguistic requirements (Kamhi and Koenig, 1985; Flavell, 1976). The last subtest of the

Auditory Comprehension section, for example tests Complex Ideational Material. In addition to linguistic comprehension per se, the listener is required to make inferences and appreciate humor. Since the focus of these tasks is the automatic meaning rather than a conscious analysis of the form of the material, however, we do not consider them metalinguistic.

COMPARISON OF THE TWO TESTS

Overall the structure of the two tests is not markedly different. Both are composed of a sizeable number of subtests, each organized so that easier items precede more difficult ones, ranging across a large number of language tasks in all four modalities. As discussed above, strictly linguistic notions of hierarchy do not determine major aspects of the tests' structures. However, differences can be seen in the basic philosophies on which the tests are based, and particularly with respect to the questions posed in this paper.

One marked difference between the two tests is that the MTDDA sees language as unitary, whereas the BDAE assumes it is organized according to subsystems. The MTDDA expects subjects to manifest an across-the-board reduction in language abilities; the BDAE expects subjects to be impaired to a different extent on different abilities. For the BDAE the Subtest Summary Profile is quite important for diagnosis and rehabilitation and the Rating Scale Profile of Speech Characteristics is critical for classification of Subjects as fluent or nonfluent. The authors suggest that one interpret that scale "in light of" the Aphasia Severity Scale. This implies that the primary questions of interest about severity interacts with the bigger questions of classification as to aphasia type.

Schuell et al. (1964, 1973) prioritize the auditory modality in ways that Goodglass and Kaplan, with their more modular presentation, do not. However, Goodglass and Kaplan do prioritize the oral/aural modalities of both speaking and comprehension. The written modalities appear less elaborated than the oral ones in both tests and are placed in the final position.

No specific linguistic level appears to be primary for the BDAE. Rather, as indicated above, interaction between levels is assumed to take place. This can be seen particularly at the discourse level, which is placed first in the test. From the MTDDA, by contrast, we inferred that the phonological level is primary in some sense, and receives the most linguistic attention.

Metalinguistic aspects of language performance per se, appear to be more a part of Schuell's conception of language abilities than that of Goodglass and Kaplan. Both sets of authors assume an interaction between certain cognitive abilities and language. For Schuell et al. (1973), memory is the primary example; for Goodglass and Kaplan cognitive aspects of discourse processing such as inferencing and appreciating humor are included.

In conclusion, we may note that, as we expected, the linguistic theories that might have influenced these tests, such as structuralism or generative transformational grammar in their fully articulated forms do not determine their structure or content. However, certain basic linguistic principles have been seen to underlie one or both tests. These principles concern characteriza- tions of language as unitary or componential and the degree to which metalinguistics or other cognitive aspects of language performance are a part of language. They also reflect assumptions about linguistic levels, structures, units and modalities.

Acknowledgements

Even when we disagreed, the authors are indebted to Houri Kalhoustian for discussions about modularity and to Dava Waltzman for discussion about metalinguistics.

References

Chomsky, N. (1957). *Syntactic structures*. The Hague: Mouton.
Chomsky, N. (1965) *Aspects of the theory of syntax*. Cambridge, Mass.: MIT Press.
Duffy, J. (1986). Schuell's Stimulation Approach to Rehabilitation. In R. Chapey (ed.) *Language intervention strategies in adult aphasia*, Second edition. Baltimore, Md.: Williams and Wilkins.
Flavell, J.H. (1976). Metacognitive Aspects of Problem Solving. In L.B. Resnick (ed.) *The nature of intelligence*. Pp. 231-235. Hillside, N.J.: Lawrence Erlbaum Associates.
Fodor, J. (1983). *The modularity of mind*. Cambridge, Mass: MIT Press.
Goodglass, H. and Kaplan, E. (1983). *The Assessment of aphasia and related disorders*, second edition. Philadelphia, Pa.: Lea and Fibiger.
Goodglass, H., Klein, H., Carey, P. and Jones, K.F. (1966). Specific Semantic Word Categories in Aphasia. *Cortex*, 2:74-89.
Hakes, D.T. (1982). The development of Metalinguistic abilities: What develops? In S. Kuczaj (ed.), *Language Development, Vol. 2: Language, thought and culture* (pp. 163-210). Hillsdale, N.J.: Erlbaum Assoc.
Hildebrandt, N. and Caplan, D. (1989). Patterns of Syntactic Comprehension Disorders. Presented at CUNY Conference on Human Sentence Processing, March 1989.
Kamhi, A.G. and Koenig, L.A. (1985). Metalinguistic Awareness in Normal and Language Disordered Children. *Language, Speech and Hearing Services in the Schools*, 16, 3:199-210.
MacMahon, M.K.C. (1972). Modern linguistics and Aphasia, *British Journal of Communication*, 7, 54-63.
Marshall, J.C. and Newcombe, F. (1973). Patterns of paralexia: A psycholinguistic approach, *Journal of Psycholonguistic Research*, 2, 175-199.
Porch, B. (1967). *Porch Index of Communicative Ability*. Palo Alto, CA. Consulting Psychologists Press.

Schuell, H. (Revised by J.W. Sefer). (1973). *Differential diagnosis of aphasia with the Mlinnesota Test.* Minneapolis: Lund Press.
Schuell, H., and Jenkins, J. (1959). The nature of language deficit in Aphasia. *Psychological Review*, **66**: 45-67.
Schuell, H., Jenkins, J. and Carroll, J.B. (1962). A factor analysis of the Minnesota Test for differential diagnosis of aphasia. *Journal of Speech and Hearing Research*, 5, 350-369.
Schuell, H., Jenkins, J. and Jimenez-Pabon, E. 1964 *Aphasia in adults; diagnosis, prognosis, and treatment.* New York: Harper and Row.
Schuell, H., Shaw, R., and Brewer, W. (1969). A Psycholinguistic Approach to Study of the Language Deficit in Aphasia. *Journal of Speech and Hearing Research*, 12, 794-806.
Sefer, J.W. and Shaw, R. (1972). The Use of Psycholinguistic Principles in Treatment of Aphasia, *British Journal Of Communication Disorders*, 7, 87-89.
Shapiro, L. (1990). Presentation at CUNY Human Sentence Processing Conference. New York, N.Y.
Spreen, O. and Risser, A. (1981). Assessment of Aphasia. In M.R. Sarno (ed.), *Acquired aphasia.* New York: Academic Press.
Swinney, D., Zurif, E. and Nicol, J. (1989). The Effects of Focal Brain Damage on Sentence Processing: An examination of the neurological organization of a mental module. *Journal of Cognitive Neuroscience*, 1: 12-24.
Yaden, D.B. and Templeton, S. (eds.) (1986). *Metalinguistic Awareness and Beginning Literacy.* Portsmouth, N.H.: Heinemann.

APPENDIX A

Sections of MTDDA

A. AUDITORY DISTURBANCES

1. Recognizing common words (18 items)
2. Discriminating between paired words (24)
3. Recognizing letter (26)
4. Identifying items named serially (6)
5. Understanding sentences (15)
6. Following directions (10)
7. Understanding a paragraph (6)
8. Repeating digits (6)
9. Repeating sentences (6)

B. VISUAL AND READING DISTURBANCES

1. Matching forms (5)
2. Matching letters (20)
3. Matching words to pictures (32)
4. Matching printed to spoken words (32)
5. Reading comprehension, sentences (12)
6. Reading rate, sentences (6)
7. Reading comprehension, paragraph (8)
8. Oral reading, words (1
9. Oral reading, sentences (30)

C. SPEECH AND LANGUAGE DISTURBANCES

1. Imitating gross movements (10)
2. Rapid alternating movements (8)
3. Repeating monosyllables (32)
4. Repeating phrases (2)
5. Counting to 20 (20)
6. Naming days of week (7)
7. Completing sentences (8)
8. Answering simple questions (8)
9. Giving biographical information (15)
10. Expressing ideas (6)
11. Producing sentences (6)
12. Describing picture (6)
13. Naming pictures (20)
14. Defining words (10)
15. Retelling paragraph (6)

D. VISUOMOTOR AND WRITING DISTURBANCES

1. Copying Greek letters (5)
2. Writing numbers to 20 (20)
3. Reproducing wheel (6)
4. Reproducing letters (18)
5. Writing letters to dictation (26)
6. Written spelling (10)
7. Oral spelling (10)
8. Producing written sentences (6)
9. Writing a paragraph (7)
10. Writing a paragraph (6)

E. DISTURBANCES OF NUMERIAL RELATIONS AND ARITHMETIC

1. Making change (8)
2. Setting clock (5)
3. Simple numerical combinations (12)
4. Written problems (8)

APPENDIX B

LIST OF BOSTON DIAGNOSTIC APHASIA EXAMINATION SUBTESTS

I. CONVERSATIONAL AND EXPOSITORY SPEECH

II. AUDITORY COMPREHENSION

    A. Word Discrimination
    B. Body-Part Identification
    C. Commands
    D. Complex Ideational Material

III. ORAL EXPRESSION

    A. Oral Agility
    B. Automatized Sequences
    C. Recitation, Singing, and Rhythm
    D. Repetition of Words
    E. Repeating Phrases
    F. Word Reading
    G. Responsive Naming
    H. Visual Confrontation Naming
    J. Animal Naming (Fluency in Controlled Association)
    K. Oral Sentence Reading

IV. UNDERSTANDING WRITTEN LANGUAGE
   A. Symbol and Word Discrimination
   B. Phonetic Association
   C. Word-Picture Matching
   D. Reading Sentences and Paragraphs

V. WRITING
   A. Mechanics of Writing
   B. Recall of Written Symbols
   C. Written Word-Finding
   D. Written Formulation

6 Pragmatics applied to aphasia rehabilitation

Lisa Perkins and Ruth Lesser

The need to consider pragmatics in aphasia rehabilitation has arisen from the concern, first voiced by Taylor (1969), that traditional assessments examine a patient's language abilities removed from the context of natural language use. The primary aim of rehabilitation is to improve the aphasic person's ability to communicate in his or her social context and not to achieve a higher score on an aphasia assessment battery. Thus it is clear why there has been so much interest from clinicians in approaches to assessment and therapy which are sensitive to communicative context, and which focus on an overall ability to communicate rather than on a decontextualised language ability.

Prutting and Kirchner (1987) state that while there is a consensus that pragmatic aspects of language should be assessed in language disordered populations, there is no agreed paradigm from which to view pragmatics. Pragmatics can broadly be defined as the study of the use and understanding of language in context. However when an attempt is made to offer a more specific definition the precise nature of what is meant by pragmatics becomes fuzzy. This fuzziness arises from the disparate and sometimes incompatible traditions of pragmatic theory and analysis drawn from many disciplines, including philosophy, psychology, linguistics and sociology. As a consequence of this mixed academic heritage, pragmatics lacks an agreed terminology and descriptive framework to an extent which creates considerable difficulty in delimiting precisely the range of phenomena which might be described as pragmatic.

It is a tradition of language pathology to take the developments of other academic disciplines and apply them in the investigation and remediation of disordered language. However in doing this it is essential that there is an awareness of the background from which the new developments come and of their strengths and weaknesses. Levinson (1983) warns that, as always in the application of academic ideas to vital practical issues, there is the very real possibility of the premature acceptance and application of untested concepts and theories. McTear (1985) suggests that much published work which attempts to measure pragmatic ability suffers from poor definition of categories and insufficient consideration of the theoretical assumptions underlying the analysis.

In writing this chapter we have tried to give consideration to such warnings and criticisms. We start with a review of some pragmatic theories that have been utilised in the aphasiology literature, examining the limitations of these theories as well as evaluating the contribution that the different perspectives have to make to the investigation of aphasia. Using the findings of investigations from the various orientations we will then address the issue of pragmatic ability in aphasia and the relationship that exists between pragmatic and linguistic impairment. The final three sections deal more directly with rehabilitation issues. We review some of the assessments that have been developed to evaluate pragmatic abilities and look at what contribution these have to make to intervention with aphasic patients. This is followed by a review of the treatment approaches that have been developed from a pragmatic orientation. We conclude by examining ways forward in a pragmatic approach to aphasia rehabilitation.

THEORETICAL PERSPECTIVES

Two broad orientations can be distinguished in the pragmatics literature; one is a top-down or theory driven perspective seen in approaches from philosophy and linguistics; the other is a bottom-up or data driven approach typified by Conversation Analysis (CA). Both approaches have been invoked in the examination of pragmatic abilities in aphasia. The central features of these theoretical perspectives and how they have been applied to aphasia will be examined, starting with an examination of "top-down" approaches.

Top-down Pragmatic Approaches

Top-down approaches characteristically employ a deductive type of reasoning, in which an abstract competence is modelled by idealised speakers in idealised situations. Organisational principles of some kind are posited and an attempt is made to fit these to the data. One of the important goals of pragmatics is the specification of how connected discourse differs from any random set of consecutive utterances. This abstract property of discourse connectedness is what is generally referred to as coherence. Several assessments of aphasic subjects' pragmatic abilities appeal to appropriacy in judgements (Penn, 1985; Prutting and Kirchner, 1987; Wirz, Skinner and Dean, 1990). It appears that it is coherence which is being appealed to in making these judgements and coherence is addressed by the theories under discussion. It is beyond the scope of this review to describe these theories in detail. We will instead refer briefly to two theories developed from philosophy (Gricean implicatures and speech act theory) and to two linguistic approaches (discourse analysis and text grammars).

PHILOSOPHICAL APPROACHES: GRICEAN IMPLICATURES

Grice (1975) addressed the fact that natural language utterances do not convey the same meanings that the corresponding logical propositions would. To account for these divergent or extra meanings that arise in natural language he proposed that they arise not from syntactic or semantic rules of languages, but instead from principles of conversation. He proposed that conversationalists are guided by a co-operative principle in which contributions to conversation are assumed to be intended as relevant and generally helpful even when they appear on the surface not to be. The co-operative principle is elaborated in terms of four basic maxims of conversation which specify what interlocutors have to do in order to converse in a maximally efficient, rational, co-operative way. The maxim of Quality is concerned with being truthful; the maxim of Quantity is concerned with providing the appropriate amount of information; the maxim of Relevance is concerned with making a contribution relevant; and the maxim of Manner involves making a contribution with clarity, orderliness and brevity.

The maxims are not arbitrary conventions but describe rational means for conducting co-operative exchanges. Grice does not argue that people follow these guidelines to the letter. His claim is, rather, that these principles are oriented to, such that when talk does not proceed according to their specifications, hearers assume that the principles are nevertheless being adhered to at some deeper level. In these cases, inferences arise to preserve the assumption of co-operation which Grice calls conversational implicatures. These are inferences beyond the semantic content of the sentences uttered, based on both the content of what has been said and some specific assumptions about the co-operative nature of ordinary verbal interaction. The way that conversationalists make implicatures can be illustrated with one of Grice's hypothetical examples:

(1) A: I am out of petrol
    B: There's a garage around the corner.

If taken literally B's response is irrelevant. It simply states that there is a certain kind of business around the corner. However, if we assume that B is abiding by the co-operative principle we can assume that B is at some level being relevant so we assume that he believes that the garage is open and that it sells petrol. Although B does not actually say these things they can be derived by implication.

Implicatures can also come about by overtly and blatantly not following one of the maxims and thereby provide evidence of the robustness of the assumption of co-operation in conversation. If someone drastically deviates from a maxim then his or her utterances are still read as being co-operative at some level if this is at all possible. Levinson (1983) provides examples of floutings of all of the maxims and the way that implicatures can be drawn from these. We will provide one example to illustrate this type of implicature (Levinson, 1983: 112). Imagine that in a musical review a statement like (2) is found:

(2) Miss Singer produced a series of sounds corresponding closely to the score of an aria from Rigoletto.

Example (2) appears to violate the maxim of manner in that it is certainly not brief. However if we assume that the reviewer is at some level adhering to the co-operative principle then the reader is likely to implicate that there was little to recommend Miss Singer's performance.

The origin of Gricean theory in the abstractions of philosophy makes direct application to situated speech problematic (Leech and Thomas, 1990). While

Grice's maxims offer a suggestion of the everyday reasoning used by communicators to extract meaning in discourse, it is not really clear how these concepts can be applied in any principled manner to the assessment of aphasia.

Indeed references to Gricean maxims in investigations of aphasic subjects' pragmatic abilities have been rather general. Both Penn (1985) and Prutting and Kirchner (1987) cite Grice in relation to appropriateness. In addition Penn suggests that aphasia appears to reduce the individual's capacity to apply the conversational maxims effectively. She does not appear to be suggesting that aphasic subjects have lost their knowledge of the maxims, rather that linguistic impairments interfere with their implementation.

To our knowledge, there has been no experimental work which measures patients' discourse comprehension in terms of their knowledge of the maxims, although the work discussed below on the relatively preserved ability of aphasic subjects with indirect speech acts indicates that, in general, inferencing ability is not impaired in aphasia. However Hawkins (1989) discusses an aphasic man, EP whose problems he describes as being at the level of discourse. He suggests that EP's utterances can be seen to be inappropriate because they violate Grice's maxim of relevance. An excerpt of the discourse is given below:

(3) T: right well I'm going to ask you a couple of questions like we've done before / would you say it was hot today
  P: er depends on where you're going / what time / from summer /they say the middle of summer
  T: today / is it very hot
  P: yes I like it / sun can be very hot er nice and breeze / it's lovely to lie back and enjoy it you know.

Hawkins points out that EP's responses are inappropriate; they can be seen to be violating the maxim of relevance in that his responses are not relevant to the question. The only explanation that might be offered for his responses, Hawkins suggests, is that he has understood the word "hot" but little else. His reply is of the sort "tell him anything you know about hot". While the use of Gricean maxims is of use in explaining the reason for the inappropriateness of EP's responses it is necessary to ask whether this problem arises from the loss of knowledge of the maxims which guide conversation (i.e., EP no longer appreciates that contributions to conversation should be relevant) or whether the problem of appropriacy is the manifestation of a severe auditory comprehension problem which EP deals with by using a strategy of picking on salient lexical items and taking these as his topic. Without information on EP's performance on assessments of auditory comprehension it is not possible to distinguish between the two. However this dichotomy is important as it distinguishes between pragmatic impairments which are independent of other linguistic impairments and pragmatic impairments which are a consequence of other underlying deficits. Further consideration is given to this question below.

PHILOSOPHICAL APPROACHES: SPEECH ACT THEORY

Originally proposed by the philosopher Austin (1962) and subsequently developed by the work of Searle (1979), the chief concern of speech act theory is to address the way in which language is used to perform actions such as warning, requesting, apologising etc. Look at example 4:

(4) This plate is hot.

It is possible to analyse (4) in terms of truth conditional semantics as a proposition concerning a plate which can be either true or false, with the assumption that its function is purely to make an assertion. However such an analysis does not explain that when uttered in an appropriate context, for example when a canteen worker is passing over a plate to a customer, (4) is likely to be interpreted and intended as a warning. For the speaker to grasp the speaker's meaning, the addressee needs not only to access the propositional meaning but also be able to make a number of assumptions about the motives of the speaker and to be sensitive to the context of the utterance.

It is possible to make the distinction between the semantic and syntactic form of an utterance and its communicative (speech act) function. While functions (or illocutionary force) may have a form traditionally associated with them (for example the interrogative form being associated with the function of asking or requesting) it is possible for a function to be realised in a number of different forms. Speech act theory deals with the issue of how conversationalists manage to assign a particular meaning or force to ambiguous utterances with the notion of felicity conditions. These are conditions which must be fulfilled in the situation in which a specific speech act is carried out if it is to be carried out felicitously or appropriately. Thus the act of requesting may be said to be felicitous only if the speaker believes that the hearer will be able to carry out the request and if the speaker wants his request fulfilled.

As with Gricean theory the philosophical basis of speech act theory makes its direct application to situated speech problematic. Searle himself has made the point that the speech act approach deals with isolated pairs of utterances and has not yet been developed to handle conversation (Searle, 1986: 7). Another problem is that there has not been much success in producing principled sets of speech act types. Even if this were achieved, another major defect of speech act theory is that it offers no account of the possible multifunctionality of utterances. In natural discourse speakers often seem to achieve several communicative goals simultaneously and often the intended illocutionary force is indeterminate and only emerges from the sequential context of the discourse.

Despite the many problems in speech act theory, however, some of the basic distinctions and concepts are very relevant to issues in aphasia remediation. One of the issues of relevance to aphasia is patients' abilities to comprehend and produce both direct and indirect meanings of utterances. Traditional assessment procedures for auditory comprehension do not make a distinction between comprehension of direct meanings which involve computing sense relations between lexical items or sense properties of sentences, and comprehension of indirect meanings which requires inferencing skills. However, studies examining aphasic subjects' abilities with indirect commands (Foldi, 1987) and indirect requests (Weylman, Brownell, Roman and Gardner, 1989) demonstrate that aphasic subjects who do not have severe linguistic comprehension deficits do not have problems in deriving inferential meaning. Rather it appears that this is a problem pertinent to right brain damage.

There have also been investigations of aphasic subjects' abilities to produce a variety of speech acts. Davis and Wilcox (1985) suggest that studies of aphasia have shown that there is a dissociation between speech acts and sentence production, in which sentence production is impaired but ability to express intentions, including requests, remains intact (Guilford and O'Conner, 1982; Prinz, 1980).

Ability in both comprehending and producing speech acts is one of the abilities examined in a number of pragmatic assessments of aphasia. Prutting and Kirchner's (1987) Pragmatic Profile contains judgements on the appropriacy of speech act pair analysis which is defined as "the ability to take both speaker and listener role appropriate to the context" (1987: 118). An appropriacy judgement is also required on the variety of speech acts produced. Five out of ten aphasic subjects in this study were judged as being inappropriate in terms of the variety of speech acts produced although none of them was analysed as inappropriate for the speech act pair analysis. It appears likely that while the aphasic subjects are able to interpret and produce speech acts appropriately, they may be limited in the diversity of speech acts produced. This may link to limitations in sentence production, or, given that speech acts are sensitive to context, it may reflect the restricted contexts in which aphasic data are usually collected.

Penn's (1985) Profile of Communicative Appropriateness also has a category concerned with speech acts. However rather than looking at the variety of speech acts, this profile asks for a judgement on the use of indirect speech acts, including the use of modal verbs and past tense in requests, as well as oblique comments, softeners and warm-ups. It appears that although both Penn (1985) and Prutting and Kirchner (1987) are drawing on speech acts in their examination of pragmatic ability, each is focussing on rather different parameters of the theory. Penn's concern with indirect requests can be seen to link with subjects' abilities to use polite forms.

A further assessment which draws upon speech act theory is the conversational assessment described by Copeland (1989) in its pilot version. This assessment involves rating performance for twenty communicative acts on a five point scale. The communicative acts were an expansion of those noted by Wilcox and Davis (1977). However Copeland found that this was still an incomplete list of the communicative acts which occurred spontaneously. This reflects the conceptual problem which has already been commented on, namely the provision of a finite and principled set of acts. It was also noted that messages were sometimes extended, a feature of natural conversation which has important implications for the sharing of the burden of conversation in aphasic conversation but which is not touched upon by a speech act type analysis. The Edinburgh Functional Communication Profile in its original and revised forms (Skinner, Wirz, Thompson and Davidson, 1984; Wirz, Skinner and Dean, 1990) is also based upon speech act theory, with ratings of efficacy being made on categories such as greeting, acknowledging, responding, requesting and initiating.

Having examined two top-down approaches to pragmatics developed from philosophical theory we will now move on to two theoretical orientations developed from linguistics, each of which has been drawn upon in the language pathology literature. These approaches essentially attempt to extend the techniques of linguistics beyond the unit of the sentence. Levinson (1983: 286) summarises the procedures employed as : (a) the isolation of a set of basic categories or units of discourse, (b) the formulation of a set of concatenation rules stated over the categories, delimiting well-formed sequences of categories (coherent discourses) from ill-formed sequences of categories. Firstly we will look at the work of the Birmingham (UK) group of discourse analysts (e.g., Sinclair and Coulthard, 1975; Coulthard and Brazil, 1979) before moving onto the work of the text grammarians (e.g., Van Dijk, 1972).

LINGUISTIC APPROACHES: DISCOURSE ANALYSIS

The Birmingham group of discourse analysts developed procedures for analysing exchange structure. In the early seventies Sinclair and Coulthard developed a linguistic framework for analysing interactive discourse to use with overtly structured types of discourse such as that seen within the classroom or interview situation. They suggest that while the ultimate aim of discourse analysis must be to account for conversational interaction, because of the "looseness" of conversation it is better to begin the development of a discourse model by analysing more formally controlled types of behaviour such as classroom interactions (Sinclair and Coulthard, 1975). As a model is developed one may then consider conversation.

Within this system, discourse is conceptualised as a further level of linguistic organisation, where elements are realised by items at the level of the grammar. Although Sinclair and Coulthard talk about speech acts their conception of speech acts is very different from the speech acts of Searle and Austin. Rather than being defined in terms of shared felicity conditions and by the intentions of the speaker, the acts are defined by their function in the discourse.

The basic unit of interaction is the exchange, consisting of at least an initiating move by one speaker and a responding move by another. While initiating moves are prospective, setting up constraints on the possible response, responses are retrospective, only being analysable in terms of what has gone before. A third type of move is a follow-up which is optional. This type of two or optionally three part structure seems to be characteristic of interviews. The three part structure is particularly common in instructional contexts, where the teacher initiates, the pupil responds and the teacher follows up with feedback. In such contexts one of the participants is responsible for controlling the discourse and it appears that in therapist-patient interaction this type of structure is operating. This can be illustrated by an excerpt from discourse between a speech and language therapist and an aphasic woman:

(5) T: can you tell me your name?
    P: er (.) right wait a minute (.) Mrs (2 seconds) I don't know (.)
    T: it's H - Hil

This structure of initiation-response-follow-up is often found in therapeutic discourse. Test questions are asked by the therapist who produces a subsequent follow-up to the response.

While this type of model of discourse is able to account for structured interactions it does not capture the nature of everyday conversation in which none of the participants have the overt obligation to control the discourse. Rather participants negotiate turns at talk. The problems of applying an exchange-type model become even more apparent if multi-party conversations are examined. Stubbs (1983) discusses the question of extending the exchange structure analysis to casual conversation. The model has indeed been extended to a more elaborate framework than that presented here (Coulthard, 1985: 120 ff). However as Lesser and Milroy (1993) have stated the issue is not whether such a model can or cannot be extended, but whether a model based on institutional discourse should be used at all as the basis for analysing casual conversation.

Pragmatic profiles have been developed which draw upon exchange structure. McTear (1985) has developed a profile, part of which at least is based upon the work of the discourse analysts, with an examination of initiation in terms of the type of initiator (question, request for action or statement) and an examination of responses. Hawkins (1989) uses this profile to examine the discourse of two aphasic patients with their therapists. However no further light is thrown on the problems than that there appears to be some problem in the appropriacy of responses; an explanation which is no more precise than that obtained using an explanation of Gricean implicatures or speech act theory. Furthermore if casual conversation is considered to be the best data on which to judge pragmatic ability, given the problems of this type of approach to conversation, it is clear that it has its limitations in assessment and remediation.

It has already been mentioned that, although the Birmingham discourse analysts refer to speech acts, these differ from the speech acts discussed in the previous section in the way that they are defined. In pragmatic profiles when speech acts or communicative acts are referred to it is not always clear whether the developers of the profile are appealing to speech acts in terms of philosophical theory or whether their conception is more in line with the discourse analysts. This is true for Holland's (1980) Communicative Abilities in Daily Living, the Edinburgh Functional Communication Profile and Copeland's

(1989) profile of conversation. This demonstrates the problem of the use of loose terminology in the applied pragmatics literature.

LINGUISTIC APPROACHES: TEXT ANALYSIS

Another linguistic approach investigating the discourse level is text-grammar. This movement originated in Germany and was concerned with extending generative grammar to the level of discourse to describe strings of sentences belonging to a discourse. Van Dijk (1972) challenged this approach by arguing that this conception of text-grammar which used a step-by-step analysis of contiguous sentences was not able to adequately explain the true nature of discourse. As well as an analysis at a microstructure level he proposed that there is a higher level or macrostructural level which must also be considered. Macrostructure refers to the overall structure of a discourse and has been equated with coherence. In contrast microprocessing deals with individual propositions and their relationships as conveyed by various syntactic and stylistic cohesion devices. As reviewed by Huber (1990), there are two competing theories on how macroprocessing takes place. While in their earlier work Kintsch and van Dijk proposed that macroprocesses generalise and summarise the contents of micropropositions, other workers (e.g., Johnson-Laird, 1983) have suggested that general world knowledge and pragmatic reasoning play a crucial role in macroprocessing. Macropropositions are not members of linguistically defined classes; their mode of processing is heuristic with lexical information such as key words and idioms being related to world knowledge and expectations about people, events and situations. The occurrence of logical relations between propositions which are conveyed not by specific words but by relationships between the facts expressed by propositions, such as cause-effect relations, gives rise to coherence. Coherence can be overt when the type of logical link can be established directly from the propositional content or covert if the link involves the drawing of inferences.

In contrast to macroprocessing, microprocessing requires extensive linguistic knowledge in order to understand cohesive relationships and consists of semantic ties between words across sentence boundaries. There are two different types of cohesive ties. The first type are grammatical and include the use of anaphora in text to refer. The second type are lexical and involve the semantic relations created across texts through, for example, word repetition, synonymy and hyponomy.

The main limitation of text grammar and research which has developed from it is that it does not pay any attention to the interactional aspect of communication. It is able to account for monologue better than dialogue and this is reflected in the type of discourse studied using this type of approach, mainly narrative story telling or procedural discourse rather than conversational discourse.

This focus of interest in narrative clearly limits the usefulness of this theory to the aphasiologist who is interested in improving the aphasic patient's functional ability as it takes no account of interactional aspects of language use, an important area of study. However certain features of the work developed from text grammar have been seen as having a useful contribution to the study of aphasia. This interest can be seen to arise because of certain parallels which can be drawn between the dichotomy of linguistic microprocessing and world knowledge macroprocessing in the theory and the dichotomy between impaired linguistic processing and the more intact cognitive abilities of aphasic subjects. It has been suggested that linguistically impaired aphasic subjects may rely on macrostructure in both comprehension and production of discourse while failing to utilise the linguistic microstructure (Huber, 1990).

Studies examining aphasic subjects' comprehension performance of narrative discourses have manipulated features in both the macrostructure and microstructure. Chapman and Ulatowska (1989) have shown that moderately impaired aphasic subjects when asked to identify referents in a text had significantly greater difficulty when the sentences were connected with pronominal co-reference than when explicitly stated noun phrase referring expressions were used throughout the text. Performance in understanding the referents of the pronouns was influenced by the plausibility of the texts, with better performance on the items that could be inferred from the communicative situation of the text.

Huber and Gleber (1982) have reported a study which investigated whether high versus low linguistic cohesion among the sentences of a text would influence aphasic subjects' performance. The task was to order the sentences of texts. Analysis of results showed that the degree of linguistic cohesion had no significant influence on any of the aphasic groups. This suggests that the aphasic subjects are relying on macroprocessing and are not using the microstructure information to obtain a precise understanding of the sequences presented in the text.

Huber (1990) reviews several studies which have examined text comprehension predominantly using picture selection to test the subjects' comprehension of narratives read to them. Several studies report beneficial effects of linguistic context on comprehension in aphasia. Whether predictive context helps aphasic subjects to understand difficult sentences (i.e. to enhance microprocessing) or whether it merely provides enough information to make those sentences superfluous (i.e. to support macroprocessing) is questionable. However on the basis of a comparison of two investigations carried out by himself and colleagues (Stachowiak, Huber, Poeck and Kerschensteiner, 1977; Kohlert, 1979) Huber proposes that the beneficial effects of redundant context on text comprehension of aphasic subjects are rather indirect. Rather than aphasic subjects showing better comprehension with greater redundancy, their results indicate that the non-language impaired control subjects obtain a lower accuracy score with the more redundant texts. It appears that normal subjects, like aphasic subjects, rely to a larger extent on macro- than microprocessing when the texts are more redundant and consequently longer and more complex.

Moving onto an examination of aphasic subjects' ability to produce discourse, the most common paradigms used are verbal descriptions of cartoon stories and the telling of narratives and procedures. Ulatowska, Allard and Chapman (1990) summarise their findings in a number of studies of different aphasic groups which examined production of narrative and procedural discourse. The tasks used include telling about a memorable experience, telling a story based on a sequence of pictures and giving a summary of it, retelling a story read by the investigator, and procedural tasks requiring an explanation of how to perform routine procedures such as making a sandwich. For a group of mildly impaired aphasic subjects it was found that they were able to produce well-structured discourse although compared to normal controls they produced less complex language, more sentential errors amd more discourse errors. In a further study examining moderately impaired aphasic subjects discourse structure was still found to be maintained despite a number of sentential errors and discourse errors. They differed from the normal subjects, with a more abrupt onset of climax rather than a steady progression of events in the memorable experience narrative. In the procedures, the aphasic subjects showed some reduction in the number of essential steps, although they were more likely to omit optional steps. In a third study of severely aphasic subjects there was evidence of a breakdown at the discourse level, with a marked reduction in information content as well as a disruption in narrative superstructure characterised by errors in sequencing of events, and also by inadequate production of setting and resolution propositions. However, studies by Gardner and his associates (Gardner, Brownell, Wapner and Michelow, 1983) indicate that right brain damaged subjects show the most striking deficit in discourse production. In recalling fables this group make sequencing errors, produce confused story renderings and have difficulty in both extracting the moral of a story and paraphrasing.

A variety of studies have examined aphasic subjects' use of cohesive devices. Aphasic subjects have been reported to produce a large number of pronouns without antecedents (Berko-Gleason, Goodglass, Obler, Green, Hyde and Weintraub, 1980; Nichols, Obler, Albert and Helm-Estabrooks, 1985). As Ulatowska et al. (1990) note, because of the complexity of the reference system, different populations may show deficits in use of reference for different reasons. Excessive use of pronouns may result from a failure to recognise the listener's need for precision of reference. However such a problem may also arise as a manifestation of a linguistic deficit. Garman and Edwards (1989) report on a fluent aphasic subject who produces a greater proportion of pronouns than nouns. They suggest that the use of proforms may be a strategic response to the difficulties of noun selection for this subject.

Bates, Hamby and Zurif (1983) examined the use of a number of cohesive devices used by Broca's and Wernicke's aphasic subjects in a three frame picture description task. The cohesive devices that they examined included probability of lexicalisation versus ellipsis, pronominalisation, definite and indefinite article use, pragmatic word order variations in dative structure, and use of connectors. Bates et al. found that both aphasic groups showed pragmatic sensitivity to at least two of these areas and from this proposed that deficits in cohesive devices may arise as manifestations of syntactic or lexical-semantic impairments rather than failure to appreciate the need to use such devices.

A Bottom up Pragmatic Approach: Conversational Analysis

Conversation analysis has been pioneered by a break-away group of sociologists called ethnomethodologists and this sociological background is reflected in the methodological principles that derive from it. Ethnomethodology arose in reaction to the quantitative techniques and the arbitrary imposition on the data of supposedly objective categories that were typical of mainstream American sociology. Instead it was proposed that the techniques that the members of a society themselves use to interpret and act within their own social worlds should be used in sociological study. This view led to the development of ethnomethodology, the study of "ethnic" (i.e. the participants' own) methods of production and interpretation of social interaction. Thus conversation analysis can be seen as distinctive from the top-down pragmatic approaches already discussed in that initially it posits no set of analytic principles but inductively seeks patterns in the bodies of naturally occurring data. Analysts confine themselves as far as possible to descriptions of observable conversational behaviour, generalising from recurring details.

As discussed by Taylor and Cameron (1987), in contrast to top-down approaches which propose rules that determine participants' behaviour, ethnomethodologists do not view rules as governing behaviour in such a deterministic way. Instead they would suggest that participants design their behaviour with an awareness of its accountability. Given that they are aware of the rules relevant to the situation that they find themselves in, they choose to follow or not to follow the rule in the light of what they expect the interactional consequences to be. If they choose not to conform to the rule they can expect their co-participants to look for the reasons why. Thus rules have a normative force without having to be seen as internalised determinants of conduct.

A feature of the relationship that exists between rules and behaviours is that it emphasises the sequential relevance of an action. A participant's behaviour is designed in light of what reaction he or she expects from co-participants, whether this is adherence to the relevant rule or violation of it (the latter leading to the drawing of conclusions as to the reason for the violation). Any component action is inevitably temporally situated in a sequential context. It will add to that context and within it will be interpreted, held accountable and be responded to in turn. Participants succeed through the sequential progression of interaction to display their understanding of its events and of the rules to which they are orienting, thus making possible the achievement of a shared interactional world. In doing this they also make that shared world publicly observable to the investigating analyst. Because the interactants' own understanding of events is displayed in their subsequent responses to those events, the analyst can obtain a clear grasp of the ways in which the participants themselves are analysing the interaction.

Conversation analysts using this empirical methodology have investigated the details of conversational organisation, including the mechanisms of turn-taking, repair, adjacency pairs and preference organisation. It is beyond the scope of this chapter to attempt to review this work; Levinson (1983) provides a comprehensive review. In order to illustrate this approach we will briefly describe the central conversational device of repair as it is a concept which, as we will see, is of relevance to pragmatic investigations of aphasia.

Conversation analysts use the term "repair" to refer to the many different ways of managing various trouble sources in the interaction. There is no one-to-one relationship between error and repair, since speakers may revise their utterances where there is no hearable error or ignore error or ambiguity on the assumption that matters will become sufficiently clear as the discourse proceeds. Thus the issue is not whether the speaker makes an error, but how and whether the mechanism for dealing with the trouble source works.

Schegloff, Jefferson and Sacks (1977) in their seminal work on repair point out that there is a distinction to be made between the initiation of the repair and its carrying out. These two phases can be carried out by different participants. Repair can be initiated by the speaker of the trouble source (self-initiated repair) or by another party (other-initiated repair). In either case the repair work itself may be done by either self or other. In relation to this Schegloff et al. emphasise that while self and other initiation of repair are distinct types they operate over the same trouble sources and their respective placements are ordered relative to each other. Other initiation is invariably withheld until the trouble source turn's possible completion; frequently it is withheld after possible completion. This allows the maximum time for self to initiate repair and thus gives rise to a preference for self repairs.

Another feature of repair examined by Schegloff et al. is the different trajectories of self and other initiation. Self initiation of repair can be, and usually is, combined with doing the candidate repair. In contrast other initiation often locates the trouble source to yield self repair in the next turn. The organisation of repair in normal speakers thus provides centrally for self repair which can be arrived at through either self or other initiation. While other repair does occur it is often modulated with the use of uncertainty markers or by the use of various types of question format. Schegloff et al. point out that when hearing or understanding is adequate for the production of a correction by other, it is adequate for the production of a sequentially appropriate next turn. Therein lies the basis for the modulation of other repair; if it were confidently held it ought not to be done. This also accounts for the paucity of other repair; those who could do them do a sequentially implicated next turn instead. From the organisation of repairs explicated by Schegloff et al. it is clear that it is an efficient mechanism which deals with trouble sources quickly in a short number of turns. Repair is either carried out within the turn in which it appears (self-

initiated self repair), within the next turn (self-initiated other repair or other-initiated other repair) or within the following turn (other-initiated self repair).
A number of workers have outlined the weaknesses of CA. While acknowledging that CA does have strengths Graddol, Cheshire and Swann (1987) report that although conversation analysts emphasise looking for evidence in the raw data (which should represent the event precisely as it happened), in reality transcription itself represents a first step in analysis, with selection of features which are considered to be most relevant.

A more serious criticism is put forward by Taylor and Cameron who discuss the application of CA principles to two pieces of data described in the literature (1987: 117-123). They state that it is not legitimate to use the conversational data as evidence for the claim that it is by orientation to organisational devices that the conversation develops as it does, when the interpretation of the data itself depends on taking for granted the speakers' orientation to these same organisational devices. Instead they argue that independent empirical evidence of how particular occurrences in the conversation are interpreted by the participants is required. They suggest that conversation analysts forego giving any criteria for determining whether a second turn displays understanding of a first turn as having a particular function, or as orienting to a particular rule. Instead they simply provide an intuitive characterisation of the understanding it displays.

Conversation analysts defend themselves against such criticisms, demonstrating that their analysis is more than a purely analytic construct by protracted examination of small fragments of data. Their claims are based not on instances from single texts but instead from examination across different discourses in order to discover the systematic properties of the sequential organisation of talk and the way in which turns at talk are designed to manage some sequences.

The painstaking examination of small excerpts of data so typical of CA may make this approach appear too time-consuming for its application to aphasia therapy. However Milroy and Perkins (1992) propose that conversation analysis has a number of hitherto under-exploited attractions for aphasiologists. The chief one of these is that its data-driven approach provides procedures for approaching intransigent data from spoken language which are intolerant of prior theoretical assumptions. It allows aphasic discourse to be approached without any prior problematic assumptions about how this relates to "normal" discourse. Furthermore, CA is the only approach which explicitly takes account of minutiae such as filled and unfilled pauses, overlaps, repetitions and repairs, all of which have interactional consequences dependent on their sequential position but which are often abstracted away from as messiness in the data. Such phenomena are generally very common in aphasic conversation. Therefore, CA addresses itself quite directly to the level of pragmatic difficulty typically experienced by aphasic patients, and it does not stress the separateness of linguistic and pragmatic impairment, an issue which is given further discussion in the next section below. The analyst is free to seek explanations for communicative success or failure in structural characteristics of a patient's language and in limitations imposed by the linguistic impairment. Overall CA provides tools for describing with some precision the observable communicative consequences of particular impairments.

A further useful feature of CA is that it explicitly treats conversation as a "collaborative achievement" (Schegloff, 1982) with successful communication seen as the joint responsibility of both the impaired and unimpaired partner. This orientation aligns with the treatment interest in indirect therapy which seeks to involve the patient's relatives and carers in intervention. Indirect therapy is given further consideration in the section below on pragmatic approaches to therapy.

There has been a limited amount of research carried out applying a CA methodology to aphasia. Schienberg and Holland (1980) examined a conversation between two Wernicke's aphasic patients in order to investigate whether they obeyed the turn-taking rules found to be operating in normal conversation (Sacks, Schegloff, and Jefferson 1974). They found that for these two patients turn-taking was not impaired. While the aphasic person may know and orient to the rules of turn-taking it is important to remember that taking a turn at talk can be competitive when there has not been allocation of next speaker by previous speaker. Given aphasic people's linguistic deficits it is possible that they may have difficulties in getting a turn, or keeping hold of it if they run into difficulty within the turn. A number of studies have shown that aphasic patients use non-verbal strategies to gain and hold onto turns. Ahlsen (1985) reports that a widely used strategy is the raising of a hand to avert interruption, the hand being dropped to signify completion of the turn. Conway (1990) reports that one aphasic patient she studied sat forward to signal movement into speakership.

A further area of investigation using a CA methodology has been repair in aphasic discourse. Given difficulties which routinely arise from language impairments such as word-retrieval problems, semantic and phonemic paraphasias, disturbances of grammatical production, and comprehension deficits it can be seen that an examination of repair organisation is particularly relevant to an evaluation of aphasic conversational abilities. Milroy and Perkins (1992) review the limited work carried out on aphasic repair and conclude that repair strategies differ from those of normals such as have been described by Schegloff et al. (1977). Specifically there is less self repair (the aphasic interlocutor not having the linguistic resources to deal with trouble sources). Furthermore, the organisation is not so easily explained in terms of "self" or "other". Instead they propose a collaborative model based on Clark and Schaefer's (1987, 1989) CA-style model of communication. The main feature of this model is that in repair the interlocutors work together to minimise the collaborative effort invested. This model allows a unitary analysis of the often protracted repair sequences observed in aphasic conversation in preference to an analysis which sees each subsequent turn as a failed repair attempt. The analysis examines how each of the partners contribute to repair sequences and offers insights into the way that repair can be achieved most efficiently; these insights might be useful in the development of communicative strategies by both the aphasic interlocutor and his or her partner. This work demonstrates the value of using the methodology of CA to examine aphasic conversation rather than simply applying the findings of the research on normal conversation and assuming that the same mechanisms are operating.

A number of assessments concerned with an examination of pragmatic ability in aphasia draw upon CA terminology. Wirz, Skinner and Dean (1990) include a rating of whether the patient is carrying out effective or ineffective repair in their revised Edinburgh Functional Communication Profile. Prutting and Kirchner (1987) in the Pragmatic Protocol include a section on turn taking which includes categories such as initiation, response, repair, pause time, overlap and feedback to listener. The therapist is required to make a judgement of appropriacy on a two point scale. These borrowings from CA do not embrace the methodology of a bottom-up approach by looking at the sequential consequences of a turn at talk in the interaction. Rather than looking at how the interaction is being managed by the interlocutors, the CA terminology is taken and created into a top-down category of analysis and then analysed in terms of appropriacy or presence or absence. This use does not capitalise on the value of a CA methodology discussed above.

One assessment which draws upon CA methodology in addition to its terminology is the Assessment Protocol of Pragmatic and Linguistic Skills (APPLS) developed by Gerber and Gurland (1989). It examines breakdown-repair sequences, including the aphasic patient's repair attempts as well as examining the strategies of his or her conversational partner to effect repair. Since this is the first published system for applying the methodology of CA to aphasia, we shall describe this in more detail shortly. Lesser and Milroy (1993) have also proposed applying a CA methodology in the assessment of pragmatic ability in aphasia, through use of a checklist of conversational abilities in turntaking, repair, embedding of sequences, routines and discourse markers.

PRAGMATIC ABILITIES IN APHASIA

Perhaps one of the most important issues to consider in the application of pragmatics to aphasia is the status of pragmatic ability. It has often been asserted (e.g., Holland, 1991; Glosser, Weiner and Kaplan, 1988) that pragmatic abilities are relatively preserved in contrast to linguistic abilities in aphasia and it is this which gives rise to aphasic subjects "communicating better than they talk". It appears that at the levels of propositional, logical and infererential structure of discourse, or larger scale "textual organisation" pragmatic ability is likely to be preserved. Evidence for this can be drawn from the studies reviewed in reference to the various pragmatic theories in the previous section. Thus although both Penn (1985) and Hawkins (1989) refer to the reduction of ability to apply Gricean conversational maxims effectively in aphasia, it appears that this can be seen as a secondary consequence of linguistic impairment rather than a primary pragmatic deficit in which knowledge of the maxims are lost. Studies of both comprehension and production of various speech acts indicate that most aphasic patients have no difficulty with indirectness, indicating that their ability to inference is still intact. Indeed this type of ability is one which appears to be compromised in right brain damage rather than left. Work looking at comprehension and production of text also provides evidence for the preservation of higher level pragmatic ability through macroprocessing, although performance is compromised by microprocessing which is influenced by linguistic deficits. Again the findings in the literature indicate that these sorts of pragmatic impairments arise primarily as a consequence of right brain damage. Studies using conversation analysis to investigate interactional aspects of language use demonstrate that while linguistic deficits may influence the interaction, aphasic subjects orient to the rules that govern conversation; they appear sensitive to turn-taking rules and the need to repair breakdowns. While the evidence supports the view that pragmatic abilities are relatively intact, in the discussion of the above investigations it becomes clear that the linguistic deficits of aphasic people are going to have an impact on their pragmatic abilities as they do not have the same resources to draw upon to communicate. It is therefore hard to argue that pragmatic ability in aphasia is unimpaired. It appears that the distinction that is required is whether pragmatic deficits are found in aphasia which are autonomous from deficits at other levels of language or whether they are manifestations of linguistic impairments. Autonomous pragmatic impairment is the type seen as a consequence of right brain damage and a review of research into the problems of this clinical population is provided by Code (1987). In aphasia it appears that pragmatic limitations are a consequence of the impact of linguistic impairment and this knowledge has consequences for the approach that we take in applying pragmatics to aphasia rehabilitation; the focus of the rest of the chapter.

Assessments of pragmatic ability in aphasia

Our review so far has incidentally mentioned a number of pragmatic assessments of aphasic individuals. These are included in Table 1. To complete this review of such assessments we also refer here briefly to the others shown in the Table.

In elicitation schedules the therapist conducts a structured interview in the clinic, using a test-like format, and the behaviour elicited from the patient is analysed against predetermined categories which are usually based rather loosely upon one or more pragmatic theories. The best known of the elicitation schedules is Holland's (1980) well standardised instrument for examining Communicative Abilities in Daily Living (CADL) which draws upon speech act theory. Within an interview patients are asked to perform a range of such acts, many of them using role play in which contextual cues are given through props such as a white coat and stethoscope to indicate that the therapist is taking the role of a doctor.

Another assessment which relies upon role play of everyday situations is the Everyday Language Test (ELT) devised by Blomert, Koster, Van Mier and Kean (1987). The tester describes a situation and the patient is required to role play the

| Elicitation schedules ("top-down") |
| Communicative Abilities in Daily Living (Holland, 1980) |
Everyday Language Test (Blomert et al, 1987)
Observation schedules ("top-down")
Pragmatic Protocol (Prutting and Kirchner, (1987)
Profile of Communicative Appropriateness (Penn, 1985)
Edinburgh Functional Communication Profile (Wirz et al, 1990)
Conversational Assessment (Copeland, 1989)
Observation schedules (scan summaries)
Functional Communication Profile (Taylor, 1969)
Communicative Effectiveness Index (Lomas et al, 1989)
Observation schedules ("bottom-up")
Assessment Protocol of Pragmatic-Linguistic Skills (Gerber and Gurland, 1989)
Checklist of Conversational Abilities (Lesser and Milroy, 1993)

TABLE 1. Some pragmatic assessments of aphasia

appropriate response. As with CADL appropriate contextual cues are provided for the patient in the form of real objects. The scoring procedure is based upon the patient's ability to produce two different types of information: necessary elements which consist of information required to make the message understood, and socially conventional elements which elaborate the message but are non-essential. Both elements are rather arbitrarily defined, and the classing of some elements as non-essential in a natural situation is questionable in this top-down approach to assessment.

The obvious advantage of elicitation schedules is that they provide a standardised assessment which allows comparison of performance between subjects. Holland (1980) provides norms for the CADL. One criticism which can be levelled at these elicited schedules is that while they are an attempt to provide a measure of naturalistic communication, the use of role play is far from natural; it is a behaviour which is not called upon in everyday interaction and which requires metacommunicative abilities which the aphasic patient may or may not have.

Both the CADL and the ELT can demonstrate that communicative performance, when the patient has the wealth of contextual cues typical of naturalistic interaction, is dissociable from linguistic performance when tested in isolation from context (as is typically measured by standardised aphasia assessments). However such assessments do not appear to offer any clear insight into the complex relationship which exists between language impairment and functional ability. These assessments appear to give a broad measure of functional communication which may be useful in measuring change but which do not provide obvious guidance for remediation.

In contrast to elicited schedules, the top-down observation schedules listed in Table 1 avoid the artificiality of role play with evaluation being based upon observation of naturalistic interaction. All these assessments involve analyses carried out on a naturalistic conversational sample collected from an interaction between the therapist and patient. As has already been mentioned all these assessments draw upon speech act theory, the limitations of which have been discussed above.

These profiles offer a finer-grained analysis of the aphasic client's ability to communicate in his or her social setting than the more global measure provided by the elicitation schedules, allowing the more specific identification of pragmatic strengths and weaknesses which could be useful in deciding on goals for rehabilitation. However the appropriacy or effectiveness judgements do not allow us to tease out precisely what is giving rise to the pragmatic problem. Prutting and Kirchner (1987) note that the Pragmatic Protocol is to be considered as a general communicative index and not as a diagnostic procedure. They state that treatment strategies for a specific patient will be based upon detailed assessment of the pragmatic parameters that have been judged inappropriate. These protocols thus appear to have a useful role in terms of a screening procedure, identifying what merits further investigation.

All of the procedures that we have reviewed require the therapist's expertise and time, but there are two other assessments which can be undertaken by relatives or carers. These have the attraction of providing a quicker, although necessarily very limited, summary of patients' communicatory success as seen by those who communicate to the greatest extent with them in a variety of settings. The Functional Communication Profile (Taylor, 1969) was the first assessment to tackle the issue of assessing functional communication. Although originally designed to be used by experienced therapists it has been used by carers and relatives of patients. It adopts a 9-point rating scale for five categories of communication: non-verbal (movement), speaking, understanding, reading and "other", the last including writing and calculating. Each category is divided into a number of activities, for example the understanding category includes recognition of family names and understanding complicated verbal directions. A conversion chart allows one to derive an overall percentage score of functional communication which can be used for test-retest puroposes.

Lomas, Pickard, Bester, Elbard, Finlayson and Zoghaib (1989) have developed an assessment specifically to be used by the "aphasic client's significant other" who spends enough time with him or her to make accurate judgements of communication performance. The Communicative Effectiveness Index (CEI) is intended to be a reliable and valid measure of functional communication which will quantitatively assess change in performance over time. After a brief explanation and training period, the carer is given a 16 item questionnaire concerned with everday communication. For each item he or she is asked to provide a rating on a visual analogue scale which runs from "not at all able" to "as able as before stroke". However Crockford (1991) in her evaluation of a number of measures of pragmatic abilities proposed that the emotional state and attitude of the spouse to his or her aphasic partner may subjectively influence ratings on CEI on different occasions. It is clearly important that clinicians are aware of such factors in the use of this tool.

BOTTOM-UP APPROACHES TO ASSESSMENT

In the assessments noted above, the focus is on a judgement of what the aphasic person does rather than looking at what arises from an interaction between patient and partner. However there is emerging realisation of the strengths that the conversation-analytic perspective has to offer to assessment of pragmatic abilities in aphasia. This is reflected in Gerber and Gurland's (1989) development of the Assessment Protocol of Pragmatic-Linguistic Skills (APPLS). Underlying their approach is the notion that an evaluation procedure should unite the assessment of pragmatic ability and the assessment of linguistic ability, recognising the synergy that exists between the two in natural language use.

The first stage of the APPLS analysis involves coding conversational turns which contain a signal to repair. The analysis embraces CA principles by using the conversational partner's response to decide what is and what is not a breakdown (i.e.,whether the partner initiates a repair sequence). Thus the success of interaction is used in preference to problematic judgements of appropriacy or communicative effectiveness prominent in other pragmatic assessments already discussed. Each breakdown is analysed in terms of whether the underlying reason for a breakdown was a "linguistic problem" (phonological, word-retrieval, or semantic-syntactic problem) or a "pragmatic problem" (contextually irrelevant, presuppositional-referencing problem, topic maintenance problem, topic shift problem, turn-taking problem or other). An analysis of the breakdown repair sequence is then made, noting both patient strategies in revision attempts and partner strategies in signalling repairs as well as examining the length of the whole process. A note is also made of both the linguistic structure of successful conversational turns which result in topic introduction and topic maintenance as well as their pragmatic function. The final part of the analysis involves quantitative and qualitative summaries of the patient's and partner's performance in the conversational sample. This includes a percentage score of the number of conversational turns in which breakdowns occurred. The APPLS is based upon observation with two different conversational partners in recognition of the potential influence of this on interaction.

One of the advantages of the APPLS procedure is that the findings of the analysis can be used to motivate remediation. This function makes such an approach a particularly valuable clinical tool which is not fulfilled by other pragmatic assessments, which appear more useful as a test-retest measure or as a screen of pragmatic ability. In particular what is laudable about this assessment is that it looks both at what gives rise to the breakdown, which could lead to direct therapy targeted at the linguistic deficits of the impairment, as well as examining the repair strategies used by both partners, which is useful for "environmental" therapy, developing useful strategies and eliminating unproductive ones.

There are clearly limitations with the APPLS. Although an integrated approach, looking at the relationship between linguistic and pragmatic impairments, is invaluable in offering remediation which has a functional impact, the division between linguistic problems and pragmatic problems is rather arbitrary. It seems likely that so-called pragmatic problems such as topic maintenance arise from so-called linguistic problems and that rather than this artificial dichotomy it is necessary to look at the possible cause-effect relationships that exist between the two. Furthermore in order to do this it is likely that much more rigorous linguistic investigation will be needed than the distinctions made by Gerber and Gurland of phonological, word-finding and semantic-syntactic problems. Such an approach is reflected in Perkins' (in progress) work investigating the relationships between aphasic subjects' cognitive neuropsychological impairments and their conversational ability using a CA methodology. The role of pragmatic assessments as measures supplementary to linguistic investigations is one which has been acknowledged by most researchers developing such tools (e.g., Taylor, 1969; Holland, 1980; Penn, 1985).

A further aspect of APPLS which could be developed is the analysis of strategies used by the aphasic client and his or her conversational partner. We have already mentioned Milroy and Perkins' (1992) model of collaborative repair in which the turns produced by the interlocutors in long repair sequences are seen as contributions to the resolution of the trouble source. Such a characterisation of repair appears to more successfully account for the jointly negotiated sequences than the treatment of each turn within a repair sequence as a strategy in isolation that has either succeeded or failed as Gerber and Gurland treat them. This underlines the problem of attempting to delineate a protocol or checklist of pragmatic behaviours; the central feature of CA methodology is that conversation is sequentially constructed and it is neccessary to look at the sequential environment to derive an analysis. This leads to problems in a checklist type protocol which does not take account of the sequential environment.

Lesser and Milroy (1993) have also proposed that a CA type analysis has much to offer to a pragmatic assessment of aphasic conversation. They have not designed a formal protocol but instead discuss an open-ended "bottom-up" procedure to approach varied and intransigent data using a CA methodology. A two-step procedure is proposed, the first step being a scan of a short conversational sample using a basic checklist in order to identify how the interlocutors are achieving conversation. This involves an examination of the conversational management procedures of turntaking, repair strategies, embedding sequences, opening and closing routines, and the use of discourse markers. The major questions to be asked are whether and how particular procedures are handled by the participants, and whether specific communicative problems can be attributed to an impairment, either directly or as a result of the interlocutors' responses. The authors propose that the detailed findings from this initial analysis will form the basis for both direct and indirect therapy.

The second stage of the informal procedure involves a comparison between speakers or between the same speakers in different circumstances. This kind of analysis may involve simple quantification, such as the proportion of major and minimal turns produced by the interlocutors in order to investigate the sharing of the conversational burden. It is at this second stage of analysis with the focus on the aphasic partner's contributions where an evaluation of effectiveness can be attempted. This evaluation relates not only to basic management procedures such as repairing and turn-taking, but to acts such as referring or requesting. Lesser and Milroy point out that while such categories are only used informally in CA, they can be treated in conversation analytic terms by examining their success or failure as evidenced by the subsequent interaction. If such acts do not result in effective communication then intervention may be focussed on developing strategies which are more effective.

As Lesser and Milroy themselves point out, their suggestions based on conversation analytic principles are intended to supplement and refine existing assessment schedules rather than to replace them. It is a principled means of approaching varied and intransigent data rather than a formal assessment procedure. Its strength is in giving an insight into the organisational strategies that are operating, thus providing a source of ideas for both direct and indirect therapy.

While the CA approaches to assessment of Gerber and Gurland and Lesser and Milroy are perhaps the strongest approaches in terms of providing information useful to the development of intervention programmes their greatest limitation is in measuring change over time. Clearly it is possible to examine the same features in conversation and see whether strategies being promoted in therapy are being harnessed in conversation. However the CA approach does not lend itself very well to quantification. A possible measure to evaluate change is an examination of the sharing of the burden of conversation by examining the number of major and minimal turns produced by the interlocutors. Gerber and Gurland's assessment includes a calculation of the percentage of conversational turns in which breakdowns occur. However a number of factors could influence whether this was a valid measure of improvement. Perkins (in progress) describes a relative who glosses over the need to repair when his aphasic partner runs into difficulties. While this avoids overt breakdowns it also results in the aphasic partner failing to make many contributions to the conversation. If a conversational partner developed such a strategy this would show up as an improvement in terms of the percentage of breakdowns although it would not be a particularly useful communicative strategy. Furthermore a more realistic goal for remediation may not be to reduce the number of breakdowns (which arise from linguistic deficits and which may be resistant to change) but rather to teach the conversational partners more successful strategies to deal with such breakdowns and therefore facilitate communication. This improvement would not be detected in terms of a percentage measure of conversational turns containing breakdowns.

A PRAGMATIC APPROACH TO APHASIA TREATMENT

The pragmatic approach to treatment represents a move away from more traditional therapy focussed on improving the aphasic patient's production and comprehension of normal and correct language structures. Instead remediation is concerned with the broader issue of optimal communication, with the patient using all resources available to maximise his or her communication ability. In practice this involves the aphasia therapist encouraging strategies already used which are successful, eliminating ones which appear to be counter-productive and teaching ones which the patient is not already using.

There are two areas developing from this broader orientation to remediation. The first of these involves direct work with the patient to help him or her develop strategies which optimise communicative ability. The second area is concerned with indirect therapy, focussed on the development of functional strategies in the patient's conversational partners. As we will see this dichotomy is not a discrete one, as work with the aphasic patient and his or her conversational partner may be run in tandem.

In order to decide on a treatment programme of communicative strategies it is necessary to assess the way in which the patient and his or her conversational partner are managing interaction. Such an assessment should yield information pertinent to the formulation of treatment plans. In our review of assessments it was proposed that approaches based on a conversation analysis methodology offered the greatest guidance for theoretically motivated therapy in this domain. An understanding of the way in which the conversationalists are managing conversation allows us to consider suitable strategies that could be used by both the interlocutors. Hence it appears that the use of a CA framework has a role in both direct and indirect therapy.

Direct Therapy

The basis for aphasia therapists to encourage development of effective strategies is identification of those used by aphasic patients who are good communicators despite their aphasia. Penn (1984) has developed a comprehensive taxonomy of compensatory strategies used by eighteen aphasic patients, the aim of which was to provide some insight into what compensatory strategies differentiate the performance of subjects viewed as overall good communicators from those who lacked efficiency in communication. In addition to strategies used by the aphasic patients' conversational partners (which are discussed in the next section in terms of indirect therapy) Penn identified six categories of strategies. The first of these was simplification strategies which could be observed primarily in syntactic terms (change in word order to bring salient aspects foremost, use of direct speech to avoid complex embedding, pronominalisation to refer back in discourse without a clear referent). It could also be seen in interactive behaviour where patients used shorter conversational turns in order to ensure effective and ongoing conversation.

The second category consisted of a range of elaboration strategies which compensated for lexical retrieval problems and included circumlocution, elimination strategies ("it's not Monday but Tuesday"), postmodification and coordination or embedding of clauses.

The third identified strategy was the use of repetition, including repetition of self and paraphrases of the proposition which the patient is trying to express. This type of repetition implies an active self-monitoring process. Other strategies in this category included stylistic repetition which was seen as a compensatory mechanism to achieve a descriptive function in the absence of adjectives and adverbs. A further repetition strategy involved repetition of all or part of the interlocutor's preceding utterance which Penn suggests seems to function as a facilitator to processing incoming information.

A further category of strategies was those serving to maintain fluency. These included the use of stereotypical chunks of language such as "you know what I mean", of filled pauses and discourse markers such as "well". Penn also identified what she called sociolinguistic strategies, devices which reflected a sensitivity on the part of the speakers to the sociolinguistic demands of the situation and to their own communicative inadequacies. Examples include self-correction, comment clauses, requests for clarification, pausing and topic shift at a point of breakdown. The final category involved the use of non-verbal strategies which Penn reports had multiple functions from merely supporting the verbal message to substituting or yielding additional information (Behrmann and Penn, 1984). Strategies included the use of alternative modes of communication such as writing as well as the use of gesture.

Penn found that some of the patients, although using identical compensatory strategies, appeared to be differentially effective in communication terms depending on the frequency and sequential placement of the devices used. Thus some dysfluencies were felt by the judges in the study to be enhancing communication in that they reflected normal searching and conversational management procedures. However if such behaviours occurred with high density or at unexpected junctures, they were regarded as inappropriate. The implications that Penn draws from the findings of her study are that for some patients it may be necessary to teach and develop new strategies, while the task may be to "undo" spontaneously acquired strategies which may in fact be hampering overall communicative competence.

While Penn's taxonomy has some useful things to say about how effective communication can be achieved by aphasic subjects, the question arises as to whether production of such devices as short conversational turns because of severely restricted syntax should be viewed as a compensatory strategy or whether it is rather better viewed simply as a symptom of the language impairment. The fuzzing of this distinction between strategies and symptoms can be seen in other work on pragmatic ability in aphasia. Holland (1991), while acknowledging that the language production errors made by aphasic patients could be argued to reflect their linguistic impairments, proposes, for instance, that the fact that in confrontation naming the most conventional errors are semantic paraphasias is indicative to some extent that pragmatic skills are retained in aphasia. While a semantic paraphasia may be more communicatively effective than the production of a neologism, it is more a reflection of the nature of the linguistic deficit than of differing compensatory strategies.

Making the distinction between strategies and symptoms has an important role to play in planning intervention, for while it may be possible to explicitly instruct a patient in the strategy of circumlocuting to communicate a lexical item that he or she cannot retrieve, it is unlikely that telling a patient that production of neologisms is not an effective strategy and that he or she should try for a phonemic paraphasia will be useful. Instead to tackle such issues it is necessary to employ direct therapy at a specific linguistic level perhaps using a cognitive neuropsychological framework to decide on appropriate therapy goals (such as is described in Lesser's Chapter 7 in this book).

Teaching of productive strategies has become the basis of a functional approach in aphasia therapy often referred to as "total communication". At the most basic level such an approach involves the use of materials in traditional therapy activities that is of relevance to the patient's everyday life and interests. Chapey (1981, 1986) developed Wepman's (1976) proposal that "indirect thought provoking therapy" is more beneficial than direct work on language into a programme of cognitive intervention in aphasia. This programme is concerned with the stimulation of cognition and memory, and of divergent and evaluative thinking as well as the convergent thinking which is so typical of the work carried out in aphasia clinics.

Green (1982, 1984) discusses the teaching of communicative strategies, suggesting that firstly it is necessary to determine whether there are any naturally occurring strategies which help communication. One might then test the usefulness of other strategies by trying them out in various situations, and then practising the strategies which have been shown to be useful. This can be done by instruction, demonstration and rehearsal in a variety of situations. While emphasising that therapists should try to develop those strategies that are as socially acceptable as possible, Green stresses that to communicate with any strategy is better than no communication at all. She suggests seven expressive strategies for the patient which include the use of an alternative communication system, giving the listener cues about the words that they are searching for through circumlocution or an initial letter, using fillers to maintain attention and requesting help. Strategies suggested for patients to aid their comprehension include requesting repetitions, telling the speaker that they do not understand, clarifying what was said by repeating it and asking questions. There is a large degree of overlap between Green's strategies and those proposed by Penn (1984).

Davis and Wilcox's (1981; 1985) PACE procedure (Promoting Aphasics' Communicative Efficiency) is a well known and widely practised technique to develop communicative strategies. The philosophy behind the technique is that therapy should resemble natural interaction. The way that this is achieved is by allowing both therapist and patient to participate equally as receivers and senders of messages with the exchange of new information between them. Feedback to the patient is based on the patient's success in communicating a message and is characteristic of receiver feedback occurring in natural settings. The patient can use any means to get the message across and the therapist models what are expected to be productive strategies for the patient in his or her turns as sender of the message. These principles are applied in an interaction in which the therapist and patient have a set of cards e.g. with object pictures; they take turns to communicate what the information is on their card, which the other cannot see. While the principles of PACE may be those of naturalistic communication, the task of communicating information such as the names of objects is much closer to didactic therapy involving picture naming than it is to imitating naturalistic communication. In a study examining the efficacy of modelling in PACE-therapy, Glindemann, Willmes, Huber and Springer (1991) concluded that the interactions underlying PACE therapy may be more determined by task and material as well as severity of aphasic impairment than by the verbal model of the therapist. Commenting on PACE, Howard and Hatfield (1987) suggest that some pragmatic therapists are confusing the aims of treatment (to improve everyday communication ability) with the means of treatment, and suggest that there is a place for direct instruction in the use of strategies.

Holland (1991), a pioneer in the development of the application of pragmatics in the assessment and treatment of aphasia, has developed a procedure in which such direct instruction in the use of strategies is provided. The procedure is called Conversational Coaching and involves the preparation of a short monologic "script" of six to eight words that is written to be slightly too difficult for the patient to produce, giving rise to the need for the use of strategies that the therapist and patient have already worked on. The script is intended to be a bridging framework to initiate a transfer of strategy use to patient-generated conversation. The script is practised and the therapist suggests how strategies might be used to get the script's content across. When this has been carried out a family member, who does not know the script, is called into the room, and the patient communicates the content while being coached by the therapist. Conversational Coaching also involves work with the family members in developing strategies, an approach considered further below. Holland suggests that the interaction is videotaped and this is viewed by all the participants to permit discussion which heightens awareness of which strategies are most effective and why.

All researchers concerned with the identification and teaching of compensatory communicative strategies refer to the use of non-verbal means of communication (e.g., Penn, 1984; Green, 1984; Holland, 1991). Garrett, Beukelman and Low-Morrow (1989) state that the focus on pragmatic competence opens the door for augmentative and alternative communication (AAC) approaches to communication for the individual who is unable to effectively generate messages through deficient modalities. The use of AAC, other than microcomputer assisted, is reviewed elsewhere in this book (Methé,

Huber and Paradis, this volume: 14-15), and here we mention briefly only two examples. Garrett et al. (1989) report on the development of an AAC system for an adult with Broca's aphasia that consisted of a package of techniques which could be drawn on to enhance communication. This included a word dictionary dealing with his favourite conversational themes which he was able to use to cue verbal output. An alphabet card was used to provide an additional cue to the conversational partner when the patient's production was unintelligible due to phonemic paraphasias and which the patient himself used to cue word retrieval. A card containing breakdown resolution cues, which guide the conversational partner through a structured form of twenty questions, was also provided to help in dealing with breakdowns in the conversation. The patient was also supplied with phrases to be used as conversational control strategies. This included phrases such as "I'm changing topic" as the patient had problems in signalling this and "we will stop" to indicate that breakdown resolution work should be abandoned. Finally the patient was encouraged to use writing and drawing as well as communicate with verbal output when possible.

The other form of AAC we will mention is the use of gesture in the regulation of conversational turns. In the review of the application of CA to aphasia it was noted that researchers had found that aphasic subjects used body position and arm movements to aid in their management of turn taking. These findings have implications for the teaching of such strategies to patients who do not already spontaneously use such techniques. The pragmatic rather than symbolic use of such gestures may make this suggestion more feasible to implement in aphasia than the use of symbol systems.

Indirect therapy

A pragmatic orientation to indirect therapy has also been called environmental intervention, in which the aim of therapy is not to work directly with the aphasic patient but instead effect change in the patient's communicative environment so that his or her potential can be maximised. The most effective way of doing this is by working with the aphasic person's most common conversational partners to develop strategies which will enhance communication. This type of approach aligns well with a conversation analysis perspective of interaction because, as we have seen from the review of the differing pragmatic theories, this is the only approach which treats discourse as a jointly negotiated endeavour.

One issue that arises in the examination of aphasic conversation as a joint endeavour is that the aphasic person's conversational partner is likely to have to take a greater part of the conversational burden in attempting to compensate for and facilitate the communication. This may arise in a number of different ways. Firstly the conversational partner will be required to take an active role in dealing with breakdowns that arise from the patient's linguistic impairments. Secondly the aphasic patient may take a passive role in the interaction, dealing with the language impairment by minimising his or her contribution to the interaction.

Silvast (1991) has reported that in the speech therapy clinic, even when engaging in conversation with the therapist, aphasic patients are forced into a passive communication role, with the therapist regulating the interaction by asking questions, while the patient responds as directed. As noted in the discussion of discourse analysis above, this is a common feature of institutional settings where one of the interlocutors can be seen to be in a position of authority. Silvast suggests that if conversation is to be used as a way to develop aphasic patients' communication then it is necessary to establish a more equal and creative conversation. Perkins (in progress) examined the sharing of conversational burden between aphasic subjects and their interlocutors by looking at the proportion of turns consisting of minimal responses (e.g., yes, mhm, right) and the proportion consisting of "major" turns. It was found for subject EN that while in conversation with a relative she relied predominantly on the use of minimal responses, with very few major turns. This strategy is effective in creating the impression of participating in conversation without requiring the use of impaired linguistic skills. However the use of this strategy was one which was influenced by the conversational partner, because when in conversation with the researcher the two interlocutors produced roughly equal proportions of major and minor turns. This is a clear demonstration of the influence of the conversational partner on the aphasic patient's communicative effectiveness. When EN ran into problems in completing a turn the need for repair work was glossed over by her relative who went onto a next turn. This resulted in the contents of EN's major turns not shaping the following interaction and gave rise to EN taking a passive role in the conversation. In contrast, in the conversation with the researcher problems in the interaction were dealt with collaboratively allowing EN to contribute to the conversation to a greater extent.

The findings of such studies make the relevance of work with the aphasic person's conversational partner clear. A lot of the research that we have already discussed in relation to teaching compensatory strategies to patients also takes account of the strategies used by patients' conversational partners. Holland (1991) includes work with the patient's partner in her Conversational Coaching technique. Penn (1984) in her taxonomy of compensatory strategies which relatives might employ includes the use of probe and yes/no questions, input simplification and modification and the use of phonemic and semantic cueing. Green (1984) discusses the strategies that could be used in this sort of work in more detail.

One of the issues which Green suggests should be tackled in assessment is collection of information from the aphasic patient's partner about pre-aphasia communication, including favourite topics of talk, style of communication, typical communication contexts and partners. Webster, Dans and Saunders (1982) proposed the use of the Critical Incident Technique (Flanagan, 1954) to collect information about the nature of spouse interactions pre- and post-stroke. They describe a study in which the spouses of aphasic patients were asked to recount situations which illustrate what communication was like before or after his or her stroke. From the transcript an analysis of who was the initiator for each of the communicative events was made. A comparison of reports of pre- and post-aphasia allows inferences to be made regarding the issue of whether interspouse communication patterns have changed with the onset of aphasia. Webster et al. propose that this technique may provide useful information which may influence the approach taken in the development of strategies that the spouse could use to enhance communication.

Green (1984) provides an inventory of strategies designed for use by a specific patient's partner to aid his understanding. It includes slowing down rate, pausing, use of less ambiguous sentences, use of strategies which alert the patient to a change of topic, placing main words at the end of the sentence, using short sentences, harnessing the use of stress and gesture and rephrasing if not understood. The techniques used to develop the use of these strategies by the partner included video feedback, audio-tape analysis, role-playing and observations of the therapist's communication.

The strategies discussed by Green focus on those that are designed to foster a patient's understanding. However, another important area involves the development of strategies which help the aphasic subject communicate despite production problems. This primarily involves the use of repair strategies to resolve any trouble sources that arise. In the same way that a useful starting point in the teaching of compensatory strategies to aphasic patients has been seen to be an analysis of the strategies already used, an analysis of the strategies used by the patient's conversational partner is also valuable. Flowers and Peizer (1984) used a referential communication task in which the aphasic subject was required to convey the contents of a picture to his or her partner. After asking an initial question, if the answer was unclear the partner could use any means to obtain the information required. Two measures were taken; task accuracy (on a two point scale dependent on the amount of information obtained) and task duration. A number of strategies to obtain information were noted, the most heavily used one being a series of guesses in which the partners tried to narrow down the answer by asking more and more specific questions. They suggest that in assessing the effectiveness of the strategies used by the aphasic partner the following questions should be asked: Does the partner make the best use of information provided by the aphasic subject? In asking successive yes/no questions does the partner proceed systematically from general to specific questions? Is the frequency of the partner's understanding checks appropriate? Does the partner prompt the aphasic partner to use other modalities which might help in conveying the message? Such observations can be used to decide on appropriate strategies to teach to the aphasic patient's partner. Milroy and Perkins' (1992) analysis of collaborative repair, already discussed above, would also provide information useful to the development of such strategies.

Some innovative approaches to the training of aphasic patients' partners in the use of strategies have been proposed by Newhoff, Bugbee and Ferreira (1981) and Miller (1989) in which one of the aims of therapy has been to develop an awareness in relatives of the communication limitations that his or her aphasic partner faces. Newhoff et al. (1981) described an eight session therapy programme based on the principles of PACE. The partner followed the PACE protocol with the clinician, with each taking turns as receivers and senders of the message. Each session was divided into three conditions. In the first section communication was limited to the use of gestures and pantomime; in the second section verbal communication was allowed but each participant had to communicate as the aphasic patient would have been expected to communicate; in the third section a combination of these two forms of communication was allowed. A comparison of a video-recording of the patient and partner in conversation before and after the programmes was made. Newhoff et al. report that the treatment affected discourse in ways not readily apparent to the statistical analysis that they carried out. They report that there appeared to be an increase in interaction in each dyad (a fact substantiated by increased communicative attempts by the aphasic patient post-treatment) and a change in the verbal strategies of each partner with an increase in discussion, disagreement and partner utterances indicating lack of comprehension.

Miller (1989) also discusses facilitating partners' communicative strategies using the central tactic of simulating the aphasic patient's predicament by placing the partner in a foreign language situation. To simulate impaired comprehension Miller suggests the use of passages and questions in which there are some recognisable or English language cues incorporated. Either nonsense words are used or languages which have certain affinities with English. In tackling the task strategies that aid comprehension can be used. The passage can be reread until the person manages to gain full insight into the meaning and each time a cue can be added such as the meaning of a key word or a non-verbal signal through the use of a real object, a gesture or tone of voice. The effectiveness of cues can be discussed with the partner as well as the significance of prior information and how meanings can be inferred. In work on strategies to deal with production impairments Miller suggests that situations are discussed where formal language knowledge has let the partners down (such as being on holiday in a foreign country) exploring their reactions and solutions. To further approximate the difficulties of the aphasic partner the relative can be asked to carry out verbal language tasks using only a limited range of words or syntactic structures. After this, discussion can be focussed on strategies which can be harnessed to overcome such difficulties and which support from the therapist proved helpful. The major aim of this novel approach is to help the partner achieve a better insight into the aphasic person's state and develop a repertoire of strategies that are potentially useful to communication for the two of them.

WAYS FORWARD IN A PRAGMATIC APPROACH TO TREATMENT

In this review we have stressed the importance of critically evaluating the theoretical principles that are applied to pragmatic intervention in aphasia. We concluded from the review of the relevant literature that most aphasic people appear to have intact knowledge of how to use language. However this is not to say that their pragmatic ability is intact. It appears that pragmatic limitations arise as a consequence of primary linguistic deficits. This led us to propose that the most useful framework to use in an investigation of pragmatic ability is one which allows the teasing out of the relationships between specific deficits and their interactional consequences. We propose that the best framework to achieve this is that of conversation analysis. It allows the investigation of the way that aphasic patients and their conversational partners manage the interaction; the cause of conversational breakdowns can be related to specific, independently identified language impairments.

Although the application of CA to aphasia is in its infancy, its focus on both partners in an interaction aligns well with the philosophy apparent in the literature of working not only with the aphasic patient but also with his or her spouse or carer. The identification and teaching of strategies is reported in the literature without explicit reference to CA (e.g., Penn, 1984; Green, 1984). However a CA-style analysis can supplement such work by enabling the therapist to identify quite precisely both facilitative strategies which need to be taught and maladaptive strategies which need to be discouraged.

This review of pragmatic approaches to treatment has focussed exclusively on the teaching of strategies and the use of alternative or supplementary modes of communication with no mention of direct work to deal with specific language deficits. While most workers acknowledge the importance of an integrated approach to treatment it is necessary to identify the most fruitful way to link these two different ways of working. We would propose that a CA methodology provides a way of approaching this. Through the identification of the impact that specific impairments have on an aphasic patient's conversational ability it is possible to identify targets of direct therapy which are likely to have the most impact on functional communication. For example if the production of phonemic paraphasias leads to protracted conversational breakdowns it may be worth selecting this as a target for direct treatment. Such therapy could be combined with "pragmatic treatment" to help the patient and partner to deal with their occurrence by discussing repair strategies and the use of other modes to communicate the word. However if phonemic paraphasias occurring in conversation do not lead to breakdowns then direct treatment may be considered as unnecessary. Perkins (in progress) is currently investigating more extensively the relationship between cognitive neuropsychological abilities and conversational abilities in aphasia and it is hoped that such research will lead eventually to the integration of pragmatic approaches to assessment and remediation with more direct therapy approaches.

References

Ahlsen, A. 1985. Discourse patterns in aphasia. *Gothenberg monographs in linguistics*, 5, University of Gothenburg.
Austin, J. 1962. *How to do things with words*. Cambridge: Harvard University Press.
Bates, E., Hamby, S. and Zurif, E. 1983. The effects of focal brain damage on pragmatic expression. *Canadian Journal of Psychology*, 37: 59-84.
Behrmann, M. and Penn, C. 1984. Non-verbal communication of aphasic patients. *British Journal of Communication Disorders*, 19: 155-168.
Berko-Gleason, J., Goodglass, H., Obler, L., Green, E., Hyde, M. and Weintraub, S. 1980. Narrative strategies of aphasic normal-speaking subjects. *Journal of Speech and Hearing Research*, 33: 370-382.
Blomert, L., Koster, C., Van Mier, H. and Kean, M. 1987. Verbal communication abilities of aphasic patients: The everyday language test. *Aphasiology*, 1: 463-474.
Chapey, R. 1981. Divergent semantic intervention. In R. Chapey (ed.) *Language intervention strategies in adult aphasia*. (pp.155-167) Baltimore: Williams and Wilkins.
Chapey, R. 1986. Cognitive intervention: Stimulation of cognition, memory, convergent thinking, divergent thinking and evaluative thinking. In R. Chapey (ed.) *Language intervention strategies in adult aphasia*, 2nd edition (pp.215-238) Baltimore: Williams and Wilkins.
Chapman, S. and Ulatowska, H. 1989. Discourse in aphasia: Integration deficits in processing reference. *Brain and Language*, 36: 651-668.
Clark, H. and Schaefer, E. 1987. Collaborating on contributions to conversation. *Language and Cognitive Processes*, 2: 19-41.
Clark, H. and Schaefer, E. 1989. Contributing to discourse. *Cognitive Science*, 13: 259-294.
Code, C. 1987. *Language, aphasia and the right hemisphere*. Wiley: Chichester
Conway, N. 1990. Repair in the conversation of two dysphasics with members of their families. Unpublished BSc. dissertation, University of Newcastle upon Tyne.
Copeland, M. 1989. An assessment of natural conversation with Broca's aphasics. *Aphasiology*, 3: 391-406.
Coulthard, R. 1985. *An introduction to discourse analysis*. London: Longman.
Coulthard, R. and Brazil, D. 1979. Exchange structure. In R. Coulthard and M. Montgomery (eds.) *Studies in discourse analysis* (pp. 82-106) London: Routledge and Kegan Paul.
Crockford, C. 1991. Assessing functional communication in aphasic adults: A comparison of three methods. Unpublished BSc. dissertation, Newcastle University.
Davis, G. and Wilcox, M. 1981. Incorporating parameters of natural conversation in aphasia treatment. In R. Chapey (ed.) *Language*

intervention strategies in adult aphasia. (pp. 169-193) Baltimore: Williams and Wilkins.
Davis, G. and Wilcox, M. 1985. *Adult aphasia rehabilitation: Applied pragmatics.* Windsor: NFER-Nelson
Flanagan, J. 1954. The critical incident technique. *Psychological Bulletin,* 51: 327-358.
Foldi, N. 1987. Appreciation of pragmatic interpretations of indirect commands: Comparison of right and left hemisphere brain-damaged patients. *Brain and Language,* 31: 88-102.
Flowers, C. and Peizer, E. 1984. Strategies for obtaining information from aphasic persons. In R. Brookshire (ed.) *Clinical aphasiology conference proceedings* (pp. 106-113) Minneapolis:BRK.
Gardner, H., Brownell, H., Wapner, W. and Michelow, D. 1983. Missing the point. In E. Perceman (ed.) *Cognitive processing in the right hemispherere.* (pp. 169-191) New York: Academic Press.
Garman, M. and Edwards, S. 1989. Case study of a fluent aphasic. In P. Grunwell and A. James (eds.) *The functional evaluation of language disorders.* (pp.163-181) London: Croom Helm.
Garrett, K., Beukelman, D. and Low-Morrow, D. 1989. A comprehensive augmentative communication system for an adult with Broca's aphasia. *Augmentative and Alternative Communication,* 5: 55-61.
Gerber, S. and Gurland, G. 1989. Applied pragmatics in the assessment of aphasia. *Seminars in Speech and Language,* 10: 263-279.
Glindemann, R., Willmes, K., Huber, W. and Springer, L. 1991. The efficacy of modelling in PACE-therapy. *Aphasiology,* 5: 425-429.
Glosser, G., Weiner, M. and Kaplan, E. 1988. Variations in aphasic language behaviours. *Journal of Speech and Hearing Disorders,* 53: 115-125.
Graddol, D., Cheshire, J. and Swann, J. 1987. *Describing language.* London: Open University.
Green, G. 1982. Assessment and treatment of the adult with severe aphasia: Aiming for functional generalisation. *Australian Journal of Human Communication.* 10: 11-23.
Green, G. 1984. Communication in aphasia therapy: some of the procedures and issues involved. *British Journal of Disorders of Communication,* 19: 35-46.
Grice, H. 1975. Logic and conversation. In P. Cole and J. Morgan (eds.) *Syntax and semantics 3: Speech acts.* (pp.41-58) New York: Academic Press.
Guilford, A. and O'Conner, J. 1982. Pragmatic functions in aphasia. *Journal of Communication Disorders,* 15: 337-346.
Hawkins, P. 1989. Discourse aphasia. In P. Grunwell and A. James (eds.) *The functional evaluation of language disorders.* (pp.183-199) London: Croom Helm.
Holland, A. 1980. *Communicative abilities in daily living: A test of functional communication for aphasic adults.* Baltimore: University Park Press.

Holland, A. 1991. Pragmatic aspects of intervention in aphasia. *Journal of Neurolinguistics*, 6: 197-211.
Howard, D. and Hatfield, F. 1987. *Aphasia therapy: Historical and contemporary issues*. London: Lawrence Erlbaum Associates.
Huber, W. 1990. Text comprehension and production in aphasia: Analysis in terms of micro- and macroprocessing. In Y. Joanette and H. Brownell (eds.), *Discourse ability and brain damage: Theoretical and empirical perspectives*. (pp. 154-179) New York: Springer-Verlag.
Huber, W. and Gleber, J. 1982. Linguistic and non-linguistic processing of narratives in aphasia. *Brain and Lnaguage*, 16: 1-18.
Johnson-Laird, P. 1983. *Mental models*. Cambridge: Cambridge University Press.
Kohlert, P. 1979. Zur neurolinguistischen Diagnose von Sprachverstandnisstorungen bei Aphasie. Medical dissertation, RWTH, Aachen. Cited in W. Huber (1990).
Leech, G. and Thomas, J. 1990. Language, meaning and context: Pragmatics. In N. Collinge (ed.) *An encyclopedia of language*. London: Routeledge.
Lesser, R. and Milroy, L. 1993. *Linguistics and aphasia: Psycholinguistic and pragmatic aspects of intervention*. London: Longman.
Levinson, S. 1983. *Pragmatics*. Cambridge: Cambridge University Press.
Lomas, J., Pickard, L., Bester, S., Elbard, H., Finlayson, A. and Zoghaib, C. 1989. The communicative effectiveness index: Development and psychometric evaluation of a functional communication measure for adult aphasia. *Journal of Speech and Hearing Disorders*, 54: 113-124.
McTear, M. 1985. *Childrens' conversations*. Oxford: Basil Blackwell.
Miller, N. 1989. Strategies of language use in assessment and therapy for acquired dysphasias. In P. Grunwell and A. James (eds.) *The functional evaluation of language disorders*. (pp. 97-124) London: Croom Helm
Milroy, L. and Perkins, L. (1992) Repair strategies in aphasic discourse: Towards a collaborative model. *Clinical Linguistics and Phonetics*, 6: 27-40.
Newhoff, M., Bugbee, J. and Ferreira, A. 1981. A change of PACE: Spouses as treatment targets. In R. Brookshire (ed.) *Clinical aphasiology conference proceedings*. (pp. 234-243). Minneapolis: BRK Publishers.
Nichols, M., Obler, L., Albert, M. and Helm-Estabrooks, N. 1985. Empty speech in Alzheimer's disease and fluent aphasia. *Journal of Speech and Hearing Research*, 28: 405-410.
Penn, C. 1984. Compensatory strategies in aphasia: Behavioural and neurological correlates. In K. Grieve and R. Griesel (eds.), *Neuropsychology III*. Pretoria: Monicol.
Penn, C. 1985. The profile of communicative appropriateness: a clinical tool for the assessment of pragmatics. *South African Journal of Communication Disorders*, 32: 18-23.

Perkins L. (in progress) The impact of cognitive neuropsychological impairments on conversational ability in aphasia: An investigation. PhD thesis, University of Newcastle upon Tyne.

Prinz, P. 1980. A note on requesting strategies in adult aphasics. *Journal of Communication Disorders*, 13: 65-73.

Prutting, C. and Kirchner, D. (1987) A clinical appraisal of the pragmatic aspects of language. *Journal of Speech and Hearing Disorders*, 52: 105-119.

Sacks, H., Schegloff, E. and Jefferson, G. 1974. A simplest systematics for the organisation of turn-taking for conversation. *Language*, 50: 696-735.

Schegloff, E., Jefferson, G. and Sacks, H. 1977. The preference for self-correction in the organisation of repair in conversation. *Language*, 53: 361-382.

Schegloff, E. 1982. Discourse as an interactional achievement: Some uses of 'uh huh' and other things that come between sentences. In D. Tannen (ed.) *Georgetown University Roundtable on language and linguistics* 93. (pp.71-93) Washington: Georgetown University Press.

Schienberg, S. and Holland, A. 1980. Conversational turn taking in Wernicke's aphasia. In R. Brookshire (ed.) *Proceedings of the Clinical Aphasiology Conference*, (pp.111-116) Minneapolis: BRK Publishers.

Searle, J. R. 1979. *Expression and Meaning*. Cambridge: Cambridge University Press.

Searle, J. R. 1986. Introductory essay: notes on conversation. In D.G.Ellis and W.A.Donahue (eds.) *Contemporary issues in language and discourse processes*. Hillsdale: Lawrence Erlbaum Associates.

Silvast, M. 1991. Aphasia therapy dialogues. *Aphasiology*, 5: 383-390.

Sinclair, J. and Coulthard, R. 1975. *Towards an analysis of discourse: The English used by teachers and pupils*. London: Oxford University Press.

Skinner, C., Wirz, S., Thompson, I. and Davidson, J. 1984. *Edinburgh functional communication profile: An observation procedure for the evaluation of disordered communication in elderly patients*. Winslow: Winslow Press.

Stachowiak, F., Huber, W., Poeck, K. and Kerschensteiner, M. 1977. Text comprehension in aphasia. *Brain and Language*, 4: 177-195.

Stubbs, M. 1983. *Discourse analysis*. Oxford: Basil Blackwell.

Taylor, M. 1969. *The functional communication profile*. New York: New York University Medical Center, Institute of Rehabilitation Medicine.

Taylor, T. and Cameron, D. 1987. *Analysing conversation: Rules and units in the structure of talk*. Oxford: Pergamon.

Ulatowska, H., Allard, L. and Chapman, S. 1990. Narrative and procedural discourse in aphasia. In Y. Joanette and H. Brownell (eds.) *Discourse ability and brain damage: Theoretical and empirical perspectives*, (pp. 180-198) New York: Springer-Verlag.

Van Dijk, T. 1972. *Some aspects of text grammars*. The Hague: Mouton.

Webster, E., Dans, J. and Saunders, P. 1982. Descriptions of husband-wife communication pre- and post-aphasia. In R. Brookshire (ed.) *Clinical aphasiology conference proceedings*, (pp.64-73) Minneapolis: BRK Publishers.

Wepman, J. 1976. Aphasia: Language without thought and thought without language. *ASHA*, **18**: 131-136.

Weylman, S., Brownell, H., Roman, M. and Gardner, H. 1989. Appreciation of indirect requests by left and right brain damaged patients: The effects of verbal context and conventionality of wording. *Brain and Language*, **36**: 580-591.

Wilcox, M. and Davis, G. 1977. Speech act analysis of aphasic communication in individual and group settings. In R. Brookshire (ed.) *Clinical aphasiology conference proceedings*. (pp.166-174) Minneapolis: BRK Publishers.

Wirz, S., Skinner, C. and Dean, E. 1990. *Revised Edinburgh Functional Communication Profile*. Arizona: Communication Skill Builders.

Part III Cognitive foundations of aphasia rehabilitation methods

The chapters in this section review the current applications of aspects of cognitive (neuro)psychological modelling to the practice of aphasia therapy, as it concerns word-retrieval difficulty (Chap. 7), naming disorders (Chap. 8), reading and writing impairments (Chap. 9), and language comprehension and production problems at the sentence level (Chap. 10). In Chapter 7, Lesser argues forcibly in favour of encouraging aphasia therapists to become practitioner-researchers, so that clinical practice may inform research in rehabilitation and vice versa. In Chapter 8, Kremin investigates the import of recent developments in anatomo-clinical and cognitive neuropsychological approaches to the rehabilitation of naming disorders. The major contribution of these approaches has been to specify the processing level at which naming has been disrupted. In Chapter 9, the same approach is brought to bear on the investigation of reading and writing impairments. In Chapter 10, Byng and Lesser make the point that, while cognitive approaches have provided a more detailed analysis of the locus and nature of the deficits, very little attempt has been made to explain how therapy may be linked to changes in the identified underlying processes. Indeed, while the application of cognitive neuropsychological modelling may have considerable potential for aphasia therapy, a theory-based rationale for planning intervention is still wanting. In fact, no new techniques have been developed; there is now only a rationale for selecting among the traditional materials available for a particular deficit. As these authors rightly point out, the contribution of cognitive psycholinguistically oriented research has been a greater specification of the deficit so that it can be addressed more precisely, rather than providing altogether new ways of dealing with aphasia.

7 Cognitive neuropsychology and practitioner-researchers in aphasia therapy

Ruth Lesser

The aim of this paper is to review the current application of some aspects of cognitive neuropsychological (psycholinguistic) modelling to the practice of aphasia therapy, in particular in the United Kingdom. The paper aims to show how cognitive neuropsychology is facilitating the integration of research and clinical practice, through the encouragement of "practitioner-researchers" i.e. aphasia therapists who regard the application of research methodology as integral to good clinical practice.

The area I have selected to demonstrate this is in remediation of one common form of aphasic impairment, word-retrieval difficulties. I also attempt to reply to criticisms which have been levelled against the cognitive neuropsychological approach to aphasia therapy. I suggest moreover that this has potential, not only as a useful addition to the resources of aphasia therapists world-wide, but also that systematic reporting of the response to such therapy can feed back into the greater understanding of the nature of language processing.

RESEARCH AND THERAPISTS

It is clear that concepts and connotations of what "research" is vary amongst therapists. For some, research appears to be considered as a professional undertaking in its own right, on which down-to-earth practitioners of aphasia therapy have no time to spend or skills to perform. This attitude is reflected in advice such as that provided in the 1991 guide-lines of the British College of Speech and Language Therapists for planners of new postgraduate programmes seeking professional accreditation. These guidelines ask the planners to make a clear distinction between research and clinical components of the programmes. They state that these two components are not necessarily easily combined, that competence in research does not always equate with clinical competence and that pressure to increase research training may be at the expense of time spent on clinical/professional development.

A similar attitude is reflected in a review of aphasia therapy from Italy:

> As in all disciplines, in speech and language therapy there is a time for research and a time for application.......The clinical

> speech pathologist['s].....task is to apply what he has learnt, to learn from his clinical experience and provide treatment for all patients who request it. It is not his task to demonstrate that what he does is effective. It would be the same as asking the general practitioner to demonstrate each time he prescribes a drug that the drug is effective. (Basso, 1989a: 79)

This quotation reflects the influence on some aphasia therapists of the current medical model for research, making a specific analogy between the aphasia therapist's treatment and the prescription of a drug. Basso, accepting this medical model for aphasia, puts aphasia therapists in a different category from researchers. Therapists are seen as receivers of recommendations for therapy, rather than contributors to its understanding; the research into therapy is conducted by people who have no current commitment to its practice but are identified distinctly as "researchers" not therapists. Furthermore the therapist has no responsibility for demonstrating the effectiveness of the intervention used with the individual patients; the presumption is that somehow this effectiveness will already have been established by the researchers who devised the programme.

A diametrically opposed view is taken by McConkey (1991) (and by the present author). He argues that the "onus rests with practitioners to demonstrate effectiveness" (p. 5) in order to avoid the perpetuation of time-honoured but perhaps time-wasting practices, and goes so far as to claim that "good practice and research are synonymous" (p. 11). He asserts that if "practitioners are committed to giving their clients the best possible service, then ongoing appraisal of present work practices is essential" (p. 6). McConkey reviews the broad spectrum of what research can consist of (from something as apparently simple as taking a case history to longitudinal professional opinion-poll taking), and points out that therapists can be practitioner- researchers without adopting the values of professional researchers whose success is measured by the number of papers they publish and the size of their research grants.

One of the merits of the application of cognitive neuropsychology to the study of aphasia is that it has opened the door to aphasia therapists becoming such practitioner-researchers by providing systematic ways of testing model-based hypotheses about individual patients' language processing. It has been accompanied by an upsurge of interest in evaluating change through experimental designs for single-case studies in language pathology (e.g. Coltheart, 1983; Pring, 1986; McReynolds and Thompson, 1986; Nicholas and Helm-Estabrooks, 1990). Both these developments have provided a practical means of implementing McConkey's advocacy that therapists should conceive of themselves as simultaneously undertaking research as an integral part of clinical practice. In contrast to the approach advocated by Basso, the hypothesis-testing method sees the work to be carried out with each patient as a mini-research exercise; a necessary part of this is evaluation of whether the intervention achieved its stated objectives, essential as a means not only of justifying clinical practice but also of acquiring insights into the nature of aphasia. The only gap between the "researcher" and the "applier" (to take Basso's distinction) is in the time each can devote to this study. The practising clinician contributes to the development of the field, and brings to it the benefit of an extending and intimate knowledge of the evolving nature of individuals' disorders, rather than receiving prescriptions formulated elsewhere. Demonstration of effectiveness or otherwise with patients who share a common feature amongst the individualisms, and refinement of our understanding of aphasia, can be achieved through compilation of these clinical studies. Hence arises the need for practising clinicians to be allowed time and facilities for publishing their observations, without which the wealth of experience and insight gained through continuing interaction with evolving patterns of aphasia is lost to a wider circle.

Such an approach distances aphasia therapy from a medical frame of reference in which randomised controlled trials are seen as the desirable tests of the effectiveness of treatments (see Howard, 1986, for a discussion of the inappropriateness of using this methodology in aphasia research). Since most aphasia patients come under medical care, and all have neurological conditions, this divorce of aphasia therapy from a medical framework may seem initially to be jarring. The development of cognitive neuropsychology (despite the "neuro" in the name) has in fact made it easier for therapists to look more critically at the current neurological dominance on interpretations of aphasia, such as that applied in the Boston Diagnostic Aphasia Examination (Goodglass and Kaplan, 1983). Although they use data from brain-damaged people, cognitive neuropsychological models of language processing describe a functional architecture of mental activities, which is essentially independent of localisation in the brain. This situation may change if it does prove possible to map language processing on to brain functions (Petersen et al., 1988), but at present the independence of psycholinguistic models from a neural substrate has its own appeal to many aphasia therapists. The reality of clinical practice in aphasia therapy in many countries is that patients are referred for therapy with minimal or no neurological assessments (in a typical provincial hospital in the UK, for example, fewer than 10% of aphasic patients are referred by neurologists); even when brain scans are available they are likely to be CT or MRI scans providing anatomical rather than functional localisation. A cognitive neuropsychological approach to interpreting a patient's language processes therefore currently has more relevance to many aphasia therapists than a neurological one.

I shall now attempt to demonstrate the usefulness of the cognitive neuropsychological approach to the aphasia therapist (and in particular to illustrate how the therapist can be a practitioner-researcher) by looking at one key trouble source for many aphasic people, word-finding (as assessed through confrontation naming). I shall restrict my examples to a few of the studies undertaken by UK speech-language therapists, although excellent applications of psycholinguistic modelling to aphasia therapy are being made elsewhere (e.g. from Baltimore, Hillis, 1989; Mitchum and Berndt, in press; and, from Brussels, Bachy-Langedock and de Partz, 1989).

MODEL-BASED THERAPY FOR NAMING DIFFICULTIES

Models of single-word processing have been acknowledged to be restricted in scope, underspecified (Ellis and Young, 1988), descriptive rather than explanatory (Seidenberg, 1988), in need of supplementation by the dynamics of working memory (Saffran and Martin, 1990) or on-line processing (Tyler, 1987), controversial in detail (Allport, 1984) and based on non-naturalistic tasks (Lesser, 1989). Despite all these accepted criticisms they have a key value for aphasiology in their origin in the study of normal subjects, rather than being entirely dependent for their development on studies of abnormal users of language. They also have an empirical justification in aphasia therapy in providing a guiding (and easily visualised) framework for the analysis of individual patients' disorders, which is so far compatible with clinical data.

The part of the model which has guided therapy for word-finding difficulties is that which involves semantics, the phonological output lexicon and the phonological assembly buffer (see Figure 1), with the equivalent output

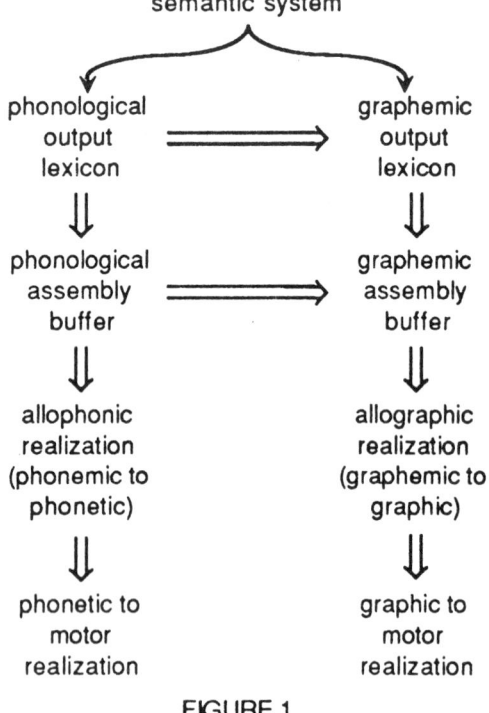

FIGURE 1

components for writing seen as potential back-ups for those patients who sometimes attempt to gesture in sky-writing their knowledge of the written form of words they cannot find orally. Proposals have been made as to how different kinds of behaviours during naming may indicate a predominance in troubles of different components of this word-finding procedure (see Figure 2).

Problem	Symptoms
SEMANTIC SYSTEM	semantic paraphasias semantic paralexias semantic paragraphias errors on semantic discrimination tests acceptance of semantic associates some categories more affected than others? (especially abstract words?)
PHONOLOGICAL LEXICON	circumlocutions exploratory phonemic paraphasias good recognition of target word strong effect of word frequency good oral reading of (regular) words
PHONOLOGICAL ASSEMBLY BUFFER	target-oriented phonemic paraphasias good recognition of target word poor imitation of target word strong effect of word length
ALLOPHONIC REALIZATION	search for phoneme shapes effect of pragmatic immediacy on articulation
MOTOR REALIZATION	consistent articulatory difficulties

FIGURE 2

A seminal group study on the treatment of anomia by a British speech therapist and his colleagues (Howard et al., 1985) has been taken as an endorsement of the value of "semantic therapy" in cases of naming difficulty. This therapy uses no overt naming by the patients, but employs tasks which emphasise the meanings of words rather than their phonological shapes. A group of British speech therapists and a psychologist (Marshall, Pound, White-Thomson and Pring, 1990) have published the results of a test of the value of "semantic therapy". As this provides an example of the application of psycholinguistic modelling to therapy "in the context of normal clinical practice (p. 178)" it is appropriate to describe it here.

The three patients studied in detail were diagnosed in terms of a version of the model shown in Figure 1. One of them, RS, had difficulties in the route connecting the semantic system with the phonological output lexicon. Two other patients, IS and FW, had disorders affecting the semantic system itself. Ten months after his stroke, RS was given a total of about three hours of therapy, aimed at reinforcing links between word semantics and phonology. The therapy consisted of reading aloud the names of sets of five semantically associated words and selecting the appropriate ones to match given pictures. On re-test, RS was significantly better at naming treated than untreated items, an effect which was maintained a month later. Pring et al. (1990) acknowledge that a possible interpretation of such results is that the improvement was due to repeated oral naming of the picture rather than to the semantic decisions involved. They believe this was not the case, however, since a substantial improvement was also noted in other patients with the same diagnosis given the same therapy, but whose reading aloud was unreliable.

The two other patients in the Marshall et al. study also improved after short periods of therapy targeted at their semantic disorder. This comprised matching semantically related written words to pictures or spoken definitions. With IS two staggered sets of treated words showed significant improvement after about two hours of therapy spread over two weeks on each, with some generalisation to untreated words. Such a result is consistent with the view that the patients' semantic difficulties were due to problems in accessing the specific verbal semantics of pictured items as distinct from their general semantics; this resulted in only partial comprehension and retrieval in which semantic associates were confused. Some generalisation to nontreated words might be expected in such circumstances, given the overlapping nature of components of meaning. After about three hours of therapy over six weeks, FW had not improved significantly, but the addition of a further equivalent period of therapy resulted in significantly improved naming on both treated and untreated words combined—again demonstrating generalisation. As Marshall et al comment, the real utility of such a procedure needs to be measured against generalisation to spontaneous speech, something which was not tested in this study. Such a lack reflects the present state of development in applying psycholinguistic modelling to aphasia rather than a fundamental restriction in it.

Marshall et al.'s study suggests that distinguishing patients' naming disorders according to whether they concern primarily semantics, or the phonological output lexicon, or access to it from semantics, should guide the selection of therapy strategies. Nettleton and Lesser (1991) have also attempted to address this question, through comparing the effects of model-appropriate and model-inappropriate therapy. Six patients were treated whose naming scores fell between 10% and 42% on the Boston Naming Test. PD and FF were considered to have a semantic basis for their naming difficulties in terms of the criteria shown in Figure 2; they resembled IS and FW in the Marshall study in this respect. PD and FF were given semantic therapy over a period of eight weeks (16 hours). Two other patients, DF and MC, were considered to have a disorder implicating the phonological output lexicon (or access to it) in terms of Figure 2. They were given eight weeks of "phonological therapy", which focused on oral repetition of picture names and judgement of rhymes, the aim being to get the patient to think about the word's phonological form rather than its meaning. The third pair of patients, MH and NC, had a naming disorder compatible with the description given in Figure 2 of trouble at the level of the phonological output buffer. As a test of the general applicability of semantic therapy which Howard et al's study had reported as effective (at least in the short term) with a group of unselected aphasic patients, these last two patients were given eight weeks of semantic therapy, although this was not appropriate in terms of the model.

Results from the period of therapy in all the cases were compared with those from the period of eight weeks before the therapy was instituted. One of the patients given model-appropriate semantic therapy, at six months post stroke, improved significantly both on treated and untreated items. The other (three years post stroke) showed a trend towards improvement which did not reach statistical significance, but now produced fewer semantic paraphasias on naming, and ones which were independently rated as being closer in meaning to the target. The two patients who were given model-appropriate phonological therapy (at one and eight years post stroke respectively) improved significantly, with one of them showing unpredicted generalisation also to untreated items. Neither of the two patients given model-inappropriate semantic therapy (at eight months and three years post stroke respectively) improved on naming.

The final example I give in this section is also by a practising UK speech-language therapist, reporting on applying a psycholinguistic model to assist in designing therapy for one of the cases seen as part of her normal workload. Cook's (1991) patient, MD, was given therapy two years and nine months after a stroke, which had left him with an aphasia characterised by nonfluency, word-finding difficulties and verbal apraxia. He had previously received 15 months of individual therapy directed at his verbal apraxia and use of verbs, and of semantic therapy in a group setting in which functionally relevant communication abilities were also emphasised. MD attempted to overcome his word-finding difficulties in spontaneous speech by circumlocuting or using appropriate gesture; he also sometimes showed the ability to sky-write the initial letter of a word he was seeking, and could sometimes give the number of letters and judge the length of the word. He responded well to phonic cues when unable to name.

Cook's psycholinguistic assessment of MD indicated a largely preserved semantic system, although with some difficulty in judging synonyms which were of low imageability. Both the output lexicons for phonology and orthography were considered to be impaired. Cook also established that MD had problems in using processes which mediate between these two lexicons, albeit indirectly, through association of graphemes and phonemes. The evidence for this was that he had difficulty in sounding out unfamiliar written words, and consequently in judging whether pairs of non-words were homophonic. Cook therefore argued that enhancement of grapheme and phoneme associations might enable the partial information in the two damaged output lexicons to be combined. If he could convert what output information he had available in the graphemic lexicon into usable form as phonemes, he could provide himself with the phonic cues which enabled him to access items in his phonological lexicon when other people provided them. Unlike the therapies described in the previous section this was, therefore, aimed at providing a cognitive relay, rather than aiming at reactivation through stimulation.

Treatment was given in two stages, the first focussing on grapheme-phoneme associations, the second on identification of the first sound in words. Six weeks were spent on each type of therapy. Cook emphasises that this treatment took place within a clinical setting, and was not part of a specific study or set up as a research design prior to therapy. It could therefore be patient-led and responsive to MD's comments; his "comments and manner of response provided evidence to support or contradict the informal working hypothesis that had been set up, and the therapy was adjusted accordingly" (p. 9). The initial psycholinguistic measures had, however, incidentally provided materials for a satisfactory design to test effectiveness. They comprised tasks which might be predicted to show change due to therapy (i.e. measures of naming, first sound identification and grapheme-phoneme association), and ones which would not be expected to show change due to therapy and could therefore act as control tasks (i.e. written lexical decision and judgement of low-imageability synonyms).

At the conclusion of the therapy period, as predicted neither of the control measures showed a significant change. There was significant improvement, however, in the activities which had been targeted in therapy. MD was significantly better in associating graphemes and phonemes after the six weeks of therapy devoted to this, with no further improvement when attention was switched to the second phase and no further therapy was directed at this. The ability to identify the first sound in words also improved significantly during the first phase, and continued to improve when this became the focus of therapy during the second phase. MD also had a lasting improvement in speed of naming, from a mean delay of 5.6 seconds before therapy to a mean delay of 1.8 seconds when tested eight months post therapy. He and his wife also reported an improvement in functional communication and increased fluency. MD described himself as using the relay strategy consciously by sounding out the word to cue himself. Cook considers that the relay strategy gave MD greater confidence in using the partial phonological information he already had. His wife also attributed the improvement to greater confidence, and confirmed that it coincided with the therapy programme instituted two and three quarter years after the onset of her husband's aphasia.

These three studies all illustrate how application of a psycholinguistic model has provided clinically useful insights to practising speech-language therapists, which have enabled them to target intervention at specific aspects of the patients' disorders and demonstrate their effectiveness. From Nettleton and Lesser's study there is also the suggestion that guiding therapy by a model may avoid the ineffective application of therapy programmes used indiscriminately for globally diagnosed "naming difficulties" identified only through use of a standard clinical test.

These studies also illustrate, however, how response to therapy may raise as many questions as it answers about the nature of language processing in aphasia. It is not clear why "phonological therapy" should generalise to untreated words, given that it is targeted at specific word forms. We also can only guess so far as to why, of two patients diagnosed as having the same quality of anomia, one should respond to semantic therapy and the other not to a significant degree. It is also difficult to explain, in terms of a simple model, why adoption of a relay strategy for naming should effect a transformation in a patient's fluency at the sentence level and the clearing of a residual agrammatism. Response to therapy is at present a relatively unused method of clarifying the nature of language processing and further modifying psycholinguistic models such as the one shown in Figure 1. Generalisation of improvements to other (untargeted) aspects of language processing may throw some light on the extent to which processing is interactive. Replication of case studies will bring into focus the differences of response and therefore of possible processing differences in individuals, and may show the extent to which a model can claim generality.

SOME CRITICISMS OF THE APPLICATION OF PSYCHOLINGUISTIC MODELLING TO THERAPY

The three examples above of applying psycholinguistic modelling to naming therapy illustrate the enthusiasm with which this approach has been adopted in the UK by many practising therapists. They provide some counter evidence to critics who believe that this approach has little to offer rehabilitationists at present.

One such critic is Caramazza (1989) who writes "I find that the promise of cognitive neuropsychology as a guide for the choice of intervention strategies is still largely unfulfilled" (p. 396). Caramazza's main reservations concern the underspecification of the models and the belief that they provide no more information than can be inferred from routine clinical observation. He also notes that no theory of cognitive rehabilitation has been developed.

Caramazza's first criticism ignores the possible contribution which aphasia therapy can make to the further specification of models, as outlined above. His second criticism undervalues the role of a theoretical framework in sharpening perceptions and in synthesizing disparate observations into an insightful pattern as a basis for action. It is also probably more realistic to expect a theory of cognitive rehabilitation to emerge from evidence of responses to model-based rehabilitation rather than to be a precondition of applying such rehabilitation.

Basso (1989b) is another critic of the application of cognitive neuropsychology to aphasia therapy. Writing as a speech therapist, she points out that her own data on recovery do not provide any support for the claim that language is processed through distinct modules, a claim which underpins present psycholinguistic models. She also feels that the strategies used in cognitive neuropsychological rehabilitation are only appropriate for the rare patients with selective impairments. A further worry is that adopting psycholinguistic modelling is a regressive step in that it ignores the pragmatic level of language.

The counter to Basso's first point is that her data were drawn from group studies and did not use measures which would reveal modular processing. Second, although patients with highly selective impairments may be rare, cognitive neuropsychological rehabilitation is in fact applied to specific aspects of the disorders of "run-of-the-mill" patients referred to clinics for help with the full range of complex problems. Third there is no reason why this type of therapy should not go hand-in-hand with intervention based on the pragmatic uses of language, such as the chapter in this book by Perkins and Lesser describes. The cognitive neuropsychological approach provides an addition to the aphasia therapist's resources, rather than diminishing them. Given the current inadequacy of measures of functional communication in aphasia, it is certainly easier at present to demonstrate effectiveness of psycholinguistic than pragmatic therapy for aphasia.

CONCLUSION

I have attempted to show that the application of psycholinguistic modelling has considerable potential for aphasia therapy, as an expression of the work of practitioner-researchers. For all its limitations aphasia therapists in the UK have welcomed what seems to them to have effected considerable changes in clinical practice. In assessment and in motivating the hypothesis testing mingle of assessment and therapy, it is proving to be a considerable addition to their resources. More fundamentally, it has contributed a theory-based rationale for planning intervention and for selecting which of the multiplicity of materials and well-tried techniques should be appropriate for a particular patient at a particular time. Therapy applying these models does not use any radically new methods (other than sometimes incidentally applying advances in technology, such as computing). The techniques used are part of classical lore. The difference is in selecting a particular technique for a particular aspect of a patient's disorder at a particular time in recovery, based on a theoretical analysis of the pattern of retained functions and dysfunctions for that individual patient, while taking into consideration the context of functional requirements for communication.

The potential of this approach provides a strong justification for not divorcing case-based research and clinical practice. The practitioner needs to be equipped with the training to integrate a research-oriented approach into clinical work (and the time to prepare publication of results). Unfortunately the profession of speech therapy still has some way to go in this respect. A recent international survey of training in 34 countries (Lesser, 1992) found that in only about half of these countries was it considered essential for initial education and training to include a research project. Only by encouraging aphasia therapists to become practitioner-researchers can we capitalise on the wealth of clinical data which daily continues to provide potential insights into the nature of aphasia and how to affect changes in this condition.

References

Allport, D.A. 1984. Speech production and comprehension: one lexicon or two? In W. Prinz and A.F. Saunders (eds.) *Cognition and Motor Processes*. Berlin: Springer. 209-228.

Bachy-Langedock, N. and de Partz, M-P. 1989. Coordination of two reorganization therapies in a deep dyslexic patient with oral naming disorders. In X. Seron and G. Deloche (eds.) *Cognitive Approaches in Neuropsychological Rehabilitation*. Hillsdale, New Jersey: Lawrence Erlbaum Associates. Pp. 211-247.

Basso, A. 1989a. Therapy of aphasia. In F. Boller and J. Grafman (eds.) *Handbook of Neuropsychology*, Vol. 2. Amsterdam: Elsevier. Pp. 67-82.

Basso, A. 1989b. Spontaneous recovery and language rehabilitation. In X. Seron and G. Deloche (eds.) *Cognitive Approaches in Neuropsychological Rehabilitation*. Hillsdale, New Jersey: Lawrence Erlbaum Associates. Pp. 17-37.

Caramazza, A. 1989. Cognitive neuropsychology and rehabilitation: an unfulfilled promise? In X. Seron and G. Deloche (eds.) *Cognitive Approaches in Neuropsychological Rehabilitation*. Hillsdale, New Jersey: Lawrence Erlbaum Associates. Pp.383-398.

Coltheart, M. 1983. Aphasia therapy research: a single-case study approach. In C. Code and D.J. Muller (eds.) *Aphasia Therapy*. London: Edward Arnold. 194-202.

Cook, K. 1991. An investigation into self-cuing techniques used by a high-level Broca's dysphasic client. In *Proceedings of Newcastle upon Tyne Symposium on Therapeutic Approaches in Aphasia*. London: British Aphasiology Society. Pp. 1-40.

Ellis, A.W. and Young, A.W. 1988. *Human Cognitive Neuropsychology*. Hove: Lawrence Erlbaum Associates.

Goodglass, H. 1988. Historical perspective on concepts of aphasia. In F. Boller and J. Grafman (eds.) *Handbook of Neuropsychology*, Vol. 1. Amsterdam: Elsevier. Pp.249-265.

Goodglass, H. and Kaplan, E. 1983. *The Assessment of Aphasia and Related Disorders*. Philadelphia: Lee and Febiger.

Hillis, A.E. 1989. Efficacy and generalization of treatment for aphasic naming errors. *Archives of Physical Medicine and Rehabilitation*, 70: 632-636.

Howard, D. 1986. Beyond randomised controlled trials: the case for effective case studies of the effects of treatment in aphasia. *British Journal of Disorders of Communication*, 21: 89-102.

Howard, D., Patterson, K., Franklin, S., Orchard-Lisle, V. and Morton, J. 1985. The treatment of word retrieval deficits in aphasia: a comparison of two therapy methods. *Brain*, 108: 817-829.

Lesser, R. 1989. Some issues in the neuropsychological rehabilitation of anomia. In X. Seron and G. Deloche (eds.) *Cognitive Approaches in Neuropsychological Rehabilitation*. Hillsdale, New Jersey: Lawrence Erlbaum Associates. Pp.65-104.

Lesser, R. 1992. The making of logopedists: an international survey. *Folia Phoniatrica*, **44**: 105-125.

Marshall, J., Pound, C., White-Thomson, M. and Pring, T. 1990. The use of picture-word matching tasks to assist word retrieval in aphasic patients. *Aphasiology*, **4**: 167-184.

McConkey, R. 1991. Practitioners as researchers. *Journal of Clinical Speech and Language Studies*, **1**: 1-15.

McReynolds, L.V. and Thompson, C.K. 1986. Flexibility of single-subject experimental designs. Part 1: Review of the basics of single-subject designs. *Journal of Speech and Hearing Disorders*, **51**: 194-203.

Mitchum, C.C. and Berndt, R.S. (in press) Verb retrieval and sentence construction: effects of targeted intervention. In G.W. Humphreys and M.J. Riddoch (eds.) *Cognitive Neuropsychology and Cognitive Rehabilitation*.

Nettleton, J. and Lesser, R. 1991. Application of a cognitive neuropsychological model to therapy for naming difficulties. *Journal of Neurolinguistics*, **6**: 139-157.

Nicholas, M. and Helm-Estabrooks, N. 1990. Aphasia. *Seminars in Speech and Language*, **11**: 135-144

Petersen, S.E., Fox, P.T., Posner, M.ii., Mintum, M. and Raichle, M.E. 1988. Positron emission tomographic studies of the cortical anatomy of single-word processing. *Nature*, **331**: 585-589.

Pring, T. 1986. Evaluating the effects of speech therapy for aphasics: developing the single case methodology. *British Journal of Disorders of Communication*, **21**: 103-115.

Pring, T., White-Thomson, M. Pound, C., Marshall, J. and Davis, A. 1990. Picture-word matching tasks and word-retrieval: some follow-up data and second thoughts. *Aphasiology*, **4**: 479-483.

Saffran, E. and Martin, N. 1990. Neuropsychological evidence for lexical involvement in short-term memory. In G. Vallar and T. Shallice (eds.) *Neuropsychological Impairments of Short-term Memory*. London: Cambridge University Press. Pp. 145-166.

Seidenberg, M. 1988. Cognitive neuropsychology and language: the state of the art. *Cognitive Neuropsychology*, **5**: 403-426.

Tyler, L.K. 1987. Spoken language comprehension in aphasia: a real-time processing perspective. In M. Coltheart, G. Sartori and R. Job (eds.) *The Cognitive Neuropsychology of Language*. London: Lawrence Erlbaum Associates. Pp. 45-162.

8 Therapeutic approaches to naming disorders

Helgard Kremin

INTRODUCTION

The anatomo-clinical approach and the psycholinguistic approach have brought to the fore a number of factors involved in the task of naming. However, these factors will not be considered within the scope of this paper, which will focus exclusively on the crucial aspects involved in rehabilitation. The complementary information presented in a few articles of a general nature (cf. Kremin, 1988; 1990; Kremin and Koskas, 1984; Lesser, 1989) is nonetheless essential in order to place the diverse attempts at rehabilitation involving naming disorders in an adequate context.

Psycholinguistic studies concerning the internal lexicon have allowed us to describe relations of form and meaning between isolated words. As regards naming disorders and their rehabilitation, the analysis of errors seems to be important for at least two reasons: on the one hand, it is possible that Broca's aphasics, Wernicke's aphasics, conduction aphasics, etc., make different types of errors. If this is the case, then we possess an important diagnostic tool. In addition to this, the hypothesis that different types of errors may be associated with distinct underlying causes is quite appealing. In the latter case, error analysis would permit identification of the level at which the process of naming has been disrupted.

Despite the clinical impression, which often seems to correspond to these hypotheses, experimental studies have not, unfortunately, confirmed either one of the hypotheses. Thus, Kohn and Goodglass (1985), while analyzing the naming errors made by different aphasic groups, noted that verbal substitution (of one word for another) was the most frequent error, followed by phonemic and other types of errors. However, this pattern was rather similar across all types of aphasia.

Similarly, it has been noted, for example, that semantic paraphasias are not necessarily indicative of a disruption in semantic organization but that, on the contrary, they may represent a problem of increased thresholds, lexicalization problems, blockage of a given word, etc. (see Kremin, in press, for a more detailed analysis). All of these deficits nonetheless occur after semantic processing, which is still intact.

Note also that the physical characteristics of the stimuli to be named (real objects, photographs, line drawings, etc.) do not crucially influence the performance of aphasic subjects (see Kremin and Koskas, 1984, for a review of the topic). On the contrary, studies have shown that pictures of objects in which the latter have been represented in a complex situational context can even hinder naming (Williams and Canter, 1982) and/or prevent semantic comprehension of the pictures in multiple-choice tasks (Pierce, Jarecki and Cannito, 1990).

Finally, even the parallelism which is often observed between semantic deficits and naming disorders does not permit us to conclude that there is a causal relationship between the two deficits. The controversy regarding the role of semantic-lexical comprehension in naming disorders appears to be settled, and to have a simple solution. A study by Gainotti et al. (1986) has shown that some anomic patients also suffer from semantic comprehension difficulties whereas other anomic patients show no such problems. Conversely, Kremin (1986) described two patients who could properly name certain pictures despite the fact that these pictures/words had not been "understood" in previous simple pairing tasks. It is to be noted, while on this topic, that these diverse deficits are in no way specific to aphasia, nor are they linked to the presence of a focal lesion. On the contrary, similar patterns have been noted in subjects with diffuse lesions and/or Alzheimer's disease—namely, a semantic deficit accompanied by anomia (Margolin et al., 1990), anomia despite the preservation of semantic analysis (Miller, Sommers and Pierce, 1990) and the preservation of the ability to name despite major disturbances of the comprehension of isolated words (Kremin, 1986).

Are we thus completely empty-handed in wishing to undertake specific rehabilitative measures for naming disorders in aphasic subjects? Rather than answer this question directly, I would like to present a summary of related work of relevance to this issue.

CHARACTERISTICS OF WORD RETRIEVAL AND/OR THE TACIT KNOWLEDGE OF WORDS IN THE NORMAL AND APHASIC SUBJECT

TOT (tip of the tongue) phenomenon in normal subjects

One of the major disturbances in aphasia consists of word-finding difficulty. However, even normal adult subjects occasionally cannot retrieve a word they are "looking for", despite the impression that the missing word is on the tip of their tongue.

Brown and McNeill (1966) studied word-finding difficulty in normal subjects. Definitions of unfamiliar words were read to the subjects, and they were then asked to produce the words or to describe the "knowledge" they had of the words when unable to name them. The authors could thus demonstrate that the subjects possessed tacit knowledge of a given target word despite an inability to name it: on the one hand, subjects correctly guessed the first letter of the word in question 57% of the time; on the other hand, subjects also had an idea of the number of syllables (and of the stress pattern) in the target word. As we shall see below, aphasic subjects do not necessarily possess the same tacit knowledge as normal subjects during naming difficulties.

Brennen et al. (1990) also explored the TOT phenomenon in normal subjects. Descriptions of celebrities were read to the subjects. When the target name could not be produced, two facilitation conditions were studied –the presentation of a picture of the person in question's face and the presentation of the person's initials. The results show that only the presentation of the initials had an effect of facilitation on the retrieval of the celebrities' names.

A study by Maylor (1990) confirms that semantic facilitation has no effect on the retrieval of a target word in a state of TOT. The author used the method of naming after verbal definition. However, another word was presented simultaneously with the verbal definition. This word was either semantically related, phonologically related, or bore no relation whatsoever to the target word. As in the aforementioned studies, subjects were requested to write the target word. The results show that only the phonological intrusions had a beneficial effect on target word retrieval, both on the complete retrieval of the name as well as on the TOT phenomenon analyzed in terms of the successful production of the initial letter, of the final letter, and of the number of syllables in the target word.

When the data were subjected to a second analysis, this time with regard to the relationship between the TOT phenomenon and/or the retrieval of the word and the age of the subject, Maylor noted that the beneficial effect of phonological intrusions decreased with age. In fact, no significant effects were found for the group of subjects who were 60 years of age and older.

Kohn et al. (1987) also explored the TOT phenomenon using the method of naming following verbal definition. Unlike the other studies cited, the authors required an oral response, as opposed to a written one, primarily to prevent the interference of one production task with another. This methodology was also used because the authors believed that oral naming represents the primary faculty with respect to written naming. When the subjects experienced the TOT phenomenon, they were encouraged to use any searching strategies or approach in order to retrieve the target word. With respect to the TOT phenomenon and/or the stages of retrieval of a word, the authors grouped the responses in several categories of which three were considered to be closely associated with the naming process: (1) the search was successful; (2) the search was unsuccessful; however the subject claimed to have searched for a specific word (which, in fact, corresponded to the target word); (3) a guessing strategy was used: the subject did not claim to have been searching for a specific word and when confronted with the target word, the subject admitted that the word was unknown.

The main conclusion to be drawn from the detailed study by Kohn et al. is that a normal subject is more likely to produce a target word if (even fragmentary) phonological and/or morphological information is available during the process of lexical search. In other words, semantic knowledge alone guarantees neither the retrieval of the lexical form of a word nor its phonological realization.

Another study relevant to this topic is that of Goldblum and Forst (1988), who used a different methodology to determine various aspects of lexical searching in normal subjects. The authors did not study the TOT phenomenon; rather, they used a "crossword puzzle" paradigm. The results show that fragments of three successive letters produced a greater facilitation effect than did series of three dispersed letters. In addition, these three letter clusters proved to be even more effective at aiding deblocking when they corresponded to a syllabic structure (and not only to pronounceable versus non- pronounceable clusters). Once again these results can be interpreted from the perspective that there are lexical sub-units which are larger than isolated letters but smaller than the word. Also, these sub-units are organized according to phonological principles.

By employing a whole series of experimental manipulations, Nelson et al. (1989) showed that word fragment completion is also facilitated by lexical cues, whereas various semantic cues proved to be completely ineffective.

Facilitation of target word retrieval by aphasic subjects using cueing

NAMING AND TACIT KNOWLEDGE OF THE TARGET WORD

As with the normal subjects (under precise experimental conditions), aphasic subjects sometimes claim to have a word on the tip of their tongue. On the basis of the work by Brown and McNeill (1966), Goodglass et al. (1976) studied this phenomenon in order to determine the partial knowledge that aphasics have of target words when they are unable to produce them. When patients failed to produce a name, or when they gave an inexact response to a given picture, they were asked whether they had an idea of the exact word; whether they could point out the first letter of the target word from a written alphabet; whether they can indicate the number of syllables in the target word (1, 2, or 3); and whether they could identify the target word from a choice of three words.

The results showed that there was a correlation between the percentage of correctly identified syllables and the percentage of correctly identified first letters within each group of aphasics. However these percentages were different from one group to the next. Conduction aphasics identified both the number of syllables and the first letter for 30% of the target words which they could not name. This contrasted with the poor results obtained from the Wernicke's aphasics and the anomic patients. The performance of the Broca's aphasics could not be differentiated from either that of the conduction aphasics or the Wernicke's aphasics. The authors concluded that naming involves the same process for both Wernicke's aphasics and anomics in the sense that either they retrieve a word well enough to produce it, or else they claim to have little knowledge of the word. When naming fails to occur, it is impossible for them to retrieve the word. In other words, these two groups of aphasics can be said to conform to the "all or none law". (However, the preserved ability to identify the target word in a multiple choice situation shows that the process of identification remains intact.)

Regarding conduction aphasics, whose performance suggests that they have retained partial tacit knowledge of a large number of words, Goodglass et al. (1976) suggested that these patients suffered from a disruption at a fairly late stage in the naming process.

In the Goodglass et al. (1976) study, clinical groups which could not be differentiated by their overall results in oral naming tasks did show significant differences with respect to their tacit knowledge of the target word. However, the notion of "tacit knowledge" as a taxonomic variable is not without contradiction. On this subject, let us consider a recent study by Le Dorze and Nespoulous (1989) which also addresses the question of tacit knowledge (relative to the form and meaning of target words) in a large series of aphasics. On the basis of neurolinguistic analysis, the authors identified three groups of patients: the first group consisted primarily of conduction aphasics, the second group consisted of subjects with (relatively pure) amnesic aphasia, and the third group was predominantly comprised of Wernicke's aphasics. Contrary to the work by Goodglass et al., Le Dorze and Nespoulous found no significant differences between the three groups of patients with respect to their tacit knowledge either of the form of the target words or of their meaning. The authors did note however that (generally speaking) semantic knowledge of the target words showed greater preservation than did the knowledge of the form of the words.

Le Bohec et al. (1989) explored the tacit knowledge of patients suffering from similar types of language disturbances. The authors studied the oral naming abilities of three Broca's aphasics as well as their performances on the following tasks when naming failed to occur: recognition of the grammatical category of the target word; recognition of the first letter (in a written alphabet); recognition of the number of syllables in the target word (the subject must point out the visual representation of the number of syllables –1, 2 or 3– following training on this type of task); recognition of the last and first syllables of the word (the subject must point out each one from a choice of three written syllables); recognition of the written word corresponding to a given picture, out of a choice of three written words (two of which are semantic distractors).

A series of control tests were used to study the performance of the patients on other tasks such as repetition, phonemic discrimination, and semantic discrimination as well as picture identification in which the names were homophones (for example: beet-beat). The results from these four control tests showed that the three Broca's aphasics performed similarly on both the phonemic and semantic discrimination tasks. On the other hand –as with the naming of pictures– they did show distinct profiles for the repetition and identification of homophone words/pictures tasks. Finally, regarding the tests exploring the tacit knowledge of unnamed items, the performances of the three Broca's aphasics differed considerably. Thus, subject H retained very little implicit knowledge of the form and/or meaning of the unnamed words. Subjects F and M, however, did obtain relatively high success rates on these form and/or meaning identification tasks. However, F and M's performances can be differentiated if we consider another, as yet unmentioned, task. In fact, the authors also subjected their patients to a judgment task, that is, identification of the correct phonemic cue (among two other cues, one corresponding to an associate of the target word and the second to a neutral, unrelated phonemic cue). Subject F was quite successful at recognizing the correct phonemic cue whereas subject M—despite good performance on the other tacit knowledge tests—encountered great difficulties in this task. In fact, on more than half of the trials, subject M wrongly accepted the phonemic cues which corresponded to semantic associates of the target word. With the task of identifying the grammatical category of the target word, M also performed at chance. The authors stressed that this pattern of performance must be interpreted with respect to M's oral naming performance, which is characterized by numerous semantic paraphasias –paraphasias which M was nonetheless capable of immediately rejecting.

The results from studies exploring tacit knowledge of the form and meaning of target words are thus doubly contradictory. In the first place, the performance of the different classical types of aphasics varies from one study to the next; in addition, this variability can also be observed within groups, such as within the group of Broca's aphasics. Thus, the type of aphasia from which a patient suffers is uninformative with respect to the tacit knowledge which that patient has of a target word.

What can deblock and/or facilitate naming in aphasic subjects?

CHANGING STRATEGY TO IMPROVE NAMING

One occasionally finds claims in the literature that a change of strategy can give rise to an improvement in performance. Johnson and Rubens (1975) presented a case in which the subject could not name when the modality of presentation was visual. In this case (of visual agnosia), the naming of visually presented objects nonetheless became possible if the subject could execute a movement typically associated with the given object, for example *"to cut"* (scissors). A similar observation regarding an aphasic subject had already been made previously by Weissenburg and McBride (1935).

Helm et al. (unpublished—cited in Albert et al., 1981: 158) undertook a more systematic study to explore this aspect of deblocking, by changing the strategy for naming, with a group of 17 aphasic subjects (fluent and non- fluent). The materials consisted of 24 pictures of which half represented easily manipulable objects (*razor*) and the other had, generally non-manipulable objects (*tree*). There were two experimental conditions: oral naming and gesturing without verbalization.

The authors found that there was a significant relationship between the two conditions: the greater the naming ability of the patients, the richer and more adequate were their gestures; conversely, those subjects who produced gestures that were very general or even without relation to the pictured object also showed greater difficulties in naming. The authors also observed –albeit in only a few subjects– a spontaneous deblocking of the oral response during demonstration of the appropriate gesture.

It is worth mentioning that Hanlon, Brown and Gerstman (1990) –who studied the effect of different unilateral gestural movements on simultaneous oral expression– found that gestures produced through activation of the proximal (shoulder) musculature of the right paralytic limb differentially facilitated naming performance in nonfluent but not in fluent (presumably posterior) aphasics. More specifically, fewer word production errors were generated by the nonfluents under the right proximal gesture condition. Note that this effect was not related to the severity of the aphasic disturbance. The authors explain their findings within the framework of a microgenetic process model.

It should be noted however that the internal operativity of the object and/or the word does not necessarily have a beneficial effect on naming in aphasic subjects. A study first done by Gardner (1973) suggested such an influence. However a second study by Feyereisen et al. (1988) in which the various variables associated with the items to be named were more strictly controlled did not confirm the beneficial effect of the operativity variable on the naming performance of aphasic subjects.

Beauvois (1982) focused on yet another completely different aspect. The author considered the possibility that normal subjects have several strategies at their disposal in order to carry out a task, one of them no longer being available to a brain-damaged subject. Beauvois tested this hypothesis during a case study. Her patient, MP, had a visuo-verbal deficit (optic aphasia for color), that is, a disturbance which was neither verbal nor visual but which concerned the interaction between the two "codes". The author thus decided to manipulate the naming strategy: in one condition, the stimulus was solely verbal, and in another, visual imagery was used. Thus, we can respond, "The snow is white" because we imagine it or because we are thinking of the verbal expression "white as snow". On the basis of these observations, MP was subjected to a color naming test following two different sets of instruction. "The first encouraged a verbal search –for example: *"You have learned what color snow is. What do people say when they are asked what color snow is?"* "The second set of instructions encouraged the use of visual imagery; for example: *"It is winter; imagine a beautiful snowy landscape; there are mountains and you can see skiers going down the slopes. Can you imagine it? Can you see it? Well, now, tell me, what color snow is?"*

The subject, MP, named 19/20 items when the instructions were of the verbal type, but only 13/20 when visual imagery was encouraged.

FACILITATING RECALL OF THE TARGET WORD

Pease and Goodglass (1978) studied the facilitation effects of several different cues on picture naming in 24 aphasic subjects. The following cues were used: a phonemic cue (the first syllable of the word in question), a description of the word's function (*knife: we cut with it. It's ...*), the rhyme, the category, the environmental context, and a linguistic context by sentence completion.

Statistical analysis of the results revealed that the various cues produced differentially beneficial effects independently of diagnostic group type or the severity of the naming difficulty during spontaneous oral naming. In fact, the phonemic cue proved to be the most effective. The latter was followed by the 'linguistic context' cue (sentence completion) in terms of effectiveness. The category, function, rhyme and environmental context cues were significantly less effective, and they could not be differentiated in terms of beneficial effect. It is thus remarkable –as the authors pointed out– that the cues which were supposed to mobilize semantic knowledge of the word, such as function, category, and environmental context (this last cue coinciding eventually with imagery), were much less effective than the phonemic cue and/or the linguistic context cue.

The facilitation effects of phonemic cues on naming performance have been confirmed by Goodglass and Stuss (1979) for several types of aphasics (Broca's, Wernicke's and anomics) and also for two different tasks: naming from pictures as well as naming following a verbal description of the target word. With regard to the latter naming condition, phonemic cues proved to be much more effective than visual cues, which consisted of presenting the correspond- ing picture.

It should be noted, however, that if the presentation of a phonemic cue constitutes an effective means of deblocking a target word, the ensuing production does not, in any way, prove that the correct response was merely blocked at the level of phonemic realization and/or articulation. Howard and Orchard-Lisle (1984) studied this problem with the help of an ingenious experimental paradigm: during a case study, the authors presented phonemic cues derived not only from the target word (for example, T(iger), but also from a close semantic associate (for example L(ion)) as well as from an unassociated word. It was found that the patient benefitted from the cue which corresponded to the target word and that he did not respond to the unrelated cue. However, the phonemic cue corresponding to a semantically related word also deblocked production, that is, it induced a semantic paraphasia. This pattern of performance seems to indicate that the patient's naming disorder was not caused by a peripheral disturbance at the level of the production of the adequate word, but on the contrary, that it reflected a more central disturbance.

The facilitation effect of phonemic cues has also been studied in prestimulation situations. Thus, Podraza and Darley (1977) examined the effect of prestimulation on the naming ability of 5 aphasic subjects. The oral prestimulations used were as follows: (1) presentation of the beginning of the word (hereafter, oral start-up), (2) sentence completion, (3) presentation of three words of which one is the target, and (4) presentation of three words semantically related to the target. The control condition consisted of the presentation of the pictures without prestimulation.

The results showed that the prestimulation technique was relatively effective: the first three conditions yielded better results than did the neutral condition, whereas the results from the fourth condition were inferior to those from the neutral condition. However, no differences were observed between the three effective types of facilitation.

Stimley and Noll (1991) also studied the effects of prestimulation cues on aphasics' picture naming. They administered phonemic (initial sound) and semantic cues. Semantic cues cues took the form of descriptive attributes (e.g., for *kangaroo* "has a pouch"), functional associates (e.g., for *bell* "you can ring"), or functional context (e.g., for *tiger* "lives in a jungle"). The results showed that aphasic naming performance varied as a function of the type of information provided by phonemic and semantic prestimulation cues. Naming performance under the phonemic condition was characterized by increased naming accuracy and an error distribution characterized by increased proportions of phonemic paraphasias. By contrast, naming performance under the semantic condition was characterized by increased naming accuracy and an error distribution characterized by decreased proportions of unrelated whole word errors.

From the difference in error distribution Stimley and Noll inferred separate mechanisms of cued word retrieval. Phonemic cues are thought to influence that stage of the naming process involving phonological aspects of selecting and producing word-level responses: they facilitate activation by the semantic system of phonological forms that share the phoneme presented in the cue and also inhibit activation of phonological word forms that do not share the intial phoneme presented in the cue. By contrast, semantic cues facilitate access of phonological representations by increasing the accuracy and completeness of semantic representations used in the access process. However, the authors also mention that not all the subjects demonstrated the effects that were reported for the group as a whole.

Love and Webb (1977) studied the effects of four facilitation conditions on the naming performance of 20 Broca's aphasics: (1) initial syllable, (2) sentence completion, (3) the corresponding written word, (4) imitation, that is, the repetition of the word. The authors noted that, on the whole, repetition of the words constituted the most effective facilitation condition. This condition was followed by the presentation of the initial syllable, which produced superior results compared with the conditions involving sentence completion and the presentation of the written word (both of which were equally least effective).

Li and Williams (1990) studied the effects of semantic and phonemic cues with regard to type of aphasia. They found that Broca's and conduction aphasics responded better to phonemic cueing while anomic aphasics were more responsive to semantic cueing. The effects of cueing with regard to two grammatical classes, nouns and verbs, was also investigated. The results showed that aphasics responded better to phonemic cueing on nouns, whereas no significant difference between phonemic and semantic cues was observed for verb retrieval.

Despite the divergence in the reported results, the deblocking effect of the phonemic cue is considered to be so important that facilitation by oral start-up constitutes part of the requirements of the Boston Naming Test (BNT, Kaplan et al., 1983). Recently, Huff et al. (1986) have shown that the successful completion of the BNT without the administration of phonemic cues does not alter the psychometric properties of the test.

It should also be noted that phonemic cueing is not necessarily effective in every case. Henaff Gonon et al. (1989) reported the case of GM, a patient suffering from severe anomia. In order to facilitate recall of the target word, the authors studied the effects of several phonemic facilitations: (1) oral start-up; (2) presentation of the first syllable; (3) presentation of the last syllable; (4) presentation of the sequence of vowels in the target word, for example /i/.../o/.../ɔ/... for *dinosaur*, (5) presentation of the consonantal sequence of the word with the wrong vowels, for example DANASAR for *dinosaur*. All of these facilitations, with the sole exception of the last one, were ineffective. Finally, the authors reported that GM's performance improved when a graphic cue (the first letter of the word) was offered.

The effectiveness of graphic cues for target word deblocking has also been observed by Bachy-Langedock and De Partz (1989). Their subject, SP, named only 25% of the pictures presented. The authors then tested the effectiveness of a number of different facilitation conditions: (1) semantico-phonemic facilitation ("beater" for *beat*); (2) final phonemic cue (last syllable); (3) initial phonemic cue (first syllable); (4) initial graphic cue (first written letter or syllable); (5) automatic linguistic context (e.g., a kettle of ...); and (6) verbal definition.

The results show that the graphic cues yielded greater facilitation than did the phonemic cues. In fact, when the patient could not name a picture, graphic facilitation via presentation of the first syllable deblocked the target word 83% of the time.

Cohen et al. (1979) considered whether or not facilitation by sentence completion could produce differential effects depending on the material used. The authors constructed two types of sentences to be completed: neutral sentences which were correctly constructed, and sentences corresponding to clichés (proverbs, children's songs, etc.). In both types of sentences, the word to be completed was in sentence-final position and was accompanied by a picture. Two groups of aphasics were examined: fluent and non fluent. The subjects were matched for age, socio-cultural level, etiology, and type and duration of aphasia.

The results show that the two aphasic groups could not be distinguished on the basis of the effects of the cues used. It also turned out that facilitation by the completion of a stereotypic phrase was significantly more effective than was facilitation of naming a target word by the completion of a neutral sentence.

Bruce and Howard (1988) studied the phenomenon of deblocking the form of a target word in Broca's aphasics. When a subject failed to name correctly after a 5-second delay, one of three conditions obtained: (1) more time was given; (2) a phonemic cue was offered; and/or (3) the subject was given a new task, that of pointing to the initial letter of the target word (from a selection of 9 letters).

The authors analyzed the success rates after the 5-second delay and found that for 7/20 subjects there was no significant difference between naming with and without facilitation; even the presentation of the initial letter had no effect.

However, a deblocking effect by initial letter presentation was observed for 10 subjects, although there was no difference between condition 1 (additional time) and condition 3 (initial letter recognition task).

Like Love and Webb (1977), Bruce and Howard noticed a certain relationship between the extent of the naming difficulties and the effectiveness of the cues: generally, patients whose naming ability was most severely disrupted tended to benefit the most from facilitation. It should be noted however that establishing the severity of the disorder does not enable us to predict that a given cue will be effective. In fact, three patients whose naming performances were extremly poor showed no facilitation following the presentation of phonemic cues.

On the basis of these results, we can assert that the retrieval of target words by aphasic subjects can be facilitated. However, it is impossible to identify specific patterns of facilitation for the different, classical aphasic syndromes. It is also impossible to conclude that a given cue –for example, the presentation of the initial letter– will be effective for every subject.

THE REHABILITATION OF NAMING DISORDERS

Immediate and long-term effects of facilitation

Evidently, the first question to be asked within a therapeutic perspective is whether or not the effects of facilitation observed for the immediate retrieval of a word (which we have just described) also represent long-term beneficial effects. This is not necessarily the case.

Let us reconsider the study by Cohen et al. (1979), concerning the two types of sentence completion. The completion of stereotypic sentences was more effective than the completion of neutral sentences for the immediate deblocking of the word. However, in a post-test after a 24-hour delay, The opposite results were obtained: after a longer delay, performance was significantly better for the pictures which had been facilitated by the completion of neutral sentences than for those facilitated by the completion of stereotypic phrases. This suggests that the deblocking of a word via an automatic expression, although immediately very effective, leaves but a faint trace over time. On the other hand, the active search for a word within the semantico-syntactic framework of a neutral sentence induces less immediate success but guarantees nonetheless the same level of performance on naming tasks after 24 hours.

With respect to the effectiveness of facilitation by oral start-up, Patterson et al. (1983) observed that the effects of retrieval of the target word with the help of a phonemic cue last no more than half an hour (and, likely, do not last more than five minutes). Phonemic cues therefore represent an effective means of immediately deblocking a word, but their therapeutic value is quite limited.

Howard et al. (1985a) studied the long-term effects of several facilitation techniques which did not involve the oral production of the word. The tasks included matching pictures to the corresponding written and/or oral word, making semantic judgments, etc. The results showed that naming was facilitated (for at least 24 hours) without prior production of the target words, before the post-test. The authors suggested that the tasks involved—matching (auditory and visual) plus semantic judgments—acted as lexical primers at two levels: at the level of the semantic system and at the level of the phonological form of the word.

In another study, Howard et al. (1985b) contrasted two therapeutic methods, one of which was based on semantic techniques and the other on phonological techniques. One week after the end of therapy, the results confirmed the effectiveness of both methods. It was also noted that the semantic condition showed a slight advantage over the phonological condition. Unfortunately, a second post-test, carried out six weeks after the end of therapy, yielded no trace of the positive effects.

Therapeutic approaches and methods for naming disorders

THE PROBLEM WITH SPONTANEOUS RECOVERY

It is obvious that when a language pathologist takes a patient into his/her care, s/he will not wait until the end of the spontaneous recovery period to begin therapy. Recall that certain authors believe that the age of the patient, the type and severity of the aphasia as well as the etiology of the lesion are likely to affect spontaneous recovery in an aphasic subject (see Sarno, 1981, for a review of the topic). Thiery et al. (1982) identified seven factors which they believe play an important role: the etiology, the location and extent of the lesion, the initial extent of the deficit and the interval between the onset of the lesion and the examination, the type of aphasia and finally, the symptomatology of the disorder. According to the authors, five factors play a more or less secondary role: the associated variables (for example, apraxia), age, sex, handedness and multilingualism.

Note that the prognosis in the case of anmesic aphasia is relatively favorable. Kertesz and Cabe (1977) reported that in fact 48% of patients with initial anomia recover completely. However, for some subjects, both naming disorders and "empty" speech persist, eventually preventing all professional activity.

Most authors believe that the disorder is more or less stable after a period of six months. Holland (1989) suggests that most of the spontaneous recovery occurs during the first two months following trauma (cf. also Basso, 1989). It should be re-emphasized that six months after trauma, one third of the subjects observed by Knopman et al. (1984) obtained normal scores despite severe initial deficits. One of the patients described by Newcombe et al. (1976) quite nicely illustrates this pattern of recovery. However the authors also describe a second case in which the naming difficulties regressed much more slowly, reaching a stable level only two years after the cerebral accident.

MULTIMODAL TECHNIQUES

Wiegel-Crump and Koenigsknecht (1973) took a multimodal approach to the rehabilitation of naming disorders. Their therapy consisted of 18, 1-hour sessions. Each patient received training on 20 (of 40) words which they had not correctly named during a selection test, which in turn consisted of the repetition of the words, their use in sentences, their deblocking by gesturing, initial phoneme, synonyms, associated words, etc. At the end of this multi- modal therapy, it was noted that the aphasics had progressed significantly on the naming of pictures, including the items which had not been a part of the training process. There was thus an effect of generalization.

A similar approach was used by Seron et al. (1979). Over the course of two months, a group of 4 patients received 20 sessions of intensive therapy. Several techniques were employed conjointly to facilitate access to the names of a limited number of words. Some of these were written and oral word associations, facilitation cues (the nature of which is not specified), sentence completions. A control group consisting of 4 aphasic subjects received the same number of therapy sessions; however a more classical therapeutic approach was used. That is, a greater number of items were "worked on". A comparison of the two therapeutic approaches showed that 3 of the 4 subjects who had followed a programmed rehabilitation made significant progress in naming, whereas the same could be said of only 1 of the 4 patients who had undergone more traditional therapy. The improvement in performance did not only apply to the studied items; again, there was an effect of generalization to the unstudied items.

REHABILITATION VIA WRITTEN LANGUAGE

Recently, there have been several therapeutic attempts to study the effect of facilitation based on written language (reading and/or writing). These methods of rehabilitation were found to be very fruitful.

Let us mention, first of all, work done by Uno et al. (1986) –cited in Sasanuma (1986) in which the rehabilitation of two Japanese patients suffering from oral naming disorders is described. The authors studied the effects of a phonological method (oral start-up; repetition of the word) and a method based on writing the words (in Kanji).

During the post-therapeutic evaluation, it was noted that the method based on writing was more effective than the phonological method. Finally, an examination carried out three and four months after therapy confirmed the superiority of the written method: both patients maintained scores of 90% and 100% correct responses for the words studied by the written method. However, the effect of the phonological method declined gradually, levelling off at only 25% and 30% correct responses after a 4-month delay. This study thus confirms the finding that gains achieved by phonological facilitation are less important and less durable. Moreover, this study suggests –finally!– that long-term (at least 4 months) therapeutic effects can be obtained.

The therapy directed by Henaff Gonon et al. (1989) also made use of the written modality. Their patient, GM, a left-hander suffering from a hemorrhage in the left temporal lobe, possessed good semantic comprehension. His performance on oral naming tasks, however, was around the 40% correct response level. This lack of words was nonetheless accompanied by appropriate comments. When pictures corresponding to polysemous words were presented to the patient, he could occasionally name one of the pictures without being able to produce the same phonological string during the presentation of the second picture bearing the same name (homograph). Even more astonishingly, when neither picture was named, the patient could sometimes describe the polysemous link between the two pictures/words which did not resemble each other at all, visually or semantically. For example, (*fraise*: strawberry/ruff). The authors pointed out that the search for the meaning of a polysemous word occurred when the patient shifted from the rare meaning toward the more familiar lexical entry.

GM's written naming was at the same level (45%) as his oral naming, however the former was characterized by orthographic mistakes. With regard to the unnamed items, GM occasionally produced the initial letter and/or some letters from the target word. If sufficient time was given (i.e., more than 1 minute) the patient sometimes had recourse to a typical strategy reminiscent of anagram completion. Thus, *antilope* was named by writing: A...I...O...P. By trying several different letters, GM finally succeeded in naming the word correctly.

It is to be noted that GM always produced isolated letters corresponding neither to syllabic structure nor to morphemes, as is the case with normals experiencing the TOT phenomenon.

Henaff Gonon et al. (1989) undertook a therapeutic treatment of their patient's naming disorders. The following facilitations were used: (1) phonological facilitation: first the initial syllable; after 10 seconds, the last syllable; (2) semantic facilitation: a picture (of an insect, for example) was presented with 6 written words (*mosquito-knitting-mill-wind-wings-late*). The patient was required to associate the words with the target picture and to explain all sorts of semantic relations; (3) facilitation by sentence completion ("*my back hurts, I have probably slipped a...*"); (4) reading and writing: a picture was presented simultaneously with its written name; the subject was required to read the word out loud and then write it.

The results show that in condition (4), involving reading plus writing, the subject's performance was still significantly better three weeks after the end of therapy. On the other hand, the phonemic cues yielded only slight improvements in performance. Both sentence completion and semantic facilitation had no significant effect.

Finally, Marshall et al. (1990) undertook an approach in their therapy which was both original and "economical": as with the approach used by Howard et al. (1985b), the subjects, consisting of 7 aphasics chosen solely on the basis of their ability to read isolated words out loud, were required to match a picture to the corresponding written word in a semantic discrimination task, for example, *harp: harp – guitare – violin – comb*. The patients were left to execute this task (a total of 60 items to match) at their own pace at home. (Ten daily work sessions were recommended.)

At the post-test, after two weeks of invidual excercises, the authors noted a significant improvement in oral naming of the studied pictures in a simple task of matching a picture to a corresponding written word from a list which included semantic distractors. These effects remained significant for the studied items for 1 month, and in some cases up to 1 year, after therapy (cf. Pring et al., 1990).

Recall once again that this rehabilitation did not directly require oral production of the target word. However, since all the subjects who underwent rehabilitation were capable of reading out loud, it is impossible to determine whether the therapeutic effect was due solely to the semantic discrimination task, or to the simultaneous intervention of verbalization during the reading of the words in the multiple choice list. It is clear, however, that only presenting the written word (i.e., the distractor) had no beneficial effect on the naming of corresponding pictures which were not part of the items to be matched.

With one aphasic subject suffering from a severe naming deficit, both oral and written, Hillis (1989: patient n.1) undertook therapy exclusively for written naming. All of the patient's oral and/or written productions were characterized by the presence of numerous semantic errors. A number of semantic errors were also observed during comprehension tests consisting of matching tasks for pictures and words. The deficit underlying the subject's poor performance was thus a central one, situated at the stage of semantic analysis.

The rehabilitation of the naming disorders was carried out in the following manner: A pre-test comprised of oral and written naming permitted the composition of two groups of stimuli: a first group of 10 pictures (which had not been named correctly by the patient in either modality) was studied in therapy. These pictures corresponded to different semantic categories. The other 40 pictures, corresponding to the same semantic categories, constituted the second group which served as a control.

During therapy, one of the items from the first group of words was presented to the patient with all sorts of facilitations: after the picture alone, the written anagram of the target word was presented with two series of unassociated letters; then the anagram alone; the initial letter; the dictation of the target word; its delayed copy). These cues were presented in a cascade until the patient correctly wrote the target word for the first time. Following the first correctly written production, the cues were re-presented and studied in reverse order until the patient correctly named the picture in writing. Only then was the patient asked to name the picture orally. The patient did not receive any feedback regarding his performance on the oral naming task.

The only scores considered were the correct responses, both written and oral, given when the pictures were presented alone (without facilitation). The results from this study are surprising. On the one hand, there was an improvement in written and oral naming with this therapy which focussed only on written language without making use of semantic tasks; on the other hand, the author noticed, as well, a generalization to the items (of the same semantic category) which had not been worked on.

It is important to recall that the therapy began only three months after the onset of the cerebral accident. Thus, we cannot be certain that the therapeutic intervention occurred in the absence of the influence of spontaneous recovery (possibly always still occurring at the time of treatment).

Yet another therapy related exclusively to written naming was undertaken by Deloche et al. (1990). They set up a multi-center study of written confrontation naming rehabilitation. It was micro-computer assisted for picture display, cue presentation and response recording (of the patients' typewritten naming performance). The study dealt with the following main issues: (i) evaluation of the global efficacy of written cueing techniques during treatment and (ii) investigation of generalizations of improvements both from trained to untrained items (in the same written modality) and from treated (written) to untreated (oral) modality.

The testing battery consisted of a set of 120 pictures selected for their high name agreement in normals controlled for age, sex and educational background (see Kremin et al., 1991). Formal testing consisted of presenting the battery twice (before and after treatment) in both naming modalities (oral and written) to 18 non-selected aphasic subjects (5 Wernicke's, 4 globals, 4 anomics, 2 Broca's, 2 conduction and one unclassified) with (at least 15%) errors in written naming pre-therapy. The experimental paradigm was the following: for each patient, a group of 40 non-treated items (20 errors, 20 hits) was constituted from his pre-test written naming of the 120 pictures; the remaining 80 pictures served for treatment in the training program for written naming with five experimental conditions (of prestimulation according to Podraza and Darley, 1977): with facilitation A, with facilitation B; with facilitation A plus a copy of the target word, with facilitation B plus a copy of the target word; without any facilitation and without copying (control condition). The facilitations used were either semantic in nature (sentence completion) or morpholexical (the first letter or first syllable or anagram of the target word).

The global results indicated specific effects of the training of written naming: drilled items improved from .46 to .65, but non-drilled items also significantly improved from .50 to .58. There was no significant effect of response modality: improvements were significant in both modalities and thus generalized from written training to untrained oral naming. Moreover, there was an immediate efficay of delivered cues on written naming during therapy since the patients' scores were higher in the cued conditions than in the uncued presentation of the 80 items. It should be stressed that the significant improvements induced by the naming programs cannot be due to spontaneous recovery since the patients participated in the study at least five months post-onset. It is also worth mentioning that the copy of the target word condition had no effect relative to the same cueing condition without copy.

However, at the level of single case analyses, Deloche and coworkers observed more hetereogeneous patterns. Indeed, in written naming, only 16 out of 18 patients benefitted from the proposed therapy (at the 95% level of confidence). Moreover, in only 4 among these 16 cases did generalization to untreated items occurr. Finally, in oral naming, only 10 out of 18 cases showed significant improvement on treated items, and generalization to untreated items occurred for only two patients. The authors conclude that the conditions of improvement and the dynamics of recovery ought to be investigated by thorough single-case analyses which take into account the patient's preserved pretherapeutic as well as impaired capacities.

While on this topic it should also be mentioned that two of the patients, when retested one year later, still showed significant improvement (M. Dordain, personal communication).

THERAPIES RELYING ON RELAYS: ALTERNATIVE STRATEGIES FOR NAMING

A therapy conducted by Beauvois and Derouesné (1982) was explicitly based on the notion of a change in strategy for naming. Their patient, RG, suffered from bilateral tactile aphasia. With the exception of difficulties in naming objects presented haptically, the subject was perfectly normal in his oral language production, and he could name without difficulty any objects which were presented visually or auditorally. Tactile identification also seemed normal since the subject could correctly mime the use of the objects presented haptically. The authors concluded that the deficit was tacto-verbal and verbo-tactile.

If this interpretation is correct and if the patient had no other difficulties, then it should be possible to modify his performance by replacing the tactoverbal route with another strategy, that is, by placing a relay between tactile identification and language. The authors decided to use visual imagery as a relay. Two conditions were thus contrasted: (1) naming objects from touch which are easy to identify haptically but difficult to visualize because they have no precise form (for example, *piece of paper, tissue, sandpaper*...); (2) naming objects from touch which are easy to visualize because they are commonly recognized visually and can be thought of when they are presented haptically. Their presentation necessitates an intermediate visual representa- tion (for example, *chair*).

In this latter condition, the examiner would strongly induce a visual strategy: the subject was asked to try to evoke the precise visual image of the object (to the extent of being able to draw it), then to name it on the basis of this visual representation, and not on the basis of the tactile sensation.

A statistically significant difference was obtained between these two types of tactile naming tasks, the performance of the patient being better when it was possible for him to use a visual relay: Words with difficult visualization were 50% correct. Words with possible and induced visualization were 75% correct.

It is readily conceivable that the construction of a mental image of an object, serving as a relay between the tactile and visual modalities, would be effective for naming that object. The possibility of changing strategies for naming is, however, not limited to naming disorders which are specific to a given sensory modality. In fact, Bachy-Langedock and De Partz (1989) showed that other forms of mental images—specifically the written form of the target word—can also serve as intermediate relays to facilitate oral naming of pictures.

The patient SP, a left-hander, suffered from a specific reading disorder, deep dyslexia, which is characterized by, among other things, the production of semantic paraphasias. In oral naming the patient had a 48% failure rate. Of his errors, 40% were non-responses, 29% were phonemic approximations, 12% were semantic paraphasias and 19% were spontaneous graphic approximations. This pattern of errors was similar for both oral naming and reading.

Since the patient was particularly responsive to facilitation by graphic cues (as noted above), Bachy-Langedock and De Partz decided to base the rehabilitation of oral naming on the principle of visualization of the written form of the word, presumably to serve as an intermediate relay. The instructions given to the patient were as follows: (1) construct a mental image of the written form of the word corresponding to the picture which is to be named; (2) decode the written word letter by letter (according to a strategy already employed during the rehabilitation of reading: it consisted of visuo-lexical associations for the isolated letters (for example, B – "*baby*") or of groups of letters (for example OO – "*hoop*"); (3) generate the word using "internal language"; (4) say the word.

The patient underwent therapy three times a week. The first evaluation, done after six months, shows that the error rate for oral naming decreased considerably (from 48% to 16%). This effect is really due to the strategy of using an intermediate mental image since the words used for the re-test had not been studied during therapy.

Note, once again, that one year after the first evaluation of the naming deficit, the patient obtained a quasi-normal score in naming (6% errors) and his retrieval of words in conversational situations was good.

THE COGNITIVE APPROACH TO NAMING DISORDERS

A reference model

The principal axiom of the cognitive approach concerns the notion that naming a picture involves a series of different stages (Morton, 1985). It is generally accepted that this succession of levels of analysis and/or processing involves serveral radically different stages which are potential sources of errors during naming (Lesser, 1989). The information processing approach postulates that the use of adequate testing procedures allows us to determine the exact level at which errors occur during naming. With regard to oral naming, Morton (1985) has identified the following: (1) the stage of perceptual analysis; (2) the stage of categorization of visual stimuli; that is, a mechanism which recognizes the 'type' of image independently of the physical realization of the stimulus; (3) after the recognition of the image, the semantic representation is accessed; (4) the semantic representation in turn activates the "logogens", that is, the corresponding entry in the output lexicon. At this level the output lexicon is conceived of as an abstract representation of the phonological form of the word;

(5) from this output logogen system, the information is sent to the phonological and articulatory mechanisms which are involved during the oral production of the response.

Note that, according to the information processing approach, the type of error—for example, a semantic paraphasia—does not constitute a sufficiently specific source of information. On the contrary, as we have already mentioned, a semantic error can be the result of several, quite distinct deficits (e.g., increased activation threshold of the word, blocking of the target word, a more general problem at the level of lexicalization, or a disturbance at the level of semantic organization).

Finally, the information processing approach treats each mode of production (oral and written) as independent. In fact, in the literature there are several examples which attest to the independence of oral naming and written naming (Bub and Kertesz, 1982; Caramazza and Hillis, 1990; Hier and Mohr, 1977; Lhermitte and Derouesne, 1974; Michel, 1979).

Let us emphasize, lastly that, according to the information processing approach, there is only one phonological output lexicon; that is, every task which requires the production of an isolated word appeals to the same output logogens. The result is that paraphasias produced during oral naming should parallel those found in spontaneous language, and indeed even those produced during reading and/or repetition in cases where the output is obtained via lexical routes. This view has received experimental support (Caramazza and Hillis, 1990; Kremin and Koskas, 1983).

A few working hypotheses

Within the framework of this model, Howard and Orchard-Lisle (1984) made the following predictions regarding the effectiveness of oral start-up cues for deblocking target words: (1) in the case of increased thresholds, oral start-up cues should be effective and all phonemic cues which do not correspond to the target word should have no effect and be rejected (including those which correspond to a close semantic associate); (2) in the case of a disconnection between the semantic system and the phonological output lexicon, oral start-up cues should be ineffective; (3) in the case of a deficit at the level of semantic processing, the subject should respond to correct oral start-up cues. However he shoud also respond to phonemic cues corresponding to semantically similar words.

These three predictions have received some experimental support. Thus Howard and Orchard-Lisle (1984) presented a detailed case study which supported prediction (3). Their patient, JCU, a global aphasic, was capable of processing conceptual information relating to non-verbal stimuli. By employing the method involving phonemic cueing, the authors attempted to determine the level of the disturbance. When a naming error occurred, phonemic cues were presented and they either corresponded to the target word, such as, for example, T(iger), to a semantically related word such as L(ion), or to a neutral and unrelated word. It was found that the patient benefitted from the correct cues, providing the appropriate names, but that he did not respond to the neutral cues. However, the cues corresponding to semantically similar words gave rise to numerous responses which, with respect to the presented pictures, were semantic paraphasias. This pattern of performance suggests that the patient's semantic representation was incomplete. This central deficit is considered to be responsible for the spontaneous production of semantic errors in naming as well as for the production of semantic errors induced by phonemic cueing, and also for problems encountered during semantic comprehension tasks.

Blanken (1989) employed the same method of deblocking by oral start-up (correct versus incorrect) in order to pin-point the naming deficit in his patient, TH. This patient could only name 10% of presented pictures, and his errors consisted mainly of non-responses. It was found that the presentation of an adequate phonemic cue deblocked 37/50 responses corresponding to the target word; moreover, 10/50 non-responses (omissions) were reported. However, during the presentation of an incorrect phonemic cue (corresponding to a close semantic approximate), the subject produced 30/50 omissions; moreover, 12/50 semantic paraphasias were reported. This pattern shows that the subject's deficit was not strictly semantic in nature but, on the contrary, was primarily linked to a problem of increased activation thresholds in the phonological output lexicon. To a certain extent, this study supports the predictions by Howard and Orchard-Lisle (1984) with respect to facilitation by oral start-up (see prediction (1) above).

This method of "controlled" oral start-up, as proposed by Howard and Orchard-Lisle in 1984, was also used by Kay and Ellis (1987) on a fluent subject who produced numerous phonemic paraphasias in spontaneous language as well as in oral naming. The patient had no problems with semantic analysis for concrete words. His performance in oral naming was greatly influenced by the frequency of the words to be produced. The presentation of the correct phonemic cue was only slightly effective and the subject did not produce any semantic paraphasias following the presentation of a phonemic cue which corresponded to a semantically related word. On the contrary, all incorrect cues irritated the subject. The authors concluded that EST suffered from a partial disconnection between the semantic system and the phonological output lexicon (see prediction (2) above).

According to the authors, the fluctuations in the subject's performance regarding the production of identical target words attests to the integrity of his phonological representations. Note that the patient's partial phonological knowledge of words as well as the numerous approaches he employed could, alternatively, be indicative of a postlexical disruption at the level of the phonological and articulatory programming mechanisms (Kremin and Ohlendorf, 1988). Either way, the published findings are not sufficiently conclusive to permit a decision to be made in favor of one or the other alternative. Recall, however, that oral start-up cues also proved to be ineffective, in this context, for the patient GM, studied by Henaff Gonon et al. (1989). It is nevertheless evident that in the case of GM, the naming deficit was, without a doubt, postlexical at the level of response blocking. Evidence for this comes from the appropriate comments made by the patient during the naming of pictures corresponding to homophones. In fact, the subject could sometimes describe the polysemous link between the two pictures/words; for example, during the presentation of (*fraise*: ruff), the subject responded: "I cannot find the word ... but it has something to do with the fruit" (i.e., *fraise*: strawberry). It is obvious, in this case, that the only link between these pictures/words, which do not resemble each other either visually or semantically, is that of an identical phonological form. The latter is thus accessible at the level of its abstract representation but cannot be realized verbally because of a specific deficit, that of oral response blocking.

These arguments seem to indicate that the predictions by Howard and Orchard-Lisle (1984) regarding "controlled" oral start-ups, have yet to be verified by other detailed case studies. Similarly it is possible, indeed probable, that there are more than three levels of processing (and thus, of disturbances) during the naming of pictures, as has been argued elsewhere (Kremin, 1988; 1990, in press).

Finally, in a recent experimental investigation Wingfield et al. (1990) reconsidered the underlying causes of the responsiveness of aphasic patients to initial sound cues. Hearing its initial sound facilitated the triggering of the target word. Such findings are commonly interpreted as reflecting an internal representation of the intended phonological form. The authors argue however that such a result may also reflect the subjects' ability to recognize a lexical stem from hearing and to provide its completion –independently of picture presentation– as in the technique of 'gating' used in studies of spoken word perception.

In order to test this ahypothesis, Wingfield et al. studied a series of 18 aphasics for picture naming in the following manner: when a naming failure did occur, the subject heard the beginning of the correct name (150 msec of word onset), followed by progressively longer word-onset duration until the subject was able to give the correct response. The main interest of the study was to compare the word-onset duration necessary to elicit the correct names under this condition with recognition times for these same words when presented in a gated format alone, without any picture presentation. Fourteen normal subjects were also tested.

The results showed that the aphasic subjects as a group were able to give the correct target word from just the gated words in the No Picture condition within an average of 368 msec of word onset (corresponding to 56% of word-onset duration) as compared to 297 msec of mean gate size duration for word recognition without a picutre by control subjects. Overall, the aphasics as a group produced the correct response with significantly less word-onset information (282 msec) when the picture was present. In absolute terms, however, this mean difference in gate size was only 76 msec across subjects, a rather small increment in word-onset information needed for completing the target word when there is no picture present, that is, prior to semantic activation.

The authors also stated that by considering only the means, important variations in the data are obscured. Indeed, in addition to a number of instances in which gate-sizes to recognition were the same in the With Picture and No Picture conditions, half of the subjects had instances in which they did better without seeing the object picture than when they had already attempted to name it. Thus word-initial sounds can be effective in yielding correct target responses for aphasic subjects even in the absence of a picture. In these cases, the sound cue alone was enough to suggest the correct response, with the picture functioning to confirm the hypothesis. Wingfield and coworkers pointed out that this circumstance "will transform what might appear to an observer as an aided word retrieval task (a tasks known to be difficult for aphasics) to a different and far easier task: the confirmation of an incipient utterance against the recognized picture. Such a name-picture matching task is in fact a process aphasics typically perfrom quite well in" (p. 388). They tentatively conclude that the facilitation of naming so often found when examiners furnish oral start ups upon aphasic naming failure may well be accounted for by the two-stage process of stem-completion followed by matching the picture name with the name so generated.

A detailed data analysis furthermore leads the authors to question the theoretical frame of reference for its interpretation. Two alternatives are considered. The first interprets the gating paradigm as entailing two discrete steps: one of recognition of an item in the phonological lexicon and one of repeating that activated item. Such a view is consistent with an application of Morton's logogen model to this paradigm. By contrast, the second interpretation dispenses with the role of the phonological lexicon and views the gating paradigm, at least in the experimental context of their study, as operating entirely within the word production system.

According to Wingfield et al. (1990) this view is consistent with a model of word retrieval on object naming in which the phonology for the naming of a picture is assembled anew upon its activation by the concept of the recognized object. In this model, word-initial phonology would play a powerful role in activating sequences that have been previously learned. This model requires also that the semantics of the pictured object act to selectively reinforce developing phonological activity that carries congruent semantic associations, and to inhibit phonological activity that does not. The failures of such semantically mediated interpretation would then account for paraphasic errors in aphasic speech (p. 388).

With regard to the effects of therapy based on explicit theoretical notions, Blanken (1989) made the following predictions: (1) in the case of a deficit at the semantic level, only treatment of the patient's semantic knowledge, following the example of the study by Howard et al. (1985b), can have beneficial effects (probably limited to items that have been practiced); (2) in the case of a (partial) disconnection between the semantic system and the phonological output system, treatment (the nature of which is not specified by the author) should have beneficial effects on the interaction between the two levels of processing with generalization to unpracticed items; (3) in the case of a deficit due to increased thresholds in the phonological output lexicon, semantic treatment should be ineffective (since this level has not been disrupted) and the effectiveness of phonological treatment should be limited to items which have been practiced without generalization to other words.

Let us now consider a few methods of rehabilitation which refer explicitly to a theoretical model, that of information processing.

Attempts at cognitive rehabilitation

Blanken (1989) attempted to rehabilitate the naming ability of a patient suffering from chronic global aphasia with major comprehension difficulty and an almost complete reduction of spontaneous language (characterized primarily by speech automatisms). Oral naming was quite disrupted (only a 10% success rate), characterized mostly by non-responses. Both dictation and reading (of names) were also very disrupted (9/30 and 1/30 correct responses). Repetition, on the other hand, showed greater preservation (20/30 correct responses). As was mentioned above, facilitation by oral start-ups was relatively effective, and the patient's specific responses to both correct and incorrect phonemic cues made it possible to identify the locus of the oral naming disorder as mainly one of increased activation thresholds in the phonological logogen system. Rehabilitation was structured in the following way: during one week (with daily, 45-minute sessions), the subject was required to work on the naming of 20 pictures. The target words (for example, *pyramid*) were comprised of one to four syllables, they were all low-frequency words and they were not a part of the patient's active vocabulary. In the event of non-responses, the target word was deblocked with the help of an oral start-up, followed by the repetition of the word. One week after the end of therapy, the patient could still correctly name 8/20 of the studied items whereas only 1/20 of the control items were correctly named.

During a second therapeutic intervention with the same patient, 10 pictures corresponding to homophones, for example *lentils* (the vegetable), were practiced (using the same method). Six days after the end of therapy, the patient could still name 8/10 items. A control test on pictures which had not been practiced but which corresponded to homonyms (for example, *lentilles*: optical lenses) showed that the subject could also produce 5/10 of the unpracticed homophones. There was thus an effect of generalization with respect to the phonological form despite the fact that the latter had been practiced with different stimuli.

Moreover, these results constitute experimental support, following the example of the Henaff Gonon et al. (1989) study, for the hypothesis that polysemous words are stored together in the phonological lexicon.

Hillis (1989, patient nr. 2) undertook the rehabilitation of the naming disorder of a 63-year-old aphasic subject. Fifteen months after the initial cerebral accident, the patient's spontaneous language was fluent but devoid of meaning. All production of words, both oral and written, was disrupted but differentially so depending on the mode of production. During oral production (for naming pictures and reading) the patient made 33% errors of which one quarter corresponded to semantic paraphasias. However, during written production (for naming pictures and for dictation), the error rate was considerably higher, at around 87%, although none of these production errors was semantic. In fact, in the written modality, the errors consisted mostly of the omission of letters (for example, *kangaroo: kagro*). This type of error (also observable in the delayed copying task), was closely linked to the length of the target word (the three letter words giving rise to 27%, the four letter words to 71% and the five letter words to 96% errors). This pattern of uniform performance for written production is interpreted as indicative of a deficit regarding the storage in short-term memory of the graphic representa- tions or "graphemic buffer" (Hillis and Caramazza, 1989) in view of their motor implementation. This specific deficit affecting writing is distinguished from a second deficit observed during oral production, which involves the notion of access to the phonological lexicon. (The published findings do not allow us to determine whether or not the problem is one of increased thresholds; the author also does not mention if oral start-ups are effective and/or whether these consist solely of correct cues according to the "controlled" paradigm.) In fact, despite the production of numerous semantic paraphasias, the subject obtained normal scores during the matching tasks involving words and pictures (with semantic distractors) which attests to the integrity of the semantic processing component.

The author decided to first undertake therapy for written naming exclusively. An oral and written naming pre-test on 50 pictures permitted the establishment of two groups of stimuli. The first consisted of 10 pictures (which had not been named either orally or in writing) which were to be practiced during therapy. These pictures corresponded to different semantic categories. The other 40 pictures (corresponding to the same semantic categories) comprised the second group which served as a control.

The therapy was carried out in the same way as it was for patient nr. 1 cited above. That is, one of the items from the first group was presented to the patient along with a series of facilitations. After presentiation of the picture alone, the anagram of the written target word was presented with two series of unassociated letters; then the anagram alone; then the initial letter. The word was then read aloud. The patient was asked to copy the word after some delay. These cues were presented sequentially until the patient managed to write the correct word for the first time. Following the first correct production, the cues were presented a second time and practiced in reverse order until the patient managed to write the name of the picture presented alone. Only then was the patient requested to name the picture orally. Note that the patient did not receive any feedback regarding his performance on the oral naming task. The only scores considered were those regarding the correct written and oral responses given for pictures presented alone.

The results show that the subject benefitted from the rehabilitation of written naming. However, the gain applied exclusively to the items which had been worked on in writing; that is, there was no generalization to the items (of the same semantic category) which had not been studied in writing, nor was there generalization to oral naming.

The author thus decided to undertake a second type of therapy, this time of the subject's oral naming abilities. In this case, the following facilitations were used: first of all, the picture alone; then dictation of the target word followed by reading it aloud; a question regarding the function of the object ("*What do you do with it?* "); the indication of its function (ex: *one cuts with this*); presentation of the target word in writing for the purpose of reading it out loud; sentence completion; then oral start-up. The procedure was the same as that mentioned above for the therapy based on writing.

The results from this second approach showed that there was a slight gain in oral naming. Once again, however, this gain applied exclusively to items which had been specifically practiced without significant generalization to the other items (of the same semantic category) which had not been studied in oral naming.

Hillis (1991) reports the method of therapy used for another subject for whom (correct) oral start-ups were ineffective. The patient had been suffering from chronic aphasia for seven years. In all production tasks, with the sole exception of reading, the subject produced many semantic errors (e.g., in repetition, dictation, both oral and written naming, as well as during matching tasks testing the comprehension of isolated words). The patient's reading (out loud) was characterized by surface dyslexia; that is, oral production was achieved by the conversion of graphemes to phonemes without recourse to the output phonological lexicon where all the words of the language are stored (see Kremin, 1989, regarding specific reading disorders).

The whole of the subject's performance can be attributed to the presence of two distinct deficits –first, a central disturbance at the level of semantic representation; and second, disturbed access to the phonological output lexicon. More support for the proposed double deficit comes from the fact that often, pictures which gave rise to semantic errors during oral naming had nonetheless been correctly named in writing, at least with respect to the access of the lexical form, short of the correct orthographic form.

On the basis of these observations, Hillis undertook the following therapy. The first phase involved semantic treatment applied exclusively to written naming. When semantic paragraphias occurred, a picture corresponding to the patient's wrong answer was drawn, for example "cherry" or "lemon". The examiner then compared the two pictures and without saying the target word, explained all sorts of differences between the two pictures (cherry, lemon), until all ten of the items from set 1 had been practiced in this way and correctly named (100%) in writing.

Then the second phase of the therapy would begin. This stage involved the phonological treatment of the words from set 1, studied exclusively by being read aloud. This stage thus consisted solely of the production of the correct phonological form of the items from set 1 (which, from a "lexicalization" perspective, had already reached 100% by the end of the first phase).

The first phase was continued throughout the second phase this time with a second set of pictures which served as control (the latter being devoid of any semantic associations with the items from set 1). When the items from this second set were correctly identified at least from the perspective of "lexicalization", rehabilitation of the phonological form of these words was undertaken by correcting the subject's reading performance.

The results show that the "semantic treatment" improved the subject's performance not only on written naming (which was practiced) but also on lexicalization during oral naming, repetition, dictation, and matching tasks. In fact, during all these tasks, the number of semantic errors decreased considerably. However, semantic treatment had no beneficial effect on either the phonological realization of these same words (set 1) during reading, or the realization at the lexical level of the unpracticed items from set 2, during oral naming.

Following the "phonological treatment" (carried out exclusively by reading), beneficial effects on reading as well as a generalization of these effects to other phonological realization tasks, such as the oral naming of pictures and the repetition of words, were observed.

Another question which Hillis addressed is whether or not semantic treatment, whose effectiveness has been demonstrated by the transfer of the learning of certain items (worked on in writing) to other modalities, also constitutes a source of generalization to items which were not practiced during therapy. In order to control the possible effects, the author composed two lists of unstudied control items. One list was taken from the same semantic category as those items which had been practiced, and another list consisted of neutral words which were semantically unrelated to the practiced items. A comparison of the subject's performance on both lists in which the words were of comparable length and frequency, shows that there was indeed an effect of generalization of the effectiveness of semantic treatment to unpracticed items, although these effects were restricted to words of the same category. As with the words studied semantically, this gain was found in all production tasks, oral and written (naming, repetition, dictation). This translates into a decrease in the total number of errors as well as a decrease in the number of semantic errors.

Note once again that this same pattern of improvement was also observed on tasks involving the matching of words to pictures with semantic distractors. Thus, the effectiveness of "semantic treatment" is compatible with the notion of (re)constitution of distinctions between items associated with the semantic level.

CONCLUSION

Our review of the various factors which come into play during naming disorders and their rehabilitation shows that the techniques used are quite similar despite the diversity of therapeutic approaches. The "catalog" of these techniques is relatively limited: oral start-ups (with one or several cues), reading out loud, repetition of words, sentence completion, and matching tasks (simple and/or with semantic distractors). In fact, treatment can be carried out with or without the (overt) production of the target word. On the other hand, the different approaches mentioned can be distinguished on the basis of their reference to a theoretical model regarding the function of naming. Only the cognitive approach refers explicitly to a model of the recognition and production of words, that of information processing. Beyond the clinical classification of language disturbances, the cognitive approach sets out to "localize" the functional naming deficit(s) in individual patients. An evaluation of linguistic functions undertaken within this framework also determines a subject's preserved abilities. During rehabilitation, a therapist can use these abilities to establish their interaction with the deficit(s) according to the modular connections proposed by the reference model. Note, however, that the model neither implies any prediction for the outcome of rehabilitation nor does it specify which strategy and/or technique ought to be selected for treatment of a precise impairment in a given patient.

Nevertheless, the comparative study of the effectiveness of various techniques, observed in a series of subjects (suffering from identical functional deficits, not only superficially similar ones), should one day allow us to make precise predictions regarding the appropriate therapeutic approaches for different patients, primarily because we will accumulate knowledge as to why a specific therapeutic technique is effective.

In this sense, the functional localization of a deficit during the course of information processing constitutes a preliminary condition necessary for all therapeutic undertakings based on explicit theoretical foundations. Conversely, the effectiveness of an experimental therapy conducted under these conditions can be considered as an argument in favor of or against the model of functional localization of naming disorders.

Acknowledgements

This research was supported in part by grant UFR 65 from Université Pierre et Marie Curie, Paris, France.

References

Albert, M.L., Goodglass, H., Helm, N.A., Rubens, A.B. and Alexander, M.P. 1981. *Clinical Aspects of Dysphasia*. Wien New York: Springer Verlag.
Bachy-Langedock, N. and De Partz, M.P. 1989. Coordination of two reorganization therapies in a deep dyslexic patient with oral naming disorder. In X. Seron and G. Deloche (eds.) *Cognitive Approaches in Neuropsychological Rehabilitation*. Pp. 211-247. Hillsdale: Lawrence Erlbaum Associates.
Basso, A. 1989. Therapy of aphasia. In F. Boller and J. Grafman (eds.) *Handbook of Neuropsychology. Vol. 2*. Pp. 67-82. Amsterdam: Elsevier Science Publishers B.V. (Biomedical Division)

Beauvois, M.F. 1982. Optic aphasia: A process of interaction between vision and language. *Philosophical Transactions of the Royal Society of London*, B**298**: 35-47.
Beauvois, M.F. and Derouesné, J. 1982. Recherche en neuropsychologie et rééducation: Quels rapports? In X. Seron and C. Laterre (eds.) *Rééduquer le cerveau. Logopédie, psychologie, neurologie*. Pp. 163-189. Bruxelles: Pierre Mardaga.
Blanken, G. 1989. Wortfindungsstörungen and verbales Lernen bei Aphasie. Eine Einzelfallstudie. *Neurolinguistik*, **2**: 107-126.
Brennen, T., Baguley, T., Bright, J. and Bruce, C. 1990. Resolving semantically induced tip-of-the-tongue states for proper nouns. *Memory and Cognition*, **18**: 339-347.
Brown, R. and McNeill, D. 1966. The "tip-of-the-tongue" phenomenon. *Journal of Verbal Learning and Verbal Behavior*, **5**: 325-337.
Bruce, C. and Howard, D. 1988. Why don't Broca's aphasics cue themselves? An investigation of phonemic cueing and tip of the tongue information. *Neuropsychologia*, **26**: 253-264.
Bub, D.N. and Kertesz, A. 1982. Evidence for lexicographic processing in a patient with preserved written over oral single word naming. *Brain*, **105**: 697-717.
Caramazza, A. and Hillis, A.E. 1990. Where do semantic errors come from? *Cortex*, **26**: 95-127.
Cohen, R., Engel, D., Kelter, S. and List, G. 1979. Kurz- und Langzeiteffekte von Benennhilfen bei Aphatikern. In G. Peuser (ed.) *Studien zur Sprachtherapie*. Pp. 350-360. München: Wilhelm Fink.
Deloche, G., Hannequin, D., Kremin, H., Metz-Lutz, M.N., Ferrand, I., Dordain, M., Quint, S., Blavier, A., Cardebat, D., Larroque, C., Naud, E., Perrier, D., Pichard, D. and Rosnet, E. 1990. Rééducation sur micro-ordinateur des troubles de la dénomination d'images chez des adultes aphasiques. *Rééducation Orthophonique*, **28**: 299-313.
Feyereisen, P., Van Der Borght, F. and Seron, X. 1988. The operativity effect in naming: A re-analysis. *Neuropsychologia*, **26**: 401-415.
Gainotti, G., Silveri, M.C., Villa, G. and Miceli, G. 1986. Anomia with and without lexical comprehension. *Brain and Language*, **29**: 18-33.
Gardner, H. 1973. The contribution of operativity to naming capacity in aphasic patients. *Neuropsychologia*, **11**: 213-220.
Goldblum, N. and Frost, R. 1988. The crossword puzzle pardigm: The effectiveness of different word fragments as cues for the retrieval of words. *Memory and Cognition*, **16**: 158-166.
Goodglass, H. and Baker, E. 1976. Semantic field, naming, and auditory comprehension in aphasia. *Brain and Language*, **3**: 359-374.
Goodglass, H., Kaplan, E., Weintraub, S. and Ackerman, N. 1976. The "tip-of-the-tongue" phenomenon in aphasia. *Cortex*, **12**: 145-153.
Goodglass, H. and Stuss, D.T. (1979) Naming to picture versus description in three aphasic subgroups. *Cortex*, **15**: 199-211.

Hanlon, R.E., Brown, J.W. and Gerstman, L.J. 1990. Enhancement of naming in nonfluent aphasia through gesture. *Brain and Language*, **38**: 298-314.
Henaff Gonon, M.A., Bruckert, R. and Michel, F. 1989. Lexicalization in an anomic patient. *Neuropsychologia*, **27**: 391-407.
Hier, D.B. and Mohr, J.P. 1977. Incongruous oral and written naming. *Brain and Language*, **4**: 115-126.
Hillis, A.E. 1989. Efficacy and generalization of treatment for aphasic naming errors. *Archives of Phys. Med. Rehabilitation*, **70**: 632-636.
Hillis, A.E. 1991. Effects of separate treatments for distinct impairments within the naming process. In T. Prescott (ed.), *Clinical Aphasiology, Vol.19*. Pp. 255-276. Austin, Texas: Pro-ed.
Hillis, A.E. and Caramazza, A. 1989. The graphemic buffer and attentional mechanisms. *Brain and Language*, **36**: 208-235.
Holland, A. 1989. Recovery in aphasia. In F. Boller and J. Grafman (eds.) *Handbook of Neuropsychology*. Vol. 2. Pp. 83-90. Amsterdam: Elsevier Science Publishers B.V. (Biomedical Division).
Helm, N.A., Kaplan, E.F. and Vercruysse, L. 1978. The role of gesture in naming. Unpublished study.
Howard, D. and Orchard-Lisle, V. 1984. On the origin of semantic errors in naming: Evidence from a case of a global aphasic. *Cognitive Neuropsychology*, **1**: 163-190.
Howard, D., Patterson, K., Franklin, S., Orchard-Lisle, V. and Morton, J. 1985a. The facilitation of picture naming in aphasia. *Cognitive Neuropsyhcology*, **2**: 49-80.
Howard, D., Patterson, K.E., Franklin, S., Orchard-Lisle, V. and Morton, J. 1985b. Treatment of word retrieval deficits in aphasia. *Brain*, **108**: 817-829.
Huff, F.J., Collins, C., Corkin, S. and Rosen, T.J. 1986. Equivalent forms of the Boston Naming Test. *Journal of Clinical and Experimental Neuropsychology*, **8**: 556-562.
Johnson, M.G. and Rubens, A.B. 1975. Case report:Visual-linguistic disturbances following left occipital lobectomy. Paper presented at the American Speech and Hearing Association Meeting, Washington, D. C.
Kaplan, E., Goodglass, H. and Weintraub, S. 1983. *Boston Naming Test*. Philadelphia: Lea and Febiger.
Kay, J. and Ellis, A. 1987. A cognitive neuropsychological case study of anomia: implications for psychological models of word retrieval. *Brain*, **110**: 613-629.
Kertesz, A. and McCabe, P. 1977. Recovery patterns and prognosis in aphasia. *Brain*, **100**: 1-18.
Knopman, D.S., Selnes, O.A., Niccum, N. and Rubens, A.B. 1984. Recovery of naming in aphasia: Relationship to fluency, comprehension and CT findings. *Neurology*, **34**: 1461-70.
Kohn, S.E. and Goodglass, H. 1985. Picture-naming in aphasia. *Brain and Language*, **24**: 266-283.

Kohn, S.E., Wingfield, A., Menn, L., Goodglass, H., Gleason, J.B. and Hyde, M. (1987) Lexical retrieval: The tip-of-the-tongue phenomenon. *Applied Psycholinguistics*, **8**: 245-266.
Kremin, H. 1986. Spared naming without comprehension. *Journal of Neurolinguistics*, **2**: 131-150.
Kremin, H. 1988. Naming and its disorders. In F. Boller and J. Grafman (eds.) *Handbook of Neuropsycholoy*. Vol. 1. Pp. 307-328. Amsterdam: Elsevier Science Publishers B.V. (Biomedical Section).
Kremin, H. 1989. Lexical access viewed from the information processing approach: Reading and writing (data from pathology). In P.G. Aaron and R.M. Joshi (eds.), *Reading and Writing Disorders in Different Orthographic Systems*. Pp. 283-304. Dordrecht: Kluwer Academic Publishers.
Kremin, H. 1990. La dénomination et ses problèmes. In J.L. Nespoulous and M. Leclercq (eds.) *Linguistique et Neuropsycholinguistique: tendances actuelles*. (pp. 47-68) Paris: Edition de la Société de Neuropsychologie de Langue Française.
Kremin, H. In press. Naming impairments and the information processing approach. In F.J. Stachowiak (ed.) *Development in the Assessment and Rehabilitation of Brain-Damaged Patients: Perspectives from a European Concerted Action*. Tübingen: Narr Verlag.
Kremin, H. and Koskas, E. 1983. Naming and spontaneous language of subjects with lesions of the left temporal lobe. *Rassegna Italiana di Linguistica Applicata*, **XV**: 189-194.
Kremin, H. and Koskas, E. 1984. Données de la pathologie sur la dénomination. *Langages*, **76**: 31-76.
Kremin, H. and Ohlendorf, I. 1988. Einzelwortverarbeitung im Logogen-Modell. *Neurolinguistik*, **2**: 67-100.
Kremin, H., Deloche, G., Metz-Lutz, M.N., Hannequin, D., Dordain, M., Perrier, D., Cardebat, D., Ferrand, I., Larroque, C., Naud, E., Pichard, B. and Bunel, G. 1991. The effects of age, educational background and sex on confrontation naming in normals; principles for testing naming ability. *Aphasiology*, **5**: 579-582.
Le Bohec, I., Perrier, D., Kremin, H. and DeGiovanni E. 1988. Trouble de la dénomination orale chez trois aphasiques de Broca: Trois processus différents. Communication à la *Société de Neuropsychologie de Langue Française*, Session de Printemps à Tours.
Le Dorze, G. and Nespoulous, J.-L. 1989. Anomia in moderate aphasia: Problems in accessing the lexical representation. *Brain and Language*, **37**: 381-400.
Lesser, R. 1989. Some issues in the neuropsychological rehabilitation of anomia. In X. Seron and G. Deloche (eds.) *Cognitive Approaches in Neuropsychological Rehabilitation*. Pp. 65-104. Hillsdale: Lawrence Erlbaum Associates.

Lhermitte, F. and Derouesné, J. 1974. Paraphasies et jargonaphasie dans le langage oral avec conservation du langage écrit. *Revue Neurologique*, **130**: 21-38.
Li, E.C. and Williams, S.E. 1990. The effects of grammatic class and cue type on cueing responsiveness in aphasia. *Brain and Language*, **38**: 48-60.
Love, R.J. and Webb, W.G. 1977. The efficacy of cueing techniques in Broca's aphasia. *Journal of Speech and Hearing Research*, **42**: 170-178.
Margolin, D., Pate, D.S., Friedrich, F.J. and Elia, E. 1990. Dysnomina in dementia and in stroke patients: Different underlying cognitive deficits. *Journal of Clinical and Experimental Neuropsychology*, **12**: 597-612.
Marshall, J., Pound, C., White-Thomson, M. and Prings, T. 1990. The use of picture/word matching tasks to assist word retrieval in aphasic patients. *Aphasiology*, **4**: 167-184.
Maylor, E.A. 1990. Age, blocking and the tip of the tongue state. *British Journal of Psychology*, **81**: 123-134.
Michel, F. 1979. Préservation du langage écrit malgré un déficit majeur du langage oral. *Lyon Médical*, 141-149.
Miller, Sommers, L. and Pierce, R.S. 1990. Naming and semantic judgements in dementia of the Alzheimer's type. *Aphasiology*, **4**: 573-586.
Morton, J. (1985) Naming. In S. Newman and R. Epstein (eds.) *Dysphasia*. Pp. 217-230. Edinburgh: Churchill Livington.
Nelson, D.L., Keelean, P.D. and Negrao, M. 1989. Word-fragment cueing: The Lexical Search Hypothesis. *Journal of Experimental Psychology: Learning, Memory, and Cognition*, **15**: 388-397.
Newcombe, F., Hiorns, R.W. and Marshall, J.C. 1976. Acquired dyslexia: Recovery and retraining. In Y. Lebrun and R. Hoops (eds.) *Recovery in aphasics*. Pp. 146-162. Amsterdam: Swets and Zeitlinger B.V.
Patterson, K.E., Purell, C. and Morton, J. 1983. The facilitation of naming in aphasia. In C. Code and D.J. Muller (eds.) *Aphasia Therapy*. Pp. 76-87. London: Edward Arnold.
Pease, D.M. and Goodglass, H. 1978. The effects of cueing on picture naming in aphasia. *Cortex*, **14**: 178-189.
Pierce, R.S., Jarecki, J. and Cannito, M. 1990. Single word comprehension in aphasia: influence of array size, picture relatedness and situational context. *Aphasiology*, **4**: 155-165.
Podraza, B.L. and Darley, F.L. 1977. Effect of auditory prestimulation on naming in aphasia. *Journal of Speech and Hearing Research*, **20**: 669-683.
Pring, T., White-Thomson, M., Pound, C., Marshall, J. and Davis, A. 1990. Picture/word matching tasks and word retrieval: Some follow-up data and second thoughts. *Aphasiology*, **4**: 479-483.
Sarno, M.T. 1981. Recovery and rehabilitation in aphasia. In M. Taylor Sarno (ed.) *Acquired Aphasia*. Pp. 485-529. New York: Academic Press.
Sasanuma, S. 1986. Universal and language-specific symptomatology and treatment of aphasia. *Folia Phoniatrica*, **38**: 121-175.

Seron, X., Deloche, G., Bastard V., Chassin, G. and Hermand, N. 1979. Word-finding difficulties and learning transfer in aphasic patients. *Cortex*, **15**: 149-155.
Stimley, M.A. and Noll, J.D. 1991. The effects of semantic and phonemic prestimulation cues on picture naming in aphasia. *Brain and Language*, **41**: 496-509.
Thiery, E., Dietens, E. and Vandereeken, H. 1982. La récupération spontanée: ampleur et limites. In X. Seron and C. Laterre (eds.) *Rééduquer le cerveau. Logopédie, psychologie, neurologie*. Pp. 33-43. Bruxelles: Pierre Mardaga.
Uno, A., Tanemura, J. and Higo K. 1985. Recovery mechanisms of picture naming (in Japanese). *Higher Cortical Function Research*, **5**: 893-902.
Weissenburg, T. and McBride, K. 1935. *Aphasia. A clinical and psychological study*. New York: The Commonwealth Fund.
Wiegel-Crump, C. and Koenigsknecht, R.A. 1973. Tapping the lexical store of the adult aphasic: Analysis of the improvement made in word retrieval skills. *Cortex*, **9**: 412-418.
Wingfield, A., Goodglass, H. and Smith, K.L. 1990. Effects of word-onset cuing on picture naming in aphasia: a reconsideration. *Brain and Language*, **39**: 373-390.
Williams, S.E. and Canter, G.J. 1982. The influence of situational context on naming performance in aphasic syndromes. *Brain and Language*, **17**: 92-106.

9 Reading and writing: Cognitive therapies of written language

Helgard Kremin

The study of reading and writing impairments dates back more than 100 years. Exner (1881) was the first to document patients with "pure" agraphia, a writing disorder without aphasic disturbances, and Déjerine (1891, 1892) distinguished "pure" alexia (without agraphia and without language disturbances) from reading impairments which occur together with an impairment of written production.

The main concerns of earlier clinical approaches to disorders of written language were (1) to circumscribe the precise localization of the cerebral lesions responsible for the observed deficits and (2) to establish classificatory schemas while adding further descriptive details. Typical, more recent work in this direction one may be found in Marcie and Hécaen (1979) with respect to writing disturbances and in Hécaen and Kremin (1976) with respect to reading impairments due to lesions of the dominant hemisphere (see also Friedman, 1989, and Bub and Chertkown, 1989, for more general reviews on acquired alexia and agraphia).

The clinical approach interpreted non-"pure" alexias and agraphias, that is agraphias and alexias in the context of aphasia, with regard to their (expected) relationship to the corresponding disorder of spoken language. Beside some noteworthy exceptions (see, for example, Luria et al., 1969), the lack of reports on specific therapies for reading and writing impairments is probably due to this "integrated" view in terms of a more general language disturbance where reading and writing are conceived as inextricably bound to the spoken language. Even in the area of recovery, little research has been conducted, although some longitudinal studies add to our understanding of the deficit. Valuable information was indeed given by Landis, Regard and Serrat (1980) and Behrmann, Black and Bub (1990) with regard to the evaluation of pure alexia, and by Glosser and Friedman (1990) and Laine, Niemi and Marttila (1990) with regard to the evolution of deep dyslexia.

However, with the recent developments in the domain of cognitive neuropsychology, the interpretation of written language disturbances has changed considerably. Indeed, under the influence of the so called information processing approach, namely in the domain of language, there has been a renewal of therapeutic approaches to written language disturbances for both reading and writing. The shared assumption of various models currently discussed is the notion of modularity. The common aim is localizing functional (rather than anatomical) lesions by describing separable subcomponents responsible for different aspects of word recognition and word production. Such an approach allows us to identify specific functional impairments which in turn may be considered as targets for specific therapeutic intervention. Research conducted within this theoretical framework has also shown that the same symptom, for example semantic errors, can arise for a number of different reasons. (We shall return to this point later.) With regard to rehabilitation strategies it is indeed valuable to have access to information which goes beyond the "surface" observation of a symptom. First, let us briefly outline the main characteristics of the model of reference which is currently so influencial with regard to the rehabilitation of reading and writing disorders.

A MODEL OF WORD RECOGNITION AND WORD PRODUCTION

From patterns to pathways

In 1973 Marshall and Newcombe proposed an original approach to reading disturbances. They distinguished three new syndromes of reading impairment according to the nature of reading errors observed in aphasic patients: (1) visual dyslexia, characterized by paralexias with visual similarity (e.g. *dug : bug; pamper : paper*), (2) deep dyslexia, characterized by paralexias of the semantic type (e.g. *little : small; entrance : exit*) and (3) surface dyslexia in which errors mainly result from failure or misapplication of grapheme-to-phoneme conversion rules (e.g. *insect : insist; guest : just*).

The impetus provided by Marshall and Newcombe's article was considerable. Research on reading in the following decade was mainly devoted to two of the described dyslexic syndromes, deep dyslexia (Coltheart, Patterson and Marshall, 1980) and surface dyslexia (Patterson, Marshall and Coltheart, 1985). Numerous cases of deep dyslexia have demonstrated that phonological output can be accessed through semantics even in the absence of any phonological reading ability (as shown by the patients' inability to read nonsense syllables). In contrast, the oral reading pattern of surface dyslexics (from left to right) demonstrated that phonology was obtained by grapheme-to- phoneme conversion. The observation of both syndromes was initially consistent with a dual processing model for oral reading.

However, as more cases were studied carefully, it became apparent that 'nonwords' and 'irregular words' are in fact the crucial variables for defining oral reading impairments within the framework of information processing: the inability of patients to read nonwords need not be associated with the production of semantic paralexias (Beauvois and Dérouesné, 1979, and others); and some patients read nonwords as well as regular words but fail with irregular ones. In spite of so many similar observations, it became apparent that the typical pattern of deep dyslexic reading –with effects of concreteness/imagery and of grammatical class– has no causal interrelation- ship. Indeed, some patients (although producing semantic paralexias and incapable of nonword reading) have no special difficulty with abstract words (Caramazza and Hillis, 1990, patient

HW; Coslett, Rothi and Heilman, 1985). Indeed, some (albeit rare) cases read abstract words better than concrete ones (Warrington, 1981) or function words better than nouns (Marin, Saffran and Schwartz, 1976).

However, the "dissolution" of the newly described syndromes does not detract from the validity of the theoretical foundations that gave rise to the observations. In accordance with Morton and Patterson's (1980) theoretical suggestions it is now commonly agreed upon that there are *three* procedures to obtain oral output for reading aloud, two of which are lexical: (1) when reading by means of the 'semantic pathway', the word is categorized in the visual word recognition system and the information is sent to the cognitive system for semantic analysis; after semantic treatment the information is sent to the phonological output lexicon where the appropriate phonological code is obtained; (2) when reading by means of the "direct route", the word is categorized in the visual word recognition system and directly sent to the phonological output lexicon by some sort of "automatic" mapping between visual and phonological wordform. Experimental evidence for a direct lexical (but nonsemantic) pathway has accumulated from several case studies of patients who correctly read words, whether regularly or irregularly spelled, in spite of severe impairment of the nonlexical transcoding route (Funnell, 1983, and others). This direct pathway is considered to be the normal route for reading virtually all words known by an individual.

The model of word recognition and production proposed by Morton (1980) and extended by others (Ellis, 1982; Miceli, 1989) hypothesizes that there are, in analogy to oral reading, three different pathways for writing isolated words from dictation: (i) nonlexical writing by acoustic-to-phonological conversion, (ii) a lexical route from auditory input lexicon via the cognitive system to the orthographic output lexicon, and (iii) a lexical route which bypasses semantics. (The interested reader may consult detailed reviews on data from pathology for both reading and writing, discussed within this theoretical framework in Kremin, 1989a; 1992).

Localizing the deficit

Knowing that a task, e.g., reading or writing, is impaired does not tell us much about the underlying deficit. The information processing approach is better suited to localizing the functional lesion responsible for the deficit. However, even within the three-route approach, the statement that one procedure—for example lexical reading—is disrupted does not yet give us enough information for the use of specific rehabilitation strategies. We should determine *why* it is disrupted: because of the inaccessibility of the visual word recognition system or because of difficulties of accessing the phonological output lexicon despite preserved visual lexical access in the orthographic lexicon. Indeed, it is standard now to distinguish two types of "surface dyslexic" readers (Shallice, 1988).

On the other hand, even when we know that a patient is reading by use of the semantic pathway—as shown, for example, by the occurrence of semantic paralexias—this information is not sufficient *per se* to localize the level of disruption within the semantic route. It should be stressed that the percentage of this error type varies from isolated productions (see Kremin, 1982, for for a review) to up to 100% of total errors (observed by De Bleser, 1987, in her case study). It seems unreasonable to think that such variations reflect the same underlying deficit.

It is inherent in the model and has been observed that semantic errors can be a "surface" symptom reflecting various underlying causes (see Kremin, in press, for a detailed discussion). Indeed, with regard to reading it has been observed that they can result from three different functional lesions.

Let us first consider problems of accessing the meaning of the target word. Warrington and Shallice (1979) presented the case AR, who showed no part of speech effect but produced predominantly visual errors and some semantic errors (5%). Among others, the patient gave responses like the following: "beaver": *"Could be an animal. I have no idea which one"*. Special testing revealed that AR suffered from 'semantic access dyslexia' since the patient preserved a striking capacity to categorize words he could not read. The patient thus displayed a selective deficit of accessing the full meaning of written words.

Other patients have shown a central comprehension deficit at the level of semantic representations. In this constellation of problems, semantic errors not only occur in production but also in comprehension and in spontaneous speech (see, for example, Nolan and Caramazza, 1982).

Finally, semantic paralexias have been observed in oral reading in spite of preserved comprehension. Such observation favors the notion of a post-semantic deficit at the level of the phonological lexicon and/or access to it. Two cases of this variety have recently been reported in detail by Caramazza and Hillis (1990). The (fluent) aphasics produced numerous semantic errors in oral reading and oral confrontation naming but such errors did not occur either in writing from dictation or in written naming. This dissociation with regard to the mode of production clearly shows that the deficit is not central in nature but limited to the production of phonological word form. Interestingly, the patients were able to give correct verbal definitions of written words they were incapable of reading aloud, e.g. **records** was read *"radio"* but defined as *"you play them on a phonograph... can also mean records you take and keep"*.

Reading impairment and aphasia

Many patients who exhibit deep dyslexia suffer from Broca's aphasia. It is thus tempting to ask whether there is any causal relationship between specific reading impairment and precise types of aphasic disturbances. Experimental evidence from numerous cases of deep dyslexia shows that the association between Broca's aphasia and deep dyslexia is rather fortuitous (probably due to some proximity of the lesioned neural substrate). In fact, even patients with fluent speech can present with deep dyslexia (Kremin, 1980; de Partz, 1986; Caramazza and Hillis, 1990), and patients with Broca's aphasia do not necessari- ly present with the typical pattern of deep dyslexia in word reading: some nonfluent aphasics read function words as well as nouns (patient BD, Caramazza, Berndt and Hart, 1981; patient RIC, Kremin, 1984); and sometimes they even read nonwords (Goldberg and Benjamins, 1982; Ross, 1983; Kremin, 1985).

Independence of reading and writing

Since there is no causal relationship between specific reading impairment and precise type(s) of aphasia one may question whether there is, at least, a constant association between impairments observed in reading and in writing from dictation. (Recall that all patients who do not present with "pure" alexia suffer from writing disturbances.) Again, the experimental evidence is entirely negative. There is no doubt that the reading performance of patients can be strikingly different from their writing performance. RG, a patient studied by Beauvois and Dérouesné (1979), wrote regular words and nonwords relatively well but produced numerous errors when writing irregular words or words with some degree of orthographic ambiguity; in contrast, in oral reading he accurately produced the vast majority of words but failed with the pronunciation of nonwords. Another patient, JC (Bub and Kertesz, 1982), was unable to write nonwords from dictation but read these items without difficulty. Kremin (1987) described a patient (with spontaneous output limited to a recurrent utterance) who read words by direct associations (in spite of his capacity to read nonwords) whereas in writing from dictation the patient initially exhibited the typical pattern of deep dysgraphia with the presence of numerous semantic errors, concreteness and grammatical class effects, and the inability to write nonwords. Finally, two patients, MK (Howard and Franklin, 1989) and GI (Kremin, 1989b) presented simultaneously with surface dyslexia and deep dysgraphia.

The observed dissociations can only be accounted for by separate word form systems for recognition and production of written and of spoken language. Further, such dissociations furnish experimental data to substantiate theoretical claims of the information processing approach (Morton and Patterson, 1980; Morton, 1980).

COGNITIVE THERAPIES OF READING AND WRITING DISORDERS

The model outlined above has no *direct* relation to the therapy of written language disorders: even the localization of the functional deficit(s) with regard to reading and/or writing impairment of an individual patient is not in itself sufficient to specify which strategy ought to be selected for treatment. Moreover, the model does not imply that *any* prediction can be made for the outcome of the rehabilitation of a precise impairment in a given patient. The cognitive approach only exerts some constraints on possible choices without specifying what these choices should be. Nevertheless, as Howard and Patterson (1989) pointed out, there are some good reasons for basing therapy on a deficit analysis in terms of information processing with the goal being to determine the impaired as well as preserved components of the patient's language system. Only with such fine-grained assessment can therapy be maximally effective. In addition, the adoption of such a theoretical approach allows for the systematic interpretation of an observed improvement in a logical manner. In order to identify changes induced by therapy, the assessment of post-treatment performance ought to be as extensive as the analysis of a patient's performance prior to rehabilitation. In this perspective post-therapy assessment should be not only quantitative but also qualitative in nature. Finally, our understanding of deficits and therapies will also depend on accumulating knowledge about therapeutic approaches that *failed* to improve performance. To date, this sort of information has only rarely been given (see however, Hatfield, 1982; Carlomagno and Parlatio, 1989; de Partz et al., 1989).

Let us now consider in more detail how these postulates have been met by the few available approaches to cognitive rehabilitation of written language impairments. We present these with reference to the current model of word recognition and production by distinguishing, within written language disorders, lexical and nonlexical impairments.

Treatments aimed at improving lexical strategies for reading and/or writing

TREATMENT OF WORD PRODUCTION USING CODES AS RELAY

Treatments focussing on homophony

TREATMENT OF PATIENTS WITH "DEEP DYSGRAPHIA"

Hatfield (1982) undertook the first therapeutic approach to specific impairments of word production which were interpreted within the framework of an information processing model. She provided data on the rehabilitation of two deep dysgraphic patients, BB and DE, who also suffered from deep dyslexia. The patients' total inability to write nonwords from dictation showed that their acoustic/phonemic transcoding procedure was inoperative. The writing of words was characterized by the presence of semantic errors and strong effects of concreteness and grammatical class. The limited goal of Hatfield's therapy was to retrain the written production of function words (which are beyond the competence of patients with performance characteristic of deep dysgraphia) by using the patients' remaining lexical processing abilities. More specifically, she made use of the homophony between a restricted set of content and function words. The training method consisted of establishing the associative link between code content words (e.g., *bean*) and target function words (*been*): the patient was instructed to first write the content word and then to correct the spelling of the key word which was elicited through sentence completion (*Barbara has /been/ to London*). (Note that, contrary to what is the case in American English, both words have the same pronunciation in the dialect used in the study.) In the final stage of the therapy "direct" writing of the target word was required, that is without prior production of the code word (other than in the patient's head). On the (necessarily) restricted set of items which were trained (pronouns, auxiliaries, prepositions) the two patients improved considerably (from 34% to 63.5% and from 49% to 67%). In spite of these limitations, Hatfield demonstrated that function could be reorganized by recourse to a spared component of the word production system. Hatfield's approach, which consisted of using a code word as relay, became one of the most influencial techniques for subsequent cognitive rehabilitation attempts –both with other patients and with regard to different rationales of therapy.

HOMOPHONIC WORDS PLUS PICTURES: TREATMENT OF A PATIENT WITH "SURFACE DYSGRAPHIA"

Behrmann (1987) also used homophony for the writing rehabilitation of a patient with surface dysgraphia. CCM, a 53-year-old well-educated woman, exhibited features of conduction aphasia due to CVA: her spontaneous speech was fluent, but a marked breakdown of repetition performance was observed together with reduced digit span (3 forward). Grammatical comprehension skills were well preserved. Written language was selectively disturbed. Oral reading of words and nonwords was possible; word reading was affected neither by imageability, word class or length, nor by orthographic regularity. In contrast, spelling regularity was the crucial variable with regard to dictation. Writing errors (both homophones and neologisms) were predominantly phonologically plausible transcriptions of the target. In spontaneous written narrative, homophone substitution errors were also observed (e.g., *threw* instead of *through*). Formal testing of homophone writing revealed 49% errors on 138 items. Note again that the patient's oral reading of the same items was flawless; she was also able to define them correctly (93%). The discrepancy between reading and writing suggests that the patient could use lexical and semantic knowledge when words were presented visually; in contrast, in writing CCM was unable to retrieve the semantic information for homophone desambiguation and was incapable of accessing the correct written form.

Homophones thus became the focus of Behrmann's writing rehabilitation. Therapy was conducted on a weekly basis over a period of six weeks. In each session, each homophone of a pair was written on a card and associated with its corresponding picture. The contrast in meaning was pointed out, and the patient was to establish mnemonic associations between differences in meaning and precise orthographic form. Finally, the patient had to write the target words without in the absence of a picture (in tasks such as sentence completion and written homophone naming of pictures which were also part of further home practice). Testing post-therapy indicated significant writing improvement exclusively on the set of trained homophones. The observed change of error pattern, from pre- to post-therapy, substantiated this finding. The absence of improvement on a set of 68 untreated homophones also suggested that the

homophone deficit was particularly stable and that each word had to be retrained individually.
It should be stressed that Behrmann also noted a statistically significant post-therapy improvement on the writing of irregular words which had never been trained (from 24% to 58%). This improvement (which was shown to result specifically from retraining and not from some nonspecific effect of therapy in general) cannot be accounted for by the cited models of information processing.

REMEDIATION OF HOMOPHONE COMPREHENSION DISORDER IN A CASE OF "SURFACE DYSLEXIA WITH SURFACE DYSGRAPHIA"

Scott and Byng (1989) undertook computer-assisted remediation of homophone comprehension disorder in a patient with surface dyslexia and surface dysgraphia. JB, a 24-year-old female student, underwent surgical intervention after a car accident. Eight months post-trauma, neuropsychological examina- tion revealed memory disturbances and aphasia. Although "hyper-fluent" in conversational speech (due to developed circumlocutory strategies) JB was found to have a severe word-finding deficit which particularly affected less frequent words. The patient also had problems following complex commands. Assessment of written language revealed very slow and laboured reading. The phonological reading procedure was largely operating. Single-word reading was affected by regularity (but self corrections were numerous). Severe problems in homophone comprehension suggested that the visual word recognition system was partially disconnected from access to semantics. Difficulties in writing from dictation were similar in nature to her reading problems.
The therapeutic intervention was directed at re-establishing –for a particular set of words, i.e. homophones– the route from the visual input lexicon to the semantic system. The computer-assisted therapy programme involved comprehension of homophonic words presented in a sentence frame. The target homophone had to be identified in a set of six words containing the other homophone of the pair as well as various visual and phonological distractors. The patient received feedback on her performance. If her response was incorrect, the same sentence item would be presented again until correct identification of the target word was achieved. One year post-onset, JB carried out the complete programme (136 sentences with homophone completion) 29 times during a period of 10 weeks. On her final session the patient scored 133/136 correct. After the end of the "homophone comprehension treatment program" the patient's written language was re-assessed and compared with baseline measures. The results are the following: (1) in a task of "homophone sentence judgement" significant improvement was observed not only for trained homophones but also for untreated items; (2) in a "homophone recognition test" improvement was significant only on treated items; (3) testing "definition of homophonic words" revealed improvement on both the treated and untreated sets.
On the other hand, no significant improvement occurred in the patient's performance in tasks which were not *directly* related to the training: (1) writing

homophones from dictation had not changed (for treated or for untreated items); and (2) there was no improvement in the writing of irregular words.

Discussing their data, Scott and Byng point out that, contrary to Behrmann's (1987) claim, homophone remediation need not be word-specific since in their own study generalizations occurred from the re-training of a specific set of words. It should be stressed however, that the observed generalization effects concerned other tasks of *written homophone comprehension*—without any benefit for *written word production*, not even for those homophonic words the patient had been exposed to many times. Such a result is incongruent with the theoretical claim that visual orthographic input lexicon and orthographic output lexicon for writing words from dictation constitute separate processsing components.

HOMOPHONY AND SYLLABIC DECOMPOSITION: TREATMENT OF A CASE OF "PHONOLOGICAL DYSGRAPHIA"

Recently Ferrand and Deloche (1991) proposed an original lexical approach to writing rehabilitation for a patient with impaired word and total inability of nonword writing. Lexical writing was dramatically influenced by length: only the writing of monosyllabic words (including homophones) was preserved. In order to remediate this specific impairment of lexical writing, probably due to problems at the level of the graphemic buffer (see Caramazza et al., 1987), the authors opted for an intervention strategy which (although relying on meaning) did not use any codes, either verbal or visual, nor was it restricted to a limited set of words.

Mr. C., a right-handed well-educated subject, had a stroke at the age of 58, resulting in initially severe Broca's aphasia, acalculia and apraxic writing. One year later, at the beginning of the therapeutic intervention, the patient was noted to have severe memory problems; digit span was 5 forward and 2 backward. He was able to repeat words and nonsense syllables. Oral reading of words was 95% correct (with nonspecific residual errors), but nonword reading was disturbed (20% correct) and oral reading of isolated letters was impossible. The patient thus presented a relatively pure pattern of "phonological dyslexia" (see Beauvois and Dérouesné, 1979; Funnell, 1983) with preserved lexical decision and semantic comprehension of written words. Assessment of writing from dictation showed that the patient could not write by phoneme-to- grapheme transcoding (13% correct for nonwords). His writing of words was dramatically sensitive to length but was not influenced by any other variables such as regularity, concreteness or word class. Semantic errors were absent and the patient was able to write (short) non-homographic homophones.

The rationale for writing rehabilitation consisted of enhancing word production by the relatively preserved lexical pathway by exploiting a peculiar feature of French: many multisyllabic words can be decomposed into meaningful monosyllabic words, for example *cartable* (school bag) consists of *CAR* (bus) plus *TABLE* (table). Since not all French words can be decomposed into meaningful subunits, the therapy developed over six different stages. In stage 1 only plolysyllabic words whose syllables correspond to meaningful words (as mentioned above) were treated. In stage 2, polysyllabic words corresponding to meaningful syllabic subunits (homophones and/or homographs) except for a final mute letter were treated (e.g. le BOIS—il BOIT; le SON, ils SONT> BOISSON). Stages 3 to 6 treated polysyllabic target words in a similar fashion by gradually integrating more complex sound-to-spelling characteristics. The whole therapy consisted of 30 half-hour sessions twice a week. A total of 500 polysyllabic words were treated by means of 300 different monosyllables.

Writing assessment after therapy showed important improvements in writing polysyllabic words (from 28% to 85%). Generalization effects were observed to the writing from dictation of plurisyllabic nonwords (53% correct after therapy) and, moreover, to the reading of nonsense syllables (from 20% to 80%) although oral reading had not directly been treated.

Ferrand and Deloche point out that phonological writing was not really re-established since the patient, in order to obtain written output for nonwords, always used a "lexical search strategy" which supposedly consists of looking for syllabic correspondence with entries in the phonological lexicon which are then addressed to the semantic system where meaning is addressed. Supposedly using this procedure (which is a proper result of the specific rehabilitation method), the patient wrote nonsense syllables which could be decomposed into meaningful units at the syllabic level much better than those which resisted such analysis. Notwithstanding these observations, the authors created an effective tool to remediate writing disorders due to problems at the level of the graphemic response buffer. Their approach is original in that it is not dependent on codes and not limited to a restricted set of words. It only depends on the patient's learning to use his own knowledge for the purpose of writing even long words. Obviously, such remediation only fits patients with preserved word comprehension.

Treatments with other mnemonic aids

PRINTED WORD PLUS PICTURE: TREATMENT OF A PATIENT WITH "SURFACE DYSLEXIA".

Coltheart and Byng (1989) attempted to improve a surface dyslexic patient's ability to read lexically. Careful assessment of the patient's reading disturbance suggested that the major source of his reading impairment was a disturbance at the level of visual word recognition: written homophone comprehension was poor and the patient exclusively relied on auditory feedback (as shown by the patient's acceptance of pseudowords (sic!), e.g., **stake**: "*dinner*"; **blew**: "*colour*"). The authors decided to enhance the patient's error-prone reading of irregular words by lexical whole-word training. They chose words containing two vowels followed by the letters GH as in *though, ought,* and *plough*. These resulted in 19/24 errors pre-therapy. For each of the words, a mnemonic aid was provided: a card contained the printed word plus a picture representing the meaning of the word (e.g., *bough* by a drawing of a tree). A total of 24 items was divided into two lists to study the effects of treatment: words of set 1 were treated while words of set 2 were not. After three weeks of treatment of set 1, there was a specific effect of treatment since set 1 was more successfully read than set 2. However, the untreated words had also improved.

In order to determine what portion of the improvement was due to treatment and what portion was due to possible spontaneous recovery, Coltheart and Byng conducted two more therapy studies with the same patient. By testing word reading twice over time before treatment they verified the absence of effects of spontaneous recovery. They also employed different mnemonic techniques using symbols (even private in nature) rather than pictures as relay. Just as with the first treatment program, there were two effects: a specific treatment effect and a nonspecific effect. That is, there was some generalization of reading success, although to a lesser extent, to untreated items. Such generalization effects are not predictable from the *item-specific* models (such as, for example, Morton and Patterson, 1980) which gave impetus to the cognitive rehabilitation approach. According to Coltheart and Byng the results are, however, compatible with *distributed* processing models.

IMAGERY CUES RELATING SEMANTICALLY TO THE MEANING OF THE WRITTEN WORD: TREATMENT OF "SURFACE DYSGRAPHIA"

De Partz and co-workers (de Partz, Seron and van der Linden, 1989; Zegiser and de Partz, 1991) undertook the rehabilitation of a patient with surface dyslexia. LP, a right-handed 24-year-old man (suffering from encephalitis) had undergone lobectomy of the left temporal point and of the lower frontal lobe. Ten months post-onset the initial transcortical sensory aphasia had regressed. At the time of testing, this patient was dysfluent with word-finding difficulties (circumlocutions and semantic paraphasias). Semantic impairments were observed in comprehension tasks that required precise differentiation of closely related items. He obtained a rating score of 3 on the severity BDAE aphasia scale. Word repetition was preserved. Dyslexia (of the agnosic type) was residual but typical: the patient generally could not understand words he could not read aloud. This deficit at the level of the visual input lexicon prevented him from judging as true or false his own writing productions. His writing impairment was severe and significantly influenced by spelling regularity. In fact, phonological plausibility was respected in 93% of the errors in written word production.

The rationale for de Partz's therapeutic approach was to re-teach the patient ambiguous and irregular words, without using reading as a back-up procedure. A first attempt at directly re-teaching conversion rules had to be curtailed (after six months of twice-a-week therapy) since this treatment procedure turned out to be totally ineffective. The second attempt at rehabilitation focussed on using an imagery technique as a mnemonic aid to re-teach specific items which were whole words. For each word incorrectly spelled by LP, an association with imagery cues was established. For example, the form of the letter H in the word *pathologie* (which constitutes its major orthographic difficulty) was used to correspond to the drawing of somebody lying in a hospital bed. Another item specific example was the double consonant MM in the French word *flamme* (flame), represented by a picture of flames. The working hypothesis was that when the patient was able to evoke the appropriate image (linked to an individual word), he automatically had access to the correct orthography.

From a baseline of 250 erroneously written words, 180 words were treated (the others serving as control items). In a preliminary stage the patient was familiarized with the visual imagery technique (in tasks unrelated to writing). In the first stage of the specific therapy the patient was trained to learn written words with embedded drawings as mentioned above. Each therapy session consisted of the retraining of a set of five written words with their respective embedded drawings. The training procedure comprised the following steps: first, the patient had to copy the written word with the drawing; then he was trained in delayed copying (after 10 seconds); finally, the patient was trained to produce the written word plus embedded drawing in response to the word spoken by the therapist.

After three months of therapy, the patient exhibited a marked change in writing from dictation, between trained and untrained items, indicating that the improvement was unrelated to spontaneous recovery. In fact, even six months after the end of the therapy the stability of the effects of treatment was confirmed.

Although the experimental paradigm does not easily allow for generalization effects the final goal of the intervention was to help to transfer the patient's knowledge on trained items from writing to dictation to spontaneous production. He was thus encouraged to track the formerly visualized words while writing spontaneously.

Lexical whole-word training without mnemonic relay

RETRAINING AMBIGUOUS PRONUNCIATIONS: TREATMENT OF "SURFACE DYSLEXIA"

Friedman and Robinson (1991) attempted direct lexical whole-word training without intermediate relay and/or mnemonic aid in a case of surface dyslexia. Patient BL, a 62-year-old left-handed engineer, was seen three years post-onset when there was no longer any aphasia. His oral reading showed the typical pattern of surface dyslexia with no effects of concreteness or of part of speech and relative preservation of phonological reading (62% correct on pseudoword reading). In contrast, his oral reading showed a marked effect of regularity. Comprehension of the written words was specifically disturbed: he comprehended written words as their homophonic counterparts and, moreover, accepted pseudowords as real words. The underlying deficit was located at the level of accessing the orthographic lexicon.

Friedman and Robinson decided to focus their treatment on words with ambiguous pronunciations of certain letter clusters. Words with a given letter cluster were grouped but separated according to their ambiguous or non-transparent pronunciations (e. g., mown, grown, etc. versus cow, clown, etc.). The words in each vowel group were divided into a larger set for training and a smaller control set. The oral reading of all 253 words, in single presentation, was tested before each therapy session. (The 32 control words were included in this check but never presented on cards for treatment.) The therapy consisted of the patient's reading the words on each card aloud. He was corrected if necessary, and he repeated the target word after the therapist. BL was seen seven times over a period of 13 weeks; he was also instructed to do home exercise (reading the words aloud to his wife).

The results of the training show a difference between trained and control words: the trained words steadily improved from session to session, beginning at 71% correct and ending at 96%. Oral reading of the control words fluctuated, but there were no elements to support the conclusion that generalization had occurred. The authors conclude that "BL's improved reading is not the result of a new 'strategy' of decoding. Rather, it appears that access to specific representations for each newly-retrained word was repaired or re-created in his orthographic lexicon" (p.526).

Moss, Rothi and Fennell (1991) trained the *speed* of reading in a case with "surface dyslexia". They attempted to decrease time of oral word production in a patient with a generally high level of accuracy in word reading. Eighteen months after trauma the patient, a 25-year-old female student, showed memory deficits for verbal and non-verbal material. Digit span was 6 forward. Problems in confrontation naming, occasional word-finding difficulties in spontaneous speech and reduction of verbal fluency (10 items per second) were observed. "Spontaneous paralexic spelling errors in writing sample" were noted by the authors but were not commented upon further.

Reading was extremely slow and generally beyond acceptable limits (initially 90% correct at a 1000-msec presentation rate). The dimension of regularity affected the patient's reading. Her accuracy decreased significantly as spelling irregularity increased. It should be stressed, however, that the patient did not show any particular difficulty in the comprehension of homophonic words. It should also be emphasized that the patient, who read slowly on all tasks, tried, while reading, to write with her finger which she reported helped her sound out the words. (In our view the patient's behaviour resembles that of an agnosic alexic rather than that of a typical surface dyslexic but, as suggested by Patterson's study (1982), the line of demarcation between these syndromes may sometimes be difficult to establish.)

The treatment program was implemented 24 months post onset and consisted of only 10 one-hour sessions. The rationale for the treatment focussed on enhancing some features of the lexical reading route by using three different tasks: (1) forced semantic analysis of words presented tachistoscop- ically at presentation times not allowing for oral production (the patient had to name the semantic category of target word); (2) reading of irregular words with same level of accuracy but decreasing presentation time; (3) correct selection of homophones within the context of written sentence pairs.

The results show that the treatment dramatically decreased processing time in all written tasks, with the original level of accuracy being maintained: on an average, in task (1) from 1000 msec to 250 msec, in task (2) from 28 to 21 seconds, in task (3) from 34 to 28 seconds. However, the patient's "reading age" stayed unchanged (13.4 years).

TREATMENTS AIMING AT IMPROVEMENTS OF NON-LEXICAL WORD PRODUCTION STRATEGIES

Treatment concerning the transcoding of single units

TREATMENT OF "DEEP DYSLEXIA"

De Partz's (1986; Bachy-Langedock and de Partz, 1989) rehabilitation of a deep dyslexic patient was aimed at reorganizing the grapheme-to-phoneme transcoding process. However, the rehabilitation strategy employed tried to make use of the patient's spared lexical knowledge as a relay between graphemes and their pronunciation. Drawing upon Hatfield's relay method, de Partz used code words as a relay to deblock single letter reading in a patient who was totally incapable of reading isolated letters.

Language assessment before treatment showed that SP (a left-handed 31-year-old man who had suffered from left cerebral hemorrhage) had fluent speech with numerous phonemic paraphasias (followed by attempts at self-correction) and occasional semantic paraphasias. Severe oral naming disturbances (49% errors with 12% semantic paraphasias on total errors) persisted but the patient was often able to outline the written words he could not express orally. He could repeat words, letters and syllables with some phonemic difficulties. Spontaneous writing was still impossible and writing from dictation gave rise to 85% No-responses and to jargonagraphia.

His reading disorder (72% error rate) resembled deep dyslexia. He could read neither nonwords (7% correct) nor letters (8%). The majority of errors in word reading were No-responses (42%); the others were semantic (12%), derivational, visual or function word substitutions. Oral reading showed the expected influence of grammatical class and concreteness. Word regularity, length and frequency had no influence on the patient's reading performance. Lexical decisions were relatively well performed (95% hits on real-word recognition) documenting the (relative) preservation of the visual word- recognition system. Semantic comprehension of written words (as judged by picture/word matching) was impaired (55% errors).

De Partz's rehabilitation of simple grapheme reading comprised three steps: (i) association of letters with code word, e.g., A "Allo", B "Bébé", etc., (ii) association of letters with the first phonemes of the code words, and (iii) from letter reading to analytical reading of short words and nonwords. It should be stressed that step 1, the association of letters with code words, was laborious and only established after 52 half-hour sessions. It should also be stressed that in step 3 the patient was asked to produce the code of the first letter before attempting analytical reading. This strategy was (re)introduced to prevent the patient from producing semantic paralexias.

In the second stage of the rehabilitation—which concerned the reconstruction of complex grapheme reading—de Partz used the "relay strategy" proposed by Hatfield (1982) for the restoration of function words homophonous with content words. Other letter groups having no word homophones were associated with frequent words which were part of the patient's sight vocabulary. In order to avoid semantic paraphasias, the relay words (e.g., *lapin* for the letter group IN) were presented together with pictures. It took six sessions for the patient to be able to match all the pictures with their corresponding letter groups. Only then was he trained to read the letter groups (in the absence of a picture or a relay word) by mental evocation of the initial relay step. The training concerned short words and nonwords.

Twelve months post-onset, that is after 9 months of intensive therapy with 5 half-hour sessions per week, the patient's pre- and post-therapy reading performance were compared. The benefits were evident: the total number of errors dropped from 72% to 14%; the word/nonword opposition was no longer significant; the effects of grammatical class and concreteness tended to disappear; and only one clearly semantic error occurred. Moreover, the patient's written word comprehension, as judged by the picture/word matching task with semantic distractors, was flawless (as compared to 55% errors pre-therapy). However, of the few oral reading errors (14%) noted on the whole testing battery, 60% of the total errors (or 8.1% of the total words read) resulted from erroneous application of grapheme-to-phoneme correspondence.

In order to prevent the patient's oral reading from developing into surface dyslexia, another 65 sessions concerning the relearning of contextual graphemic conversion rules were held (Bachy-Langedock and de Partz, 1989). After these training sessions, the patient's error rate in reading aloud dropped to 2%. These residual errors were atypical; the effect of regularity (which had appeared before this latter stage) had disappeared. At this point, the patient was able to read correctly but slowly. The authors comment:

> His reading procedures were mixed. While reading phrases, the patient had recourse simultaneously to analytical reading with grapheme-phoneme codes and to reading procedures controlled by direct access to meaning. The former was used more frequently for verbs, functional words, abstract words, and words with affixes, and the latter aided by the former, when he read concrete words and very common function words. This was indicated by differences in his reading. When SP read in syllabic fashion and used phonemic approaches, he was

probably apprehending the word analytically, but concrete words were most often produced in one single oral emission, which, if incorrect, he would correct syllable by syllable. Thus one may presume that the analytical procedure was used conjointly with the global process as a means of checking it.
(p. 238)

TREATMENT OF A CASE WITH "DEEP DYSLEXIA AND DEEP DYSGRAPHIA"

One of our students in speech pathology, Guillotte (1988), undertook a similar rehabilitation of another patient with deep dyslexia. The objectives of her work were well defined: she wanted to verify (1) whether de Partz's (1986) rehabilitation method was replicable with another patient, (2) whether the therapeutic approach is efficient for a nonfluent patient (as compared to SP who was fluent), (3) whether progress would obtain, with a severely aphasic patient, twelve years post-onset (as compared to SP whose treatment started three months post-onset), and (4) whether possible benefits in oral reading would have any influence on the patient's writing from dictation, characterized by deep dysgraphia.

Pre-treatment language assessment in 1987 showed that PM, a right-handed well educated 40-year-old man, showed the same disturbances as in 1984 (when "traditional" rehabilitation stopped): agrammatism in spontaneous speech, difficulties in confrontation naming (omissions and semantic paraphasias which could be deblocked by oral start ups), preserved repetition of (short) sentences, reduced fluency (7 items per category in 1 sec), preserved comprehension of picture/word associations and even of complex commands. Motor deficits of the right limbs were still present, but by developing the skills of his left hand he was able to draw in a rather sophisticated manner. (Drawing had been one of his professional activities before his cerebral accident in 1976.)

Oral reading (and writing from dictation) of isolated words was characterized by a deep pattern of performance. Of a total of 250 stimuli 169 were read correctly (67,6%): concrete words were read better than abstract words (100% vs. 75%), and nouns were read better than verbs (55%) and better than function words (45%). Oral reading of letter names was 50% correct and the reading of nonwords was 33%. Reading errors were visual (10/47), phonological (9), derivational (8), semantic (5), function words substitutions (5) and one content word substitution. Lexical decisions, including pseudohomophones, were fairly well executed with only three false alarms on 40 items.

The treatment program was mainly conceived for reading rehabilitation, and it lasted from January through June 1988 in two one-hour weekly sessions. It dealt with four main stages: (1) learning of (simple and complex) grapheme-to-phoneme transcoding by use of code words, e.g., P "Paris"; (2) reading of short nonsense syllables, phoneme by phoneme, then syllable by syllable; (3) oral reading of pseudowords in a similar fashion; (4) reading aloud of lists of words which were either visually (e.g., *Rêve, Trêve, Vert*) or derivationally (e.g.

Bouillon, Bouillir, Bouillant) or semantically (e.g., *Ouragan, Tempête, Vent*) related. The patient had to judge his own productions as correct or incorrect; when an error was produced he was invited to resort to analytical phonological reading. The treatment of oral reading was accom- panied by similar (although less frequent) exercises for writing from dictation. Care was taken to avoid using words from the Assessment battery of reading and writing.

Assessment of reading and writing post-therapy showed significant effects for oral reading and for writing words from dictation. Overall, reading improved from 67.5% to 82% and writing from 31% to 88%. In reading, significant changes were observed only for isolated letters (from 50% to 90%) and nonwords (33% to 70%). The improvement for abstract words (from 75% to 92.5%) and for function words (from 45% to 75%) was not statistically significant. The differences in the performance on concrete and abstract words and between content and function words were no longer significant. In contrast, the difference between noun and verbs persisted, whereas the difference between words and nonwords became less significant. In writing from dictation (words, nonwords and letters), significant changes were observed for all classes of items (with the exception of high frequency concrete nouns: 80% to 90%).

Guillotte observed the same "perverse" effects of therapy as de Partz (1986): like patient SP, patient PM showed some characteristics of surface dyslexia. More precisely, in spite of eventually correct oral reading, PM sometimes had difficulties in understanding the meaning of written items (especially of pseudowords but also of some words). In light of the patient's quasi-normal verbal comprehension (already observed before treatment) this finding further substantiates the claim that "understanding via auditory feedback" cannot be taken to be automatic or even "normal" (see Kremin, 1985, for more detailed discussion).

Notwithstanding these observations and reflexions, Guillotte's study showed, that the method of "key words as relays" –as proposed by Hatfield (1982) and extended to phonological reading by de Partz (1986)– is efficacious for the remediation of reading *and* writing disturbances not only in fluent (de Partz, 1986) but also in agrammatic patients, even when administered twelve years post-onset.

TREATMENT OF A PATIENT WITH "ALEXIA WITHOUT AGRAPHIA" IN A CASE OF VISUO-VERBAL DISCONNECTION

Beauvois and Dérouesné (1982) resorted to a *gestural* relay for single letter "reading" in a case of alexia without agraphia. The patient could identify words and even comprehend them but was unable to read them aloud (30% correct with frequent concrete words; 40% correct for isolated letters; 2% correct for nonwords).

The patient, MP, a 61-year-old woman, suffered from optic aphasia and hemianopia subsequent to CVA. Since Beauvois (1982) had experimentally shown that MP suffered from a specific deficit, i.e., optic aphasia in terms of visuo-verbal disconnection (and not from colour agnosia), it was hypothesized that her reading impairment might be due to the same underlying deficit. That is, oral reading of words may have been impossible because it entailed a visuo-verbal process. Beauvois and Dérouesné thus decided to attempt remediation of the patient's reading impairment by preventing her from directly going from vision (the written word) to language (verbal label of the target word). They trained the patient to use letter-by-letter decoding (instead of possible whole-word reading) by means of a gestural relay. The method was the following: (1) MP was to copy a given letter by gesture, (2) she had to look at her gesture, (3) she had to associate her gesture with an arbitrary verbal code to prevent letter naming, and finally (4) she had to learn the associations between (arbitrary) verbal code and letter name. When this procedure became automatized, sequences of letters were presented. (In order to avoid any spontaneous verbalization only nonwords were treated initially.)

Treatment started approximately two years post-onset and lasted for five months with eight sessions per week. At the end of this period the patient's oral reading had improved significantly in both accuracy and reading speed: letters from 45% to 100% and 18 seconds to 3 (pre/post-therapy), words from 30% to 85% and 25 seconds to 10, nonwords from 2% to 55% and 30 seconds to 12.

Treatment at the syllabic level

TREATMENT OF "SEVERE NON-SPECIFIC WRITING IMPAIRMENT" IN A PATIENT WITH SURFACE DYSLEXIA

Carlomagno and Parlato (1989) undertook the rehabilitation of a severely agraphic Italian patient. Aphasia testing showed no spoken language impairment except for a few self-corrected phonemic paraphasias in repetition and oral naming. Speech was fluent, well articulated, and apparently normal. Digit span was 4 forward. The patient, a 60-year-old right-handed man, complained only of reading and writing disturbances.

Formal assessment of the patient's written language revealed that he read words better than nonwords (80% vs 30%) with no consistent effects of concreteness, word class, frequency or length on word reading. Qualitative analysis of oral reading permitted classification of the patient's impairment as surface dyslexia which involves a deficit of lexical reading. Namely, stress assignment errors on irregularly stressed words suggested that the patient was using nonlexical grapheme-to-phoneme conversion. Moreover, he relied on the phonological form of the written stimulus for reading comprehension. However, the patient did not produce errors of context rules or misapplication of parsing procedures. The authors concluded that the "patient carried out grapheme-to-phoneme conversion by segmenting the letter string at the syllabic level and then mapping these multigraphemic sequences onto their phonological counterparts" (p. 185). The patient's inferior reading of nonwords as compared to words shows that the patient also suffered from some damage to the peripheral reading route. Since nonword reading was sensitive to homophony with real words, Carlomagno and Parlato suggested that a lexical aid was spontaneously used by the patient in order to attain phonological output. Writing was more severely disturbed in a non-specific manner in all writing tasks. In writing from dictation there were no true effects of word-class, concreteness and frequency. A slight length effect was observed. The severity of the writing impairment and the lack of a particular pattern led the authors to conclude that both the lexical and non-lexical route were seriously damaged. Carlomagno and Parlato's rationale for treatment was to attempt to improve the patient's writing by phoneme-to-grapheme translation by using *syllabic* treatment instead of isolated letters—a strategy supposedly familiar to the patient since his reading was shown to operate at the syllabic level. Also, in the Italian writing system, which is very transparent as far as conversion rules are concerned, almost all orthographic rules operate on the syllabic level.

Therapy with this patient lasted about five months, with two sessions per week. The first stage consisted of searching for verbal codes as a relay. Since the patient had worked as a railway clerk, Italian towns or Christian names were chosen which contain in the first syllable about 30 consonant-vowel clusters to be treated, e.g., mito (myth) = MI (from Milano) plus TO (from Torino). The authors state that the procedure was as follows: "The patient was trained to match spoken syllables uttered by the therapist with its code name, then code names with words, always uttered by the therapist, containing the relative syllable in first position. Finally he was requested to write the syllable without a model" (p. 191). After a few sessions, the patient mastered this task. Then multisyllabic words were introduced and, this time, the patient had to write the target word (through syllabic decomposition and association with code words). In the second stage the patient had to automate the procedure by applying the strategy to nonwords. The last month of treatment was spent in training the patient on consonant cluster and diphthong decomposition as well as the double consonant rule.

After five months, because of reasons unrelated to the therapeutic intervention, the treatment had to stop. Immediate and delayed evaluation of the patient's writing performance showed, even two months after the end of treatment, a substantial gain from treatment. Moreover, none of the few errors made were inconsistent with the application of the learned strategy. In consequence, the final result "could be considered with our prediction of 'surface dysgraphia' whose grapheme-to-phoneme system operates at the syllabic level" (Carlomagno and Parlato, 1989: 199). Note that, in Italian, simple transcoding allows for about 80% correct word production whereas for French it is estimated to allow for only 50%. In fact, in French 30 phonemes result in about 130 graphemic transcriptions.

Finally, documents on intermediate checks permitted Carlomagno and Parlato to observe significant learning transfer from written treatment to oral reading: the patient improved on *nonword* reading at the same rate as on word writing. In summary, the direct treatment of approximately 100 code names (which were easily learned and used by the patient) resulted in this positive outcome in a patient who, before therapy, probably had some nonlexical orthographic knowledge. However, he had been unable to access this knowledge without a relay via the phonological output lexicon.

TREATMENT OF IMPAIRMENT OF KANA PROCESSING

Sasanuma (1986) reports on kana writing and reading programs developed by Kashiwagi and Kashiwagi (1978). The cardinal feature of these programs is to use a kanji character as a key word to form a link with each kana so as to facilitate relearning of kana-syllable correspondences. These programs thus seem to share the same theoretical framework with those therapeutic approaches described for dyslexic and dysgraphic users of alphabetical orthographies.

Indeed, impairment of kana processing poses a major problem for Japanese patients since it is a part of reading and writing ability per se: function words and predicate inflections as well as most loanwords are represented in kana. An impairment of kana processing thus hampers daily communication more dramatically than the loss of phonological processing by grapheme-to- phoneme and/or phoneme-to-grapheme transcoding in alphabetic languages. It must be noted that the sole loss of phonological processing in alphabetic languages does not necessarily interfere with the preserved reading of all sorts of words, including function words and inflections (see Funnell, 1983, for the example of an English patient).

In Kashiwagi and Kashiwagi's treatment program for kana writing (1978, in Japanese) the patient is helped through the following steps (cited from Sasanuma, 1986) with initially 46 kana characters: *step 1*: Say the key word (/kaki/ or "persimmon" in this example) on hearing the initial syllable of the word, i.e., the pronunciation of the target kana /KA/ > /Kaki/; *step 2*: Write the kanji key word; *step 3*: Write the kana linked with the kanji key words (i.e., the initial syllable of the key word; *step 4*: Gradually phase out the strategy of using kanji key words as intermediary steps until this is done internally or omitted altogether; *step 5*: Expand the ability to write each kana character not only individually but also in words and sentences.

The program for reading is based upon a similar approach: *step 1*: Write the kanji key word after the written presentation of the target kana; *step 2*: Read the key word; *step 3*: Separate the initial syllable of the key word and say it: /kaki/ > KA. *Steps 4 and 5* are the same as in the writing program.

Three patients (whose spontaneous speech was severely impaired, either limited to one- or two-word utterances or, in the third case, characterized by empty speech) went through this program. Each patient was given 4 to 5 45-minute therapy sessions per week lasting from 6 to 14 months (with supplementary homework). All patients significantly improved in writing and reading not only individual kana characters but also kana words. However, from this study it was not clear "whether the key word kanji with its semantic representation was in fact playing an intermediary role in facilitating the

reorganization of kana-syllable correspondences, as intended by the therapy program" (Sasanuma, 1986: 161).
Another work by Kashiwagi et al. (1985) seems to answer this question. The authors studied a patient with selective impairment of kana writing.

> He was a 68-year-old college-educated man exhibiting fluent aphasia at 3 months post-onset of a CVA. After 2 months of intensive therapy based on the kana writing program just described, the patient underwent a series of 14 testing sessions over a period of 4 months, during which a total of 611 kana characters were dictated to him and his responses were recorded and analyzed. Findings of immediate relevance to our question were: (1) the patient achieved a mean of 78% correct responses (479/611) throughout the test sessions, indicative of a significant gain over his pre-treatment performance; (2) in 89% of these correct responses he also succeeded in retrieving the key words; (3) on the other hand, he failed to retrieve the key words in about half of his incorrect responses; (4) there were instances of retrieval of wrong key words which were semantically related to the target key words (semantic paraphasias) (...) and on those occasions the patient's response consisted of writing the first syllable of the paraphasic key word instead of the syllable of the target key word. Based on these findings, the authors concluded that the word meaning of kanji in fact played a crucial role in the retraining program.
>
> (Sasanuma, 1986: 162)

Taken together, the findings with Japanese patients indicate that the patients have in fact learned to use a strategy for exploiting the lexical semantic system in (re)establishing kana-syllable correspondences.

CONCLUSION

Assessment and therapy of written language impairments show some universal neurospychological mechanisms underlying these performance deficits: on the one hand, there is a more or less selective inability to derive word phonology from the word's orthographic representation which results in (partly faulty) word production by means of transcoding procedures; on the other hand, there are (more or less preserved) lexical production strategies—even when groups of patients are considered (see Carlomagno et al., 1991).

These core structures of reading and writing override script-specific features. Nevertheless script-specific features—whether in alphatical languages (like English and French as opposed to Italian or Spanish) or in non-alphabetical languages like Japanese—exert an important influence on the outcome of therapies based on *similar* rationales.

The "theory of remediation of cognitive deficits" postulated by Caramazza and Hillis (1991) surely is a final goal of the whole enterprise we have begun. But, for the moment at least, there seems to be a greater urgency in achieving understanding—within the framework of the model of word recognition and production now relatively well established. That is, *why* do generalizations sometimes extend from treated to non-treated items and/or modalities, and why so often do they not?

References

Bachy-Langedock, N. and de Partz, M.P. 1989. Coordination of two reorganization therapies in a deep dyslexic patient with oral naming disorder. In X. Seron and G. Deloche (eds.) *Cognitive Approaches in Neuropsychological Rehabilitation*. Hillsdale, New Jersey: Lawrence Erlbaum. Pp. 211-247.
Beauvois, M.F. 1982. Optic aphasia: A process of interaction between vision and language. *Philosophical Transactions of the Royal Society of London*, B **298**: 35-47.
Beauvois, M.F. and Derouesné, J. 1979. Phonological alexia: Three dissociations. *Journal of Neurology, Neurosurgery and Psychiatry*, **42**: 1115-1124.
Beauvois, M.F. and Derouesné, J. 1982. Recherche en neuropsychologie et rééducation: Quels rapports? In X. Seron and C. Laterre (eds.) *Rééduquer le cerveau. Logopédie, Psychologie, Neurologie*. Bruxelles: Mardaga. Pp. 163-189.
Behrmann, M. 1987. The rites of righting writing: Homophone remediation in acquired aphasia. *Cognitive Neuropsychology*, **4**: 365-384.
Behrmann, M., Black, S.E. and Bub, D. 1990. The evolution of pure alexia: A longitudinal study of recovery. *Brain and Language*, **39**: 405-427.
Bub, D. and Chertkown, H. 1989 . Agraphia. In F. Boller and J. Grafman (eds.) *Handbook of Neuropsychology. Vol. 1*. Amsterdam: Elsevier Science Publishers. Pp. 393-414.
Bub, D. and Kertesz, A. 1982. Deep agraphia. *Brain and Language*, **17**: 146-165.
Caramazza, A., Berndt, R.S. and Hart, J. 1981. Agrammatic Reading. In: F.J. Pirozzolo, and M.C. Wittrock (eds.) *Neuropsychological and Cognitive Processes in Reading*. New York: Academic Press. Pp. 297-317
Caramazza, A. and Hillis, A.E. 1990. Where do semantic errors come from? *Cortex*, **26**: 95-122..
Caramazza, A. and Hillis, A.E. 1991. For a theory of remediation of cognitive deficits. *NIDCD Workshop on Treatment of Aphasia*. Bethesda, MD, June 6-7, 1991.
Caramazza, A., Micelli, G., Villa, G. and Romani, C. 1987. The role of graphemic buffer in spelling: Evidence from a case of acquired dysgraphia. *Cognition*, **26**: 59-85.

Carlomagno, S. and Parlato, V. 1989. Writing rehabilitation in brain damaged adult patients: A cognitive approach. In X. Seron and G. Deloche (eds.) *Cognitive Approaches in Neuropsychological Rehabilitation*. Hillsdale, New Jersey: Lawrence Erlbaum. Pp. 175-209.

Carlomagno, S., Colombo, A. Casadio, P, Emanuelli, S. and Razzano, C. 1991. Cognitive approaches to writing rehabilitation in aphasics: evaluation of two treatment strategies. *Aphasiology*, 5: 355-360.

Coltheart, M. and Byng, S. 1989. A treatment for suface dyslexia. In X. Seron and G. Deloche (eds.) *Cognitive Approaches in Neuropsychological Rehabilitation*. Hillsdale, New Jersey: Lawrence Erlbaum. Pp. 159-174.

Coltheart, M., Patterson, K.E. and Marshall, J.C. 1980. *Deep Dyslexia*. London: Routledge and Kegan Paul.

Coslett, H.B., Rothi, L.G. and Heilman, K.M. 1985. Reading: Dissociation of the lexical and phonologic mechanisms. *Brain and Language*, 24: 20-35.

De Bleser, R., Bayer, J. and Luzzatti, C. 1987. Die kognitive Neuropsychologie der Schriftsprache —Ein Überblick mit zwei Fallbeschreibungen. *Linguistische Berichte Sonderheft* 1: 118-162.

Déjerine, J. 1891. Sur un cas de cécité verbale suivi d'autopsie. *Mémoires de la Société de Biologie*, 197-201.

Déjerine, J. 1892. Contribution à l'étude anatomo-pathologique et clinique des différentes variétés de cécité verbale. *Mémoires de la Société de Biologie*, 4: 61-90.

De Partz, M.P. 1986. Re-education of a deep dyslexic patient: Rationale of the method and results. *Cognitive Neuropsychology*, 3: 149-177.

De Partz, M.P., Seron, X. and Van Der Linden, M. 1989. Reeducation of a surface dyslexic patient with a visual imagery strategy. *Seventh European Workshop On Cognitive Neuropsychology: An Inter- disciplinary Approach*. Bressanone, Italy, January 1989.

Ellis, A.W. 1982. Spelling and writing (and reading and speaking). In A.W. Ellis (ed.) *Normality and Pathology in Cognitive Functions*. London: Academic Press.

Exner, S. 1881. Untersuchungen über die Lokalisation der Funktionen. In: *Die Grosshirnrinde des Menschen*. Wien: Wilhelm Braunmüller.

Ferrand, I. and Deloche, G. 1991. Thérapie expérimentale de l'écriture dans un cas d'atteinte de la voie phonologique avec préservation de la production des monosyllabiques. *Société de Neuropsychologie de Langue Française*, Paris, December 1991.

Friedman, R. 1988. Acquired alexia. In F. Boller and J. Grafman (eds.) *Handbook of Neuropsychology. Vol. 1*. Amsterdam: Elsevier Science Publishers. Pp. 377-391.

Friedman, R. and Robinson, S.R. 1991. Whole-word training therapy in a stable surface patient: it works. *Aphasiology*, 5: 521-527.

Funnell, E. 1983. Phonological processes in reading: New evidence from acquired dyslexia. *British Journal of Psychology*, 74: 159-180.

Glosser, G. and Friedman, R.B. 1990. The continuum of deep/phonological alexia. *Cortex*, **26**: 343-359.
Goldberg, T. and Benjamins, D. 1982. The possible existence of phonemic reading in the presence of Broca's aphasia: A case report. *Neuropsychologia*, **20**: 547-558.
Guillotte, N. 1988. A propos de la thérapie cognitive d'une dyslexie profonde. Mémoire d'Orthophonie, Faculté de Médecine de Tours.
Hatfield, F.M. 1982. Diverses formes de désintégration du langage écrit et implications pour la rééducation. In X. Seron and C. Laterre (eds.) *Rééduquer le cerveau. Logopédie, Psychologie, Neurologie*. Bruxelles: Mardaga. Pp. 135-156.
Hécaen, H. and Kremin, H. 1976. Neurolinguistic research on reading disorders resulting from left hemisphere lesions. Aphasic and "pure" alexias. In H. Whitaker and H.A. Whitaker (eds.) *Studies in Neurolinguistics, Vol 2*. New York: Academic Press. Pp. 264-271.
Howard, D. and Franklin, S. 1989. *Missing the Meaning? A Cognitive Neuropsychological Study of Processing of Words by an Aphasic Patient*. Cambridge, Massachusets: M.ii.T. Press.
Howard, D. and Patterson, K. 1989. Models for Therapy. In X. Seron and G. Deloche (eds.) *Cognitive Approaches in Neuropsychological Rehabilitation*. Hillsdale, New Jersey: Lawrence Erlbaum. Pp. 39-64.
Kremin, H. 1980. Deux stratégies dissociables par la pathologie: Description d'un cas de dyslexie profonde et d'un cas de dyslexie de surface. *Grammatica*, **VII**: 131-156.
Kremin, H. 1982. Alexia. Theory and Research. In R.N. Malatesha and P.J. Aaron (eds.) *Reading Disorders - Varieties and Treatments*. New York: Academic Press. Pp. 347-367.
Kremin, H. 1984. Comments on pathological reading behavior due to lesions of the left hemisphere. In R.N. Malatesha and H.A. Whitaker (eds.) *Dyslexia. A Global Issue*. The Hague: Martinus Nijhoff Publishers. Pp. 273-310.
Kremin, H. 1985. Routes and strategies. Data on acquired surface dyslexia and surface dysgraphia. In K.E. Patterson, J.C. Marshall and M. Coltheart (eds.) *Surface Dyslexia. Neuropsychological and Cognitive Studies of Phonological Reading*. London: Lawrence Erlbaum Associates. Pp. 105-137.
Kremin, H. 1987. Is there more than ah-oh-oh? Alternative strategies for writing and repeating lexically. In M. Coltheart, G. Sartori and R. Job (eds.) *The Cognitive Neuropsychology of Laguage*. London: Lawrence Erlbaum Associates. Pp. 295-335.
Kremin, H. 1989a. Lexical access viewed from the information processing approach: Reading and writing (data from pathology). In P.G. Aaron and R.M. Joshi (eds.) *Reading and Writing Disorders in Different Orthographic Systems*. Dordrecht: Kluwer Academic Publishers. Pp. 283-304.

Kremin, H. 1989b. Case study of a patient with surface dyslexia and deep dysgraphia with special attention to the noun/verb distinction. *International Conference On Cognitive Neuropsychology*, Harrowgate, Great Britain, 24-27 July 1989.

Kremin, H. 1992. La neuropsychologie de la lecture: de la définition de syndromes à la description de structures mentales. In P. Lecocq (ed.) *La Lecture. Processus, apprentissage, evaluation, troubles*. Presses Universitaires de Lille.

Kremin, H. In press. Naming impairments and the information processing approach. In F.J. Stachowiak et al. (eds.), *Development in the Assessment and Rehabilitation of Brain-Damaged Patients: Perspectives from a European Concerted Action*. München: Narr Verlag.

Laine, M., Niemi J. and Marttila, R. 1990. Changing error patterns during reading recovery: A case study. *Journal of Neurolinguistics*, 5: 75-81.

Landis, T., Regard, M. and Serrat, A. 1980. Iconic reading in a case of alexia without agraphia caused by a brain tumor: a tachistoscopic study. *Brain and Language*, 11: 45-53.

Luria, A.R., Naydin, V.L., Tsveskova, L.S. and Vimarskaya, E.N. 1969. Restoration of higher cortical function following local brain damage. In P. Vinken and G.N. Bruyn (eds.), *Handbook of Clinical Neurology, Vol.3*. Amsterdam: North Holland. Pp. 368-433.

Marcie, P. and Hecaen, H. 1979. Agraphia: Writing disorders associated with unilateral cortical lesions. In K.M. Heilman and E. Valenstein (eds.) *Clinical Neuropsychology*. New York: Oxford University Press. Pp. 92-127.

Marin, O.S.M., Saffran, E.M. and Schwartz, M.T. 1976. Dissociations of language in aphasia: Implications for normal reading. *Annals of the New York Academy of Sciences*, 280: 868-884.

Marshall, J.C. and Newcombe, F. 1973) Patterns of paralexia: A psycholinguistic approach. *Journal of Psycholinguistic Research*, 2: 175-199.

Micelli, G. 1989. A model of the spelling process: Evidence from cognitively impaired Subjects. In P.G. Aaron and R.M. Joshi (eds.) *Reading and Writing Disorders in Different Orthographic Systems*. Dordrecht: Kluwer Academic Publishers. Pp. 305-328.

Morton, J. 1980 The logogen model and orthographic structure. In U. Frith (ed.) *Cognitive Processes in Spelling*. London: Academic Press. Pp. 117-134.

Morton, J. and Patterson, K.E. 1980. A new attempt at an interpretation or an attempt at a new interpretation. In M. Coltheart, K.E. Patterson and J.C. Marshall (eds.) *Deep Dyslexia*. London: Routledge and Kegan Paul. Pp. 91-118.

Moss, S., Rothi, L.G. and Fennell, E.B. 1991. Treating a case of surface dyslexia after closed head injury. *Journal of Clinical Neuropsychology*, 6: 35-47.

Nolan, K.A. and Caramazza, A. 1982. Modality-independent impairments in word processing in a deep dyslexic patient. *Brain and Language*, **16**: 232-264.

Patterson, K.E. and Kay, J. 1982. Letter-by-letter reading: Psychological descriptions of a neurological syndrome. *Quarterly Journal of Experimental Psychology*, **34A**: 411-441.

Patterson, K.E., Marshall, J.C. and Coltheart, M. 1985. *Surface Dyslexia. Neuropsychological and Cognitive Studies of Phonological Reading*. London: Lawrence Erlbaum Associates.

Ross, P. 1983. Phonological processing during silent reading in aphasic patients. *Brain and Language*, **19**, 191-203.

Sasanuma, S. 1986. Universal and language-specific symptomatology and treatment of aphasia. *Folia Phoniatrica*, **38**: 121-175.

Shallice, T. 1988. *From Neuropsychology to Mental Structure*. Cambridge/ New York: Cambridge University Press.

Scott, C. and Byng, S. 1989. Computer assisted remediation of homophone comprehension disorder in surface dyslexia. *Aphasiology*, **3**: 301-320.

Warrington, E.K. 1981. Concrete word dyslexia. *British Journal of Psychology*, **72**: 175-196.

Warrington, E.K. and Shallice, T. 1979. Semantic access dyslexia. *Brain*, **102**: 43-63.

Zegiser, P. and de Partz, M.P. 1991. Rééducations cognitives des troubles de l'orthographe et/ou de l'écriture. In M.P. de Partz and M. Leclercq (eds.) *La rééducation Neuropsychologique de l'adulte*. Paris: Edition de la Société de Neuropsychologie de Langue Française. Pp. 53-77.

10 A review of therapy at the level of the sentence in aphasia

Sally Byng and Ruth Lesser

For many people with aphasia, the inability to produce language to express their needs, desires and thoughts represents a major loss, so that the restoration of the ability to use language in these respects is particularly important. For some people these aspects of communication will never be accomplished within connected language and alternative means of expression will need to be sought non-verbally or through the use of fragments of language. However for many people the production of connected language as a vehicle through which they can express themselves represents a desirable and feasible goal in therapy.

In this chapter we review the literature that describes attempts to provide therapy for people with problems with language comprehension and production at the level of the sentence. The purpose of reading a literature review in any domain would be to discover what the body of accumulated knowledge reveals about the subject being considered, and what theoretical orientations are employed in the studies. In the case of therapy studies, however, there is an additional expectation; that is that the body of knowledge reviewed should be able to provide some evidence about how people with aphasia can improve their ability to produce and interpret connected language through therapy. Clinicians know from experience and observation that these aspects of language can improve over time and through therapy. The body of knowledge on therapy at the level of the sentence might therefore be expected to reflect how this improvement is accomplished. Our review below in fact shows that, in most cases, very little attempt is made in many of these studies at an explanation of how therapy may be linked to changes in underlying processing. This may be because few of the existing therapy studies discuss or specify the aspect of sentence comprehension or production being addressed either by the analysis of the impairment or by the therapy. That is, is the therapy directed at a syntactic deficit, a semantic deficit, a phonological deficit or at an interaction of deficits underlying the sentence comprehension or production problem? This makes the results of the therapy hard to interpret; we cannot know the potential of a therapy for remediating the disorder of sentence comprehension or production if we do not know the aspect of language at which it is directed, and therefore for what type of disorder in what type of patient it is appropriate.

An assumption might also be made that the aims of therapy at the sentence level would relate to the basic function of connected language suggested above, that is that therapy studies would be seeking to facilitate the restoration of the ability to use connected language as a means of expressing needs, desires and thoughts, and as a basic tool for communicating with other people. Again this review of the literature reveals that this expectation is not met substantially by the existing studies. In addition, considering that these factors represent a fundamental aspect of impaired communication for many people with aphasia, the number of studies addressing them is small.

If the literature does not address these issues we need to consider what it does address. Examining the aims of the existing studies, a variety of different reasons for carrying out therapy is revealed which seem to fall into two main groups. In one group the focus of the studies seems to be on examining the outcome of a specific therapy for an observed symptom, with respect to the effect of that therapy on treated stimuli and the generalisation of the effects to various types of untreated stimuli. This group of studies in general uses the terminology of "syndromes", and of Broca's aphasia and agrammatism in particular. In the other group the focus is on establishing the nature of the problem underlying the difficulty in comprehending or producing sentences, and then implementing and evaluating therapy for that specific difficulty. In general this group of studies uses psycholinguistic models of sentence processing as its frame of reference. We discuss these two differing orientations below as "symptom focussed" and "processing focussed" studies.

THEORETICAL ORIENTATIONS TO THE LANGUAGE IMPAIRMENTS

Like the pragmatic approach to aphasia therapy discussed in Chapter 6, therapy at the sentence level draws eclectically on snatches of clinical, psycholinguistic and linguistic theories, without necessarily relating them to a consistent body of work.

The syndrome-based approach to therapy at the sentence level is only loosely connected to a theory base. It generally makes the assumption that problems in producing sentences are due to difficulties with grammar, or with the use of function words and inflections; the focus of treatment is therefore on the production of complete sentences of varying grammatical forms, or on the use of subsets of function words or inflections which encode time relationships or mood. The studies classified under the heading of the "symptom focussed" grouping, therefore, almost all share a set of assumptions which might be summarised as follows: (a) there is no need to specify the type or nature of the deficit beyond assigning a syndrome label to the patient or providing a very short description of the patient's language functioning; (b) this lack of specification is probably a result of another assumption that patients who are classified within the same syndrome will have the same type of deficit; and (c) if the patients who are the subject of the study are Broca's aphasics then they have a syntactic deficit. Examples of studies which fall in this category are those by Shewan (1976), Helm-Estabrooks and Ramsberger (1986) and Kearns and Salmon (1984); we describe these in more detail below.

In contrast to the "symptom focussed" grouping, a common feature of the studies grouped under the heading "processing focussed" is that each patient's language impairment needs to be interpreted within the framework of knowledge about the normal language processing system (see for example Coltheart, Sartori and Job, 1987). The assumption underlying this approach is that similarity of surface symptoms might mask different underlying impairments. The notion of homogeneity within syndrome groups has been severely challenged during the last decade, particularly with respect to agrammatism in Broca's aphasia (see for example Badecker and Caramazza, 1985). A large number of studies on sentence processing in aphasia have been carried out during the last decade so that there exists now a plethora of empirical evidence about the range of deficits, dissociations and coocurrences of symptoms and their relationship to normal language processing. Berndt (1991: 263) in a comprehensive review of these studies suggests that

> symptoms that once appeared to be quite similar across patients and consistent within a patient have been shown to be dauntingly variable when performance is carefully scrutinized. This variability may indicate that most if not all of these symptoms can arise from several distinct functional disorders.

Therefore a further assumption shared by these studies is that, having derived some working hypotheses about the specific nature of the sentence processing impairment that an individual patient might have, psycholinguistic concepts about normal sentence processing can inform the clinician about relevant aspects of sentence processing to focus on in therapy.

The process of producing a sentence, like any other psycholinguistic task, is made up of a number of components (Byng and Black, 1989; Black, Nickels and Byng, 1991). Each component contributes a particular type of information to the process as a whole, and these different kinds of information have to be translated into one another. Thus to convey an event, say, through language we have first to know what message is to be conveyed. Then we have to construct a representation at a conceptual or semantic level which specifies what role each person or thing plays in that event, in relation to the other people or objects involved. This kind of representation is referred to as the "predicate-argument structure". This information has not yet been translated into a syntactic framework. Precise models for the translation, or mapping, of the predicate-argument structure into syntactic structure have yet to be developed, but current linguistic and psycholinguistic theories of sentence processing lay emphasis on the role of lexically based information and procedures (in addition to general principles and procedures) particularly with respect to verbs and other lexical items that determine both the type of phrases that can cooccur with them and how those phrases are interpreted (see Bresnan and Kaplan, 1981; Garrett, 1982; Chomsky, 1985). Once the conceptual information has been translated into syntactic structure, that syntactic structure must be mapped on to a phonological representation and then on to a form for translation to motor-articulatory commands. A sentence production deficit could arise at any of these different levels (Caplan, 1987) and/or from an impairment in the procedures that translate, or map, between these levels (Schwartz, Linebarger and Saffran, 1985).

THEORETICAL ORIENTATIONS TO THERAPY

With neither of these theoretical orientations is it made explicit generally how the therapy devised relates to the specific wishes or needs for language and communication of the individuals concerned. This is not to say that the patients themselves were not involved necessarily in devising therapy, but rather that it is not made clear how the therapy related to them specifically. The origins of many of the studies seem to stem from observations of characteristics of performance at the sentence level and therefore the therapy is devised to remediate that observed characteristic of impaired performance, rather than to remediate the underlying deficit at a specified level of language processing.

The tasks used in the therapies deriving from these different orientations are not necessarily very different. Many therapies are based on input tasks; that comprehension of language is adequate to sustain the therapy or as a basis for the therapy tasks is implicit in most studies rather than having been established a priori. Many of the tasks across all the different orientations to therapy use multiple modalities of input, such as spoken and written sentences, with or without pictures, focussing on a set of specific sentence constructions.

Where the major difference between the two groups of studies lies is in how the tasks are presented, what it is intended that the patient gets from the task and the descriptions of the feedback provided, that is, what the clinician did when the patient made an incorrect response. The symptom focussed group studies, in general, make the presentation of the stimuli and the action of the clinician very explicit. In the processing focus group the tasks might be described, along with what the patient is required to do, but the feedback used is generally much less explicit. This is because the emphasis in these studies is not on elicitation per se, but rather on facilitation of processing of language. The interaction between the therapist and the patient is intended to allow the patient an opportunity to understand some particular aspect of how language works, such as how sentences are constructed, and so the actions of the therapist relate to conveying that information in relation to the specific responses of the patient. The interactions—what the therapist says when the patient's response is incorrect— could involve many different techniques, such as an explanation of why a particular error was made, or a strategy for the patients to implement to facilitate their sentence production for themselves.

The difference between the therapeutic procedures in the two groups could, with some exceptions, be described as prescriptive versus experiential methods. The symptom focussed studies tend to focus on providing a model of the specific response that a patient is meant to try to produce—in effect patients have to practise what they cannot do (McCrae Cochrane and Milton, 1984, would be an exception here). Patients have to demonstrate item specific learning. The therapy methods in the processing focussed group, on the other hand, could be described as emphasising skill-building and underpinning concepts about language rather than practising what is not there. The patient is learning something about how language works, with strategies for solving a specific problem and facilitating production of language. There is, usually, less emphasis on the learning of a set of specific items, since the identity of those items is less important than the nature of those items and their function in the language. As an example, therefore, in a therapy aimed at increasing the combination of nouns and verbs in sentences it would be less important to demonstrate learning of a specific set of verbs and nouns than to show that structured sentence production has improved.

There is as yet, however, no theory underlying the application of any of these therapies. In contrast to the theories available about the origin and nature of the deficit, we do not have theories available about how or why certain tasks used in certain ways might bring about change. Looking at the results of the therapy it is clear that we are in urgent need of such theories to improve the outcome of our therapy. This review serves to highlight the variation in the aims and objectives of therapy, the differences in interpretation of the deficits, and the range of different outcomes of the therapy. We will attempt to suggest some reasons for the restricted nature of much of the improvement measured after therapy.

A summary of all the studies to be reviewed is included in Table 1, in order to provide a quick, albeit incomplete, reference to the contents of the studies. A striking observation from this table is that nearly all the studies addressing therapy at the level of the sentence involve people with a non-fluent, Broca's type aphasia. This is not surprising perhaps given that Broca's aphasics demonstrate so obviously an impairment at that level. However that is not to say that people with fluent production of language do not also have problems at the level of the sentence, but it seems that they have not been the target of as much research interest. In addition it seems that most of the studies comprise either single case studies or small groups of single cases, where data are provided for each subject so that it is possible to track the effects of the therapy for each of the people treated.

The section entitled "description of the deficit" reveals that the majority of studies do not provide details of the language disabilities that the people to be treated have. This means that it is hard to know how similar these patients are. In comparing the patients described across the therapy studies being considered here, a range of different degrees of severity or of impairments is apparent amongst the patients classified as Broca's aphasics. For example all the patients treated in the studies by Thompson and McReynolds (1986), Kearns and Salmon (1984) and Helm-Estabrooks and Ramsberger (1986) are described as being agrammatic Broca's aphasics and yet they are all reported as being able to produce verbs prior to the therapy. The agrammatic Broca's aphasic patients studied by Jones (1986), Byng (1988) and Nickels, Byng and Black (1991) are all unable to

		SYMPTOM FOCUSSED STUDIES											PROCESSING FOCUSSED STUDIES					
		Holland & Levy (1971)	West (1973)	Holland & Sonderman (1974)	Shewan (1976)	McCrae, Cochrane, and Milton (1984)	Kearns and Salmon (1984)	Thompson and McReynolds (1986)	Helm-Estabrooks and Ramsberger (1986)	Naeser et al. (1986)	Davis & Tan (1987)	Doyle, Goldstein, and Bourgeois (1987)	Doyle et al. (1989)	Beyn & Shokhor-Trotskaya (1966)	Jones (1986)	Byng (1988)	Loverso et al. (1988)	Nickels, Byng, and Black (1991)	Le Dorze et al. (1991)	Mitchum & Berndt (in press)
Purpose of Study	Efficacy of a Specific Therapy Procedure	√	√	√		√	√		√	√	√		√	√	√	√	√		√
	Comparison of Two Procedures				√			√											
	Implementation of an Existing Procedure												√				√	√	
Orientations	Behavioral (a)	√	√	√			√	√	√			√	√		√				
	Stimulation							√			√								
	Processing														√	√	√	√	√	√
	Undefined				√	√				√									
Size of Study	Single Case				√				√			√				√	√	√	
	Group Study	√	√										√						
	Group of Single Case Studies (b)			√		√	√	√		√	√	√	√		√	√			
	Number of Patients	7	5	24	1	4	2	4	6	16	1	4	4	25	1	2	2	1	1	1
Purpose of Therapy	Aspect of Sentence Production	√		√		√	√		√	√	√			√				√	√
	Sentence Comprehension				√														
	Spoken Language						√		√							√		√	
	Auditory Comprehension	√						√											
	Sentence Processing														√	√			
Duration of Therapy	Number of Sessions	DK	20	DK	8	DK	45	30-40	24-113	DK	18	40-75	55-75	DK	120	2-12	36-48		20	14
	Number of Weeks	DK	6-8	DK	DK	12-28	15	10	DK	4-32	6	5-24	6	2-24	40	2-12	12-16	9	4	DK
Characteristics of patients	Months Post Onset	DK	42-3	4	21	1.5-144	30	15-26	60	1.5-132	8	30-177	28-195	13	72	60-27	8-84	3	14	84
	Type of Aphasia (c)	DK	B FL	DK	B	B W	B	B	B	W u a	B	B	B	B	B	B	FL	B	W	M
Descriptions of the Deficit	Descriptive Account				√	√				√									√	
	Described on Standardized Test						√	√	√	√										
	Syndrome Label Only		√							√	√	√	√		√				
	Psycholinguistic Account													√	√		√		√
	No Description	√	√																
Description of the Task	One Task Described												√						
	Series of Tasks Described	√		√	√	√	√		√					√	√	√		√	√	√
	Hierarchy of Tasks Described in Detail		√	√					√	√	√				√				
	Stimulus Material Described in Detail	√	√	√			√	√	√	√		√						√		√	√

TABLE 1. Summary of contents of sentence-level therapy studies

Key to Table 1.
(a) The orientation to therapy category seeks to describe the type of approach to therapy taken. The behavioural category is used for studies which describe a consistent method of presentation, response and reinforcement by the clinician. The stimulation category is used for those studies that describe the therapy implemented as such. The processing category is used for studies which have a framework of normal language processing as their basis and do not have a specifically behavioural presentation of the terapy. Studies in which the therapy does not clearly belong within a specific "approach" are described as undefined.

(b) A group of single cases describes a study in which data for individual patients is provided. It does not imply that the single cases are described in detail.

(c) B– Broca's; FL– fluent; W– Wernicke's; M– Mixed; G– Global; DK– don't know.

produce verbs, and indeed verbs become a focus of the therapy. This suggests that the therapies carried out with one group of patients would not be appropriate for the other group necessarily. If little information is supplied to describe language impairments, as in the 'symptom focus' studies, it is hard for a clinician to be able to judge whether a new patient would make an appropriate candidate for a particular therapy or not.

The time post onset at the start of therapy, in nearly all cases, is considerable, which is very different from the position that most patients are in. The duration of the therapy is very variable from a few hours to several months, a variable which must have a profound effect on the outcome of treatment, yet few authors account for why a particular duration was selected. Most studies aim to investigate the effects of a specific therapeutic procedure, with only three attempting to apply a published programme and only two comparing two different procedures. It is striking that there are very few studies that report successive attempts at therapy in an effort to find the most successful way of improving language. Some of the studies reviewed here had only very limited effects, but there are few follow up studies reported attempting to remediate the same problem through a different technique, to see whether more extensive effects of therapy are possible (but see Mitchum and Berndt (in press) and Byng (1988) for alternative therapies). It seems that if the patient fails to improve, or makes only limited improvement through one therapy, then it is assumed that no further improvement can be made. Whether this is an accurate reflection of clinical practice is doubtful.

The studies included in this review represent the main body of literature concerning therapies at the level of the sentence. A decision was made to include only those studies which were in press or published at the time of writing, though we are aware of a number of therapy studies which have been carried out recently and which meet the criterion of being at the level of the sentence, but are not yet reported. In addition, descriptions of the implementation of Melodic Intonation Therapy have not been included, as it is a technique designed more to improve ability to use some language accurately, rather than specifically to restore language at the level of the sentence.

The next section of this review will consider the origins of the different therapeutic procedures and the relationship between the impairments described and the therapy. Then the relationship of the therapy to the outcome and the quality of the outcome will be discussed.

THE RELATIONSHIP BETWEEN THE LANGUAGE DEFICIT AND THE THERAPY

The two groupings of studies, those that focus on the processing and those that focus on the symptom, have a demonstrably different basis for the relationship between the deficit and the therapy. In order to demonstrate this in detail, we will consider the basis for the therapy in some of the studies from the two groupings, looking at the aim of each study (in order to provide a context for the treatment), the nature of the patients' sentence level impairments and the basis for or origin of the therapy. In order to clarify the different bases of the therapy, the studies have been further grouped under different headings reflecting different origins of the therapy. This arises because the relationship between what the deficit is and where the idea for the therapy tasks comes from is not necessarily either predictable or constant. There seem to be two basic origins for the focus of the therapy and the tasks to be implemented. In the first case the focus comes from observation of something that the patient can already do, and the therapy is devised to capitalise upon that ability. In the second case the idea for the therapy is motivated by a consideration of what it is supposed that the patient needs to work on. This might be derived from observation of a specific symptom or from analysis of the patients' sentence processing impairments. In the following section we will consider first the studies grouped as "symptom focussed" studies, looking at those where the focus of the therapy is generated by observation of the patient's preserved abilities, and then at those where the focus or task is motivated by the symptom to be worked on. Then we will discuss the studies grouped as "processing focussed" studies and consider the origin of the therapy in the same way.

Symptom focussed studies

TREATMENT BASED ON OBSERVATION OF PRESERVED ABILITIES

Each of the studies included in this section describes a therapy method derived from observation of some retained abilities that the patients demonstrated.
Shewan (1976) set out to compare two training methods and assess generalisation of a technique to "improve sentence formulation skills and use" with an aphasic person who had a moderate Broca's type aphasia. His spontaneous speech was characterised by short phrase length, he could not integrate the various aspects of a picture he was describing, and he had word retrieval problems. He comprehended general conversation, but had more difficulty with increasing complexity and length of discourse. Shewan used active declarative sentences of SVO or SVPP structures for a variety of reasons as follows: a) SJ had shown himself able to produce these himself and could

repeat them; b) this form is "easy" linguistically, requiring no embeddings; c) the form has a high rate of frequency usage by normal adult speakers; d) the form appears in early language recovery in Broca's aphasia.

Another single case study by Davis and Tan (1987) investigated "the effects of a Schuellian stimulation approach upon progress in verbal expression". JS, the patient treated, had a mild auditory comprehension deficit and a severe deficit in spontaneous verbal production. Speech was agrammatic when utterances did occur. The treatment emerged from the observation that JS was "stimulable by producing fluent utterances upon careful cueing." She produced "fluent utterances upon sentence completion and subsequent chaining procedures. Treatment during the study was designed to capitalise on this limited capability."

A very different technique called "conversational prompting" used with severely aphasic patients is described by McCrae Cochrane and Milton (1984). The technique was devised to capitalise upon retained imitation abilities in these patients in order to facilitate more verbalisations through conversational-like exchanges in therapy. Four patients are characterised by a brief description. Three of the patients were Broca's aphasics and one was more like a Wernicke's aphasic. All four patients displayed an ability to repeat single words, and this is seen as a prerequisite for the technique to work. Two patients are mentioned briefly who had to be taught to repeat before they could use the technique.

Finally a programme set up by Naeser, Haas, Mazurski and Laughlin (1986) aimed to facilitate improved sentence level auditory comprehension. They observe that discrimination of phonemes is often retained when the ability to comprehend sentences is impaired, so they attempt to use this ability to facilitate auditory comprehension. Given their rationale that the ability to do some phoneme discrimination tasks is not related to the ability to comprehend sentences, then it is not clear why giving patients practice in discriminating phonemes, which they can already do, will improve sentence level auditory comprehension. The nature of the patients' deficits is unspecified; there are three groups of patients with different kinds of aphasia but no Broca's aphasics. It is assumed that they have auditory comprehension impairments.

TREATMENT BASED ON OBSERVED SYMPTOMS

Each of the studies in this section incorporates a programme devised for an impaired feature of performance such as omission of a type of word from spontaneous speech or reduced phrase length. The studies have in common that, once this observation has been made, the treatment represents an attempt to practise that observed feature. The amount of description of the original deficit varies; some authors establish that a type of word is absent from spontaneous speech whereas others assume a difficulty with certain sentence types.

One of the first experimental studies of treatment for sentence level deficits was carried out by Holland and Levy (1971). It represented a preliminary attempt to determine if syntactic generalisation occurs as a function of training an aphasic to use an active sentence through programmed instruction. The study aimed to discover if any generalisation to other lexical items can be effected by training a single sentence, and then using a minimally varying vocabulary in the equivalent syntax. Holland and Levy used an active sentence for the training programme "on the basis of experimental data suggesting that the obligatory transformations resulting in the active sentence constitute the simplest structure for normal speakers to comprehend." Non-reversible sentences were used so that "the operations involved in the training, as well as the end product of the training, would be at a basic level of syntactical ordering." No differentiation of type of aphasia was made amongst the patients, but a 30 item test was used before and after the treatment programme, so that if patients failed more than four items they were given the programme. The test included active, interrogative, negative, and passive sentences which were tested through repetition, oral reading, writing to dictation, comprehension and production of each of the sentence types.

Holland and Sonderman (1974) devised a sentence comprehension training programme to determine whether training on the types of tasks in the Token Test could change comprehension as measured by that test, and to determine whether such training would change performance on other measures of comprehension. The rationale for this was that the Token Test was considered to be a sensitive indicator of degree of aphasic deficit, in particular looking at comprehension, which Holland and Sonderman considered to be the most basic language skill. A study with a similar purpose was devised by West (1973), but included a comparison of a group of patients treated with a retraining programme including items similar to those in the Token Test, and a control group who received 'conventional' speech therapy.

A comparison of the effects of two treatments of different theoretical origin on agrammatic subjects' use of WH-interrogatives was undertaken by Thompson and McReynolds (1986). WH-interrogatives were selected because earlier studies suggested that "agrammatic patients lack the linguistic behaviours necessary to initiate conversation and to request information." Specifically Thompson and McReynolds examined whether learning to use WH-interrogatives and generalisation and maintenance of learning would be different depending on the treatment approach selected. The two treatment approaches compared were an auditory-visual stimulation treatment derived from principles associated with a stimulation approach and a direct-production treatment, using principles associated with a behavioural approach.

Two agrammatic Broca's aphasic patients who never produced any form of the verbal auxiliary and copula during a spontaneous language sample were trained to produce "is" in auxiliary sentences to see if it would generalize to production of "is" in copula contexts (Kearns and Salmon 1984). Generalisation from trained to untrained exemplars of auxiliary "is" was also investigated. The rationale for this basis for the therapy was that, given their common surface forms (note that this means phonological forms), training auxiliary or copula verbs may result in generalised responding to the untrained reciprocal form. Although copula and auxiliary verbs serve different grammatical roles in sentences, Kearns and Salmon suggest that a "generative response class" may exist between them. The patients were also unable to produce functors and grammatical endings so that their speech consisted primarily of nouns and verbs.

Helm-Estabrooks and Ramsberger (1986) carried out a study to report the effect on speech output of training six Broca's aphasic patients on the Helm Elicited Language Program for Syntax Stimulation (HELPSS) (Helm-Estabrooks, 1981). All six patients were diagnosed as Broca's aphasics on the Boston Diagnostic Aphasia Examination (BDAE) (Goodglass and Kaplan, 1983) and all six had relatively good auditory comprehension. Phrase length in language production averaged about three words, consisting almost exclusively of nouns and verbs. The hierarchy of syntactic difficulty of different sentence structures for Broca's patients identified by Gleason, Goodglass, Green, Ackerman and Hyde (1975) acts as the basis for the design of the HELPSS. This hierarchy is described as providing a precise analysis of the linguistic deficits. Given the plethora of studies reviewed by Berndt referenced above, the assertion that this hierarchy represents a precise analysis is hard to maintain. In addition, it is a linguistic hierarchy, not a psycholinguistic hierarchy, which might be more relevant since this task is concerned with how patients process incoming material to convert it into output for themselves. Doyle, Goldstein and Bourgeois (1987) also implemented HELPSS with four agrammatic Broca's aphasics, in order to "examine the training and generalisation effects of the HELPSS programme and to socially validate the treatment effects".

Doyle, Goldstein, Bourgeois and Nakles (1989) investigated whether a training procedure could increase Broca's aphasics' use of requests for information about trained and untrained topics in conversations with trainers and unfamiliar partners. Requests were specifically targetted because "previous research has found this behaviour to be frequently omitted in the conversational discourse of Broca's aphasic subjects" (possibly because such research has typically studied clinical interviews between therapist and patient, rather than natural conversation between equal partners in a domestic setting).

Processing focussed studies

Studies included in this section all share the same basis for treatment, which is an attempt to describe the specific deficits in sentence processing demonstrated by individual patients within a psycholinguistic framework of normal sentence processing and then basing therapy on that analysis.

TREATMENT BASED ON OBSERVATION OF PRESERVED ABILITIES

Jones (1986) reconsidered the sentence construction problems of a longstanding non-fluent patient, BB, in the light of theoretical hypotheses about sentence processing, to determine why he had severe problems in structuring sentences. Therapy was then designed in the light of these hypotheses. BB's speech was characterised by single word output in the presence of a severe word finding deficit. Output was restricted to nouns, no verbs being used spontaneously. Comprehension was 'functional' but he relied on contextual and pragmatic cues. Jones suggests that whilst he had some knowledge of verbs and nouns he had little concept of how these should be ordered to mark underlying meaning relations, and this constituted his main problem. Despite trying to improve sentence production, Jones worked only on input in therapy (a) because it was felt that this would allow him to concentrate more fully on meaning relations attached to the verb, and (b) because he had had a lot of therapy previously for production of grammatical morphemes which he had become preoccupied with. The therapeutic strategy emerged from an attempt to explain to BB what was entailed by the meaning of each verb. Jones found that he was able to use a question word to identify meaning relations in the sentence and she developed this observation into a treatment strategy. The intent of the therapy was to bring the processes involved in mapping meaning relations up to conscious processing.

A partial replication of Jones' therapy study was implemented by Le Dorze, Jacob and Coderre (1991) to determine whether a similar patient would also improve when similar therapy was provided. Therapy was directed at "focusing the patient's attention on verbs and the meaning relations attached to them by making these more explicit, thereby facilitating access to sentence structure". MG, the man they worked with, had very limited spontaneous speech and word-finding difficulties were severe. He produced single word phrases consisting of nouns and grammatical morphemes were omitted. Auditory comprehension was relatively preserved although some more "linguistically complex" items were difficult. He had poor verb retrieval in narrative speech and a near absence of sentence structure.

TREATMENT BASED ON AN ANALYSIS OF THE SENTENCE PROCESSING IMPAIRMENT

The sentence processing impairments of two people with agrammatic Broca's aphasia were interpreted as resulting from impairments to the procedures which map thematic relations in sentence comprehension and production (Byng, 1988). This hypothesis was made on the basis of specific features of sentence comprehension and production; these were that the patients were unable to comprehend simple reversible sentences and additionally, their sentence production was structurally limited, both patients relying on producing simple, either holophrastic or at least minimally structured utterances (Schwartz, Linebarger and Saffran, 1985). Both patients also made errors assigning thematic roles in a comprehension task for single words, suggesting that their problems in comprehending sentences were not because of failure to parse the syntax. With the first patient, BRB, a therapy task was devised which focussed only on enhancing the comprehension of one set of thematic roles and in input only. The hypothesis underlying this therapy was that if the hypothesis about the mapping deficit was correct then remediation of the mapping deficit in one modality should bring about improvement in the untreated modality.

A different therapy programme was devised for the second patient, JG, as the procedure that BRB followed did not prove successful for JG. The second therapy programme aimed to teach JG about mapping thematic relations in a more direct fashion than the first therapy had done. It aimed to improve sentence comprehension of reversible active declarative sentences by encouraging perception of thematic roles of agent and theme, related to their position in the sentence, and thereby also to improve production of simple sentences. An attempt to replicate the therapy programme for JG with another patient, AER, is reported in Nickels, Byng and Black (1991). The motivation for this study was to observe the extent to which the pattern of results that had been obtained with JG could be repeated with another patient.

An unusual approach to therapy was adopted by Beyn and Shokhor-Trotskaya (1966) in which they tried to prevent agrammatism from occuring. They suggest that the observed features of agrammatism are secondary symptoms of a primary deficit to the patients' inner speech, which they describe as the intermediate link of verbal thought, so that the "predicative system of inner speech" disintegrates. This interpretation might not be incompatible with the interpretations offered by Jones and Byng described above. Thus Beyn and Shokhor-Trotskaya aim to reconstruct the patient's inner speech very early on, when expressive speech is absent, in order to prevent the telegraphic style from developing. They suggest that agrammatic speakers use nouns to represent whole predicative utterances, so in developing their rehabilitation programme they exclude the use of any nouns and focus only on those words which could function as a whole sentence.

Mitchum and Berndt (in press) describe their study as an application of an approach to treatment built upon an understanding of how sentences might normally be produced, in which the patient's symptoms might reflect a breakdown of that normal functioning. Therefore the assessment of symptoms and the prediction of generalisation of treatment effects are derived from this approach. ML's speech was fluent but poorly structured and marked by frequent paraphasias, fragmented utterances and impaired word retrieval. Analysis of sentence production revealed poor verb retrieval both as single words and in sentences, limited and often incorrect use of auxiliary verbs and inflections, and incorrect realisation of the logical relations between verbs and nouns in a sentence. The therapy Mitchum and Berndt implemented was carried out in two stages. The first therapy was based upon the hypothesis that impaired verb retrieval caused or at least contributed to ML's poor sentence production and motivated an intervention to facilitate verb retrieval, "production of verbs being a crucial prerequisite to successful sentence production". The second therapy aimed to "test the hypothesis that sentence construction was undermined by poor accessibility to the grammatical elements that form a verb phrase."

A treatment based on a theory about language processing, but not specifically related to the patients, was devised by Loverso, Prescott and Selinger

1988 (see also Loverso, Selinger and Prescott, 1979, and Prescott, Selinger and Loverso, 1982). The programme entitled CVT is described as being based on "sound theoretic and behavioural evidence" that the verb is the predicate core of all simple sentences. This theory is further elaborated: "Events of an individual's database (knowledge and memory) are represented by centering all language procedures around the action. Therefore, action, in our opinion, is the central node." Loverso et al., citing Bever (1970), describe the actor-action-object framework as the primary internal structure of language, and this represents the basic structure of their treatment programme. However this interpretation limits the representation of the predicate to an association with action. Not all events necessarily involve action or actors; some events might be states, and the entities involved in those events could have a variety of roles depending on the nature of the event. Loverso et al do not develop this theory into a psycholinguistic account of the patients' deficits, but rather use it to emphasise the necessity of treating verbs in language production. Two patients with moderate, fluent aphasia were given the programme in order "to provide initial behavioural evidence about the effectiveness of treatment in improving productive language skills and the actual treatment procedure utilized." Loverso et al. suggest that both patients had reached "maximum benefit from treatment, as indicated by stable speech/language performance." The therapy is described as a technique that builds on the input skills of auditory and visual processing and enhances verbal and graphic output abilities.

Having described the rationale underlying the construction of these studies, in the next section we consider what happens to the sentence comprehension or production of the patients included in the studies described above. We look first of all at the general patterns of outcome and the quality of the changes, and then we group the studies in relation to the patterns of outcome obtained.

THE RELATIONSHIP OF THE THERAPY TO OUTCOME

What is the outcome?

Tables 2 and 3 summarise the main effects of the therapies reviewed here. Table 2 describes the improvements that have occurred in sentence production and comprehension across not only treated sentence types but also across untreated sentence types, and including spontaneous, narrative speech. If we assume that what most people with aphasia would want to get from therapy at the level of the sentence is some improvement in spontaneous speech, then Table 2 is not encouraging. Improvements in spontaneous speech seem either not to be reported, or else do not occur widely. Working on the assumption that had there been such improvements they would have been reported, it appears that these therapies have not had much effect on the ability to produce quantitatively more,

Therapy at the sentence level

SYMPTOM FOCUSSED STUDIES	Improvement on treated stimuli/ sentences/ condition in comprehension or production	Improvement on untreated stimuli/ sentences/ condition of the same type as the treated sentences	Improvement on untreated stimuli/ sentences/ condition of different type from treated sentences	Improvement in spontaneous speech	Improvement on standardized test
Holland & Levy (1971)	√	x	x	–	–
West (1973)	√	√	x	–	x
Holland & Sonderman (1974)	√	–	–	–	x
Shewan (1976)	√	x	–	–	–
McCrae, Cochrane and Milton (1984)	–	–	–	√	–
Kearns and Salmon (1984)	√	√	√	x	–
Thompson and McReynolds (1986)	√	√	x	x	–
Helms-Estabrooks and Ramsberger (1986)	–	–	–	–	√
Naeser et al. (1986)	–	–	–	–	√
Davis and Tan (1987)	√	x	–	–	√
Doyle et al. (1987)	√	√	x	x	√
Doyle et al. (1989)	√	x	–	–	–

PROCESS FOCUSSED STUDIES					
Beyn & Shokhor-Trotskaya (1966)	–	–	–	√	–
Jones (1986)	–	√	–	√	–
Byng (1988)	√	√	√	√	–
Loverso et al. (1989)	–	–	–	?	√
Nickels, Byng and Black (1991)	–	√	√	√	–
Le Dorze et al. (1991)	√	√	–	√	–
Mitchum & Berndt (In Press)	√	√	–	x	–

TABLE 2. Effects of therapy on measures of sentence production or comprehension
√ = Improvement observed/measured
X = No improvement observed/measured
— = Information not provided

or more useful, spontaneous speech. It does not follow, however, from this that the spontaneous language production skills of people with aphasia cannot be enhanced; rather, it suggests that the studies carried out so far do not reflect what is observed in clinical practice, namely, that production of connected language can improve.

The predominant method of measuring the results of therapy is to test production or comprehension of treated sentences both before and after therapy. Sentences of the same construction as those used in therapy but involving different lexical items are generally tested as well, to discover whether improvement has been construction and item specific, or whether some rule, procedure or strategy has been learnt or reaccessed which permits production or comprehension of similar sentence types using different lexical items. Less often tested are sentences of different types from those treated in therapy and using different lexical items. The ways in which these sentences are different from the treated sentences depend on the assumptions underlying the therapy. For example Thompson and McReynolds (1986) investigated the effects of treating one type of WH-interrogative sentence on the production of other types of WH-interrogative sentences, but no rationale was provided prior to the therapy for why such generalisation should or should not take place. Kearns and Salmon (1984) measured the effects of producing "is" as an auxiliary on the use of "is" as a copula, on the assumption that they are both members of the "same generative response class" (presumably on the basis that they share a phonological form). Byng (1988) considered six alternative predictions about the possible outcome of therapy. One of these six alternatives was that if the patients had learnt through therapy about mapping of thematic relations, then any tasks involving such mapping should demonstrate improvement. Therefore although the first of the two therapies that was carried out involved only mapping thematic roles in locative sentences, pre and post-therapy testing investigated comprehension of a variety of other sentence types and also aspects of sentence production, where the ability to map thematic relations would be predicted to improve performance in specific ways.

Fewer predictions and assumptions are generally made about the effects of therapy on spontaneous speech. Many studies which aim to investigate improvement in "verbal expression" (Davis and Tan, 1987) or "sentence formulation skills and use" (Shewan, 1976), for example, do not specifically investigate effects on spontaneous speech. There are few methods available for documenting and analysing changes in spontaneous speech. Few of the studies reported in this review have developed methods of investigating changes in spontaneous speech which might relate to the changes that therapy is trying to bring about. One exception would be Byng (1988) where a specific structural analysis, described in Byng and Black (1989), was used to document changes in the number and type of simple structured utterances produced before and after therapy.

It is also relevant to note, however, the limited way in which spontaneous speech is usually sampled; the tasks through which speech samples are obtained nearly always involve the use of descriptive language, in narrative or picture description tasks, whereas someone with aphasia may not be as concerned about producing descriptive language as about the conversational use of language, for example. Therapies where multiple aspects of language use are considered and measured are rare. Table 3 illustrates the quality and type of changes in sentence production that the different therapy procedures have brought about. These classifications have been based on quantitative data where supplied, or on observation of raw data.

Therapy at the sentence level

SYMPTOM FOCUSSED STUDIES	Increase in length of utterance/ MLU	Increases in production of structured utterances (e.g. Unouns + Verbs in one)	Increase in number of content words/ constituents produced	Increase in use of bound and free grammatical morphemes	Increase in number of grammatically correct utterances	Increase in semantic appropriateness of production
Holland & Levy (1971)	-	-	-	-	-	-
Shewan (1976)	-	-	√	-	√	√
McCrae, Cochrane and Milton (1984)	√	-	-	-	-	-
Kearns & Salmon (1984)	-	-	-	√	-	-
Thompson and McReynolds (1986)	-	-	-	-	√	-
Helms-Estabrooks and Ramsberger (1985)	-	-	√	√	-	-
Davis and Tan (1987)	-	-	√	-	-	-
Doyle et al. (1987)	√	-	-	-	√	-
Doyle et al. (1989)	-	-	-	-	x	-

PROCESS FOCUSSED STUDIES						
Beyn & Shokhor-Trotskaya (1966)	√	√	√	√	-	-
Jones (1986)	√	√	√	√	-	-
Byng (1988) *	√ √	√ √	x x	x x	√ x	-
Loverso et al. (1989)	-	-	-	-	-	-
Nickels, Byng and Black (1991)	√	√	√	x	√	-
Le Dorze et al. (1991)	√	√	√	√	-	-
Mitchum & Berndt (In Press)	√	x	x	√	√	√

TABLE 3. Quality of changes in production of language
√ = Improvement observed/measured
X = No improvement observed/measured
— = Information not provided

* Two entries are provided because the two patients described had different outcomes.

It demonstrates that the least commonly measured aspect of spoken language is the semantic content of the production. .More emphasis is placed on measuring the number of words produced, the number of grammatically correct utterances and the number of function words produced. None of these is particularly informative about quality of change nor pertinent to the communicative needs of people with aphasia. It might be more appropriate to measure quality of content or accuracy/ease of interpretability of content. These aspects would involve measures related to specific semantic properties of words used, the nature of the utterances they are used in and also the context. An integration of specific structural and semantic measures with pragmatic measures would provide far more information about the quality of outcome.

Despite the paucity of extensive data on the effects of therapy, we shall examine the existing studies in more detail to look not only at what kind of change there has been but also at how that change has been brought about. In summarising the effects of therapy above we noted that in few of the studies was improvement in spontaneous speech brought about, and furthermore, the improvement across treated and untreated sentence types was variable. Therefore in the following sections, rather than grouping the studies being reviewed in relation to the type of approach to therapy adopted, we have grouped them on the basis of the pattern of outcome that has been achieved, in order to try to consider how that pattern of outcome might be related to the therapy. We attempt some preliminary conclusions about the features of the studies which might be relevant to the pattern of outcome achieved. For each study we will consider the nature of the therapy that was carried out in terms of how the task was presented, what the patient was required to do and the feedback used, where possible.

How has change been effected?

STUDIES WHERE IMPROVEMENT IS MEASURED ONLY ON STANDARDISED TESTS

The outcome of the studies included in this section is described only in terms of performance on standardised tests rather than specifically related to the types of sentence implemented in the treatment procedures, so that the direct effects of the treatment are harder to discern. The Helm Elicited Language Program for Syntax Stimulation (HELPSS) implemented by Helm-Estabrooks and Ramsberger (1986) employs a hierarchy of sentence types (based on the hierarchy of difficulty suggested by Gleason et al., 1975), ranging from imperative intransitive sentences to embedded sentences, in two levels. In the first level, the response that the patient must produce is modelled in a 'story' completion task, in which the patient repeats the last statement of the story that has been read aloud. Each story is accompanied by a simple line drawing. In the second level, the patient must make an inference about the end of the story, but it must be the target sentence produced before in the first level. For example the patient might hear "Rob's grandchild is bored. Rob gets a book and what does he do?" The correct response is "He reads his grandchild a story." The relationship of the target to the picture is a consideration at this point; for some sentence types the intended target is probably more predictable from the picture than for others.

The results were expressed in terms of change to scores on the Northwestern Syntax Screening Test (NSST) and by a count of the number of content units and grammatical morphemes produced on the Cookie Theft picture description from the BDAE. A content unit here is a word or phrase carrying important information about the people, events and setting depicted. All patients improved on the NSST expressively, by between 4 and 12 points. On the BDAE measures production of content units increased by between 2 to 9 points. Morphological changes were also measured, but were of a much smaller magnitude than on the NSST, but no account for this difference is offered. In particular, one patient, JC, made a gain of 39 points on the NSST but did not increase his morphology score on the BDAE. The change in content unit production suggests that the patients produced more words, but does not suggest anything about the structure, content or appropriateness of the output. It may be that some or all of the patients had changed in these respects but this is not made clear by the methods of measurement used. In addition, it is not clear whether a gain of two, or even nine, content units on the Cookie Theft description should be regarded as significant improvement. The output of these patients does seem to have improved, but it is not clear how, or in what respect this improvement has taken place.

Helm-Estabrooks and Ramsberger suggest that "treatment programmes based on the linguistic assets and deficits of specific aphasia syndromes may prove effective with groups of patients who demonstrate these specific behavioural features (i.e., agrammatism)." This assumes that the same hierarchy will be beneficial for all patients. Since we have already seen that the specific behavioural features of agrammatism can vary extensively (Berndt, 1991) and we do not know about the different types of linguistic deficits that these patients might have, it is hard to see how this programme can be based on linguistic assets and deficits since it is not based on individual strengths and weaknesses.

Differential outcome on two different standardised tests is also evident in the outcome of the programme set up by Naeser et al. (1986) to facilitate improved sentence level auditory comprehension (SLAC). The programme consists of three levels. The first level trains discrimination between word pairs. Then a target word has to be associated with a written word from a selection of visually and phonologically similar written words. No semantic information is provided. The third level gives some sentential context which might be more semantically informative, but it is not clear from the examples given how uniform the amount of semantic information is provided across stimuli. Again the patients have to select from the same set of visually and phonologically similar written words. It is unclear how access to meaning is being facilitated through this programme. Two of the three groups of patients evaluated were receiving other therapies simultaneously, so their results should be treated with caution. Of the group that used SLAC five out of seven had a "good" response. However, this seems to be considered only on the basis of their performance on the Token Test, as their improvement on two subtests of the BDAE was either minimal or negative. Two patients made quite big gains on the Token Test, but for the other three it is not clear whether their improvement is significant. There is no discussion of the difference between the Token Test and BDAE results. Of the patients who carried out this programme alone for six months, two out of five made no progress.

Measurement of the effects of the CVT programme (Loverso, Prescott and Selinger, 1988) is confined to the PICA (Porch, 1967). This programme takes as its basic premise that the verb is the predicate core of all simple sentences. Loverso et al. claim that CVT is constructed on the structure of language, while reinforcing and teaching self-cueing strategies. Verbs are presented as pivots and WH-questions are provided as strategic cues to elicit sentences in an actor-action, actor-action-object framework. Thirty verbs controlled for frequency and imageability are included. Selection criteria for verb and thematic role type, however, is less clear. Loverso et al. base their programme on "actor action object ", but include verbs which do not represent actions but states e.g., have, live, like.

The programme includes six hierarchical levels, which patients enter at different points depending on their abilities. Initially the patient is given a verb and a cue for "who" or "what". The patient has to produce an appropriate subject, then a subject and a verb and then write it down exactly. As some of the verbs would yield ungrammatical utterances if produced in this structure e.g., "the man gives", it is not clear whether all verbs were used at each level. If the patient cannot do this, then a verb is provided with a WH cue, a subject and the whole sentence, written and spoken. The patient then has to repeat the verb, the WH cue for the subject, the subject, the subject and the verb together and then copy it. The next step is to give the verb, written and spoken, a WH cue, and a card giving a choice of subjects. The patient has to repeat the verb, repeat the WH cue, choose a correct subject, generate the subject plus verb combination and then copy it.

The next level uses an SVO structure in a similar procedure (again, some of the three place predicates would be ungrammatical at this point, and some of the verbs would take a PP not an NP). At the first stage a verb, a subject cue and an object cue are provided. The patient must provide a subject after the subject cue and an object after the object cue, then say the "full grammatical utterance" and write it down exactly. If an error is made then a verb, a subject cue, a subject, an object cue, an object, and the complete sentence are provided. The verb, the subject cue, the subject, the object cue, the object and then the whole utterance are each repeated and then copied. The next step provides the verb, the subject cue, a card giving a choice of subjects, the object cue and a card giving a choice of objects.

There are some features of this procedure in common with the procedure devised by Jones described above, although the patients and the underlying purpose of the tasks are very different. Both procedures focus on the verb and use questions to focus on appropriate constitutents. In Jones' therapy no output was required, whereas in Loverso et al's procedure the emphasis is on facilitating production. Loverso et al. provide explicit details about how a "correct" sentence was elicited. In contrast, because no output was required, the details of Jones' therapy relate to the range of sentence types and the tasks in which they were used.

The effects of the CVT programme for two patients with fluent aphasia are described. Both had stable baselines over three occasions before treatment was initiated. After the treatment the overall, verbal and graphic PICA performance scores improved and the improvement was maintained. Loverso et al mention that they improved on two specific subtests—'1' and 'C'. 'C' examines writing the names of common objects to dictation, and this improved significantly, but does not specifically relate to Loverso et al.'s aim to improve productive language. '1' assesses ability to describe the function of common objects, which also improved significantly and is more in line with the aims of their therapy. We do not know what kinds of changes to spontaneous speech were measured, so the quality of the improvement is not clear. In addition, it is uncertain whether all the improvement measured on the PICA is in productive language skills, and if not, why not. If this treatment had some non-specific effects, how are they accounted for?

In general the authors suggest that "CVT works because it emphasises the verb or action versus the traditional noun-centred approaches used. ...The cueing strategies implemented from the verb or pivot allow the patient to build on his/her language skills rather than being concerned with one word NPs. ...Use of the verb in this pivot role results in the direct, systematic application of strategies (WH-questions) by the aphasic patient. By specifying function, the patient is able to narrow the semantic field referring to the ideas he/she wishes to communicate." (Note that it is not clear how function is being used here, as many of the verbs used do not specify functions per se.) Loverso et al. claim that, by using this approach, one can enhance the potential for functional communication, but data have not been presented in this case to support this claim.

Loverso, Prescott, Selinger and Riley (1989) compared presentation of this therapy procedure administered by clinician alone or by "computer-clinician assisted". They found that more time was taken to reach criterion levels of performance when computers were also involved than when the clinician alone carried out the treatment. Prescott, Selinger and Loverso (1982) demonstrated some generalisation to untrained verbs and maintenance of these effects, although the numbers are small.

STUDIES RESULTING IN IMPROVEMENT ONLY ON TREATED SENTENCES

In an early study concerning treatment for sentence level deficits Holland and Levy (1971) used a teaching machine to train patients to use an active sentence through programmed instruction. Spoken and written material was presented using picture-word matching tasks, reading, repetition, writing and speaking and the patient would push an appropriate "response window" for each item. If the patient was correct the programme would continue. If the patient made an error a buzzer sounded and the item was replayed. As the study aimed to discover if any generalisation to untrained lexical items can be effected by training a single sentence and then using a minimally varying vocabulary in the equivalent syntax, "the man" was taught first, then "the door" was taught in a similar manner. "Open" was taught next, then combined with "the door", then the whole sentence was put together.

The major effect of the programme was to "improve sentence production orally and to a lesser extent in writing". Results were measured by sentence type and modality of response. Improvement on comprehension of active sentences was measured, for all seven patients, although individual data are not supplied. Six subjects improved on production of interrogative sentences also, but improvement on passive and negative sentences was not found. Generalisation to untrained items was not statistically reliable. In considering the effects of the treatment, Holland and Levy consider less why active sentences improve and more why interrogatives should also have been affected. They suggest that the form of the interrogative that they used—"did the man open the door"—does not "disturb" the ordering for active sentences, so that this may have facilitated comprehension of the interrogative. In addition they used interrogative forms extensively in the administration of the programme, so they suggest that they may have reinforced them that way also. (This factor influenced the decision to include this study in this section.) They suggest that the other sentences did not improve because of "their relative structural complexity". Given the limited number of lexical items trained, the restricted outcome of this procedure is not surprising. What it does suggest is that some people with aphasia can improve on their ability to produce specific types of sentences quite rapidly. Of course, the caveat here is that we do not know what sort of language impairments these patients had, nor whether their pre-therapy baselines were stable.

Similarly restricted results of therapy were observed in studies by both Holland and Sonderman (1974) and West (1973) in therapy programmes for comprehension of sentences, based on the Token Test (de Renzi and Vignolo, 1962). In each case the same sentence types as those in the Token Test were used in training, but with a variation in the lexical items. Holland and Sonderman used geometric shapes, as in the Token Test, but West used everyday objects. In both studies, patients were presented with a spoken sentence and asked to carry out the instruction given in the sentence with the coloured objects or shapes provided. If they were unable to carry out the instruction, it was repeated once or twice. If performance was still incorrect, the therapist demonstrated the correct answer. Holland and Sonderman told the patients what the error was that they had made. In both studies, change in performance was measured on both the Token Test and on the MTDDA (Schuell, 1965), and in both cases performance on the Token Test improved whereas on the MTDDA it did not, or did only to a lesser extent. The improvements in the Holland and Sonderman study were also confined to treated stimuli with little generalisation to untreated stimuli of the same type as used in the therapy. Generalisation to untreated stimuli did occur in the West study, however. Additionally, Holland and Sonderman found that patients who were performing relatively better on the Token Test prior to therapy made better progress than patients who had a greater degree of initial severity on the test.

A training procedure to increase Broca's aphasics' use of requests for information about trained and untrained topics in conversations with trainers and unfamiliar partners was applied by Doyle, Goldstein, Bourgeois and Nakles (1989). Each patient was given an initial prompt to start asking questions about a specified training topic such as personal information, leisure activities or health. A 20 second interval was allowed during which the trainer stayed silent and maintained eye contact with the patient. If during that time the patient responded with an adequate request that met their criteria, then the trainer praised the patient and provided the information asked for. The criteria to be met included that an unambiguous message was communicated which was intelligible, on a specified topic, and contained a question morpheme and a content word, or ended with rising intonation. Then another 20 second interval was allowed for the patient to ask another question. If an inadequate request was made, the trainer "praised and acknowledged the attempt" but allowed another 20 seconds for the patient to make the request meet their criteria. If no adequate response was provided then the trainer provided a specific content prompt about the topic. If no adequate response was obtained the trainer modelled an adequate request and the patient had to repeat it.

The subjects provided increasing numbers of unprompted requests over time during training. There was no generalisation from a trained topic to an untrained topic during the training period for all the subjects. Generalisation of requesting with trainers to requesting with unfamiliar volunteers was assessed. It took place gradually for all four patients, although subject 4 showed quite a lot of variability. Use of requests was generally lower with unfamiliar volunteers than with trainers, but it did increase from the baseline. Maintenance on the trained topics was good while other topics were being trained. With respect to type of the requests, subjects 2, 3 and 4 at the start of therapy used a WH-word for a substantial proportion of their requests, but by the end of therapy they were relying much more heavily on yes/no requests, using fewer requests containing question morphology and the requests used were less grammatically complete. Subject 1 relied on using yes/no requests pre-therapy, and this does not change. By the end of the training the patients' level of requesting was within the normal range for all the topics, as compared to a set of non-aphasic controls who were also studied. A subjective evaluation by students rated the subjects more talkative, requesting more information than prior to the training and conversations were more successful. Looking at the type of speech acts used by the patients and volunteers, it is clear that the conversation became more shared and the patients less passive and purely responsive.

The implications of this study are that patients do get better at making (certain types of) requests for information, with a variety of interlocutors, but only about topics that have been practised, a rather restricted result. Doyle et al. attribute the generalisation to using requests with unfamiliar volunteers as a consequence of using multiple trainers, employing functional rather than structural response criteria (i.e., accepting any kind of request), encouraging subject initiated requests and using natural reinforcers. How natural it is to receive praise every time you ask a question may be questioned. The lack of generalisation from one topic to another is suggested to be a product of the strategy employed during training of not responding to "off-topic" requests, which may have discouraged any attempt to generalise, a serious implication for this therapy procedure. Imposing such severe restrictions on the patients, use of subject matter for communication must discourage them from exercising control over their own communication. Doyle et al. do not discuss the change in type of request used, other than to say that structurally deficient requests functioned adequately. Since different types of requests will affect the kind of response that is obtained from the communicative partner, there may be another important issue to consider here, that although "requesting" has increased, the nature of that requesting may not involve sufficient variety of types of request, to serve different communicative purposes.

STUDIES RESULTING IN IMPROVEMENT ONLY ON TREATED AND SIMILAR-TYPE SENTENCES

The therapy in the studies described in this section either only affects sentences of the type used in the therapy or, alternatively, no data are provided about other sentence types, so that we do not know whether or how other sentence types benefitted.

Two techniques to "improve sentence formulation skills and use" were implemented by Shewan (1976). Thirty pictures eliciting simple active declarative sentences which were described incorrectly at pre-test were divided into three groups, one control group and two for the two different treatments to be implemented. An utterance with any semantic or syntactic error was deemed incorrect. The first treatment involved the presentation of one picture and two possible Subject noun phrases, two verbs and two Object noun phrases, written on cards. The patient, SJ, selected the phrases appropriate for the picture, sequenced them and read them aloud orally. The clinician provided prompts until he corrected his errors. The written cards were removed and SJ attempted an unaided description. If he was incorrect, the clinician provided a correct model. The second treatment involved providing both the spoken and written form of the verb. SJ made up a sentence about the picture using the verb. Then the written verb was removed and SJ reformulated the sentence. If it was incorrect, the clinician provided a correct model. Both therapies were administered in the same session.

SJ improved significantly in his ability to produce grammatically correct sentences to trained items, but he did not show any generalisation to untrained items. In addition, it seems that the correct sentences produced were in the main not more semantically appropriate but only syntactically correct. However after the training he did produce more syntactically and semantically appropriate linguistic constituents necessary to construct simple sentences, although it is not clear whether this was true for both trained and untrained items. There was no significant difference between the two therapy methods, although it looks as though giving the verb might be more facilitative. However as both modality of presentation and type of cue were compounded in this cueing condition this benefit may have arisen because the verb was given in both spoken and written forms, or because provision of the verb is more facilitative. There were only ten items in each set and the therapy period was very short (six sessions), neither being factors which facilitate monitoring the effects of therapy. The outcome of the therapy suggests that SJ can produce more semantically appropriate constituents, across all grammatical functions, but is not able to combine these into "grammatically correct" sentences unless specifically trained.

On reflection about how the therapy might have worked, Shewan wonders whether SJ may "have abstracted a strategy or strategies", but these potential strategies are not considered in detail. Alternatively she suggests that the trained items were facilitated by providing practice and repetition, which assumes that aspects of the language system can be restored through repeated practice. Another alternative that Shewan suggests is that the auditory and visual stimulation treatment (the second treatment described above) might have facilitated word retrieval, but that because of the lack of generalisation to sentence construction, processes involved in the recovery of these aspects of language might be different. With only ten items per treatment condition, however, there is really insufficient evidence to support a conclusion about the relative effects of the different treatments.

Another stimulation approach aimed at promoting "progress in verbal expression" is described by Davis and Tan (1987), using three sets of pictures of simple events. The first stage of therapy required the patient to repeat a heard sentence which went with a picture, and as repetition improved the sentences got longer. It is not clear whether that means that the number of arguments increased or the length of individual phrases or whether the sentences became embedded, for example. The second stage of the therapy focused on retrieving single words representing any part of the event being described, in order to practise the production of a propositional utterance without having to attempt an entire sentence. Questions were used to "encourage" a response, but it is not clear what the questions were cueing—the subject, the object, or the verb? The relationship between this stage and the one before is not specified. The final stage of therapy was designed to elicit more spontaneous and complete sentence production. A reverse chaining procedure was used to elicit first of all the last word of the sentence, then the last two and so on. It is not made explicit whether this elicitation of the "last word" really means the last word, e.g., a noun without its determiner, or the last phrase, e.g., a determiner plus noun. The rationale for eliciting determiners and nouns separately would not be obvious. Finally the patient described the picture again without cues. Davis and Tan describe this therapy as being a "Schuellean" stimulation programme, but the justification for this is uncertain since the focus of the therapy does not seem to have been on auditory stimulation specifically nor on multi-modality stimulation.

After treatment was instituted, performance increased on the treated set and some improvement was maintained with evidence of a small amount of generalisation. The quality of the improvement is unclear; was the patient saying more words or attempting to produce more structured utterances? There are some examples of combination of NPs but it is uncertain whether JS produced any NP V NP structures. In scoring the changes in production, an element realised with a functor achieves a higher score, but since the purpose of the therapy was to promote progress in verbal expression, it is doubtful whether it is relevant to score production of all functors. Post-therapy scores on the PICA are presented but are difficult to interpret since the patient seems to get better on most subtests, except in the visual modality. Improvements on the verbal and auditory scores by no means represented the biggest increase and yet are the aspects of the test where most change might have been anticipated. These results are not further interpreted.

Thompson and McReynolds (1986) compared the effectiveness of a stimulation approach to a behavioural approach in eliciting WH-interrogatives. The stimulation part of the training was designed to provide multi-modal stimulation of target responses. The patient was presented with a picture stimulus "corresponding to the interrogative being trained". (Does this mean that the stimulus picture represented the response to the question, or somehow conveyed the question itself?) A written sentence corresponding to the target response for that picture was also provided. The patient then listened to the clinician saying the sentence three times, stressing the initial WH-word. Then the written sentence was removed and the patient was instructed to "ask a question about the picture". The patient was told whether the response was correct. Correct in this case meant that the sentence had to be syntactically well-formed. If the response was incorrect the whole training trial was repeated again for the same item, until three consecutive trials had been completed. In the behavioural treatment the patient was presented with the same picture stimulus and was instructed to "ask a question about the picture". If a correct response was produced within 15 seconds the clinician provided verbal praise. If an incorrect response or no response was produced then the clinician indicated that the response was not correct and modelled the correct response. If a correct response was still not forthcoming then a forward chaining procedure was implemented, in which the first two words of the target response were modelled, and then the remaining words. The patient had to repeat each portion as it was modelled. Finally the whole sentence was modelled.

The results of this study suggested that direct-production treatment (i.e., the behavioural treatment) was consistently more effective than auditory-visual stimulation in facilitating production of WH-interrogatives. Generalisation from trained interrogative forms to untrained forms did not take place, that is, training on eliciting sentences such as "What is he reading?" did not improve production of sentences such as "Who is the nurse?" or "Why is he hot?". Generalisation from treated to untreated items within the same interrogative form did take place, for example training on items such as "What is he reading?" did facilitate production of items such as "What is he sweeping?". Improvement on the trained items was maintained at a reasonable level of accuracy. Trained responses did not generalise to use of correct questions in other contexts where spoken language was elicited, such as in describing the cookie theft picture (although this picture would not normally elicit questions, the patients were encouraged to ask questions about the picture).

Thompson and McReynolds suggest that "the present data provide empirical support for the use of a direct-production approach because such training facilitated more rapid acquisition, better generalisation, and better maintenance of trained responses than did the auditory-visual stimulation approach. These data suggest however, that generalisation to untrained WH morphemes and to situations beyond training may not occur." The reason given for this lack of generalisation is that

> these data suggested that WH-interrogatives may not be members of the same response class, even though they belong to the same linguistic or structural class, for a functional relationship was not observed between them. ... The present findings provide little support for the theoretical notion that aphasia results from a disruption of language processing or access mechanisms because improved access to some WH responses did not improve access to others. Instead, these data suggest that aberrant interrogative production may result from a "loss" of rules or information about how to use them appropriately to generate grammatically complete and accurate utterances, and that because rules for production of different WH forms are different, it is necessary to train each rule individually. (p. 203)

What Thompson and McReynolds do not make clear is why these WH-interrogative sentences *should* be part of the same class. They seem to be concerned with these forms only as single words, that is as one constituent of the sentence. They do not consider the sentences in which these forms occur which have very different structures and interpretations. Since the patients are producing whole sentences, then the WH-interrogatives cannot be discussed separately from the sentence contexts in which they occur.

This is not the only issue of concern in this study; there are others. First of all the use of a direct-production approach is barely supported because the effects of the treatment were very limited, with no generalisation to other types of interrogative or to any generalised use. Thus the efficacy of the approach has not really been established; it just seemed to achieve a little more than the stimulation approach did. Secondly, to suggest that generalisation amongst interrogative forms may not occur is also an unreasonable conclusion—the only conclusion that can be drawn is that generalisation from the application of these specific techniques or tasks may not occur. That is quite different from suggesting that generalisation is never possible. Thirdly, the difference between the two treatment approaches is more to do with what they require of the patients than something about the approach. In the stimulation treatment, the patient simply has to repeat what the clinician has said and what they have read on a card. They are not involved in thinking through how to put the construction together for themselves. In the direct-production treatment the patients have first of all try to think out how to formulate the questions for themselves, then, if they are unable to do this, they can repeat a model provided. Repeating when the patients have already tried actively to process the sentence for themselves may be more facilitative than when they passively repeat. The behavioural treatment also allowed the patients to build up the sentences from provision of the question word and the verb (similar to Jones, 1986), thereby providing some information from which the patients could construct their own sentences. Therefore the differential effects of the therapies might have less to do with the "approaches" and more to do with the specific nature of the tasks involved. Finally, considering the method of scoring for this study, both pre- and post-therapy, Thompson and McReynolds score lexical errors as correct, but grammatical errors as incorrect. Therefore after the training it is not clear whether patients could still be making lexical errors but scoring correctly because they have produced a grammatical utterance—the criteria for success in training. It would seem more useful to a patient to be able to produce relevant lexical items than to put them into an entirely grammatical construction.

More detailed evidence about the nature of the outcome of the HELPSS programme described earlier was the aim of a study by Doyle, Goldstein and Bourgeois (1987). Instead of training the whole set of sentence exemplars from the programme, only the first five sentence types to which each subject provided consistent "low levels" of response were included. Five different sentences for each sentence structure were also created, one of which was used for training, and the others for measurement of baseline and generalisation. In addition feedback was incorporated, such that an incorrect response was responded to first by providing a spoken model. If that was not successful in eliciting a correct response, a forward chaining procedure was implemented, so that the clinician modelled 'portions' of the sentence for the patient to repeat until the whole sentence was modelled if necessary. Whether the portions corresponded to phrasal constituents or single words is not stated. In order to obtain a correct score, all free and bound morphemes had to be produced in a correct sequence.

In general there seemed to be good learning of the trained sentences, and reasonable generalisation to untrained sentences of the same type as the trained sentences. There was some variation within patients; generalisation from trained to untrained items was better for some patients in some particular sentence types. For example subjects 1, 2 and 3 had less generalisation to untrained sentences within the declarative transitive condition. Subject 4 had difficulty generalising from trained to untrained items within passives and direct indirect object sentences. There was no generalisation from training of one sentence type to another, across all patients. Maintenance was variable across subjects and sentence types. Subject 1 maintained all the sentence types tested reasonably well, but subject 2 showed steadily declining performance. Subject 3 maintained gains for WH-interrogatives, but not for any of the other forms trained. Subject 4 maintained effects for WH-interrogative and declarative intransitive forms.

Ability to produce the same sentence types in stimulus conditions other than the HELPSS was also measured. The generalisation measured was patchy with subjects 1 and 3 showing limited generalisation of ability to produce 4/5 sentence types, and subjects 2 and 4 on 1/5. Western Aphasia Battery aphasia quotients improved for all subjects, but this does not seem to be a function of improving spontaneous speech (one of the patients even drops a point on spontaneous speech) but rather to do with improved naming or repetition, and in one case comprehension. Two of the subjects showed some improvement in the number of content units produced, as in the Helm-Estabrooks and Ramsberger study. Doyle et al. carried out a 'social validation' study by asking naive judges to rate pre- and post-therapy samples of the untrained items and the trained sentence type material for adequacy, accuracy and grammaticality. The findings suggest that accuracy (an utterance containing all the grammatical elements of the target form and appropriate to the verbal and visual context) did improve as did grammaticality (getting the constituents in the right order), but adequacy did not improve across the board. Doyle et al. suggest that HELPSS was sufficient for the patients to learn verbal production of their respective trained sentence types, but given the patchy nature of the generalisation and maintenance this "learning" has to be considered to be very restricted in nature. The efficacy of the programme in facilitating the production of sentences has not been established clearly.

In considering the effects of the therapy Doyle et al suggest that the problem that subject 4 had with generalisation may be "that beyond a certain level of structural complexity, the strategy of teaching multiple exemplars is not an efficient means of programming for generalisation." This suggests that unsatisfactory effects are attributed to procedural effects rather than prompting the re-examination of the content of the therapy or the relationship betweeen the therapy and the nature of the deficit. Discussing the lack of generalisation to other "stimulus conditions" Doyle et al. suggest that "there may be few stimulus characteristics common to both situations, and/or the opportunities to respond with a particular trained form may be significantly reduced." This suggests that patients are having to rely on trained forms, so that what they are learning through therapy are prototypical sentences that they have to reproduce.

Most of the studies described in this section so far fall within the "symptom focus" grouping. However, the study by Mitchum and Berndt (in press) belongs to the 'processing focussed' group as their therapy was rooted in an interpretation of their patient's deficit within a framework of normal language processing. The goal of the first therapy they carried out with a man with fluent aphasia, ML, was to determine whether or not improvement in verb retrieval would affect the production of verbs in sentences and of verb-relevant grammatical morphemes, resulting in production of grammatical structures. The focus was on enhancing the lexicalisation and production of a small set of eight different verbs, controlled for a variety of psycholinguistic variables, with eleven different depictions of each verb (to avoid repeated presentation of the same materials for pre and post-therapy testing and training). ML practised naming the seven depictions of two verbs, with the two verbs at first presented separately and then mixed together. Errors were corrected by providing the target.

The second therapy was designed to "test the hypothesis that sentence construction was undermined by poor accessibility to the grammatical elements that form a verb phrase" and focused explicitly "on the production of verb morphology within the limited domain of tense/aspect markers". Fourteen common actions were selected for which three points in the sequential time course of the activity (future, present and past) could be depicted easily. ML was asked first to order the triads in the correct sequence and then to describe each sequentially ordered set of pictures using the auxiliary verb and appropriate inflection on the verb to denote if the action "is already done", the action "is right now" or the action "is about to happen". If he could not name the action he was given the verb. Where necessary, he was reminded to provide the correct grammatical morphemes by a tense indicative cue e.g., "about to ... right now ... already." Larger parts of the sentence were modelled if necessary. ML was practising a specific sentence frame to represent each tense marker, by remembering what the specific form of that tense marker was and trying to produce it to the relevant picture. If he could not do it he was reminded of the temporal element that he was trying to represent.

Despite much improved ability to name action pictures using specific verb targets, ML was no better at producing full sentences to describe pictures using the same verb targets after the first therapy. However the second therapy succeeded in "reinstating ML's ability to use tense/aspect markers productively, which had far-reaching effects on his ability to formulate sentences". He showed generalization of use of the trained morphological elements to untrained verbs and was also able to produce morphologically complete and semantically specific verb phrases in sentence formulation without pictures and without temporal clues/ constraints, but given a target word to include in a sentence. That is the changes were not bound to specific stimuli or specific forms of elicitation. No improvement in production of sentences in narrative speech comparable to improvement in tasks eliciting single sentences was measured, although the structures produced were longer and better formed.

Mitchum and Berndt suggest that the first therapy did not facilitate sentence production for two possible reasons: one, that the processes used in retrieving a verb to name a pictured action are different from the processes needed to retrieve a verb to construct a sentence, or, two, that other impairments in addition to the verb retrieval deficit prevented sentence construction. Additionally it is possible that item specific learning of the type used in this first therapy was not facilitative of production of other forms. Therefore they carried out the second therapy. From observing ML's performance during the second therapy they suggest that he deliberately extracted the central action from the picture by focussing on naming it, before attempting to construct a sentence. They also suggest that the explicit use of temporal information in the treatment may have provided him with a clearly specified message at an early level of processing which facilitated later levels of production. The lack of generalisation to narrative speech may have been due to increased demand that recalling the details and sequence of a story imposed, such that processing resources were diverted from structural and morphological processing.

STUDIES RESULTING IN IMPROVEMENT IN SENTENCES OF SIMILAR AND DISSIMILAR TYPES, BUT NO IMPROVEMENT IN SPONTANEOUS SPEECH

The study described in this section describes a treatment in which there has been some generalisation to treated sentence types using different lexical items and also to an untreated sentence type, but no improvement measured in spontaneous speech.

Production of "is" in auxiliary sentences was trained by Kearns and Salmon (1984), to see if it would generalise to production of "is" in copula contexts. The treatment was based on ten stimulus line drawings used in two treatment phases. In the imitation phase the "trainer" pointed to the picture and said, for example, "say 'girl is washing'." If the sentence was repeated correctly the patient was given a redeemable token. If it was incorrect the trainer said so and repeated the model. If the patient still did not get it right, the trainer went on to the next item. In the second spontaneous phase the trainer pointed to the picture and said, for example, "tell me about the girl". A correct response, "girl is washing", elicited a token. After an incorrect response the trainer modelled the target for the patient to repeat. No further action was taken if the response was still not correct, but if it was correct a token was presented. Once criterion performance was reached of 9/10 correct responses on three consecutive ten trial sets, the reinforcement schedule changed to one token for every two correct responses. The facilitation procedure therefore involved repetition of the whole sentence. In order to examine the effects of the therapy, a reversal training was also used to train the patients not to use the forms they had just learnt, and then they were retrained to produce the auxiliary and copula forms that they had just been trained not to produce.

In considering the results of this treatment it should be noted that there are only five items in each of the copula categories, so that the numbers being considered were small. There was also some variation at baseline of one or two items out of the five. Subject 1 improved on ability to produce auxiliary "is" after training on a restricted set of items. Generalisation to copula + adjective forms took place during training of auxiliary "is". Reversal of use of copula + adjective forms was achieved through direct training. Reversal training on auxiliary "is" was necessary as only partial reversal took place on these items after reversal on copula + adjective forms. Reversal training applied to auxiliary "is" was successful after eight sessions. Retraining was successful as ten correct responses were achieved after training on only three items. Retraining of auxiliary "is" led to generalisation again to copula + adjective forms. No generalisation took place to copula + NP structures at any time, and Subject 1's responses to these remained "off target" e.g., for "he is a fireman" he said "puts out fire". The communicative effectiveness of this 'error' response is not considered. Maintenance of treatment gains was observed but there was no carryover to spontaneous speech. The effects for subject 2 were very similar except that he showed generalisation to copula + NP structures as well as to copula + adjective structures. Neither subject generalised to production of copula

+ PP structures. Subject 2 also showed no generalisation of the use of auxiliary "is" or the copula to spontaneous speech.

Kearns and Salmon attribute the generalisation from auxiliary "is" to copula + adjective as supporting the notion that the auxiliary "is" and copula "is" (which share only a similar phonological form) are related through response class membership. But they have to explain why only one out of the three copula contexts actually shows that generalisation. To find an account they look to procedural manipulations suggesting that it might be necessary to train multiple exemplars of the auxiliary construction simultaneously, although no rationale is provided about why this should be more effective at obtaining generalisation. They do not have an account for why auxiliary "is" and copula + adjective should show a consistent relationship. The lack of generalisation to spontaneous speech is, they suggest, because generalisation is not automatic but should be actively planned for. Thus these results are hard to interpret, both in terms of the pattern of improvement and also in terms of how the training might have worked.

It is not clear that this therapy would be considered relevant by all patients. If someone can already say "girl washing" it is hard to know of how much benefit it is to produce "girl is washing" unless focus on this aspect of language production is part of a wider context of goals that a patient has set, to work on a number of different aspects of accurate sentence production. In this case this might be relevant if a patient was concerned to focus on representation of temporal aspects of language. In addition, the clinician might want to find out whether a patient understood the implications for meaning of use of the present progressive as opposed to the present tense for example, to ensure that the form will be used appropriately to convey the intended message. Reversing the effects of therapy, then retraining, might also be challenged as a relevant therapeutic procedure.

STUDIES RESULTING IN IMPROVEMENT ON SENTENCES OF SIMILAR AND DISSIMILAR TYPES AND IN SPONTANEOUS SPEECH

The pattern of improvement demonstrated in the studies grouped in this section suggests that whatever the patients have learnt in therapy they have been able to apply in contexts other than those used in therapy, specifically in spontaneous speech. Additionally, where it is possible to extract the data, it seems that whatever they have learnt through therapy has generalised to facilitate the production of types of sentence construction other than those used in the therapy. There could be a problem here in establishing that these more widespread effects are not due to spontaneous recovery, but of the five studies cited, there seems to be good evidence that the results represent a specific effect of therapy for nearly all the patients described. Most of the studies that fall within this group were previously described as 'processing focus' studies, with the exception of McCrae Cochrane and Milton (1984).

In reconsidering the sentence construction problems of a longstanding non-fluent patient, BB, Jones (1986) discovered that whilst he had some knowledge

Therapy at the sentence level 351

of verbs and nouns he had little concept of how these should be ordered to mark underlying thematic relations, and she suggested that this constituted his main problem. Therapy was designed in the light of this hypothesis and was based on making judgements about sentences. The therapy, which involved only input tasks, consisted of the seven principal stages shown in Table 4.

1	Given simple written sentences, BB had to identify the phrasal constituents of the sentence. Having done this he was asked to identify the verb. In all subsequent stages of the therapy he had to identify the verb before doing anything else.
2	The concept of Agent was established by explaining that it answered the question who or what undertook the activity. BB selected the phrase that went with WHO (animate Agents) or WHAT (inanimate Agents). All verbs used at this stage were intransitive. BB identified the verb and then labelled the Subject with the correct question word.
3	The concept of theme was introduced, by explaining that it represented the entity which undergoes the event. Again, BB selected the phrase that went with WHO (animate themes) or WHAT (inanimate themes). Themes were distinguished from Agents initially by using verbs which took obligatorily animate Agents and inanimate Themes.
4	Verbs which demanded the question WHERE were introduced. Again, BB identified the verb, and determined which were the phrasal constituents. Obligatory three place predicates like 'put' were used first, then verbs where the location was optional such as 'sit' were included.
5	Other phrases were added answering the questions WHEN, WHY and HOW. BB was given a chart listing all the question words and their relation to the verb.
6	BB was required to detect omitted or misordered constituent phrases. Given an incomplete sentence, BB had to retrieve the relevant question word to be supplied with the missing phrase. When BE and HAVE were introduced as main verbs initially, BB was confused. However, he quickly grasped the concept when IS as copula was introduced with an inanimate Agent.
7	More complex sentence structures were introduced, including the passive and embedded clauses. BB had to identify all the verbs first and determine the actor for each verb (including the use of pronouns).

TABLE 4. Jones' programme for patient BB.

With every new stage, initially lexical items were selected so that sentences could be interpreted by use of pragmatic knowledge rather than relying on having to interpret the sentence itself fully; then that support would be taken away, and for example, reversible Subjects and Objects used. What is not clear, however, is what happened at any stage when BB did not make the correct response on first selection. The feedback procedures are not made explicit.

The results of the therapy are reported descriptively rather than numerically, but the considerable improvement in quality and quantity of spontaneous speech is obvious from the speech samples provided. BB grasped the concept of mapping meaning relations both in input and output, despite all the therapy focussing on input, and after the therapy he used more sentence structure in output. From the speech samples provided it appears that the majority of the structured utterances he produced did not include embedding or the use of clauses. He used appropriate verbs correctly inflected for both tense and subject agreement, and there was also a marked improvement in the production of function words, neither of which aspects of production were worked on directly. Tense agreement improved although he still made errors. Errors were rarely made in using pronouns, prepositions, auxiliaries and determiners and "is" as copula could be used. Single words were still produced as complete utterances in spontaneous speech but BB could attempt a sentence if a single word did not suffice. BB also improved in sentence comprehension of reversible sentences.

In considering how the therapy worked, Jones suggests that the question word strategy appeared to give him the 'tag' upon which to construct planning frames. "The concept of each question word seeking information about the activity—the verb—seemed to clarify for him the central role of the verb in any sentence. It seemed to afford him the opportunity to access consciously verb information about predicate-argument assignment previously inaccessible." These predicates include not only verbs but also prepositions. The emphasis in the programme on input rather than output focussed his attention on the fundamental problem he had and "not practising" sentences overcame his obsession with functor words. It seems that what BB had gained from this therapy procedure was a method or strategy of facilitating his own language production. He had not learned just one response to a particular stimulus, but rather a means of helping himself to formulate and structure his own output. Jones comments that "Aphasic patients have lost access to aspects of language processing automatic in normal speakers. If they were able to reaccess these automatically by some 'absorption' techniques during language stimulation then they would do so. BB had been exposed to plenty of stimulation over the past six years and yet little improvement had taken place in spoken output."

The results of a partial replication of Jones' study by LeDorze et al. (1991) largely support the original findings. Another agrammatic aphasic man, MG, who had not improved during six months of previous 'traditional' therapy, went through stages 1-4 and 6 of Jones' therapy programme. Because of his difficulty with written language, all the materials used were pictorial. This meant having to represent the subject, verb and object in pictorial form, by representing the subject and object on separate pictures as entities not engaged in any action, and the verb as a "solid stylised outline of a figure executing an action". All therapy tasks were input tasks as in Jones' study and no production was required. After the therapy the number of NPs produced did not improve but the number of verbs did, on treated and untreated stimuli, in both picture description and narrative tasks. In addition there were more attempts to combine noun phrases and verbs in production after therapy, also on treated and untreated stimuli, in picture description and narrative tasks. Some of these attempts represented production of verb plus two argument structures. The quality of the improvement appears similar to that of BB in that MG is also producing simple structured utterances at the end of the therapy. All of the treatment effects were statistically significant. A measure of maintenance of effects on narrative speech one month after therapy had ceased demonstrated that performance was deteriorating slightly. It should be noted that MG's therapy was of a very much shorter duration than that of BB.

This replication of the outcome of a therapy procedure is important because it shows that the effects of this way of doing therapy are specific neither to one therapist nor to one patient. Both patients worked only through input modalities, in the case of MG only through spoken input and pictorial material, and yet both improved in their ability to produce structured language in narrative or spontaneous output.

The following study describes two therapies (Byng, 1988) one of which was also replicated in a second study (Nickels, Byng and Black, 1991). The therapy carried out aimed to facilitate the mapping of thematic relations which, according to the mapping deficit hypothesis in agrammatism, should improve both sentence comprehension and production. The first therapy involved only comprehension of mapping in locative sentences, and used only written sentences which were reversible (i.e., the order of the noun phrases could be reversed plausibly), for example "the bucket is in the box." In order to interpret this sentence, it has to be established that "the bucket" is the thing (theme) which is inside "the box" (the location). Because the box could as well be in the bucket, guessing at which interpretation is correct would lead to an incorrect interpretation fifty percent of the time. The thematic relations, that is the role of the entities in both subject and object position, have to be established to comprehend the sentence. BRB was given written reversible sentences accompanied by two pictures, a picture representing the correct interpretation of the sentence and a picture representing the reversed interpretation of the picture. BRB had to select the correct picture and was provided with clues to help him work out the thematic relations. For each of the five prepositions included in the therapy he was given a "meaning" card which explained diagrammatically the meaning of a sentence containing that preposition, describing the relationship between the two NPs in the sentence. Colours were also used such that the same colour not only represented the NP in the subject position in the written sentence but also the theme in the picture and also the first entity in the diagram. A different colour was used for the NP in the PP in the written sentence, for the location in the picture and for the second entity in the diagram. Using these materials, BRB could work out for himself how to interpret these reversible locative sentences. At a later stage in the therapy the colours were no longer included, requiring BRB to interpret the sentences without the use of any cues. These materials were designed for him to take home to practise. He reached errorless rapid performance within two weeks on the therapy materials, and was then retested on the range of tasks requiring mapping of thematic relations that had been tested prior to the therapy.

The results of this testing demonstrated that he had made statistically significant improvement in comprehending all sentences where he had to map thematic roles onto syntactic relations in order to interpret the sentence correctly. This included not just the sentence types used in therapy but also reversible simple active sentences, and passive sentences. Therefore BRB had not just learnt a strategy for interpretation of sentences, such as the most agent like entity will be in the subject NP position, which would facilitate comprehension of active sentences. Since application of that strategy would not have allowed him to interpret locative or passive sentences any better, he must have learnt some principle about mapping thematic relations.

Mapping of thematic relations in production of sentences was also investigated after the therapy. It is considered that an inability to do this mapping for the purposes of production leads to a reliance on structurally impoverished speech, so that constituents are produced as isolated phrases. After the therapy, a statistically significant shift was observed in the production of narrative speech away from producing single words or phrases towards combining nouns and verbs to form structured utterances.

The same therapy was attempted unsuccessfully with a second patient, JG, which also prompted a further therapy procedure to be implemented. This therapy was then replicated with another severely agrammatic patient, AER, (Nickels, Byng and Black, 1991) to investigate its effects in more detail. The basic purpose of the therapy task was to facilitate structuring conceptualisation of an event (as depicted in a picture) through increasing the patients' awareness of the linguistic structure that can describe that event, at first through a sentence ordering task and then by describing the event themselves.

The therapy programme had two principal stages. In the first stage, the aim of the task was to make explicit the relationship between the roles of the participants in an event and the relative position in the sentences of the NPs expressing those participants. Action verbs in the active voice were used in a sentence ordering task in relation to pairs of pictures which contrasted the subject, verb or object. A colour cueing system was used initially to distinguish between nouns and verbs and to give an indication of ordering of nouns and verbs. Having established the principles of assigning agent and theme to subject and object positions within the therapy task, the patients were encouraged to apply the same principles in production of sentences in the second stage of the therapy. In this stage, to begin with, the same pictures were used as in the first stage. The patients were asked to describe each picture in whatever way they could. As they produced each constituent they were asked to indicate where that constituent should go in the basic sentence structure they had been working with. The therapist facilitated the patients in specific ways to enable them to see how they could use what words they could produce within a structured utterance (see Byng, 1992). In the replication study, a further stage of therapy was introduced to assist in using the strategies learnt in spontaneous speech tasks. The tasks used were basically the same, but the materials were more familiar to the patient, involving pictures of himself, family and friends engaged in different activities, as well as current events.

Both JG and AER demonstrated statistically significant gains in both comprehension of simple active agentive sentences and in production of structured utterances. This improvement in production was not confined to picture description tasks but was also measured in production of narrative speech and, according to observations and reports from relatives, in day-to-day spontaneous speech. Additional tasks were performed pre- and post-therapy, which used processes that were predicted not to be involved in therapy, and therefore served to act as controls for the processes that were predicted to be positively affected by the therapy. In both patients, these control measures showed that the improvements were specific to those processes involved in the therapy and not to other unrelated processes. This result was taken to reinforce the specific effects of the therapy.

A very different "prophylactic" approach to treatment was devised by Beyn and Shokhor-Trotskaya (1966) and administered to 25 patients very early post onset, who had no expressive speech. Comprehension was relatively intact, but they seem to have had problems in comprehending some thematic relations. The basis of their therapy was to "regulate" the type of words introduced into the patient's speech from outside, specifically excluding the use of any nouns, on the basis that these could only serve to aggravate the "disintegration of the predicative system", which the authors consider to be the basis for the development of agrammatism. Although few explicit details of this therapy are provided, it is possible to discern the basic principles and progression through the therapy. The authors selected single words which could "function as a whole sentence" by expressing a complete idea, including verbs and interjections, citing examples such as "Here!, Give!, No!, Ugh!". The lexical composition of the sentences used gradually became more complex through the introduction of pronouns, adverbs, modal verbs and "compound predicates". Nouns were not introduced until spontaneous words began to appear in the patients' speech. They describe this therapy as advancing from the "sentence to the word". At the same time as this therapy was being carried out other methods of "stimulation and disinhibition using automatised forms of speech" were also being implemented.

Again few details are provided about the outcome, and there is little experimental methodology to permit the specific effects of the treatment to be ascertained. However, it does seem that all the patients made some progress, 22 out of 25 achieving speech at the level of the sentence. Nine of these reached a stage of being able to produce sentences actively, which seems to mean without having a picture as a prompt. The most important outcome for Beyn and Shokhor-Trotskaya is that none of the patients developed a telegraphic style (a reliance on producing nouns or verbs as single word utterances), which they say would otherwise be inevitable. They comment that by the second month of treatment patients were beginning to introduce their own words, but they note that these words were "verbs, adverbs, and auxiliary parts of speech". More detailed data on four patients are provided, two of whom had the preventive treatment and two did not. The two who did not have the treatment used a large majority of nouns (80-90%) as compared to verbs (10-13%) and almost no other parts of speech, whereas the two patients who received the preventive treatment produced a much wider variety of parts of speech and had no predominance of nouns over verbs. Although it is hard to discern exactly what the therapy consisted of, what the influence of spontaneous recovery was, and exactly how the patients improved, this therapy appears to be promising. Given how long ago this paper was published it is disappointing that a more detailed study of some of the principles involved has not been undertaken.

Another different approach to therapy is described by McCrae Cochrane and Milton (1984). Their conversational prompting technique involves "creating a therapy atmosphere and a conversational context which attempts to personalise treatment content, guide expression of the individual's communicative intent and maximise the opportunity for spontaneity during ongoing interchanges." The technique was devised to capitalise upon the imitation abilities of severely impaired agrammatic patients in order to facilitate more verbalisations through conversational-like exchanges in therapy. Linguistic expansion of spoken responses took place through "skillful and creative clinician "orchestration" of such conversational interactions". A detailed description of the approach is given to provide an explicit description of how the technique was applied in addition to a method for charting patients' performance whilst therapy was in progress.

Given the detailed description of this approach, and the interesting vignettes of case histories, the method of measurement of outcome does not capture the essence of the effects. Since few adequate measures of qualitative change in spoken language, especially in spontaneous conversation, have been developed, as noted above, this is perhaps not surprising. McCrae Cochrane and Milton use mean length of utterance which, although it is not very satisfying, does demonstrate that all the patients seem to make considerable gains (although figures and examples are not given for each of the patients). It seems that a major effect is for the patients to become more participatory and communicative, in addition to saying more, and in some cases the effects are reported to have extended beyong the clinical setting to use at home. It is not clear to what extent the patients are still relying on cues from a communicative partner to effect repairs, or how much language they can produce spontaneously. It is also hard to know how much development there is in the production of propositions, in structural complexity or type of content.

McCrae Cochrane and Milton suggest not only that modelling and repetition are important ingredients in the success of this approach but also, interestingly, that the grammatical forms are presented in a context where meanings can be adduced, and where "enriched contextual information" can be derived. This might mean that the patients are being provided with a consistent and meaningful framework in which to both interpret and produce language, in a wide variety of forms. In addition, each attempt to produce an utterance seems to be worked through with a variety of different types of input, so that the patients' experience of certain linguistic forms is varied, but derived from a meaningful and explicit context.

The explicit nature of the contexts in which the pertinent aspects of language are demonstrated is a unifying feature across these studies. In each of the therapies described in this section, in which there are more extensive effects on spontaneous or narrative output, the common thread amongst the therapies seems to be one of enabling the patients to establish a strategy or to understand an aspect of language formulation within a meaningful context. The strategies and concepts being developed could pertain to any communicative situation, not just to the specific treatment context or to a specific set of lexical items.

DISCUSSION

In this chapter we have tried to cluster these studies on therapy at the level of the sentence not only from the point of view of the type of approach used, but also in relation to the basis for therapy and to the outcome of the treatment. We have seen that there are two ways of approaching the language problem to be treated: one is to observe surface symptoms of that problem and one is to analyse aspects of the deficit in relation to theories about normal language processing. Within these two approaches the origins of the therapy tasks to be implemented are different. We have seen that they may be generated to capitalise on retained features of performance or they may be constructed by the therapist to practise either what the patients cannot do or to find a way of facilitating an impaired process.

Considering the outcome of these studies, it is striking to see how little improvement in spontaneous speech there is overall as a result of the therapy carried out. As we suggested earlier we suspect that this is not because sentence production in spontaneous speech does not improve, but rather that the kind of studies that report therapy at the sentence level do not bring about that kind of change. Moreover, when one therapy has not been successful, researchers rarely attempt another therapy (although this must be relatively common in clinical practice). Therefore we do not know whether patients have not improved because they are unable to improve or because therapy was inappropriate. Assumptions are made on the basis of the outcome of one therapy attempt with one set of patients, and often these assumptions are then related to patients in general rather than just to the patients in question.

The restricted outcome from these studies could be attributable to a number of factors. First, the methods of measurement of changes in production are, as we have seen, inadequate, so that it is possible that changes which have occurred have not been monitored. Measures are usually confined to numbers of words produced or to the production of grammatically correct specific structures. Little attention is paid to changes in the quality of either the content or the structure of the utterances produced. Moreover, spontaneous language is nearly always measured in relation to descriptive tasks, and yet descriptive language is not necessarily a part of everyday communication. Development of measures of both the structure and content of language, relevant to pragmatic and functional use of language, is urgently needed in order to investigate the real value of therapy for enhancing spontaneous language. Some suggestions in this direction have been made in Chapter 6 of this book.

Secondly, many of the studies, particularly those classified within the "symptom focussed" grouping, emphasise learning of a specific set of items. An assumption is made that patients should be able to generalise from that set of items to similar items, and such generalisation is often examined in detail, as we have seen. There are two issues here. One is that there are different definitions of what a "similar item" might be. In some cases it might be examplars from within the same word class as the treated items (e.g., Mitchum and Berndt, in press). In others it might be sentence constructions hypothesised to require similar psycholinguistic processes to the sentences being treated (e.g., Byng, 1988). In yet other cases items are assumed to belong to the same "response class" (e.g., Kearns and Salmon, 1984; Thompson and McReynolds, 1986). Whilst the basis for the similarity between items within the same word class and between items sharing psycholinguistic processes can be defined, the similarity between items within the same "response class" is much less clear, as we discussed earlier. Thompson and McReynolds suggest that "generalisation may not be a natural process resulting from behaviour change. At best, when retraining interrogative responses (and possibly many other aspects of language in neurologically stable aphasic subjects), we can expect generalisation to occur only to a very limited set of responses and incompletely to spontaneous language". The implication of this statement by Thompson and McReynolds is that recovery of language in aphasia requires training item by item. This cannot be a realistic option. Moreover, from the results of studies reviewed here, we also know that this is unnecessary. We have seen that where there are good psycholinguistic reasons for generalisation to spontaneous speech to take place, and when these are made explicit and incorporated within therapy, then such generalisation can take place. If the therapy is aiming to change the processing underlying the language, then effects of the therapy beyond the treated items can be anticipated.

A psycholinguistic rationale, however, is not sufficient on its own to bring about a more extensive change. The design and implementation of the therapy procedures must also facilitate the application of the rationale. If we consider the studies that did effect more extensive change, not all of them are based on a psycholinguistic rationale. McCrae Cochrane and Milton (1984) did not have such a rationale, and yet their therapy technique seems to have brought about substantial change. We suggested that a feature of these studies, including that of McCrae and Milton, was the relation of the therapeutic procedures and stimuli to meaningful contexts, which facilitated the patients not only in making their own deductions about how language works, but also in developing strategies to use to facilitate their own language production, from a rich but consistent source of contexts.

By contrast many of the studies reviewed here used modelling and repetition as a basic technique in therapy. Modelling might provide a consistent source of language experience, but the information provided by modelling a sentence is limited simply to the provision of an exemplar with no explanation or demonstration about the nature, construction or use of that exemplar. How the technique is meant to work to enable patients to produce those sentence types on other occasions, when they are not repeating them, is never specified. Why should hearing a sentence enable patients to produce that sentence for themselves in another context? Is it intended that the sentences heard should remain encoded in memory and then serve as prototypical sentences? Or should patients be able to extrapolate from the sentence they have heard some features about how sentences are structured to allow them to generalise to produce other sentences of that form? Does this assume that all patients need is plenty of exposure to language which they can repeat, to allow them to relearn or reestablish sentence production? As Jones (1986) pointed out, many years of exposure to language during treatment did not facilitate BB's sentence production.

We have to explain, however, why patients *do* make item specific and sentence specific improvements after therapies which employ a predominantly modelling technique. It is possible that during these therapies patients are learning to listen for key constituents, which they can then hold onto in relation to a specific context, often a picture, which suggests which entities are relevant. This might provide them with a means of determining the type of content words to use in a specified context, which they then have to work out how to put together. This might help them to produce the trained response, as in the HELPSS programme, but how patients could use this strategy if the constituents were not provided by their interlocutor, and for a specific context, is not clear.

We would suggest that the limited effects of many of these therapies must be taken seriously, so that a different way of approaching the remediation of these problems is sought, to meet the needs of the patients. Howard and Hatfield (1987), in the context of describing treatments for word retrieval deficits, suggest that "...many of the contemporary schools of aphasia therapy assume that simply eliciting a response from an aphasic, by whatever means available, is itself therapeutic. This assumption is clearly no longer tenable... Effects depend on *how* a word is elicited" (p. 129). The results of our present review suggest that the same statement could be made about therapy for sentence level deficits; both the rationale underlying the therapy and the methods employed to bring about that change must be specified to determine both *why* and *how* they should be effective. In this way we believe that our therapy at the level of the sentence could improve; it needs to.

Acknowledgements

Maria Black provided helpful comments on earlier drafts of this chapter and Antoin Smith contributed to its preparation.

References

Badecker, W., and Caramazza, A. 1985. On consideration of method and theory governing the use of clinical categories in neurolinguistics and cognitive neuropsychology: the case against agrammatism. *Cognition*, 20: 97-126.
Berndt, R.S. 1991. Sentence processing in aphasia. In M.T. Sarno (ed.) *Acquired Aphasia*, second edition. Academic Press. Pp. 223-270.
Bever, T. 1970. The cognitive basis for linguistic structures. In J.R.Hayes (ed.) *Cognbition and the Development of Language*. New York: John Wiley and Sons.
Beyn, E.S. and Shokhor-Trotskaya, M.K. 1966. The preventive method of speech rehabilitation in aphasia. *Cortex*, 2: 96-108.
Black, M., Nickels, L. and Byng, S. 1991. Patterns of sentence processing deficit: processing simple sentences can be a complex matter. *Journal of Neurolinguistics*, 6: 79-101.
Bresnan, J.W. and Kaplan, R.M. 1981. Lexical-functional grammar: a formal system for grammatical representation. In J.W. Bresnan (ed.) *The Mental Representation of Grammatical Relations*. Cambridge, Mass: MIT Press.
Byng, S. 1988. Sentence processing deficits: theory and therapy. *Cognitive Neuropsychology*, 5: 629-676.
Byng, S. 1992. Testing the tried: issues in replicating therapy for sentence processing in agrammatic aphasia. *Clinics in Communication Disorders*, 2: 34-42.
Byng, S. and Black, M. (1989) Some aspects of sentence production in aphasia. *Aphasiology*, 3: 241-263.
Caplan, D. 1987. *Neurolinguistics and Linguistic Aphasiology*. Cambridge: Cambridge University Press.
Chomsky, N. 1985. *Knowledge of Language, Its Nature, Origin and Use*. New York: Praeger.
Davis, G. and Tan, L. 1987. Stimulation of sentence production in a case with agrammatism. *Journal of Communication Disorders*, 20: 447-457.
de Renzi, E. and Vignolo, L.A. 1962 The token test: a sensitive test to detect receptive disturbances in aphasia. *Brain*, 85: 665-678.
Doyle, P., Goldstein, H., and Bourgeois, M. 1987. Experimental analysis of syntax training in Broca's aphasia: a generalization and social validation study. *Journal of Speech and Hearing Disorders*, 52: 143-155.
Doyle, P., Goldstein, H., Bourgeois, M. and Nakles, K. 1989. Facilitating generalized requesting behaviour in Broca's aphasia: an experimental analysis of a generalization training procedure. *Journal of Applied Behavioural Analysis*, 22: 157-170.

Garrett, M.A. 1982. Production of speech: observations from normal and pathological language. In A.W. Ellis (ed.) *Normality and Pathology in Cognitive Functions.* London: Academic Press.
Gleason, J.B., Goodglass, H., Green, E., Ackerman, N., and Hyde, M. 1975. The retrieval of syntax in Broca's aphasia. *Brain and Language,* **24**: 451-471.
Goodglass, H. and Kaplan, E. 1972. *Boston Diagnostic Aphasia Examination.* Philapelphia: Lea and Febiger.
Helm-Estabrooks, N. 1981. *Helm Elicited Language Program for Syntax Stimulation.* Austin, Texas: Exceptional Resources Inc.
Helm-Estabrooks, N. and Ramsberger, G. 1986. Treatment of agrammatism in long term Broca's aphasia. *British Journal of Disorders of Communication,* **21**: 39-45.
Holland, A.L. and Levy, C. 1971. Syntactic generalization in aphasics as a function of re-learning an active sentence. *Acta Symbolica,* **1**: 34-41.
Holland, A.L. and Sonderman, J.C. 1974. Effects of a program based on the Token Test for teaching comprehension skills to aphasics. *Journal of Speech and Hearing Research,* **17**: 589-598.
Howard, D. and Hatfield, F.M. 1987. *Aphasia Therapy: Historical and Contemporary Issues.* Hove: Lawrence Erlbaum Associates.
Jones, E. 1986. Building the foundations for sentence production in a non-fluent aphasic. *British Journal of Disorders of Communication,* **21**: 63-82.
Kearns, K., and Salmon, S. 1984. An experimental analysis of auxiliary and copula verb generalization in aphasia. *Journal of Speech and Hearing Disorders,* **49**: 152-163.
Le Dorze, G., Jacob, A., and Coderre, L. 1991. Aphasia rehabilitation with a case of agrammatism: a partial replication. *Aphasiology,* **5**: 63-85.
Loverso, F.L., Prescott, T.E. and Selinger, M. 1988. Cueing verbs: a treatment for aphasic adults. *Journal of Rehabilitation Research and Development,* **25**: 47-60.
Loverso, F., Selinger, M., and Prescott, T. 1979. Application of verbing strategies to aphasia treatment. In R. Brookshire (ed.), *Clinical Aphasiology Conference Proceedings.* Minneapolis: BRK Publishers. Pp. 229-238.
Loverso, F., Selinger, M., and Prescott, T. and Riley, L. 1989. Comparison of two modes of aphasia treatment: clinician and computer assisted. In R. Brookshire (ed.), *Clinical Aphasiology Conference Proceedings*:**18**. Boston: College-Hill. Pp. 297-317.
McCrae Cochrane, R. and Milton, S. 1984. Conversational prompting: a sentence building technique for severe aphasia. *Journal of Neurological Communication Disorders,* **1**: 4-23.
Mitchum, C.C. and Berndt, R.S. In press. Verb retrieval and sentence construction: effects of targeted intervention. In G.W. Humphreys and

M.J. Riddoch (eds.) *Cognitive Neuropsychology and Cognitive Rehabilitation.*

Naeser, M., Haas, G., Mazurski, P., and Laughlin, S. 1986. Sentence level auditory comprehension treatment program for aphasic adults. *Archives of Physical Medicine and Rehabilitation,* **67**: 393-399.

Nickels, L., Byng, S., and Black, M. 1991. Sentence processing deficits: a replication of therapy. *British Journal of Disorders of Communication,* **26**: 175-199.

Porch, B.E. 1967. *Porch Index of Communicative Ability.* Palo Alto, CA: Consulting Psychologists Press.

Prescott, T., Selinger, M., and Loverso, F. 1982. An analysis of learning, generalization and maintenance of verbs by an aphasic patient. In R. Brookshire (ed.), *Clinical Aphasiology Conference Proceedings.* Minneapolis: BRK Publishers. Pp. 178-182.

Schuell, H.M. 1965. *Differential Diagnosis of Aphasia with the Minnesota Test.* Minneapolis: University of Minnesota Press.

Schwartz, M.F., Linebarger, M.C. and Saffran, E.M. 1985. The status of the syntacic theory of agrammatism. In M-L. Kean (ed.) *Agrammatism.* New York: Academic Press

Shewan, C. 1976. Facilitating sentence formulation: a case study. *Journal of Communication Disorders,* **9**: 191-197.

Thompson, C., and McReynolds, L. 1986. WH-interrogative production in agrammatic aphasia: An experimental analysis of auditory-visual stimulation and direct production treatment. *Journal of Speech and Hearing Research,* **29**: 193-206.

Thompson, C., McReynolds, L. and Vance, C. 1982. Generative use of locatives in multiword utterances in agrammatism: a matrix training approach. In R. Brookshire (ed.), *Clinical Aphasiology Conference Proceedings.* Minneapolis: BRK Publishers. Pp. 289-297.

West, J.A. 1973. Auditory comprehension in aphasic adults: improvement through training. *Archives of Physical Medicine and Rehabilitation,* **54**: 78-86.

Part IV Neurological foundations of aphasia rehabilitation methods

Chapters in this section examine the neurological foundations of aphasia rehabilitation. Chapter 11 stresses the fact that spontaneous recovery from aphasia takes place in stages and is a function of lesion size and, to some extent, location. Initial severity is considered one of the most important factors in recovery. Time post-onset is also important as a predictor of further recovery. Etiology may also be a major factor. Since contralateral homologous cortical substitution is probably not a major mechanism in the recovery of speech, it may legitimately be concluded that, in the right hemisphere, compensatory mechanisms, rather than reacquisition of implicit grammar, are to be expected and possibly enhanced. (For a review of the application of neurobehavioral research to aphasia rehabilitation, see Helm-Estabrooks [1988]).

Chapter 12 also acknowledges the superiority of the left hemisphere for the processing of verbal material. Experiments by the authors have shown that in left- hemisphere-damaged aphasic patients unilateral visual field defects cause bilateral involvement of the entire visual perceptual system, and present some subclinical neglect in the hemifield contralateral to the lesion. These experiments also support the notion that words tend to be processed by the right hemisphere as simple visual patterns, not as written material, and that words are processed by the left hemisphere as authentic verbal material. Holistic approaches to the processing of verbal material should provide the alexic aphasic patient with a way to have access to reading.

Reference

Helm-Estabrooks, N. 1988. The application of neurobehavioral research to aphasia rehabilitation. *Aphasiology*, 2: 303-08.

11 Neurobiological foundations of aphasia rehabilitation

Andrew Kertesz

INTRODUCTION

The biology of recovery has taken an important position in neuroscience. Initially the central nervous system was regarded relatively finite and damage to it was not considered to be recoverable. However, in the last 80 years an increasing amount of information has become available concerning the extent and mechanisms of recovery, not only in animals but also in clinical populations. Therapists trying to help brain damaged individuals need to know the extent of spontaneous recovery and its biological limitations. A solid foundation in neurobiology is necessary to construct efficacious methods of treatment. In addition to the practical implications for the biology of recovery, there are many theoretical issues that can be addressed through such studies. In this chapter, the neurobiology of recovery will be summarized with specific emphasis on the recovery of language in humans.

FIRST-STAGE RECOVERY

Recovery from cerebral injury in man takes place in several stages. The first stage of recovery is related to the absorption of hemorrhage, cellular debris, and edema. Cellular reaction and chemical and transmitter factors initiate tissue repair. The re-establishment of the circulation in the "ischemic penumbra" (Kohlmeyer, 1976; Astrup et al., 1981) or reperfusion after thrombolysis (Zivin et al., 1985) are possible early mechanisms of recovery. The ischemic threshold of cerebral perfusion for membrane failure is around 8 ml/100 g/min. Therefore damage can be reversed if blood flow can be elevated. The rapid recovery from neurological deficit after a stroke is often attributed to the recovery of function of the cells in this area of partial ischemia or the ischemic penumbra. Alterations in tissue water and electrolytes and edema in trauma and ischemia are responsible for some of the early changes in first-stage recovery. Some of the pharmacological treatment of initial trauma is an attempt to control cerebral edema. Corticosteroids are controversial in trauma, and are contraindicated in stroke edema, although undoubtedly valuable in brain tumor. Hyperosmolar agents such as mannitol and glycerol reduce brain volume and increase cerebral blood flow, but most of the acutely effective drugs unfortunately have only a temporary effect, and some even produce a rebound edema.

One of the important electrolyte changes in acute ischemia and trauma is the increase of intracellular calcium, which inhibits mitochondrial respiration. Calcium activates phospholipase, and other lyososomal enzymes destroying mitochondria and the cytoskeleton. Impaired membrane permeability increase calcium influx, which seriously interferes with neuronal functioning. Some excitatory aminoacids, particularly glutamate, influence calcium channels and since calcium influx is considered a major mechanism of injury, pharmacological agents inhibiting the glutamate cascade and calcium channels may promote early recovery (Gelmers et al., 1988; Stevens and Yaksh, 1990). Calcium channel antagonists not only inhibit calcium influx, but also prevent the post-ischemic reduction of cerebral blood flow and reduce mortality in experimental animals.

Another mechanism of damage is the accumulation of free radicals and new agents, called "Lazaroids", that mop up these, may promote early recovery (Tazaki et al., 1988). This restoration of high energy phosphates depleted by ischemia also contributes to early recovery (Argentino et al., 1989).

The first few days and weeks after the onset of a stroke may be a time of critical regrowth, when somewhat damaged neurons start to regenerate injured parts, and other neurons form new connections to compensate for ones that have been lost. Feeney et al. (1979) suspected that a depression of the catecholamine neurotransmitter system might contribute to behavioural deficits after a stroke. Amphetamines, which increase catecholamine transmission, produced a lasting improvement.

SECOND-STAGE RECOVERY

After the first stage, which takes anywhere from a few days up to a month, recovery continues for months and even years but this second stage is a largely unexplained process. Active repair of damaged structures includes axonal sprouting, axonal regrowth and collateral sprouting. All of these have been shown to take place in the mammalian CNS, even though originally they were thought to occur only in the peripheral nerves. Synaptogenesis, the formation of new synapses also occurs, but the extent to which this underlies recovery of cognition is not established. Recently nerve growth factors have been discovered in the CNS that may contribute to recovery. The association between recovery of cognitive function and the regrowth of axons in the human CNS however is not well established. It is likely that in higher level nervous systems, in complex networks subserving cognitive processes, axonal regrowth or collateral sprouting are not the only or not even the major mechanisms of recovery.

Goldstein (1948) considered that most recovery is a readjustment by which the organism manages to get along with the functions which have been lost. He contrasted "true recovery of function, that comes only as result of restoration of the anatomical substratum or by relearning with the help of the remnant of the substratum which participated in the original". Lashley (1938) also thought that some part of the original system must be preserved in order for recovery to take place. In contrast to this, the principle of vicarious functioning emphasizes the

Neurophysiological foundations 367

plasticity of the nervous system and states that functions may be taken over by structures not previously involved in their function. This assumes a redundancy that will provide elements that can substitute for the damaged functions. The theory of vicarious functioning was first proposed by Fritsch and Hitzig (1870) to account for the recovery of dogs from a cerebral paralysis produced by removal of the motor cortex of one hemisphere. This theory has played a role in almost every subsequent attempt to account for the restitution of function. Pavlov also argued for vicarious functioning and a large factor of redundancy saying that there are many potential conditioned reflex paths which are never used by the normal organisms.

Diaschisis, which was Monakow's (1914) explanation for second-stage recovery, implies that the damaged brain deprives the surrounding, or functionally connected areas from a trophic influence causing the initial deficit and subsequently these areas recover by acquiring innervation from somewhere else or becoming autonomously functional. The phenomenon of diaschisis has been widely accepted and subsequently elaborated, with biochemical and physiological supporting evidence. Denervation hypersensitivity may explain why some central structures become more responsive to stimulation after damage (Stavraky, 1961). The remaining fibers may have a greater effect on the denervated region, which is hypersensitive to neurotransmitters, thereby promoting recovery. The opposite effect, however, has also been argued. The initial hypersensitivity to inhibitory neurotransmitters could induce inhibition of function (diaschisis), and the appearance of collateral sprouting might reduce the denervation and the accompanying inhibition (Goldberger, 1974).

Even before Monakow's diashisis theory, it was noted that sudden lesions produced a more serious deficit than slowly growing ones. Dax (1865) suggested that left-hemisphere lesions may not result in aphasia if the lesion develops gradually. Modern animal experimentation confirmed this "serial lesion" effect (Ades and Raab, 1946).

Spontaneous reorganization which is independent of regeneration or redundancy may occur. Even the associationistic or connectionistic theory of cerebral organization could accommodate the possibility of the development of new associative connections, therefore the restitution of function seems to be related to learning. However, it is evident from clinical experience and experimental data that recovery from a higher level cortical deficit is limited by certain factors. It is useful, as well as theoretically important, to examine what influences recovery and the limits to what is usually observable in the evolution of various deficits.

Rather than a general and widespread substitution of functions, it appears that compensation occurs only in specific areas. For instance, Bucy (1934) showed that in monkeys, motor recovery depends on the adjacent anatomical areas and only the combined destruction of precentral and premotor gyrus results in permanent paralysis. In rodents it is impossible to produce paralysis with cortical lesions only. However, combined destruction of the motor cortex and the corpus striatum produces a permanent spastic paralysis (Lashley, 1938). In the hierarchically organized visual system, visual discrimination is relearned quickly after striatal cortex destruction but secondary destruction of the tectum prevents relearning. Such experiments suggest that restitution of function depends upon the preservation of some part of a limited system which is normally concerned with the function. However, the system which allows compensation cannot always be limited to an anatomical neighbourhood. In many instances, not necessarily the adjacent tissue but connected areas, which were functionally involved to some degree, could play a part.

Functional compensation explains recovery with a behavioral rather than a neural model. Instead of rerouting connections, the brain-damaged organism develops new solutions to problems using residual structures. Substitute maneuvers or "tricks" have been observed in various experimental situations (Gazzaniga, 1974). Luria (1970) promoted the theory of retraining, which claims that the dynamic reorganization of the nervous system is promoted by specific therapy.

There are some cortical areas in man that are primary in the sense that their destruction permanently abolishes function, such as the striate cortex, which Munk (1881) called the "cortical retina". Hemianopsia due to striate cortex lesions are not compensable. It appears that the language cortex also represents a primary cortex in this context, because of the large number of permanently disabled aphasic individuals. However, to what extent and what areas of the language cortex can be destroyed that is still compatible with recovery and what other areas are taking part in compensation, are topics for ongoing investigation.

RECOVERY OF LANGUAGE

The recovery of human cognitive function, especially recovery of language, can provide an important behavioural model which sheds light on the rate and biological determinants of second stage recovery. The study of these factors has more than just practical prognostic value for the clinician. It also provides an important theoretical framework for understanding cerebral reorganization following damage to CNS. The fact that aphasic patients recover considerably has been long recognized, and Von Monakow (1914) used recovery from aphasia as a model for diaschisis. Unfortunately, aphasic syndromes are often presented as if they were static, and many linguistic or therapeutic studies disregard the major influence of biological recovery.

The multiplicity of prognostic factors in aphasia has been recognized and some quantification has been achieved by several advances in methodology. Firstly, aphasia tests became better standardized and more specific for a language-disordered population (Goodglass and Kaplan, 1983; Kertesz, 1979; Kertesz, 1982). Secondly, the methods of follow-up and the statistical evaluation of change have become more sophisticated. Thirdly, neuroimaging in vivo with computerized tomography (CT) and magnetic resonance imaging () have enabled us to localize and quantify lesions in the same patients who are documented cognitively. In recent years, factors such as initial severity, etiology, age, aphasia type and the effect of therapy have been extensively investigated. Initial severity is one of the most important factors in recovery. Early investigators considered initial severity to have a highly predictive value (Godfrey and Douglass, 1959; Schuell et al., 1964; Sands et al., 1969; Sarno et al., 1970a; Sarno et al., 1970b; Gloning et al., 1976; Kertesz and McCabe, 1977). The severity of deficit at onset has a considerable effect on comparing recovery rates because mildly affected aphasics do not have much potential "room" for recovery (i.e., they experience a "ceiling effect") while severe aphasics often have more potential. Patients considered suitable for treatment tend to be selected from the less severe groups and hence they can bias the results. Unless initial severity is considered a major factor to be controlled, studies of treatment should not be considered reliable. There are various methods of controlling for initial severity, such as analysis of covariance or using outcome measures instead of recovery rates or the change expressed as percentage of initial severity.

The time from onset when patients are studied is also an important factor. When patients are entered into recovery studies at various stages in their progress, comparison among treatment groups becomes very difficult (Basso et al., 1978). In our studies, we took care to start our evaluation within the acute period, between 10 to 45 days after a stroke. Since most of our patients were examined at exactly 14 days post-onset, this provided a rather homogenous sample. Only the more severely affected patients, who could not be examined at that time because of concurrent medical illness or altered conciousness, were kept until the upper limit of the acute period.

Etiology is the third major factor that needs to be considered. Traumatic aphasic patients, for instance, recover quickly if their lesion is related to closed head injury. Persisting dysarthria, however, is common in severe trauma and this often disrupts communication to such a degree that the extent of post-traumatic aphasia is difficult to determine. Penetrating head injury often affects a different age group and yields different behaviour because of the variation in the speed and path of the missiles and the associated concussion. Therefore, post-traumatic aphasia is biologically different from the vascular type. There are many similarities, nonetheless, indicating that the recurring patterns of aphasia types are not necessarily related to the distribution of vascular lesions. A recent study by Ludlow et al. (1986), on Vietnam veterans, showed that the lesions that produce a persisting asyntactic or Broca's aphasia are large, involving the subcortical structures and the parietal area, in addition to Broca's area. They have thus reached very much the same conclusions that have been obtained studying stroke recovery.

LESION SIZE AND LOCATION

Lesion size and location have also been recognized as interrelated and complex factors. Until recently, clinicians relied on autopsy correlations but modern neuroimaging has provided an opportunity to study lesion characteristics in vivo. We found, in our first study of lesion size measured on computerized tomography (CT) and recovery from aphasia, that the larger the lesion the poorer the outcome; in other words, outcome correlated negatively with lesion size (Kertesz et al., 1979). This has been known to clinicians since Jackson's time and supports the principle of the so-called "mass effect" (Lashley, 1938). Recovery rates also showed a trend of negative correlation with one unexpected exception. The recovery rate of comprehension was found to be correlated positively with lesion size! This can best be understood if we look at another study of ours in which the best recovered modality was found to be comprehension (Lomas and Kertesz, 1978). Patients with large lesions and who have global or severe Broca's aphasia often show greater improvement in comprehension. Patients with smaller lesions, such as anomics, already have good comprehension. Therefore they have less room for recovery. The large lesions with more recovery and small lesions with relatively less recovery give rise to a consistently positive correlation, unless the initial severity is covaried (as was done in our subsequent studies).

Since then, various other studies have been conducted dealing with localization of the lesions in a somewhat different, symptom-oriented approach (Selnes et al., 1983; Knopman et al., 1983). Knopman et al. (1983) examined lesion size and location and found that, in the language area, about 60 cm^2 was a critical mass, and larger lesions resulted in relatively less recovery. They concluded that speech output has less redundancy and, if the surrounding areas are involved, less recovery will take place. The symptom-oriented approach generally leads to less focal localization of deficits as some of these functions are widely distributed in the brain.

In a recent study (Kertesz, 1988), the recovery of Broca's aphasics was correlated with lesion size, age, the degree of atrophy, and initial severity. The most significant correlations were obtained in the outcome measures which showed a negative correlation with lesion size. Cerebral atrophy, as measured by the ratio of fontal horn width and brain diameter on the horizontal section at the level of the pineal, did not correlate significantly, with recovery, or outcome measures.

Lesion location was evaluated in Broca's aphasics and divided at the median for poor and good recovery. The structures with significant involvement (more than 50%) were the inferior frontal gyrus, especially the pars opercularis and triangularis, and the insula in both groups. Supramarginal, angular and superior temporal gyri were not involved in those cases where recovery was good. The subcortical regions showed significant differences in the involvement of the putamen and the caudate, which involvement was twice as frequent in the persistent cases.

The most consistently involved structure in patients with Wernicke's aphasia was the superior temporal gyrus, confirming previous studies (Kertesz et al., 1979; Naeser et al., 1987). In cases of poor recovery, the middle temporal gyrus and the supramarginal gyrus were significantly more frequently involved, in addition to the isthmus of the temporal lobe and the insula whichwas twice as frequently involved in the unrecovered cases (Kertesz et al., 1989). Subcortical structures did not seem to be as significant for persisting Broca's aphasia. The involvement of isthmus and posterior insula is similar to the findings of Naeser et al. (1987), but our emphasis on posterior structures in compensation is different (Kertesz et al., 1989). In recovery from Wernicke's aphasia, lesion size was also a major contributor. Those lesions associated with low recovery rates were over 60 cc, and this is compatible with the finding of Selnes et al. (1983) who studied recovery of comprehension.

It is interesting to note that Wernicke's aphasics did not show as high recovery rates, as global and Broca's aphasics, confirming our previous study which was done independently from lesion localization that Broca's and global aphasics recover best (Kertesz and McCabe, 1977). There is some controversy in the literature about this, mainly because some authors who claim that expressive aphasic patients do not recover as well take a large number of global aphasics with poor outcome and contrast them to the relatively good outcome of Wernicke's and anomic aphasics. Therefore, when outcome measures are contrasted with recovery rates, a discrepancy may result.

The idea of right-hemisphere substitution remains a favored explanation for language recovery even though the evidence in favour of it is often complex and contradictory. Wernicke (1886) postulated that the contralateral homologous cerebral cortex is responsible for the return of language function. Jackson (1873) considered the right hemisphere capable of automatic utterances and stated that it was responsible for the residual output in global aphasia. The idea of contralateral hemispheric compensation has been reemphasized by Henschen (1920-22) in his monumental monograph of aphasic cases with localization. Subsequently this was called the "Henschen principle" by Nielsen (1946), who also believed that it is the right hemisphere that compensates for language deficits. Nielsen also thought that the right hemisphere was able to substitute mainly for comprehension but not for expressive functions.

Initially, the principle of contralateral homologous cortical substitution was based on large left-hemisphere lesions with good recovery where there was very little left hemisphere remaining to assume functions. More recently, CAT scan studies have enabled researchers to arrive at the same conclusions (Cummings et al., 1979; Landis et al., 1980). In addition, in some patients who became aphasic with a single left-hemisphere stroke, but who recovered, a second right-hemisphere stroke produced a language deficit again (Nielsen, 1946; Levine and Mohr, 1979; Cambier et al., 1983). These cases, however, may have represented bilateral language organization to begin with, rather than a commonly operating mechanism of functional transfer to the contralateral hemispheres.

The idea of compensation through right-hemisphere function, even after partial left-hemisphere damage, was also supported by studies of sodium amytal given to aphasics who had recovered (Kinsbourne, 1971; Czopf, 1972). These studies indicated that, even though the aphasic disturbance occurred from a left-hemisphere lesion, it was the right hemispheric injection that increased the language disturbance, implying that the right-hemisphere compensated for the previous deficit produced by the left side. Cerebral blood flow (CBF) and positron emission tomography (PET) studies of cerebral metabolism provide methodologies that add functional information to structural or lesion studies of recovery. Recent studies of cerebral blood flow with xenon 133 have also revealed a right-hemisphere hypometabolism in aphasic strokes, the extent of which has correlated with recovery to a modest degree (Knopman et al., 1984). Positron emission tomography (PET) studies of cerebral metabolism have shown a great deal of hypometabolism surrounding, but also remote from cerebral infarcts, thus suggesting that not only surrounding areas but also homologous areas in the contralateral hemisphere play a role in compensation (Metter et al., 1981). However, one CBF study showed no significant change while clinical recovery occurred in severe aphasics (Demeurisse et al., 1983). Patients who improved more showed more activation in the left hemisphere. This appears to be the consequence of the size of the lesion, which correlates with the CBF changes. Another CBF study showed better than 60% hemispheric flow in patients with good recovery (Nagata et al., 1986). We had experience with four aphasics whose distant hypometabolism (ipsilateral cortex outside this CT lesion) persisted, despite of significant recovery. Recent unpublished studies by Cappa (1990) have suggested improved CBF at distant (contralateral) hypometabolic sites, but this requires further study.

HEMISPHERIC LATERALIZATION

The variable degree of recovery that cannot entirely be explained by the extent and location of lesions has been postulated to relate to differences in language laterality due to handedness and gender. Subirana's (1969), Gloning's et al. (1969), and Geschwind's (1974) suggestion that left handers and right handers with a family history of left handedness recover better from aphasia because of greater bilateral language distribution is based on anecdotal evidence. There is also a recently popularized, but yet to be proven, theory of greater bilateral distribution of language in women (McGlone, 1980). Studies of anatomical asymmetry on CT scans, inspired by the finding of a larger *planum temporale* in the left (Geschwind and Levitsky, 1968), have correlated better outcome with atypical asymmetry (Pieniadz et al., 1983). We have studied the factor of anatomical asymmetry on CT, as measured by occipital width, frontal width, and protuberance (petalia) and could not confirm that it played a role in recovery in any of the aphasic groups. Neither did we find any sex differences in recovery (Kertesz, 1988). It could be that anatomical asymmetries relate more to handedness variables than language distribution, as suggested by some of our studies in normals (Kertesz et al., 1986); therefore, we have not observed an effect on language recovery.

PHARMACOTHERAPY

It was hoped that pharmacotherapy for aphasia would promote recovery through providing missing neurotransmitters or by decreasing inhibitory mechanisms. Pharmaceuticals and chemicals used to treat aphasia include dexamethasone, sodium amytal, priscoline, meprobamate, and hyperbaric oxygen—with unimpressive results (Darley, 1975). Recently, bromocriptine has been found to improve spontaneous speech in a case of transcortical motor aphasia (Albert, 1987). Apparently, language deteriorated to baseline after cessation of therapy.

CONCLUSIONS

The complex interaction of size and location of lesions, in addition to time from onset, and etiology and initial severity, are the main factors in the recovery of cognitive and language loss. Other biological factors, such as age, sex and handedness play a less significant role in an adult stroke population.

Lesion size is undoubtedly a significant factor in the extent of recovery. However, the important exceptions are certain crucial areas in the left hemisphere that are more important for prognosis than others. Motor and premotor phonemic assembly mechanisms are elaborated by a cortical/ subcortical network that can be damaged partially, and followed by good recovery. However, if both cortical and subcortical components of the network are impaired, good recovery is much less likely. Therefore, the role of hierarchical structural organization is important in the recovery of speech and in recovery from Broca's aphasia. Contralateral, homologous, cortical substitution is probably not a major mechanism for the recovery of speech.

Certain aspects of hemispheric specialization may vary according to individuals even though anatomical asymmetries, as measured on CT scan, do not seem to play a role in recovery, according to our preliminary findings. The individual variations in the intra- and interhemispheric distribution of various functional components may contribute, to an important extent, to the ability of the mature brain to compensate after a single nonprogressive lesion. Other pathological variations, such as repeated stroke insults, cerebral atrophy, intercurrent latent dementia, etc., are factors to be considered or even examined directly, although they have been controlled by exclusion in our studies.

A complex network of various structures is also needed for the processing of language comprehension. However interhemispheric connections may play a larger role in comprehension than in motor output. It seems that, when there is a restricted deficit in the dominant hemisphere auditory association area, the posterior superior temporal gyrus and the *planum temporale*, the deficit can be compensated for by surrounding structures in the inferior parietal regions, and the insula. However, when most of these compensating structures are affected and the lesion is large, recovery may take place. Some argue that large lesions also preclude right-hemisphere access.

References

Ades, H.W., and Raab, D.H. 1946. Recovery of motor function after two-stage extirpation of area 4 in monkeys. *Journal of Neurophysiology*, 9: 55-60.
Argentino, C., Sacchetti, M.L., Toni, D., et al. 1989. GM_1 ganglioside therapy in acute ischemic stroke. *Stroke*, 20: 1143-1149.
Astrup, J., Siesjo, B.K., and Symon, L. 1981. Thresholds in cerebral ischemia—the ischemia penumbra. *Stroke*, 12: 723-725.
Basso, A., Capitani, E. and Vignolo, L.A. 1979. Influence of rehabilitation on language skills in aphasic patients. *Archives of Neurology* 6: 190-196.
Bucy, P.C. 1934. The relation of the premotor cortex to motor activity. *Journal of Nervous and Mental Disorders*, 79: 621-630.
Cambier, J., Elghozi, D., Signoret, J.L., and Henin, D. 1983. Contribution of the right hemisphere to language in aphasic patients. Disappearance of this language after a right-sided lesion. *Revue Neurologique*. 139: 55-63.
Cappa, S.F. and Vallar, G. 1992. Neuropsychological disorders after subcortical lesions: implications for neural models of language and spatial attention. In G. Vallar, S.F. Cappa and S.-W. Wallesch (eds.) *Neuropsychological disorders associates with subcortical lesions*. Oxforx: Oxford University Press.
Cummings, J.L., Benson, D.F., Walsh, M.J., and Levine, H.L. 1979. Left-to-right transfer of language dominance: A case study. *Neurology*. 29: 1547-1550.
Czopf, J. 1972. Role of the non-dominant hemisphere in the restitution of speech in aphasia. *Archiv für Psychiatrie und Nervenkrankeiten*, 216: 162-171.
Dax, M. 1865. Lésions de la moitié gauche de l'encéphale coïncidant avec l'oubli des signes de la pensée. *Gazette Hebdomadaire Medédicale Chirurgicale*, Paris.
Demeurisse, G., Verhas, M., Capon, A., and Paternot, J. 1983. Lack of evolution of the cerebral blood flow during clinical recovery of stroke. *Stroke*, 14: 77-81.
Feeney, D.M. and Wier, C.S. 1979. Sensory neglect after lesions of substantia nigra or lateral hypothalamus: differential severity and recovery of function. *Brain Research*, 178: 329-346.
Finger, S. and Stein, D.F. 1982. *Brain Damage and Recovery*. Plenum Press, New York.
Fritsch, G.T. and Hitzig, E. 1870. Über die elektrische Erregbarkeit des Grosshirns. *Archiv für Anatomie und Physiologie*, 300-332.
Gazzaniga, M.S. 1974. Determinants of cerebral recovery. In D.G. Stein, J.J. Rosen and N. Butters (eds.), *Plasticity and Recovery of Function in the Central Nervous System*. New York: Academic Press. Pp. 203-217.
Gelmers, J.J., Gorter, K., DeWeerdt, C.J. and Wiezer, H.J.A. 1988. A controlled trial of nimodopine in acute ischemic stroke. *New England Journal of Medicine*, 318: 203-207.

Geschwind, N. 1974. Late changes in the nervous system: An overview in plasticity and recovery of function in the central nervous system, in D. Stein, J. Rosen, and N. Butters (eds.). *Plasticity and recovery of function in the central nervous system* New York: Academic Press. Pp. 467-508.
Geschwind, N. and Levitsky, W. 1968. Human brain, left-right asymmetries in temporal speech regions. *Science*, **161**: 186-187.
Gloning, I., Gloning, K., Haub, G. and Quatember, R. 1969. Comparison of verbal behavior in right-handed and nonright-handed patients with anatomically verified lesion of one hemisphere. *Cortex*, **5**: 43-52.
Gloning, K., Trappl, R., Heiss, W.D. and Quatember, R. 1976. *Prognosis and speech therapy in aphasia in neurolinguistics*, Volume 4, *Recovery in aphasics*. Amsterdam: Swets & Zeitlinger, B.V.
Godfrey, C.M. and Douglass, E. 1959. The recovery process in aphasia. *Canadian Medical Association Journal*, **80**: 618-824.
Goldberger, M.E. 1974. Recovery of movement after CNS lesions in monkeys, in *Plasticity and Recovery of Function in the Central Nervous System*. D. Stein, J. Rosen, and N. Butters (eds.). New York: Academic Press.
Goldstein, K. 1948. *Language and Language Disturbances*. New York: Grune and Stratton.
Goodglass, H. and Kaplan, E. 1983. *Boston Naming Test*. Lea and Febiger, Philadelphia.
Henschen, S.E. 1920-22. *Klinische und anatomische Beitrage zur Pathologie des Gehirns*, Vols. 5-7, Stockholm: Nordiska Bokhandel.
Jackson, J.H. 1873. On the anatomical and physiological localization of movements in the brain. *Lancet*, **1**: 232-234.
Kertesz, A. 1979. *Aphasia and Associated Disorders: Taxonomy, Localization and Recovery*. New York: Grune and Stratton.
Kertesz, A. 1982. *The Western Aphasia Battery*, New York: Grune and Stratton,.
Kertesz, A. 1988. What do we learn from recovery from aphasia? in S. G. Waxman, (ed.) *Advances in Neurology, Vol. 47: Functional Recovery in Neurological Disease* . New York: Raven Press. Pp. 277-292.
Kertesz, A., Black, S.E., Polk, M., and Howell, J. 1986. Cerebral asymmetries on magnetic resonance imaging. *Cortex*, **22**: 117-127.
Kertesz, A., Dennis, S., Polk, M. and McCabe, P. (1989). The structural determinants of recovery in Wernicke's aphasia. *Neurology*, **39** (Suppl. 1): 177.
Kertesz, A., Harlock, W. and Coates, R. 1979. Computer tomographic localization, lesion size and prognosis in aphasia. *Brain and Language*, **8**: 34-50.
Kertesz, A. and McCabe, P. 1977. Recovery patterns and prognosis in aphasia. *Brain*, **100**: 1-18.
Kinsbourne, M. 1971. The minor cerebral hemisphere as a source of aphasic speech. *Archives of Neurology*, **25**: 302-206.

Knopman, D.S., Rubens, A.B., Selnes, O.R., Klassen, A.C., et al. (1984). Mechanisms of recovery from aphasia: evidence from serial xenon 133 cerebral blood flow studies. *Annals of Neurology*, **15**: 530-535.
Knopman, D.S., Selnes, O.A., Niccum, N. and Rubens, A.B. 1983. A longitudinal study of speech fluency in aphasia: CT scan correlates of recovery and persistent nonfluency. *Neurology*, **33**: 1170-1178.
Kohlmeyer, K. 1976. Aphasia due to focal disorders of cerebral circulation: some aspects of localization and of spontaneous recovery. In *Neurolinguistics*. *4. Recovery in Aphasics*. Amsterdam: Swets & Zeitlinger B.V.
Landis, T., Cummings, J.L., and Benson, D.F. 1980. Passage of language dominance to the right hemisphere: Interpretation of delayed recovery after global aphasia. *Revue Médicale de Suisse Romande*, **100**: 171-177.
Lashley, K.S. 1938. Factors limiting recovery after central nervous lesions. *Journal of Nervous and Mental Disease*, **88**: 733-755.
Levine, D.M., and Mohr, J.P. 1979. Language after bilateral cerebral infarctions: role of the minor hemisphere. *Neurology*, **29**: 927-938.
Lomas, J. and Kertesz, A. 1978. Patterns of spontaneous recovery in aphasic groups: A study of adult stroke patients. *Brain and Language*, **5**: 388-401.
Ludlow, C., Rosenberg, J., Fair, C., Buck., D., et al. 1986. Brain lesions associated with nonfluent aphasia fifteen years following penetrating head injury. *Brain*, **109**: 55-80.
Luria, A.R. 1970. *Traumatic Aphasia*. Hague: Mouton.
McGlone, J. 1980. Sex differences in human brain asymmetry: a critical survey. *Behavior and Brain Sciences*, **5**: 215-264.
Metter, E.J., Wasterlain, C.G., Kuhl, D.E., Hanson, W.R., and Phelps, M.E. 1981. FDG positron emission computed tomography in a study of aphasia. *Annals of Neurology*, **10**: 173-183.
Monakow, C. von 1914. *Die localisation im Grosshirn und der Abbau funktionen durch corticale Herde*. Bergmann, Wiesbaden, F.R.G.
Munk, H. 1881. *Über die Funktionen der Grosshirnde, Gesammelte Mitteilungen aus den Jahren 1877-1880*. Berlin: Hirshwald.
Naeser, M.A., Helm-Estabrooks, N., Haas, G., Auerbach, S., and Srinivasan, M. 1987. Relationship between lesion extent in Wernicke's area on computed tomographic scan and predicting recovery of comprehension in Wernicke's aphasia. *Archives of Neurology*, **44**: 73-82.
Nagata, K., Yunoki, K., Kabe, S., Suzuki, A., and Araki, G. 1986. Regional cerebral blood flow correlates of aphasia outcome in cerebral hemorrhage and cerebral infarction. *Stroke*, **17**: 417-423.
Nielsen, J.M. 1946. *Agnosia, Apraxia, and Aphasia*, New York: Hoeber.
Pieniadz, J.M., Naeser, M.A., Koff, E., and Levine, H.L. 1983. CT scan cerebral hemispheric asymmetry measurements in stroke cases with global aphasia: atypical asymmetries associated with improved recovery. *Cortex*, **19**: 371-391.

Sands, E., Sarno, M.T., and Shankweiler, D. 1969. Long-term assessment of language function in aphasia due to stroke. *Archives of Physical Medicine and Rehabilitation*, **50**: 202-222.

Sarno, M.T., Silverman, M. and Levita, E. 1970b. Psychosocial factors and recovery in geriatric patients with severe aphasia. *Journal of the American Gereatric Society*, **18**: 405-409.

Sarno, M.T., Silverman, M. and Sands, E. 1970a. Speech therapy and language recovery in severe aphasia. *Journal of Speech and Hearing Research*, **13**: 607-623.

Schuell, H.M., Jenkins, J.J. and Pabon, J. 1964. *Aphasia in adults*, New York: Harper Row.

Selnes, O.A., Knopman, D.S., Niccum, N. and Rubens, A.B. 1983. T scan correlates of auditory comprehension deficits in aphasia: A prospective recovery study. Annals Neurology 13: 558-566.

Stavraky, G.W. 1961. *Supersensitivity following lesions of the nervous system.* Toronto: University of Toronto Press.

Stein, H.D., Rosen, J.J. and Butters, N. 1974. *Plasticity and Recovery of Function in the Central Nervous System.* New York: Academic Press.

Stevens, M.K. and Yaksh, T.L. 1990. Systemicc studies on the effects of the NMDA receptor antagonist MK-801 on cerebral blood flow and responsivity, EEG, and blood-brain barrier following complete reversible cerebral ischemia. *Journal of Cerebral Blood Flow Metabolism*, **10**: 77-88.

Subirana, A. 1969. Handedness and cerebral dominance, In P.J. Vinken and G.W. Bruyn (eds.), *Handbook of Clinical Neurology..* Amsterdam: North Holland.

Tazaki, Y., Sakai, F., Otomo, E. et al. 1988. Treatment of cerebral infarction with a choline precursor in a multi-center double blind placebo-controlled study. *Stroke*, **19**: 211-216.

Wernicke, C. 1886. Die neueren Arbeiten über Aphasie. *Fortschr. Med.* **4**: 371-377.

Zivin, J.A., Fisher, M., DeGirolami, J. et al. 1985. Tissue plasminogen activator reduces neurological damage after cerebral embolism. *Science*, **230**: 1289-1292.

12 Interhemispheric participation in recovery from aphasia

Josep M. Vendrell, Pere Vendrell and Daniel Ibáñez

The asymmetrical performance of the right and left cerebral hemispheres in the processing of verbal and non-verbal material is well documented in normal subjects. Superiority of the left visual field in the identification of visually presented spatial stimuli (Gross, 1977; Berlucchi et al., 1979; Berrini et al., 1982) and physiognomical material (Rizzolatti, Umiltà and Berlucchi, 1971) indicates a right hemisphere advantage in contradistinction to a right visual hemifield superiority for letters and words, indicating a left hemisphere advantage in alphabetical and written-word processing (Bryden and Rainey, 1963; Rizzolatti, Umiltà and Berlucchi, 1971; Hines, 1972; Gross, 1977; Hay, 1982). The neurophysiological approach supports the superiority of the left hemisphere in written-word identification tasks, showing asymmetrical visual event-related potentials dependig on the hemifield where the stimulus-word is located: the N410 component obtained in the anterior temporal area is more negative from the left than the right hemisphere and this asymmetry is largest for right visual hemifield presentations (Neville, Kutas and Schmidt, 1982).

Therefore, we can maintain the basic assumption that the left hemisphere is more accurate in letter and written-material recognition, whereas the right hemisphere is more accurate in visual spatial recognition (Berrini et al., 1982). This implies a predominantly analytical left hemisphere processing in contrast to a predominantly holistic right hemisphere processing. The word "predominantly" means that both hemispheres can process information analytically and holistically (Bagnara et al., 1982) although in normal circumstances each subserves its respective specialized tasks. Following an exhaustive revision of studies with verbal stimuli, Beaumont (1982) concludes that words, letters, digits, letter strings and nonsense words are generally associated with a right visual field advantage for tasks of identification, recognition and nominal matching, although with physical (or simple identity) matching this advantage is not always found.

Studies of reaction time in aphasic patients by randomly localized luminous stimuli in both visual hemifields provides a general insight into the visuo-perceptive performance following a brain lesion, although this research in itself does not provide specific infomation on language processing.

REACTION TIME IN APHASIC PATIENTS STUDIED BY RANDOMLY LOCALIZED STIMULI IN BOTH VISUAL HEMIFIELDS.

Since the work of Poffenberger (1912) the measuring technique of visual reaction time (RT) has been considered as a useful procedure in the study of normal brain organization. RT techniques have also been applied to the study of brain-damaged patients. DeRenzi and Faglioni (1965) carried out a study on 166 brain-damaged patients confirming the usefulness of RT in the evaluation of cerebral involvement. They found that the patients with a right hemisphere lesion, responding with their preferred hand, presented a more severely affected RT than the left-hemisphere patients responding with their non-dominant hand. DeRenzi and Faglioni attributed this more severely affected RT in right-hemisphere lesions to a non-specific effect depending on the extension or severity but not on the site of the lesion (mass effect). Howes and Boller (1975), however, studied the RT in 49 brain-damaged patients and confirmed the asymmetric effect of the hemispheric lesions in the RT experiment, concluding that right-hemispheric lesions slowed the RT more than the lesions of the dominant hemisphere. It may thus be said that the lesions of the non-dominant hemisphere have a specific effect on RT as opposed to a non-specific effect of the dominant hemisphere lesions (Howes and Boller, 1975).

These data correspond to the study of Anzola et al. (1977) in normal subjects, the results of which showed that the RT is shorter for stimuli presented in the left visual hemifield, regardless of the responding hand.

On the other hand, it is well known that spatial neglect caused by unilateral cerebral lesions is more closely related to right hemisphere lesions (with neglect of the left hemispace) than to the left hemisphere lesions (Brain, 1941; Gainotti, 1968; Hécaen and Albert, 1978). Heilman (1979) states that each cerebral hemisphere, as well as being responsible for the movements of the contralateral extremeties and processing contralateral sensory input, is also responsible for mediating the behavior in the contralateral spatial field independently of which extremity is working in this hemifield. The hemispatial neglect does not therefore consist of a contralateral visual field defect but of a decreased capacity to act in the hemispace contralateral to the lesion. This may be related to an attention-arousal defect which could depend on a dysfunction in a cortico-limbic-reticular loop (Heilman and Valenstein, 1972; Watson et al., 1973; Heilman, Schwartz and Watson, 1978).

A great part of research into visual RT is carried out by means of stimuli from constant sites on either side of a fixed central point. The present work was designed to investigate the RT performance in brain-damaged patients by means of stimuli presented at random from different points in the two visual hemifields around a central fixation point. By means of random presentations we eliminated the factor of foreseen responses determined by the constancy of localization of stimuli, and we were able to see to what extent we could obtain information concerning visual field defects either of a purely perceptive character or due to neglect. Furthermore, we wanted to see if we could detect subclinical problems of the visual fields, that is, those problems undetected by means of routine clinical examination.

The aims of this work were therefore to obtain the following in aphasic patients:
a) a visual RT ("global RT");
b) an RT for all stimuli from each hemifield ("partial RT");
c) a graph representing the time of response (or absence of) for each stimulation point; and thus all the responses of the patient for both visual hemifields;
d) complementary measurements which can contribute to the reliability and analysis of these results. Such complementary measures were: RT for the central stimuli (these being the 14 stimuli appearing nearest to the central fixation point, at a distance of 7 cm. around this, each at a visual angle of 5 degrees); RT for the peripheral stimuli (the 14 stimuli appearing in the sites furthest from the central fixation point, at a visual angle of 38 degrees); the number of stimuli not responded to in each visual hemifield; the total number of stimuli to which the subject gave an "anticipated" response.
e) the correlation of RT with the size of the lesion.

MATERIALS AND METHODS

Subjects: Twenty aphasic patients were selected from an initial group of 33 left-hemisphere damaged patients visited in the Section of Neuropsychology of the *Hospital de la Santa Creu i Sant Pau* in Barcelona, Spain. Selection was based on pre-established criteria of reliability, based on the pre-analysis of the data obtained (see Procedure for detailed description). The range of severity of aphasia varied from mild to severe. Only in three cases were the language disturbances very slight. Global or Wernicke's aphasics unable to cooperate were excluded. Etiology was stroke in all cases. All subjects were assumed to be right handed (right hemiplegia).

A non-matched group of sixteen normal subjects was used, selected from an initial group of 20, and all were right-handed.

Apparatus: Stimuli were produced by means of 94 red super-LEDs (Light Emiter Diode, High Bright), radially placed from a central fixation point, with a total of 14 radii covering a surface of 1 m. wide by 0.8 m. high, situated at a distance of 0.7 m. from the subject's eyes. The LEDs were covered with a transparent plastic screen on a black background making them invisible when not lit, even though the room was illuminated.

The central fixation point consisted of a yellow LED, clearly visible at all times.

The apparatus was controlled by a specially designed microcomputer with CPU Zilog Z80-A, with a driver to light the LEDs.

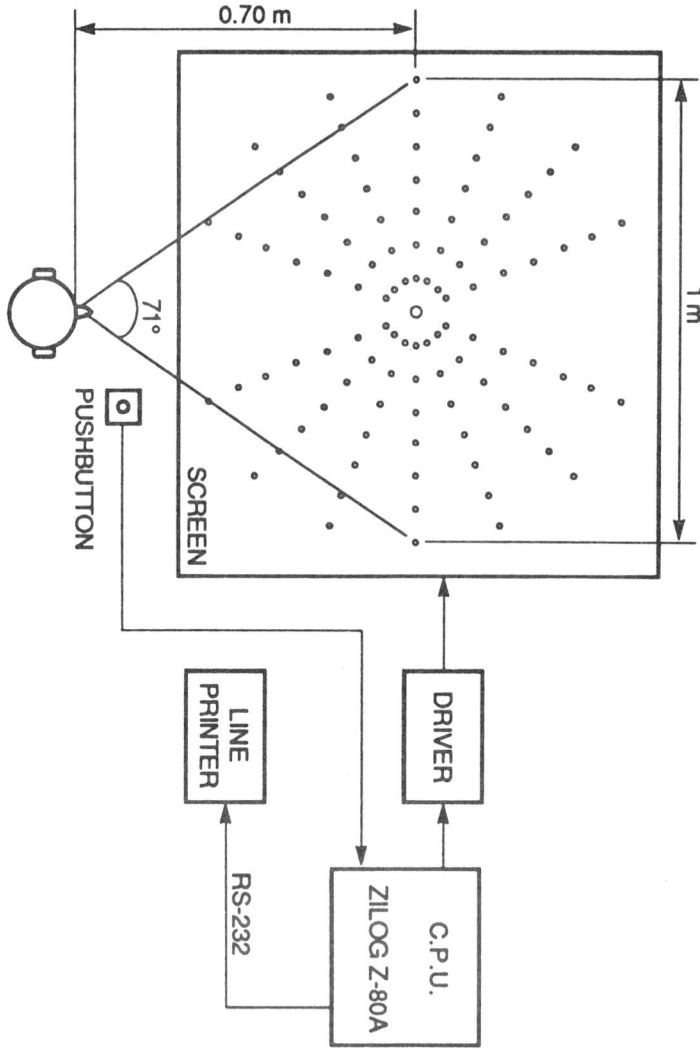

FIGURE 1. Diagram of apparatus.

Procedure: The test was carried out in a sound-attenuated room in darkness, the only light being that of the On/Off control on the printer, which was situated behind the subject. The subject was seated in front of a screen with eyes centered in respect to the fixation point at a distance of 0.7 m. The hand homolateral to the lesion rested on a table placed between the subject and the screen, and was kept in a supine position on the manual control, at the anteposterior level of ipsilateral shoulder (at a distance of 20 cm from the medial line). The pressbutton on the manual control was in a horizontal position and was operated by the thumb. Subjects in the control group operated the press-button with the same hand as did the patients (the left hand).

Subjects were instructed to press the button as quickly as possible after the illumination of any red point in any part of their visual field, following the warning of the central fixation point. Once the instructions were given, 42 practice stimuli were presented to ensure that the subject understood the procedure. When necessary, more than one series of practice stimuli were given. Once the test itself was underway, no additional instructions were given.

The onset of each stimulus was preceded by the illuminaton of the central fixation point which had the dual purpose of fixing the gaze and announcing that, after a variable interval., one of the 94 stimuli would randomly appear. Each stimulus appeared on the screen for 100 msec. If no response was given, the maximum time between two consecutive onsets of the central fixation point was three seconds (two seconds on and one second off). When a response was given, the next stimulus appeared (with the onset of the central fixation point) one second after the press-button was released.

When no response was given, the microcomputer recorded NR (No Response). Responses made earlier than 150 msec after the onset of stimulus were regarded as "anticipated" (this value was decided on the basis of previous tests in control subjects) and the same stimulus was presented again at the end of the 94 randomly-sequenced stimuli. RTs of responses produced after the first 150 msec –and within the maximum available time– were recorded to the nearest one-hundredth of a second. Except in case of "anticipatons", no stimulus was presented twice.

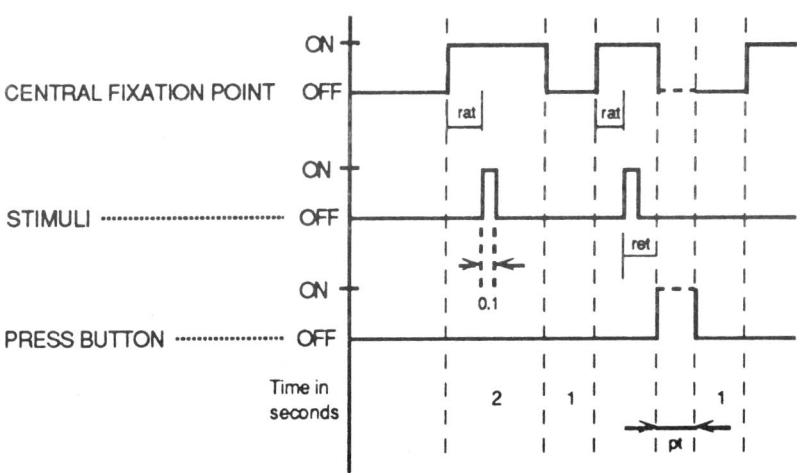

FIGURE 2. Graph of the stimuli;
rat: randomized time; ret: reaction time; pt: press button time

Total and partial results were recorded on a printer once the series of stimuli was completed. The total duration of the test was less than 5 minutes.

The main criterion of reliability was based on the different sensitivity of the retina depending on the central or peripheral incidence of the stimulus: it is well known that the foveal stimulation has an excitability threshold inferior to that of the peripheral regions of the retina (Sloan, 1961; Marks, 1968). In agreement with this, Berlucchi et al. (1971) found a progressively longer RT when the angular position of the visual stimulus was increased. Vendrell-Brucet et al. (1983) showed that RT tests with random stimuli in normal subjects were highly significant in the differences between results of central and peripheral stimuli. Accepting that the responses to central stimuli should yield faster values than those of peripheral stimuli, we therefore had a method for discarding subjects with poor cooperation.

A second criterion of reliability was the number of "anticipated" responses, which indicated that the test had not been well understood, or that the subject's capacity to cooperate was poor. The limit of anticipations allowed in the control group was 10. Among the left-hemisphere damaged patients, the highest number of anticipated responses in any one patient was 11.

A third criterion for excluding subjects was based upon the observation during the test itself: any behavior showing a lack of attention was recorded. In general., the clinical impression corresponded to the number of anticipations, so that both criteria (as well as the central/peripheral sensitivity criterion) were combined in the decision to exclude patients from the final analysis. As mentioned earlier, 13 subjects were thus excluded.

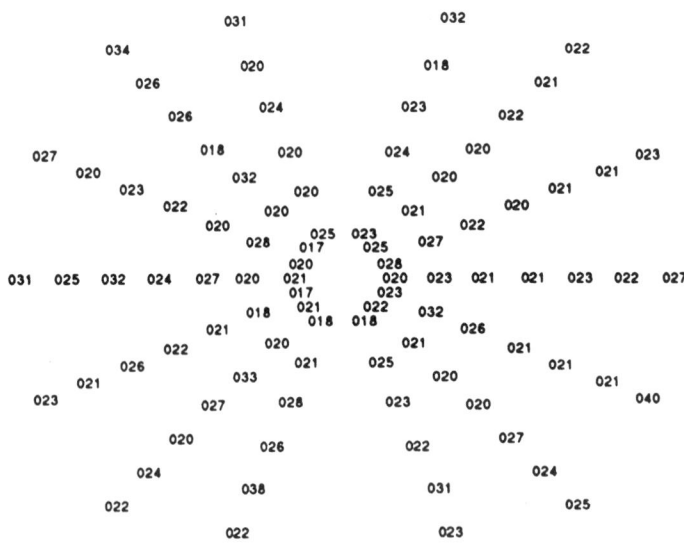

FIGURE 3. Graph of a normal subject.

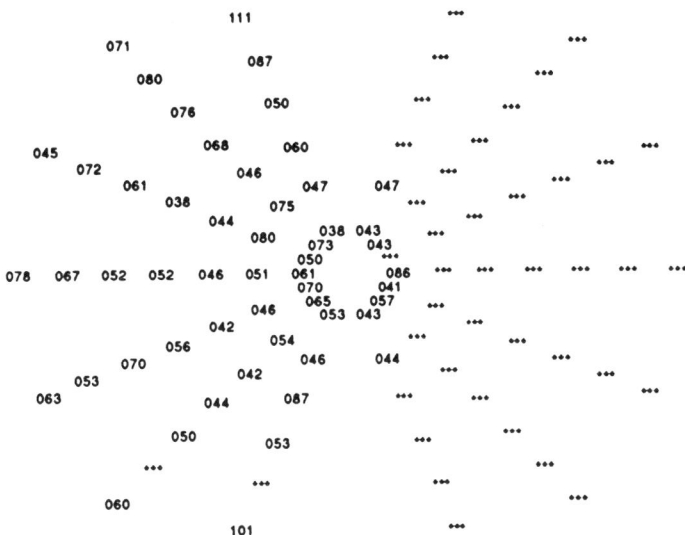

FIGURE 4. Graph of hemianopsia.

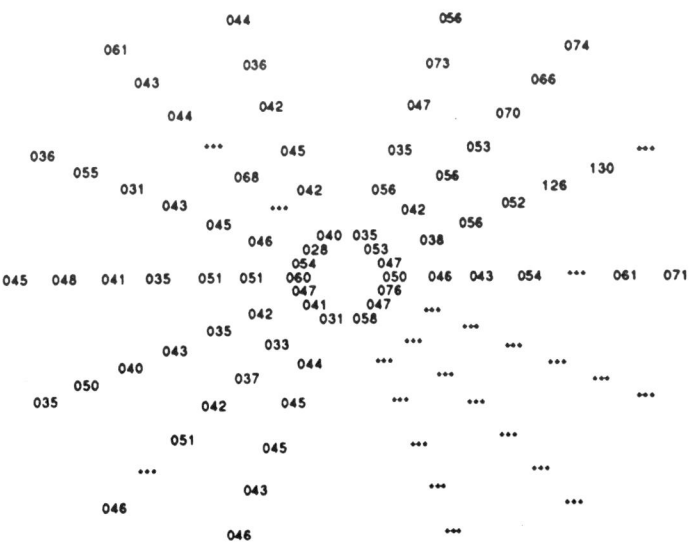

FIGURE 5. Graph of right visual field defect.

Brain scans: We used the control CT of the residual lesion, but not the CT obtained during the acute process. The extent of the lesion indicated by the scan was calculated by relative units (RU). That is, using a graphics table on a microcomputer Apple-II-Plus, the value of 100 RU was attributed to the A-P distance of the slice W (Wernicke's slice, Naeser and Hayward, 1978). The 100th part of this value is the RU (relative unit). From then on, the microcomputer worked on RU, and all measures were taken using this unit. Once the relative unit of measure was determined for each patient, we outlined the lesion which appeared in the standard slices at constant distances from the slice taken as pattern, and its extension was measured in RU. Obviously, the attribution of the RU value based on the A-P distance of the slice taken as a pattern had to be carried out in each patient before beginning any other calculation. When the scan was defective or too diffuse to be outlined, or the preestablished standard cuts could not be completed, the CT was not considered in the study. Finally the partial values were added together: the total value was considered a valid figure for statistical comparison, although it could not be used as an absolute value.

The values were calculated without our kowledge of the reaction-time data.

	Max–Min	Mean	SD
AGE	72–34	57	9.1
EXT. Lesion	3113–62	1225.2	964.7

TABLE 1. Age and extension of lesion in selected patients.

RESULTS

1. Reaction times for the left-hemisphere-damaged patients and the control groups were 465 msec and 291 msec, respectively. The RTs were subjected to a one-way analysis of variance (ANOVA). Results revealed a highly significant group difference [$F(1,34) = 52.42$; $p<0.0001$]. The global RTs were longer in brain damaged patients than in the control group (Fig. 6).

2. Reaction times for the right and left visual fields were 494 msec and 443 msec, respectively. Comparison of RTs between both visual hemifields by a repeated measures ANOVA [$F(1,19) = 12.99$; $p<0.003$] revealed that visual reaction time was significantly longer in the hemifield contralateral to the lesion (Fig, 7).

3. Patients were also divided into two groups (Fig. 8): those with and those without visual field defect (VFD). The comparison of global RT between VFD patients (540 ms) and non-VFD patients (415 ms) was done by the use of a one-way ANOVA: [$F(1,18) = 16.42$; $p<0.002$]. This indicated a significantly more affected RT in VFD patients when compared to non-VFD patients. This greater

RT seemed to depend on the VFD, and it would be logical to think that in a study of partial RTs we would basically find the deficient visual hemifield involved.

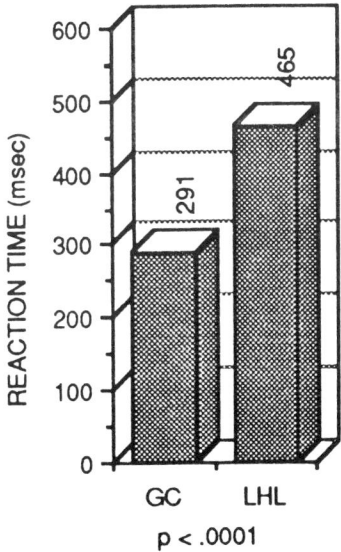

FIGURE 6. Reaction time (in milliseconds) obtained in normals (CG) and in brain-damaged patients. CG: Control group; LHL: Left hemisphere lesion

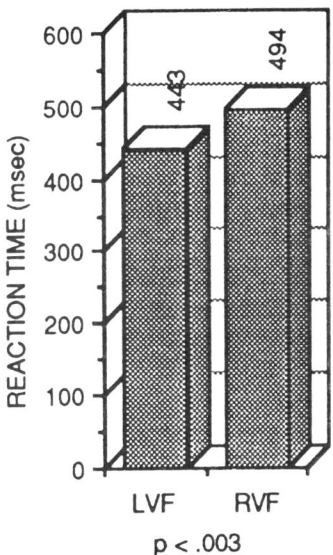

FIGURE 7. Partial results in each hemisfield in left-hemisphere patients. LVF: Left visual field; RVF: Right visual field.

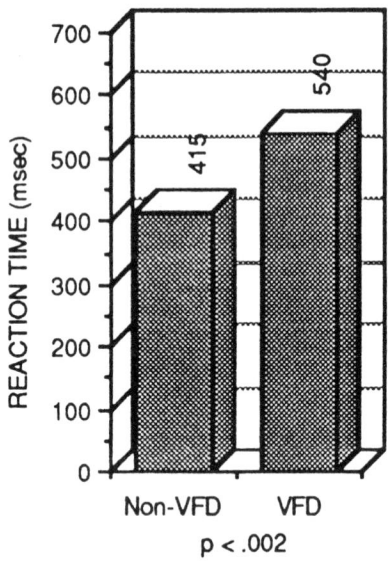

FIGURE 8. Left hemisphere damaged patients, with and without visual fild defect (VFD).

Statistical analysis of the partial RT obtained in the right visual field of both subgroups of patients (Fig. 9) yielded these results: [F (1, 18) = 22.47; p<0.0004. That is to say, the partial RT in the affected hemifield of VFD patients (591 ms) was greater than in the same hemifield of non-VFD patients (430 ms). This finding seems logical due to the perceptive difficulties caused by the visual field defect itself in the hemifield under study.

Nevertheless, an interesting fact emerged upon further analysis of the partial results obtained by both subgroups of patients in the hemifield ipsilateral to the lesion (Fig. 9) that is to say, in the hemifield which had no visual field defect in either of the two groups: [F (1,18) = 11.24; p<0.004]. This result indicated that, besides the longer RT in the affected hemifield, there was also a significantly greater RT in the non-affected hemifield of the VFD patients (505 ms) in relation to the RT obtained in the same hemifield –ipsilateral to the lesion– of non-VFD patients (401 ms). This finding seems to suggest that a unilateral VFD causes bilateral involvement of the entire visual-perceptual system, at least regarding the visual RT.

4. *Correlation between reaction time and lesion size.* The strength of the relationship between RT and the extent of the lesion was evaluated by means of the Pearson product moment correlation test, by calculating the correlation between the different values of RT (mean, left-hemifield RT, right hemifield RT)

and the extent of the lesion in the CT scan measured by RU. None of the correlations were strong or statistically significant.

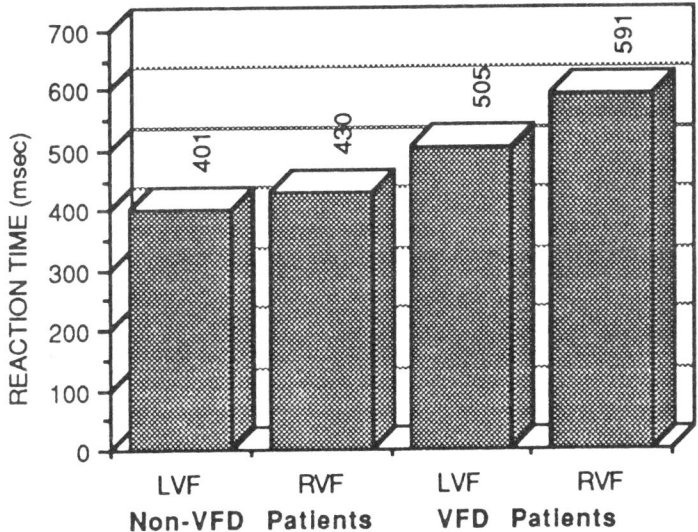

FIGURE 9. Left hemisphere damaged patients, with and without visual field defect (VFD). Analysis of RT to stimuli proceeding from each hemisfield (see text).

5. *Correlation between RT and age.* Results of another Pearson test revealed no statistically significant correlation between RT and age, and again, the correlation obtained was weak.

6. *Correlation between RT measurements.* The correlation between the different measurements carried out are summarized in Table 2. Analysis of this table indicates a high positive correlation with strong statistical significance ($p<0.001$) for all parameters studied, showing a strong correlation between partial RTs.

DISCUSSION

The fact that RT is sensitive to cerebral lesions has already been established in previous studies such as those by DeRenzi and Faglioni (1965) and Howes and Boller (1975). Our approach, in which randomly presented stimuli were directed at both visual hemifields, has emerged as an adequate method for measuring visual RT, with significantly affected values in left-hemisphere- damaged patients.

The possibility of obtaining partial values for stimuli originating from each visual hemifield has proven to be useful, as it has allowed us to show a significantly longer RT in the hemifield contralateral to the lesion. This fact seems to indicate an effect of the cerebral damage related to a deficit for the integration of sensory input from the contralateral hemispace. The data obtained

	L + R	LVF	RVF	C
P	.934 *	.936 *	.811 *	.851 *
C	.95 *	.891 *	.901 *	
RVF	.923 *	.81 *		
LVF	.968 *			
* p < .001				

L + R: Global RT (Left & Right)
LVF: Left visual field RT
RVF: Right visual field RT
C: RT for central stimuli
P: RT for peripheral stimuli

TABLE 2. Correlation between global RT, partial RT in each hemifield and RT for central and peripheral stimuli. Total group of patients selected (n=20).

are consonant with a decreased capacity for visual exploraton of space in each hemifield produced by the contralateral hemispheric lesions (DeRenzi, Faglioni and Scotti, 1970), and they agree with the hypothesis that each hemisphere mediates behavior in the contralateral hemispace (Heilman, 1979). Moreover, our data in left hemisphere patients favor the presence of a subclinical neglect contralateral to the lesion that may show up by means of random stimuli in both visual fields. The existence of this subclinical neglect may indicate that lesions of the left hemisphere give rise to a disturbance of contralateral attention-arousal response. If this phenomenon was also demonstrated in lesions of the right hemisphere, it may suggest that any unilateral hemisphere lesion causes neglect in the contralateral hemispace whether this is of a clinical or subclinical nature.

The finding of a significantly more affected visual RT in VFD patients when compared to non-VFD patients is of particular interest. This fact may be related to the studies that indicate that patients with visual field defects have more problems handling sensory input when compared with brain-damaged patients with full visual fields (DeRenzi, Faglioni and Scotti, 1970; Chedru, Leblanc and Lhermitte, 1973). On analyzing these results in greater detail it seems logical to suppose that the poorer performnce in the visual RT in these patients was a consequence of the visual field defect itself; and thus, comparing the results between the right hemifield of both groups of left-hemisphere-damaged patients (with and without VFD) the existence of VFD in the hemifield under study seems to justify in itself the observation that partial RT for the defective hemifield was more severely affected in VFD patients than it was in the homologous hemifield of non-VFD patients.

However, the most interesting result from our study is perhaps the finding of a significantly longer RT in the non-affected hemifield of VFD patients in relation to the homologous hemifield of non-VFD patients. This finding seems to suggest that a unilateral VFD causes bilateral involvement of the entire visual perceptual system; this would at least seem to be the case in relation to the visual RT, and it supports the claim made by DeRenzi, Faglioni and Scotti (1970) and Chedru, Leblanc and Lhermitte (1973) that brain-damaged patients with visual field defects have particular difficulty handling sensory input. Moreover, our data can be related to the work of Smith (1972), who considers prognosis to be poor in the rehabilitation of aphasic patients with sensory loss, including visual field defect (VFD).

The lack of a significant correlation between RT and the extension of the lesion measured in RU from the TC agrees with the results obtained by Howes and Boller (1975) and does not favor the hypothesis of mass-action given by DeRenzi and Faglioni (1965), as it does not seem that the RT is systematically related to lesion size.

Our series should be contrasted with studies of right hemisphere lesions, and more thorough investigation should be carried out to further explore these findings which may have interesting physiopathological implications.

A marginal but interesting result has been the possibility of carrying out a quick visual field screening in patients with a lesion in the speech dominant hemisphere, as the graphic representation of particular results for each point of stimulation constitutes in itself clear evidence about the state of the visual fields. This is very useful for providing complementary exploratory data if the intensity of the aphasia hinders patient cooperation in conventional exploration of visual fields.

Another marginal but also interesting result is the high concordance between the results for central and peripherial stimuli, as well as concordance of these results with the clinical impression of the patient's cooperation (and also with the number of "anticipated responses"). Such findings, which are related to the neurophysiology of retinal perception, give us a useful method for discarding patients with a poor level of cooperation, thus making excessively sophisticated methods (TV monitoring of patient's eyes, infrared rays) or methods requiring patient verbalization (reading letters, numbers or signs at the center of the screen) unnecessary for this type of examination.

Thus, the two main conclusions of this reaction-time study in left-hemisphere-damaged aphasic patients were that even unilateral visual field defects cause bilateral involvement of the entire visual perceptual system (at least in relation to visual reaction time), and that left-hemisphere-damaged patients present some subclinical neglect in the hemifield contralateral to the lesion. Following these two conclusions, a further point of interest arises, which is the study of differential performance of each hemisphere in the processing of verbal and non-verbal material in aphasic patients. Moreover, such study should be useful for providing bases for later rehabilitation exercises and objective periodic revisions to follow the patient's change.

Therefore, an additional study was undertaken. The aims of this work were to examine the following in aphasic patients: (1) The performance differences between the two hemispheres in the processing of non-verbal material (pictures) in a picture-matching test. (2) The performance differences between the two hemispheres in the processing of verbal material (words) in a word- matching test. (3) The performance differences between the two hemispheres in the ability to match a picture with the corresponding written word in a picture/word-matching test.

MATERIALS AND METHODS

In short, the apparatus we designed to carry out our research consisted of a combination of microcomputers with programmes on which the keyboard characters and/or colored pictures could be presented to either hemifield or to the center of the field of vision. Presentation time was controlled, mean reaction times were calculated, and the number of correct and erroneous responses were recorded.

We used one computer for permanent storage of the drawings to be presented (with CPU Z80, CP/M operating system and a 24 Megabytes hard-disk), and another which consisted of a personal home computer (designed for home games) able to generate these drawings (with CPU Z80A, ROM resident monitor program of 8 K, RAM static memory of 4 K and a screen memory of 16 K). Dialogue between both computers was assured through a serial interface RS232. The specific program was written in Assembler language, compiled, recorded in an EPROM memory and incorporated into the CPU Z80A microcomputer. The output screen of testing materials consisted of a 14" PAL-Color Monitor.

The three groups of tests used were as follows: The patient was seated in a dark, sound-proof room, with his eyes situated at the same level as the computer screen, at a distance of 550mm from its center. A four-letter word was then presented with an angle of three degrees and, in lateralized presentations, the center of the word formed a visual angle of five degrees from the fixation point.

The test began with an asterisk flashing four times in the center of the screen; each time it appeared, a bell rangs. The sample was presented immediately; it consisted of a picture or a written word, depending on the test which was being carried out. Presentation of the sample lasted one second.

On continuation, a warning/short-central-fixation point appeared once, and the stimulus was presented, in either hemifield or centered. Depending upon which test was being used, the stimulus could consist of a picture or a written word.

The stimulus picture or word remained on the screen for 80 msecs. The average error due to the scanning of the screen itself (the time spent by the screen in completing the full presentation of the display) was approximately 10 msecs. We (practically) eliminated the after-image due to the TV screen, by illuminating the screen with a hallogene light. Therefore, the stimulus presentation time was well below 150 msecs, which assured that the stimulus was not within reach of the displacement of the eyes, in case this occured when a stimulus was presented in one hemifield (Pirozzolo and Rayner, 1980; Young, 1982; Beaumont, 1982).

In order to rule out the use of language in the responses, and to provide a simple responding device, we constructed a system for responses which consisted of two levers, placed perpendicular to each other in a horizontal and vertical position. The first was pressed by means of downwards pressure (YES response), and the second, beside it, was pressed by lateral pressure (NO response). The movement of the lever interrupted the circuit of a photoelectric cell, thus recording the response and calculating the reaction time.

Stimuli and procedures

To carry out testing, twelve two-syllable words with the stress on the first syllable and all with a CVCV structure were selected. Appendix 1 contains the pictures (in black and white here) and the list of written words. We tried to include a wide variety of letters within the words.

To avoid artificial predominances in any of the three presentations (left hemifield, centre, right hemifield) we created specific criteria:
  a) each item had to appear in all three positions; thus each test contained 36 balanced presentations;
  b) order of presentation had to be random; thus, we designed a random sequence of presentation which meet all the specific criteria we needed; this sequence was used with all patients;
  c) in half the presentations (18) the stimulus had to be the same as the sample, and different in the other half;
  d) there were not to be more than three consecutive presentations in congruent conditions (e.g., the picture of a cloud and the word 'cloud'), so as to avoid monotony in responses;
  e) to avoid a possible decrease in performance due to loss of attention, the order of presentation was balanced between each localization;
  f) to the extent possible, we avoided semantic association between sample/stimulus pairs. There were no consecutive presentations of pictures or words, and sample/stimulus pairs were not repeated.

The tests were carried out as follows: The patient first had a trial run with three short lists of 4-letter words, different from those in the test itself. Each list consisted of 12 presentations of words. During this trial run, the patient became familiarized with the response levers and the three types of tests.

Then a picture/word matching test of 24 presentations was administered. It contained words of different lengths, all centered. Once this test had been completed, the patient was left to rest for approximately 10 minutes.

Finally, the actual tests (see Appendix 2) were carried out in a counter-balanced presentation across patients.

Subjects:

Seventeen right-handed aphasic patients with a dominant (left) hemisphere lesion were studied. Their mean age was 56.8 years. Twelve patients were males, and

5 were females. Cause of aphasia in all cases was stroke. Mean interval between stroke and testing was 25.6 months.

Before carrying out the tests, patients were asked to identify some written materials on the screen at the adequate distance, in order to rule out visual accommodation problems.

Patients with occipital or bilateral lesions were not included, as the study was limited to patients with lesions in the classical language areas. Severe Wernicke's and global aphasics were also excluded, due to their inability to learn the required task.

RESULTS

Total results. Correct responses (Fig. 10)

Comparison between the three tests was carried out by a repeated measures ANOVA. The results showed that the three tasks were not equivalent [F (2,30) = 28.34; p<0.0001]. Post-hoc Scheffé analysis revealed significant differences for all comparisons (p<0.01 for each comparison). The order of differences, from greater to lesser scores in number of correct responses, was this: picture-matching test, word-matching test, and picture/word-matching test.

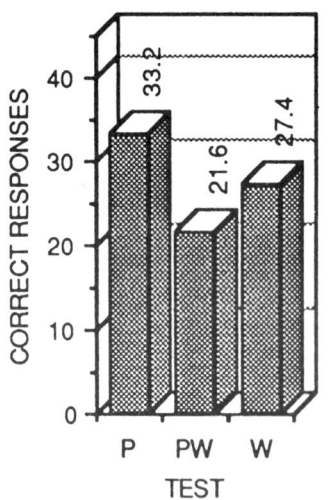
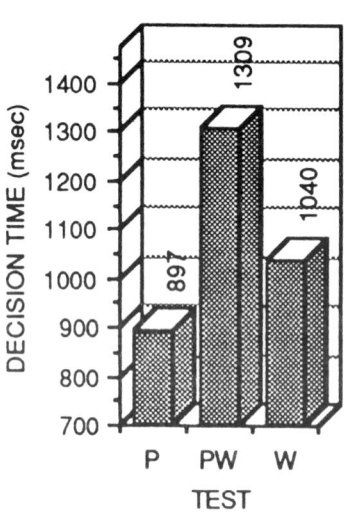

FIGURE 10. Total results. Correct responses.   FIGURE 11. Total results. Time.

P: Picture matching test.
PW: Picture/written word matching test
W: Written word matching test.

Total results. Reaction time (Fig. 11)

The results of the analysis of the RTs are similar to those of the correct responses. By comparing Figs. 10 and 11, we can see that the scores are inverse to each other: as the number of correct responses increases, the reaction time decreases.

These results indicate that the three tests we have designed are different, and that their level of difficulty increases from the easiest (pictures-matching test) to the most difficult (picture/word-matching), with the word-matching test being somewhere in between.

Analysis of performance on each test:

Picture-Matching

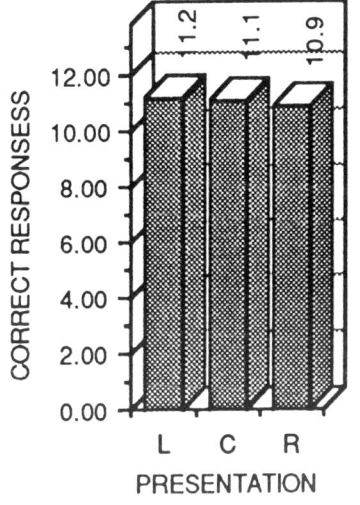

FIGURE 12. Picture matching (P).
Correct responses

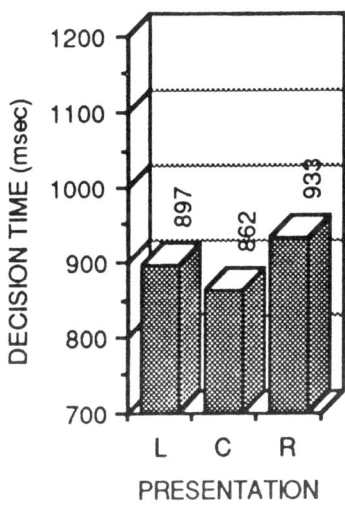

FIGURE 13. Picture matching (P).
Time.

L: Left hemifield presentations
C: Centred presentations.
R: Right hemifield presentations.

Analysis of the scoring of correct responses (Fig. 12), as well as analysis of the reaction time (Fig. 13) using ANOVA repeated measures revealed no significant difference between each stimulus site of presentation (left hemifield, center, right hemifield).

These results reveal that there was no relatioship between the site of the stimulus and the capacity to recognize it. This finding is consonant with the possibility that both cerebral hemispheres can carry out gnosic processing of colorful pictures.

Picture/Word-Matching

Both for the number of correct responses [F (2,30) =20.85; p<0.0001] (Fig. 14) and for decision time [F (2,28) = 4.42; p<0.03] (Fig. 15) ANOVA repeated measures revealed a significant contrast between centered and both lateral stimulus presentations.The source of differences comes from centered presentations, which resulted in a higher number of correct responses, with lower reaction times.
There was no significant difference between hemifield presentations.

Word Matching (Fig. 16; Fig. 17)

The results are very similar to those obtained in picture/word matching.

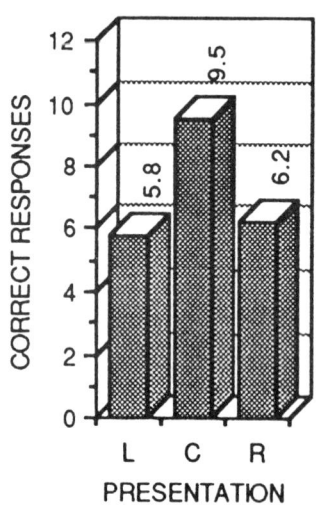
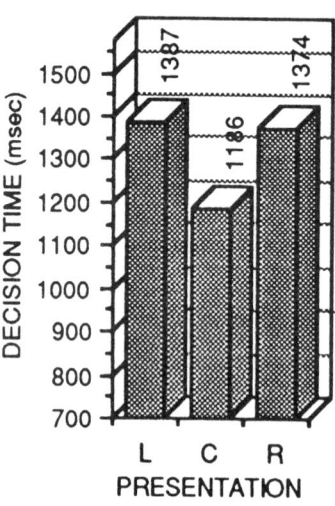

FIGURE 14. Picture/written word matching (PW). Correct responses.

FIGURE 15. Picture/written word matching (PW). Time (p<0.03)

Analysis of performance according to presentation site

CENTRED PRESENTATIONS.

Both analyses, that of correct responses (Fig. 18) and that of reaction time (Fig. 19), show similar results.

ANOVA repeated measures revealed a significant difference between the three tests (correct responses [F (2,30) = 5.51]; p<0.01; time, [F (2,30) = 23.25]; p<0.0001). Scheffé post-hoc analysis indicated that the source of differences comes from the picture/word matching test, while results in the picture matching test and word matching test are similar. That is to say, the analysis of correct responses and reaction time indicate that there are significant differences between the picture/word matching test and the picture matching test (correct responses, p<0.05; time, p<0.01) and significant differences between the picture/word matching test and the word matching test (correct responses, p<0.05; time, p<0.01).

These results suggest that the mechanisms involved in solving the picture/word-matching test are different from the mechanisms involved in solving the other two tests.

The explanation for this finding comes from the analysis of results in the hemifield presentations.

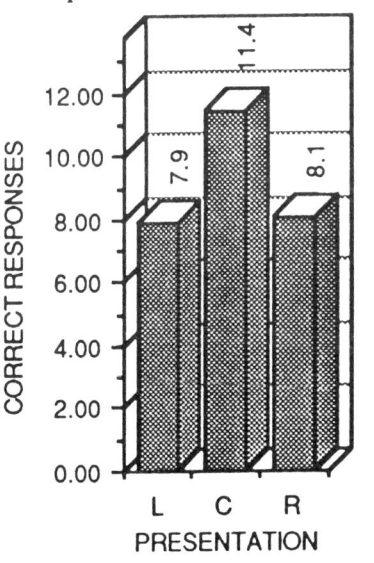

FIGURE 16. Written word matching (W). Correct responses (p<0.0001)

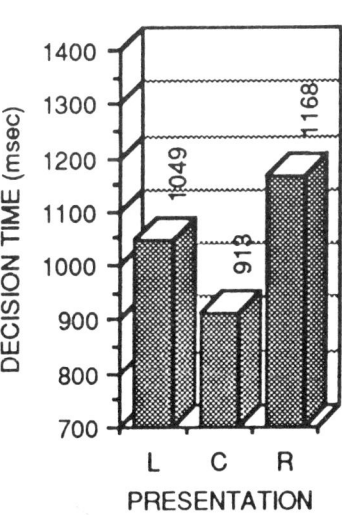

FIGURE 17. Written word matching (W). Time (p<0.0003).

LEFT HEMIFIELD PRESENTATIONS.
Correct responses (Fig. 20)

ANOVA repeated measures showed a high significance in the differences between the three tests [F (2,30) = 30.17; p<0.0001]. Scheffé post-hoc analysis showed once more that the picture/word-matching was the most difficult and picture-matching was the easiest (p<0.01).

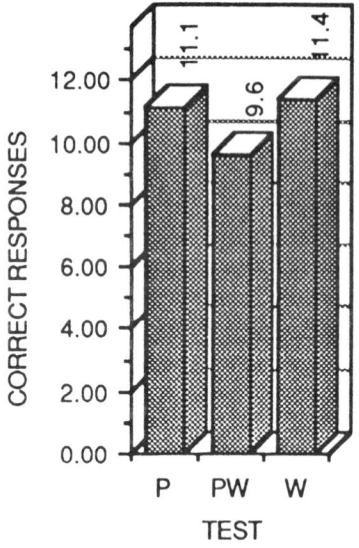

FIGURE 18. Centred presentations. Correct responses (p<0.01).

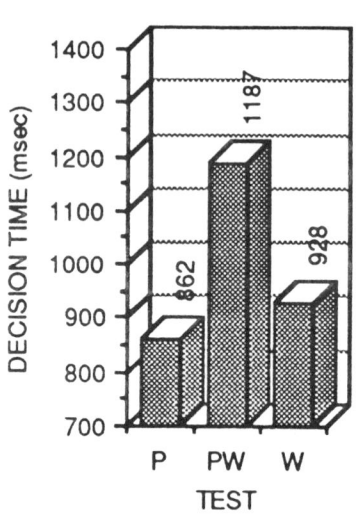

FIGURE 19. Centred presentations. Time (p<0.0001)

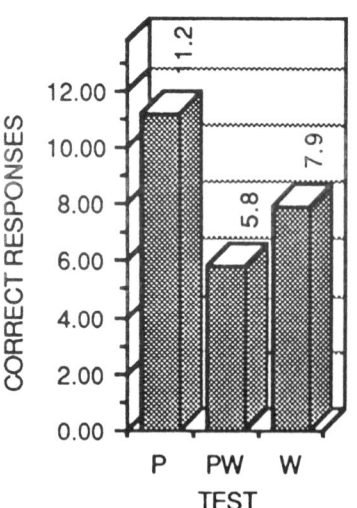

FIGURE 20. Left hemifield. Correct responses (p<0.0001).

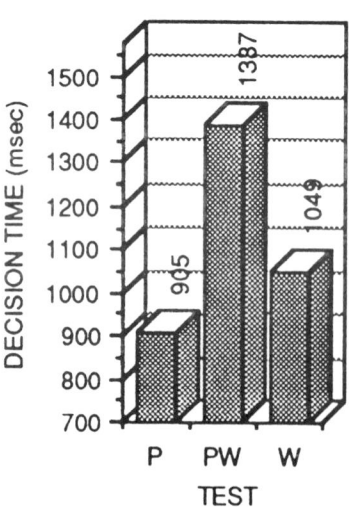

FIGURE 21. Left hemifield. Time (p<0.0001). Partial results: P/PW p<0.0001; P/W p<0.002; PW/W p<0.006).

Reaction time (Fig. 21)

Significant differences were also observed in the reaction time analysis [F (2,28) = 31.21; p<0.0001]. Scheffé post-hoc analysis revealed that the source of differences is due to patients' performance on the picture/word- matching test, which yielded significantly longer reaction times than the other two tests (p<0.01). However, there were no significant contrasts between results of the picture-matching test and the word-matching test. This lack of significance in the differences between these two tests might suggest that the word-matching test tends to be performed (resolved) in the same manner as the picture-matching test.

In other words, in the word-matching test, words seem to be processed as if they were a global shape or a visual pattern. This is not surprising, as the presentation was always in the same type of letters, i.e., uppercase letters.

RIGHT HEMIFIELD PRESENTATIONS

Correct responses (Fig. 22)

As can be seen in Figs. 22 and 23, results were very similar to those obtained in left hemifield presentations [F (2,30) = 25.29; p<0.0001. Post-hoc Scheffé analysis: p<0.01 for the three tests].

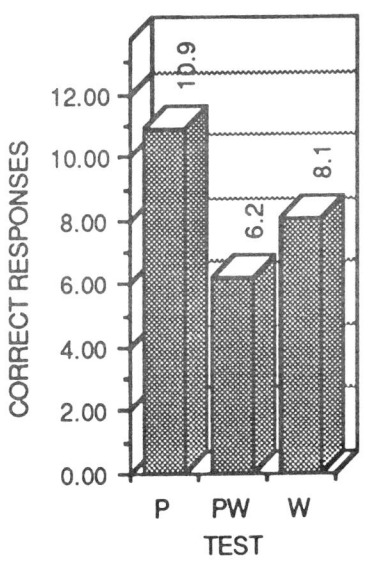
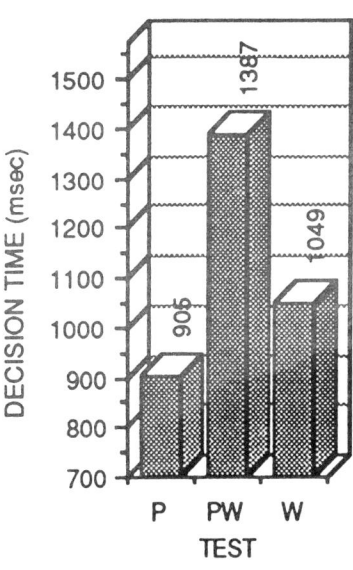

FIGURE 22. Right hemifield. Correct responses (p<0.0002).

FIGURE 23. Right hemifield. Time (p<0.0001). Partial results: P/PW p<0.0001; P/W p<0.002; PW/W p<0.006).

Reaction time (Fig. 23).

ANOVA repeated measures showed significant differences between the three tests [F (2,28) = 29.36; p<0.0001], and Scheffé post-hoc analysis revealed significant differences for all comparisons (p<0.01). We therefore found that the differences became more obvious when the stimuli were directed to the right hemifield.

By comparing scores in the left and right hemifields, we may make the following hypothesis: The word-stimuli coming from the right hemifield can only be processed as words by the left hemisphere, in contrast with the word-stimuli coming from the left hemifield, which can be processed as simple designs. Verbal-analytical processing of words by the left hemisphere increases the reaction time, thus increasing the difference between test scores. In short, these results confirm that written words coming from the right hemifield are processed as truly written verbal material.

A further conclusion that we may reach as a consequence of this refers to analysis of centered presentations of the three tests (Fig. 18 and 19). We have suggested that the mechanisms involved in solving the picture/word-matching test were different from those involved in the other two tests. We now have a better understanding of this difference.

Specifically, the picture/word-matching test is the test which truly analyzes the reading ability of the aphasic patient (in our study). The word-matching test, as designed (always in uppercase letters), can be solved by means of holistic procedures, and thus avoids the more specific linguistic-related reading mechanisms. It is for this reason that no significant difference was seen with the picture-matching test. Further experiments are needed in order to confirm this hypothesis; for example, words could be alternately presented in lowercase and uppercase letters.

CONCLUSIONS: Analysis of tests

1. The three tests require different cerebral procedures, and are of varying degrees of difficulty (Figs. 10 and 11). The picture-matching test was the easiest and the picture/word-matching the most difficult. Such a difference can be seen both in the number of correct responses and in the reaction time.

2. Results from the picture matching test agree with the concept that the brain can utilize bilateral mechanisms to carry out gnosic processing for pictures of concrete objects (Figs. 12 and 13).

3. In the tests containing written material., this is better processed in the central rather than the lateral presentations (Figs. 14, 15, 16 and 17).

4. The picture/word-matching test is responded to quite differently from the other two tests, and requires actual reading, while the word-matching test can be solved by processing the words as designs (Fig. 18 and 19).

5. The word-matching test may possibly be solved by holistic strategies (Figs. 20 and 21). For this reason, the stimulus words proceeding from the left

hemifield tend to be processed by the right hemisphere as simple visual patterns, and so their role as written material is somewhat lost. On the other hand (Figs. 22 and 23), when the stimulus words come from the right hemifield, both in the picture-word matching and the word matching tests, these words are processed by the left hemisphere as authentic verbal material.

6. Finally, in terms of rehabilitation, the possibility that verbal material can be interpreted by holistic processing would provide the alexic aphasic patient with a way of access to reading.

Analysis of aphasic groups

Another set of interesting results appears when we attempt to identify different groups of readers among our aphasic patients. We distributed the patients into three groups, according to the topography of the lesion analyzed by means of the CT scan. In Group 1, areas 44, 6 and the insula were involved: lesions predominantly in areas of motor significance. In Group 2, with lesions in areas 22 and 37, and the upper part of 21, damage was predominantly in Wernicke's area. In Group 3, the Supramarginal and Angular gyri were involved. (One patient was excluded due to a doubtful evaluation of the locus of damage).

Results

We shall now discuss some of the most significant results observed.

WORD MATCHING (Fig. 24):

Analysis of Fig. 24 shows that groups B (Broca's area damaged patients) and A (Angular gyrus area damaged patients) had similar performance. However, group W (Wernicke's area damaged patients) performed differently. A two-way repeated measures ANOVA (aphasic group by presentation site) using the correct responses as dependent variable reveals that the behavior for centered presentations was alike in the three groups [$F(2,13) = 1.54$, n.s.]. But in hemifield presentations, the different performance of the groups indicates that Group W did not have alexic problems, while Groups B and A did (aphasic group by stimulus site interaction: $F(4,26) = 2.80$; $p<0.05$). Obviously, these alexic disturbances in B and A groups were related to anterior and posterior lesions, respectively. On the other hand, the good performance of W patients has to be attributed to the fact that W patients with severe involvement were excluded from testing because of poor cooperation. Therefore, the W patients included in our study had only mild aphasic disturbances, and the alexic phenomena, if present, did not interfere with the test.

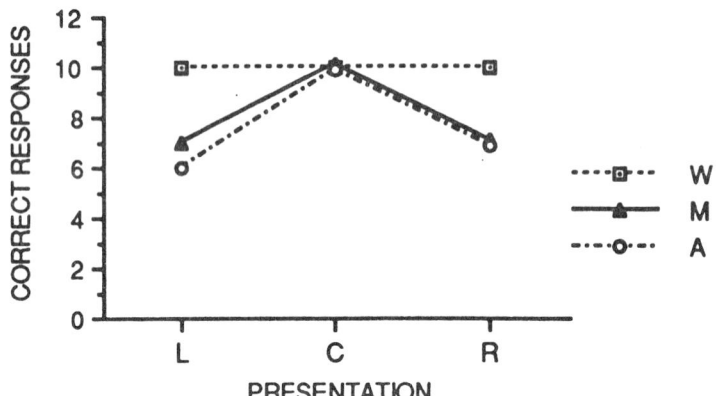

FIGURE 24. Written Word matching. Correct responses.
Group x Stimulus Site Interaction (p<0.05).

P: Picture matching test.
PW: Picture/written word matching test.
W: Written word matching test.

L: Left hemifield presentations.
C: Centred presentations.
R: Right hemifield presentations.

M: Brain lesion in areas of motor significance.
W: Brain lesion predominantly in Wernicke's area.
A: Brain lesion in supramarginalis and angularis areas.

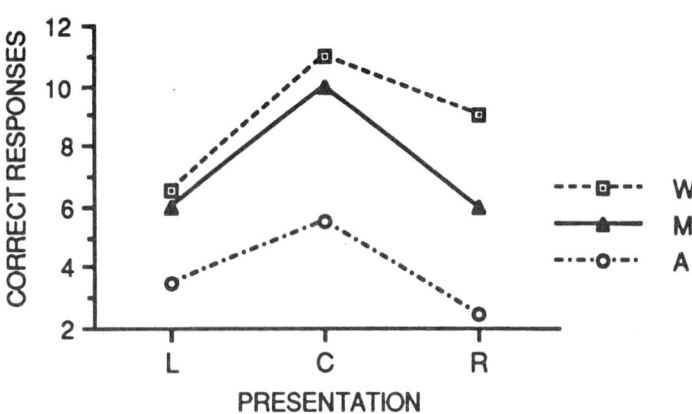

FIGURE 25. Picture/Written word matching. Correct responses. (p<0.02).

PICTURE/WORD MATCHING (Fig. 25)

A two-way repeated measures ANOVA (aphasic group by presentation site) using the correct responses as dependent variable showed significant differences between the three groups [F (2,13) = 6.68; p<0.02). As post-hoc analysis revealed, the source of differences is attributed to the Angular gyrus damaged group, which obtained much lower scores than the others (p<0.05). These results support the previous finding, showing that this test requires the reading of the written word, if subjects are to be able to recognize the matching or non-matching of the word with the corresponding picture. For this reason, patients with posterior brain damage obtain the worst scores. The analysis of central and lateralized presentations in the picture/word-matching test sheds more light on the subject of interhemispheric differences:

CENTRAL PRESENTATIONS (Fig. 26):

A two-way repeated measures ANOVA (aphasic group by test) using the correct responses in central presentations as dependent variable revealed no significance in the differences between the three groups [F(2,13) = 3.65; p<0.06], but analysis revealed a significant group by test interaction [F (4,26) = 4.90; p<0.002). Post-hoc Scheffé analysis showed that the angular-gyrus-damaged patients performed poorly compared to the other two groups (p<0.05). These results were due to the very low scores of angular-gyrus-damaged patients in the picture/word-matching test, that is to say, in the true reading test.

Conversely, this group of angular-gyrus-damaged patients has scores very similar to the other two groups in the other tests, which, as previously seen, may be solved by holistic mechanisms.

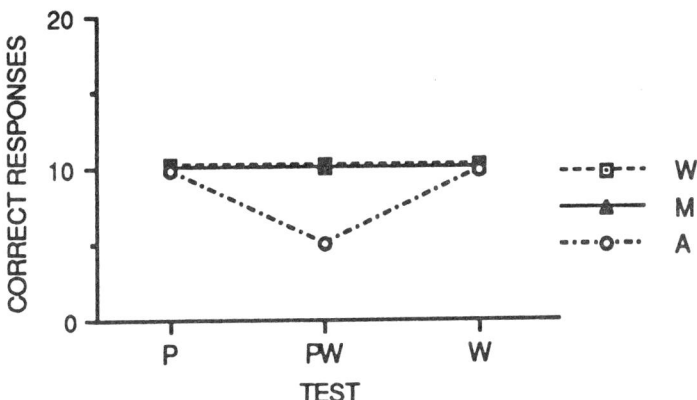

FIGURE 26. Centred presentations, correct responses. Group x Test Interaction (p<0.002).

LEFT HEMIFIELD PRESENTATIONS (Fig. 27).

A two-way repeated measures ANOVA (group by test) using the left hemifield presentation correct responses as dependent variable revealed no significant differences between groups [$F(2,13) = 2.46$, n.s.] and no interaction between groups and tests [$F(4,26) = 1.04$; n.s.].

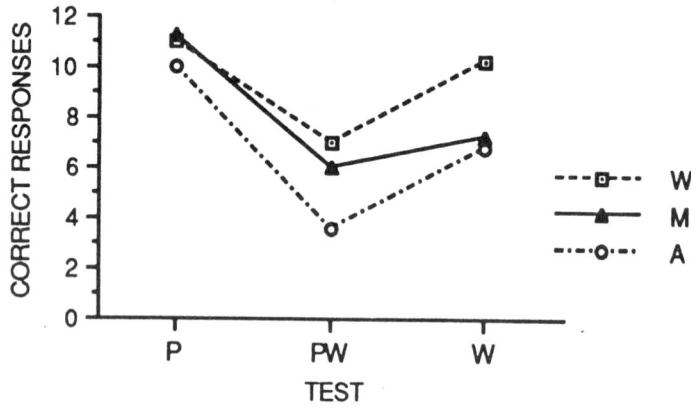

FIGURE 27. Left hemifield presentations, correct responses. Group x Test Interaction N.S. ($p<0.4$).

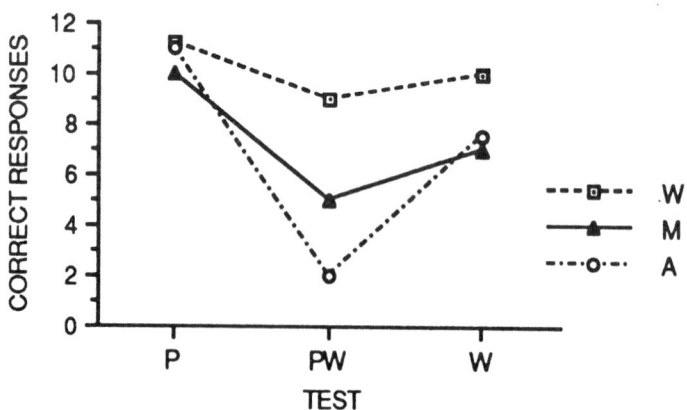

FIGURE 28. Right hemifield presentations, correct responses. Group x Test Interaction ($p<0.02$)

RIGHT HEMIFIELD PRESENTATIONS (Fig. 28)

A two-way repeated measures ANOVA (group by test) using the right hemifield presentation correct responses as dependent variable revealed a significant difference between the three groups [$F(2,13) = 4.21$; $p<0.04$], and a significant aphasic group by test interaction [$F(4,26) = 3.57$; $p<0.02$]. Sheffé post-hoc analysis indicated significant differences between the three groups ($p=0.01$), as shown in Fig. 28. The group by test interaction may be seen as follows:

The three groups of patients had the same pattern of picture processing. In contrast, the picture/word-matching test, which required reading processing, showed clearly defined differences between the groups. By comparison of Figs. 28 and 26, it can be seen that the differences observed in centered presentations (Fig. 26) have increased sharply in right hemifield presentations (Fig. 28).

And again we find that the written-word-matching test was better solved than the picture/word-matching test, thus supporting the hypothesis that holistic strategies were used in our word-matching test, in spite of the fact that the stimulus went to the left hemisphere.

But here, in the right hemifield presentations, we observed significant differences between the two tests picture-matching and written-word-matching, in contrast with the non-significant differences found when the same stimuli in both tests came from the left hemifield. This leads us to think that the differences are greater because the reading material was received by the left hemisphere; this would compel the brain to read, thus increasing the differences in the groups' performance. This is also apparent when results from the three presentation sites are compared (Figs. 26, 27 and 28).

On encouraging the patients to use right-hemisphere strategies (presentation in left hemifield), our group of Wernicke-area-damaged patients, who proved to be good readers, tended to perform like the Broca's group (who were not good readers). This may indicate that aphasic patients who are good readers do not use right-hemisphere strategies. Instead, poor readers (the Broca's and Angular Gyrus groups) utilize strategies which are more right-hemisphere-like; for this reason, when a written stimulus comes from the right hemifield, the performance of the group with more severe reading difficulties (Angular Gyrus group) worsens, thus contributing to a significance in the statistical differences between groups. In other words,

a) analysis of responses to reading material coming to the right hemisphere showed no statistically significant differences among aphasic patients;

b) analysis of centered presentations (combined processing by the two hemispheres) showed significant differences, but there were two groups of patients (Broca's and Wernicke's) with the same pattern of reading behavior;

c) analysis of responses to reading material presented to the left hemisphere, besides showing significant differences between patients, also pointed to the existence of three separate groups of readers among aphasics: frontal., temporal and Angular gyrus-damaged patients.

CONCLUSIONS: Analysis of Groups

1. Two groups of patients with different reading abilities were found: good readers and poor readers.
2. The group of good readers presented mild aphasic disturbances due to the exclusion of severely impaired patients. Brain damage in this group was mainly located in Wernicke's area.
3. The group of poor readers may be subdivided into two: one group with anterior lesions affecting the motor areas; and another group with posterior lesions affecting the Angular and Supramarginal Gyri areas.
4. Aphasic patients with good reading ability used primarily left-hemisphere-based strategies.
5. The two groups of aphasic patients with alexic problems tend to use holistic strategies to read.
6. The more severe the alexia, the more the remaining reading possibilities depended on the use of holistic strategies.
7. From our study we can reach a provisional conclusion regarding the theory of interhemispheric transfer for compensation in aphasia.

It seems that the group of aphasic patients studied utilized left-hemisphere-based strategies for reading, although with holistic mechanisms. According to the theory of interhemispheric transfer for compensation, the reading task in aphasic patients may shift towards the right hemisphere. However, in our group of patients, our tests seem to suggest that the left hemisphere retains the basic reading mechanisms; what happens is that the holistic mechanisms are more prominent in aphasic patients. It is a well-known fact that these holistic mechanisms are not exclusive to the right hemisphere, but the holistic mechanisms of the left hemisphere are usually subordinated to sequential analysis, which is a prominent feature of the left hemisphere. When this sequential analysis fails (as in aphasia) the holistic mechanisms become more important.

8. Even if the only valuable result of this study was that many predictable findings can be obtained, it would suggest that the technique used, was reliable, thus showing the way for further research in applications of computer technology for evaluating the aphasic patient and for developing programs for the treatment of aphasia.

Further experiments with greater numbers of patients and with a wider range in the severity of the aphasic disturbances will be needed in order to provide more insight into this subject.

References

Anzola, G.P., Bertolini, G., Buchtel, H.A. and Rizzolatti, G. 1977. Spatial compatibility and anatomical factors in simple and choice reaction time. *Neuropsychologia*, 15: 295-302.

Bagnara, S., Boles, D.B., Simion, F., and Umiltà, C. 1982. Can an analytic/holistic dichotomy explain hemispheric asymmetries? *Cortex*, 18: 67-78.

Beaumont, J.G. 1982. Studies with verbal stimuli. In J.G. Beaumont (ed.), *Divided Visual Field Studies of Cerebral Organisation*. Academic Press, London.

Berlucchi, G., Brizzolara, D., Marzi, C.A., Rizzolatti, G., and Umiltà, C. 1979. The role of stimulus discriminability and verbal codability in hemispheric specialization for visuospatial tasks. *Neuropsychologia*, 17: 195-202.

Berlucchi, G., Heron, W., Hyman, R., Rizzolatti, G. and Umiltà, C. 1971. Simple reaction times of ipsilateral and contralateral hand to lateralized visual stimuli. *Brain*, 94: 419-430.

Berrini, R., Della Sala, S., Spinnler, H., Sterzi, R. and Vallar, G. 1982. In eliciting hemisphere asymmetries which is more important: the stimulus input side or the recognition side? A tachistoscopic study on normals. *Neuropsychologia*, 20: 91-94.

Brain, R. 1941. Visual disorientation with special reference to the lesions of the right cerebral hemisphere. *Brain*, 64: 244-272.

Bryden, M.P. and Rainey, C.A. 1963. Left-right differences in tachistoscopic recognition. *Journal of Experimental Psychology*, 66: 578-581.

Chedru, F., Leblanc, M. and Lhermitte, F. 1973. Visual searching in normal and brain-damaged subjects. *Cortex*, 9: 94-111.

DeRenzi, E. and Faglioni, P. 1965. The comparative efficiency of intelligence and vigilance tests in detecting hemispheric cerebral damage. *Cortex*, 1: 410-433.

DeRenzi, E., Faglioni, P. and Scotti, G. 1970. Hemispheric contribution to exploration of space through the visual and tactile modality. *Cortex*, 6: 191-203.

Gainotti, G. 1968. Les manifestations de négligence et d'inattention pour l'hémiespace. *Cortex*, 4: 64-91.

Gross, M.M. 1977. Hemispheric specialization for the processing of visually presented verbal and spatial material. *Perception and Psychophysics*, 12: 357-363.

Hay, D.C. 1982 Cerebral asymmetries in processing proper names: Evidence of an efficiency difference. *Cortex*, 18: 385-393.

Hécaen, H. and Albert, M.L. 1978. *Human Neuropsychology*. New York, John Wiley.

Heilman, K.M. 1979. Neglect and related disorders. In K.M. Heilman and E. Valenstein (eds.), *Clinical Neuropsychology*. New York, Oxford University Press.

Heilman, K.M., Schwartz, H.D. and Watson, R.T. 1978. Hypoarousal in patients with the neglect syndrome and emotional indifference. *Neurology*, 28: 229-232.

Heilman, K.M. and Valenstein, E. 1972. Frontal lobe neglect in man. *Neurology*, 22: 660-664.

Hines, D. 1972. Bilateral tachistoscopic recognition of verbal and non-verbal stimuli. *Cortex*, 8: 315-322.

Howes, D. and Boller, F. 1975. Simple reaction time: evidence for focal impairment from lesions of the right hemisphere. *Brain*, 98: 317-332.

Marks, L.E. 1968. Brightness as a function of retinal locus in the light-adapted eye. *Vision Research*, 8: 525-535.

Naeser, M.A. and Hayward, R.W. 1978. Lesion localization in aphasia with cranial computed tomography and the Boston Diagnostic Aphasia Exam. *Neurology*, 28, 545-551.

Neville, H.J., Kutas, M. and Schmidt, A. 1982 Event-related potential studies of cerebral specialization during reading. *Brain and Language*, 16: 300-315.

Pirozzolo, F.J. and Rayner, K. 1980. Handedness, hemispheric specialization and saccadic eye movement latencies. *Neuropsychologia*, 18: 225-229.

Poffenberger, A.T. 1912. Reaction time to retinal stimulation with special reference to the time lost in conduction through nerve centers. *Archives of Psychology*, 23: 1-73.

Rizzolatti, G., Umiltà, C. and Berlucchi, G. 1971. Opposite superiorities of the right and left cerebral hemispheres in discriminative reaction time to physiognomical and alphabetical material. *Brain*, 94: 431-442.

Sloan, L.L. 1961. Area and luminance of test object as variables in examination of the visual field by projection perimetry. *Vision Research*, 1: 121-138.

Smith, A. 1972. Diagnosis, intelligence and rehabilitation of chronic aphasics: Final Report. Ann Arbor, University of Michigan.

Vendrell-Brucet, J.M., Vendrell-Gomez, P. and Lacorte-Pi, T.M. 1983. Temps de reaccio i hemicamps visuals. Estudi diferencial aplicant un programa de micoordinador. *Sant Pau*, 4: 190-195.

Watson, R.T., Heilman, K.M., Cauthen, J.C. and King, F.A. 1973. Neglect after cingulectomy. *Neurology*, 23: 1003-1007.

Young, A.W. 1982. Methodological and theoretical bases of visual hemifield studies. In J.G. Beaumont (ed.), *Divided Visual Field Studies of Cerebral Organisation*. Academic Press, London.

Appendix 1. Pictures and words used.

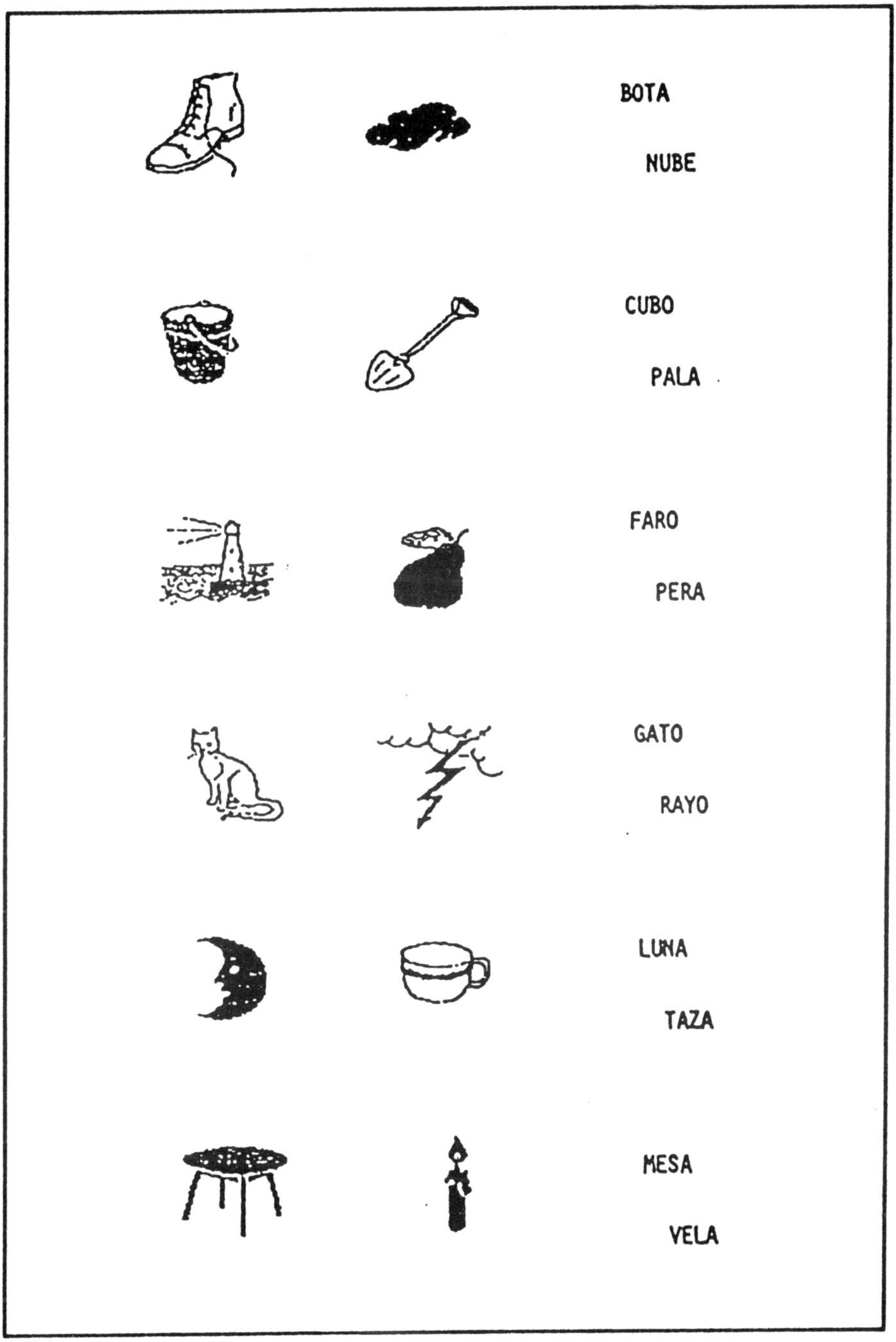

BOTA

NUBE

CUBO

PALA

FARO

PERA

GATO

RAYO

LUNA

TAZA

MESA

VELA

Appendix 2. Tests used. Each test contains 36 balanced presentations.

P : PICTURE MATCHING, i.e. :

PW : PICTURE-WRITTEN WORD MATCHING, i.e.:

W : WRITTEN WORD MATCHING, i.e. :

Part V Foundations of rehabilitation for special populations

It might be argued that bilinguals are not a special population at all, but rather that they constitute the norm. Indeed, there may be more bilinguals and multilinguals in the world than unilinguals (Grosjean, 1982). However, until recently, the bilinguality of the patient was—and currently still is in most hospitals around the world—too often ignored. Most often, the fact that a patient speaks a language other than the one routinely used in the hospital is simply recorded in the patient's file, and nothing else is done about it. (Sometimes the fact that the patient is bilingual is not even mentioned in the file.) Yet, when the language of the hospital environment is partially or completely unavailable to the patient, it is important to determine whether another language may serve as a means of communication. Moreover, some deficits may be observable in only one of the patient's languages. These deficits would go unnoticed if the better preserved language happened to be that of the hospital environment and the other language(s) were not tested. In both cases the results may help one decide the language in which the patient should receive therapy. These concerns are discussed in Chapter 13.

Another growing population is, unfortunately, that of adult aphasic patients with a history of alcohol and drug abuse. Contrary to most studies of aphasia that are generally conducted on right-handed males whose aphasia is secondary to a circumscribed cerebral vascular accident, in the absence of any other pathologic condition, Chapter 14 relates a pilot experiment on such a special population. There are reasons to expect that individuals with substance abuse histories may benefit from aphasia rehabilitation in the same manner as the usually assessed typical aphasic population.

13 Bilingual aphasia rehabilitation

Michel Paradis

Before research into bilingual aphasia therapy may proceed, the questions to be addressed must be identified. They are numerous and varied and, so far, unanswered. Not all have even been formulated. One multifaceted question is whether therapy should be provided in both languages simultaneously (and if so, why? if not, why not?), or successively, and if in only one of them, which one (the native language, the one most fluent premorbidly, the best recovered, the language of the hospital or that of the home environment). Another type of question is whether there is transfer of therapeutic benefits (if any) from the treated to the nontreated language, and if so, whether the transfer is a function of (1) structural distance between the languages, (2) order of acquisition, (3) dominance pre-onset, (4) dominance post-onset, (5) type of aphasia, (6) pattern of recovery, and/or (7) type of therapy. If there is an effect, is it directional (e.g., from the native language to the second language—or the reverse; from the most to the least fluently recovered—or vice versa; from the most fluent to the least fluent pre-onset—or vice versa). Should translation be used as a deblocking device, or should translation be avoided for fear of increasing inhibition? Are various therapy techniques equally efficient in different languages, or is their efficiency a function of the structural type of language (e.g., morphology-rich/poor; configurational/nonconfigurational; or with/ without tones)? Is degree of transfer covariant with efficiency of a particular techinque in the treated language, or is transfer greater with a technique that obtains fewer gains in the treated language?

Even though a number of papers on rehabilitation in bilingual aphasia have been published over the years, it is not known at present whether therapy in one language has beneficial effects on the nontreated language and, if it has, in what proportion, in what manner (specific or general improvement), and in what circumstances (whether linked to the history or type of bilingualism, structural distance between the languages, or the type of aphasia (site and size of lesion) or the pattern of recovery (whether the degree and pattern of impairment is equal in both languages, or whether one language is recovered sooner or better than another; whether the two languages are being uncontrollably mixed, or whether each language is accesible for alternating periods of time).

Most authors seem to agree that therapy in both languages exerts an inhibitory influence upon speech restitution in general (Chlenov, 1948) and can hinder the recovery of all the languages used premorbidly (Wald, 1958; 1961) Hence they insist that therapy be provided in only one language, at least initially

(Chlenov, 1948; Wald, 1961; Hemphill, 1976, Lebrun, 1988). Chlenov (1948) even goes so far as to suggest that it is essential to forbid patients to practice on their own languages other than the one involved in therapy. Of course, once the restoration of the basic language has progressed to a significant degree, even Chlenov (1948) agrees that therapy in a second language is acceptable.

There is less consensus when it comes to deciding *which* language is to be treated. Krapf (1961) suggests that it should be the mother tongue. Chlenov proposes that one must assist the restoration of the language which appears spontaneously: the tendency for preferential recovery displayed by the patient should not be overridden. Hilton (1980) agrees that, as a general rule, therapy should focus on the language currently exhibiting the strongest spontaneous recovery. Lebrun (1988) then suggests that where the patient lives in a bilingual environment, it would be desirable that his interlocutors use only the patient's best recovered language. Hilton (1980) recommends that, if it is determined that the patient's language preference is other than that of the clinical staff, every effort should be made to recruit a family member or friend to help in therapy in the patients's preferred language.

Little actual evidence has been published so far. Watamori and Sasanuma (1976; 1978 case 1) report only partial transfer from the treated to the nontreated language in a 69-year-old right-handed male English-Japanese bilingual patient with Broca's aphasia. The degree of improvement in the treated language (English) was clearly greater than that in the nontreated language (Japanese) across all modalities, and particularly in writing, which seems to have benefited most from therapy. Two factors may have interacted here: (1) the patient had preferred English from the start, and (2) the writing systems of English and Japanese differ radically. There was some limited transfer of oral abilities from English to Japanese, but hardly any in writing. Hence it would appear that in this case, the amount of transfer of the benefits of language therapy was proportional to the degree of similarity between the various aspects of the two languages. Later, therapy was provided in Japanese along with therapy in English. After 8 months, the patient's writing in Japanese had improved considerably, which suggests that improvement was directly related to therapy for each specific task. To the extent that the nature of the task differed in the two languages, transfer did not occur.

Therapy, even when provided in both of the patient's languages, may have effects on only one of them. The gap between English and Japanese in the oral production modality in Watamori and Sasanuma's (1976) patient did not narrow even after therapy in Japanese had been provided. The patient's oral ability improved in English subsequent to 12 months of therapy in that language, even though he had no occasion to speak English outside the clinic, but his Japanese did not improve in spite of 31 months of therapy in that language, and in spite the fact that he was living in a Japanese unilingual environment. This case desmonstrates that it is possible for a bilingual patient to be selectively responsive to therapy in only one of his languages.

Furthermore, in Watamori and Sasanuma's (1978) patients, similar improvement was observed in the nontreated language and in the treated language *only for auditory comprehension*, irrespective of the type of aphasia.

Watamori and Sasanuma (1978) interpret their first patient's better performance on naming tasks in English than in Japanese (with unilateral interference from English words in the Japanese naming task) as an indication "that the patient premorbidly had a stronger association with English words than Japanese words."(137). This could equally reflect a preferential recovery of English in the context of equal premorbid competence.

Therapy in the second language (also the language of the environment) has been reported to have a positive effect on the nontreated native language as well as on the treated language. Weisenburg and McBride (1935) rehabilitated aphasic polyglots mostly in one language and observed notable progress in the rest of the languages as well. Voinescu et al. (1977) report that intensive therapy exclusively in Rumanian (after 11 months without spontaneous progress) had a significant, albeit lesser, effect on each of their patient's other three languages, including his mother tongue. Some report an even greater effect on the nontreated native language (Durieu, 1969; Linke, 1979). Therapy only in German resulted in substantial (*deutlich*) improvement in Italian, the patient's native language, though minimal (*gering*) improvement in the treated second language even though for the previous 16 years the patient had spoken German exclusively. Durieu (1969) reports the patient's wife's observation that his native Spanish had improved to a greater extent than French, subsequent to therapy in French only. The patient had lived and worked in France with his Spanish wife for over 20 years and had spoken French since adolescence. Fredman (1976) reports that speech therapy in the second language (Hebrew, the major language of the environment) of her 40 patients had a positive effect on their nontreated mother tongue as well as on their treated second language.

Some authors predict transfer of the benefits of therapy in one language to the nontreated language on grounds of the commonality of neurophysiological or linguistic structures. Peuser (1978) predicts concomitant improvement in the nontreated languages known premorbidly by the patient. Voinescu et al. (1977) postulate a level of "deep structures" which are involved in meaning (i.e., semantic structures) and "basic deepest structures" involving "cerebral psychophysiologic processes" where "the linguistic meets the psychologic and the physiologic, which are the basis of linguistic communication in any language whatsoever." (p. 175) They hypothesize that "the deep psycholinguistic structures" are "more or less the same for all languages" (p. 174) and, therefore, according to them, therapy in one language would transfer its effects to all other languages at these levels, though not at the level of "superficial structures". This can be interpreted to mean that therapy would transfer its effects for whatever aspects two languages have in common (universal grammar as well as whatever parameters they share) but not those aspects for which parameters differ. It could also mean that languages are more similar at the semantic than at any other level, and that gains at the semantic level are therefore more likely to transfer than those at the syntactic, morphological or phonological levels.

Other authors report a lack of transfer in the nontreated language. Denès (1914) reports that his patient's French improved after therapy in that language but that his Italian did not. Byng et al. (1984) attribute the discrepancy between their patient's relatively worse production in the mother tongue (Nepalese) than in the second language (English) to the fact that the patient had received intensive speech therapy exclusively in English.

On the other hand, Bond (1984) suggests that *bilingual* stimulation should be the most effective approach to language rehabilitation among bilingual aphasics. Several different means of language processing could be stimulated. In particular, translation may be an appropriate means of rehabilitating a patient's second language, especially when translation skills are less impaired than other language skills. She discusses the case of an 83-year-old male patient born in Mexico where he had lived and had spoken only Spanish until the age of 23, at which time he emigrated to the United States and acquired English gradually. He used English exclusively at work for 30 years before retiring 19 years pre-onset, using both Spanish and English every day since retiring, the Spanish translation equivalent was a successful cue to the retrieval of the English word. However, the reverse did not occur —i.e., an English translation equivalent did not facilitate retrieval of a Spanish word. This underscores the unidirectionality of translation skills available to some bilingual aphasics (see Paradis, Goldblum and Abidi, 1982; Paradis, 1984). In this case, unidirectionality is explained by the fact that the subject probably learned the names of English lexical items by reference to their Spanish equivalents, as suggested by the author (Bond, 1984: 133). Also, Falk (1973) reports having successfully retrained an aphasic translator in two languages.

Thus, in the literature, we find reports of partial transfer of benefits from the treated to the untreated language, benefits only or greater in the untreated mother tongue subsequent to therapy in the second language, no transfer to the untreated language, as well as treatment in both languages resulting in improvement in only one language. We may therefore conclude at the least, that it is possible for therapy to be effective in only one of a patient's languages and not in the other, even though at the onset of therapy both languages present the same qualitative and quantitative picture (i.e., the same symptoms to the same degree). Hence, if therapy shows no effect in one language, it does not mean that it should not be attempted in the other.

In order to assess the effects of therapy on the nontreated language of bilingual aphasic patients, a world-wide research project was initiated in 1989. Ultimately, this project will enable us to compare results from various centres where different languages are spoken in different contexts. In Barcelona, for instance, bilingual aphasic patients who showed no evidence of spontaneous recovery six months prior to testing are being assessed with the Bilingual Aphasia Test (BAT) in Catalan and in Spanish. Though all patients were fluent in both languages, having acquired them both before the age of 5, half are considered somewhat Catalan-dominant and half somewhat Spanish- dominant before insult. Half the patients received therapy in their dominant language, the other half in their weaker language. A language program has been designed for each patient on the basis of his/her performance on the BAT. Each patient's specific deficits are being treated for a period of 13 weeks at the rate of four 45-minute sessions per week. Patients are then assessed on the BAT again in both languages to ascertain whether (1) there has been an effect of therapy on the treated items, (2) whether that effect, if any, has transferred to other aspects of the language, and (3) whether the improvement in the treated language, if any, has transferred to the nontreated language.

Thus this study documents a series of single case studies (not a group study) in which each patient is his/her own control. The possible contamina- tion of spontaneous recovery of the patients' scores has been eliminated by selecting only chronic aphasics having demonstrated no improvement over the six months preceding therapy.

Our hypothesis is that there will be positive transfer to the extent that (and in the areas where) the two languages are similar. The more similar the two languages, the greater the amount of transfer to be expected; the less similar the languages for any aspect of language structure (e.g., phonological, morphological, syntactic, lexical) the lesser the amount of transfer is to be expected. Transfer may be expected only in those aspects of language structure that are similar between the two languages. For instance, the syntax and lexicon of Basque are very different from those of Spanish. On the other hand, European Spanish posesses only two phonemes that do not exist in Basque, and Basque possesses only 4 phonemes that are not part of the Spanish repertoire. To the extent that the systems are similar, one may expect transfer in articulatory phonetics and phonology from Basque to Spanish (and vice versa) but not in syntax or in the lexicon —to the extent that these differ between Spanish and Basque. By contrast, improved syntax may be expected to transfer readily from Catalan to Spanish (and vice versa), but not necessarily in phonology.

In addition, one may expect an interaction between structural distance and type of aphasia, in that articulatory problems in anterior aphasia might transfer only to the extent that the two languages share the same phonemes, and syntactic problems might transfer only to the extent that the structures in the two languages are the same, whereas in posterior aphasia the transfer may be more extensive, inasmuch as the deficits are of a higher-order (semantic) nature. The greatest amount of transfer would thus be in patients with Wernicke's aphasia speaking two closely related languages, whose semantic systems would greatly overlap; the least degree of transfer would be in patients with Broca's aphasia speaking two languages which have radically different phonological and/or syntactic systems.

Because all patients in the Catalan study acquired their second language before the age of 6, the results cannot be generalized to late bilinguals (i.e., to those who acquired or learned their second language later in life). Another similar study is necessary to verify whether results obtained with early bilinguals are also obtained in late bilinguals, and to ascertain whether therapy in the native language has the same effects on later acquired/learned languages as therapy in the second language has on the native language.

Because all patients in the Barcelona study (Peña and collaborators) and in the Vigo study (Juncos and collaborators) spoke two very closely related languages, the findings cannot be generalized to bilinguals whose two languages differ considerably in surface syntax and lexicon. A similar study with Basque-Spanish bilinguals (Bidegain and collaborators) and Finnish- Swedish bilinguals (Kukkonen, Tuomainen and collaborators), as well as Farsi-Armenian, Farsi-Azari and Farsi-Kurdish (Nilipour and collaborators), and hopefully in other language combinations from contributors from around the world, will soon determine whether structural distance is a factor in the transfer of rehabilitated language skills.

In these studies, the same general method of language therapy is used, to the extent possible. If therapy is the major variable in the pattern of transfer from the treated to the untreated language, then results should be different from center to center, where methods are more likely to vary, irrespective of similarities between patient groups, such as type of bilingualism, structural distance, aphasic symptoms or site of lesion.

When all the results are in, we will be in a better position to determine whether both languages should be treated concurrently or whether only one should be treated, and if so, which language should be treated first.

Acknowlegment

This research is funded by Grant 410-91-1864 of the Social Sciences and Humanities Research Council of Canada.

References

Bond, S.L. 1984. Bilingualism and aphasia: Word retrieval skills in a bilingual anomic aphasic. Unpublished M.A. thesis. North Texas State University.

Byng, S., Coltheart, M., Masterson, J., Prior M. and Riddoch, J. 1984. Bilingual biscriptal deep dyslexia. *Quarterly Journal of Experimental Psychology*, **36A**: 417-433.

Chlenov, L.G. 1948. Ob afazii u poliglotov. *Izvestiia Akademii Pedagogicheskikh NAUK RSFSR*, **15**: 783-790.

Denès, P.-A. 1914. *Contribution à l'étude de quelques phénomènes aphasi- ques*. Thèse de Doctorat, Paris: Ollier-Henry.

Durieu, C. 1969. *La rééducation des aphasiques*. Bruxelles: Dessart.

Falk, K. 1973. Die aphasischen Störungen aus der Sicht des Logopäden. *Die Sonderschule*. **18** (2). Volkseigener Verlag.

Fredman, M. 1975. The effect of therapy given in Hebrew on the home language of the bilingual or polyglot adult in Israel. *British Journal of Disorders of Communication.* **10**: 61-69.

Hemphill, R.E. 1976. Polyglot aphasia and polyglot hallucinations. In S. Krauss (ed.), *Encyclopedic Handbook of Medical Psychology.* London: Butterworth.

Hilton, L.M. 1980. Language rehabilitation strategies for bilingual and foreign-speaking aphasics. *Aphasia, Apraxia, Agnosia,* **3** (2): 7-12.

Krapf, E.E. 1961. Aphasia in polyglots. *Proceedings of the VIIth International Congress of Neurology.* Rome, Vol. **1**: 741- 742.

Lebrun, Y. 1988. Multilinguisme et aphasie. *Revue de Laryngologie.* **109**: 299-306.

Linke, D. 1979. Zur Therapie polyglotter Aphasiker. In G. Peuser (ed.) *Studien zur Sprachterapie.* München: Wilhelm Fink Verlag.

Paradis, M. 1984. Aphasie et traduction. *META: Translators' Journal.* **29**: 57-67.

Paradis, M., Goldblum, M.C. and Abidi, R. 1982. Alternate antagonism with paradoxical translation behavior in two bilingual aphasic patients. *Brain and Language.* **15**: 55-69.

Peuser, G. 1978. Vergleichende Aphasieforschung und Aphasie bei Polyglotten. *Folia Phoniatrica.* **12**: 123-128.

Voinescu, I., Vish, E., Sirian, S. and Maretsis, M. 1977. Aphasia in a polyglot. *Brain and Language.* **4**: 165-176.

Wald, I. 1958. Zagadnienie afazji poliglotow. *Postçpy Neurologii Neurochirurgii Psychiatrii*: **4**: 183-211.

Wald, I. 1961. Problema afazii poliglotov. Moskva: *Voprosy Kliniki i Patofiziologii Afazii.* Pp. 140-176.

Watamori T. and Sasanuma, S. 1976. The recovery process of a bilingual aphasic. *Journal of Communication Disorders.* **9**: 157-166.

Watamori T. and Sasanuma, S. 1978. The recovery process of two English-Japanese bilingual aphasics. *Brain and Language.* **6**: 127-140.

Weisenburg T.H. and McBride, K.E. 1935. *Aphasia, a clinical and psychological study.* New York: Commonwealth Fund.

14 Aphasic adults with premorbid heroin and alcohol abuse: Rehabilitation research strategies

Camille Bushell, Loraine K. Obler and Patricia Kerman-Lerner

INTRODUCTION

As a rule in research on aphasia, subjects are selected according to standard criteria. These established criteria target right handed aphasics with CT-scan documented single, focal lesions within the left hemisphere. Moreover, the presence or presumed presence of multiple lesions and/or other diffuse cellular changes is typically considered grounds for exclusion from a subject pool. Thus patients with a history of drug or alcohol abuse have been excluded, along with patients with previous neurologic histories.

Several rationales may be suggested for excluding aphasic adults with histories significant for alcohol or heroin abuse (ETOH and IVDA respectively) from subject pools. Aphasia syndromes resulting from the complex etiologies of unilateral focal lesion in conjunction with diffuse damage might be expected to deviate from the traditional aphasias. Moreover, diffuse neurologic conditions such as those caused by ETOH or IVDA may contribute effects to the clinical profile which are impossible to control. Theoretically, language disturbances are expected to be grossly similar in any two patients with a single, focal lesion in the same language zone, (but see Basso, Lecours, Moraschini and Vanier (1985); and Naeser, (1978)). Diffuse cellular dysfunction is more difficult to document than circumscribed lesions, and is likely to vary greatly across subjects with diagnosis of IVDA or ETOH, thus experimental replicability is endangered.

One consequence of excluding aphasic individuals with neurologically diffuse lesions from neurolinguistic research is that commonly used diagnostic tools (e.g., Goodglass and Kaplan, 1983 and Kertesz, 1982) and treatment programs (e.g., Albert, Sparks, and Helm 1976 and Helm-Estabrooks, Fitzpatrick and Barresi, 1982) are normed solely on aphasic subjects with strictly focal lesions. As a result, we have little idea of the range of aphasic symptoms in non-standard populations, or of the rehabilitative techniques that may be successfully employed with them.

MOTIVATION FOR THE RESEARCH

Consider the population of alcohol abusers, even those without Korsakoff's syndrome. One might predict that the ability to learn is likely to be impaired after years of excessive heroin or alcohol consumption. However, it appears that alcohol-induced neural changes may resolve with time to some degree (Carlen, 1978), resulting in relatively spared learning capacity. At least with respect to alcohol abuse, we are justified in expecting, on the basis of the neurobiological literature on humans and even rodents, that aphasic patients may benefit from aphasia rehabilitation. Chronic cerebral changes in humans associated with nutrition and hydration secondary to long-term alcohol abuse appear to show some degree of recovery after months of abstinence. The common changes associated with alcohol abuse are cerebral atrophy due to underlying loss of neurons, glial cells, and supporting vasculature, and enlargement of the ventricles and cortical sulci (Ron, Acker, Shaw and Lishman, 1982). These CT documented features were reported to progressively diminish, but not completely resolve after months of abstinence (Carlen et al., 1978; Ron et al., 1982).

Further changes associated with ethanol abstinence have been studied on a cellular level in rodents. In rats, neuronal degeneration is known to be characterized by loss of dendritic spines and axonal arborization, particularly affecting those neurites of the pyramidal cells of the cortex and the hippocampus, granule cells of the hippocampus, and Purkinje cells of the cerebellum (see Carlen et al., 1978 for review). However, morphological plasticity has been demonstrated in rats after weeks of abstinence, showing reinnervation of partially deafferented neurons (Carlen et al., 1978). Additional cellular changes of a metabolic nature have been studied in mice. A 50-percent reduction in protein syntheses was recorded after long-standing ethanol ingestion. However, after weeks of abstinence, protein synthesis increased significantly (Carlen et al., 1978).

Some chronic neurologic effects, such as transverse myelitis and encephalitis, are shown to be associated with intravenous heroin abuse (Grant and Mohns, 1975). However, it is not clear whether these sequelae reflect a direct effect of heroin or rather an effect of the adulterants in which heroin is embedded. Moreover, the reversibility of these effects is not known. However, cognitive changes subsequent to long-term heroin usage were not apparent on neuropsychological testing of attention, memory, intelligence and new learning. Abstinent heroin users could not be distinguished from normal controls in a neuropsychological study by Fields and Fullerton (1974).

Given these findings of spared cognition seen in heroin users and reversible cerebral effects in alcohol abusers, one would want to characterize learning abilities in aphasic patients with such premorbid histories. Our aim is to determine whether aphasic subjects with complex etiologies respond to the same treatments given to aphasics without these histories.

SUBJECT SELECTION

Subject selection is difficult in studies of individuals whose behavior is illegal or otherwise stigmatized. Naturally the most difficult feature in subject selection for studies of aphasic adults with histories of substance abuse is to determine the amount of substance abuse and the ages at which it occurred. Self-report is likely to be the best evidence available, but, perhaps particularly in a study of substance abuse, it is not highly reliable.

Ideally, the control group would be aphasic subjects drawn from the same population but without history of substance abuse, who receive the same therapy programs as the experimental group. Clearly, both groups must have equivalent focal aphasia-producing lesions. Also, both groups should be an equivalent amount of time post onset in attempting to control for the extent of spontaneous recovery that might have occurred. In addition, severity level of aphasia at onset must be balanced between the two groups, as well as the duration and type of treatment. Experimental and control groups must be matched as closely as possible for age, education, and socioeconomic status. All subjects would be native speakers of the same dialect of the language of rehabilitation. Also, ideally, motivation to succeed would be equal in all groups. Aphasic individuals living in the community or to be discharged to the community may be more engaged in the rehabilitative process than those bound for placement in nursing facilities. Finally, the speech/language pathologists involved in rehabilitation of the patients would be blind to whether the patients they treated were part of the experimental group or the matched control group.

PILOT STUDY

In a retrospective study designed to determine whether patients with histories of heroin abuse or alcohol abuse would benefit at all from speech/language therapy, we studied sixteen such aphasic patients. The subjects had aphasia as a result of lesions in the dominant hemisphere. Their ages ranged from 33 to 66 years old at the time of onset. Educational levels varied from grade school to two years of college, with five patients being high school graduates and one illiterate. Fifteen subjects were male and one was female. Six subjects were diagnosed with fluent aphasia, and 10 with nonfluent aphasia. These patients were at a comparable chronic severity level, with 14 to 16 patients being in the moderate to severe range. The remaining 2 patients fell into the mild-moderate range.

Mean months post-onset at the time of initiating treatment was on average approximately six months, ranging from 1 to 23 months. Duration of treatment ranged from 3-15 months with a mean of 10 months. Documentation regarding nature and length of treatment prior to admission to the current facility was virtually unavailable. Therefore, initial severity and previous course in treatment had to be disregarded in this study.

Because this was a retrospective study and because of the great variation in the amount and type of treatment given to our subjects, we confined ourselves to a particular set of learning conditions. Specifically, we investigated the ability to learn the following compensatory strategies: gestures, writing, drawing, circumlocutions, enhanced intonation, requests for repetition and/or clarification, and signaling yes/no. We relied on speech language pathologists' reports of whether patients achieved the ability to learn these strategies with clinician cues, and whether they advanced to be able to use them independently.

Seventy-five percent of all patients achieved maximum gains in utilizing at least one strategy independently. The remaining 25 percent of patients still required a clinician cue to initiate and/or fully execute trained strategies. Indeed, 100 percent of the aphasic subjects presented here experienced success in this clinician-cued, less challenging condition for at least one strategy. Although our numbers were too small to conduct statistics, we looked at the extremes and the group with respect to months post onset, chronic aphasia severity, and education; none of these appear to have influenced the extent to which these aphasic patients succeeded in mastering the compensatory strategies. Nor did the type of substance abused make a difference. Both subjects with alcohol and heroin premorbid histories ranged in their successes from 0 to 100 percent.

It was the case that certain strategies resulted in particular success for these subjects, namely, circumlocutions, yes/no responses and requests for repetitions and clarification. By contrast, the remaining strategies (gestures, enhanced intonation, drawing, and writing) resulted in markedly fewer successes. Because we did not have a control group of non-abusers available, however, it is unclear the extent to which this pattern of success might hold for any aphasic population studied.

In our study, all subjects succeeded in the cue-dependent condition and some subjects, on one or more strategies, in the independent condition. Our main conclusion was that learning did occur in varying degrees, independent of treatment duration, months post onset, aphasic severity, education, and intoxicant type.

We may account for spared learning abilities observed in our patients with premorbid ETOH by reference to the neurology and neurobiology literatures which demonstrate the reversibility of alcohol-related neuropathological effects. The reported improvements in neuropathological conditions seen in humans and rodents are consistent with our observation of spared learning abilities seen in humans after years of excessive alcohol consumption. With respect to patients with histories of IVDA, the intact capacity to learn a strategy also follows from the neuropsychology literature cited in the Introduction. Our finding of functional learning abilities for compensatory strategies is consistent with those studies discussed above.

Our report of substance-abusing aphasics' success in learning compensatory communication strategies may be viewed in the light of several studies that provide evidence for the abilities of the nonaphasic, non-Korsakoff, abstinent alcoholics to learn and use compensatory strategies. Cermak (1975) and Binder and Schreiber (1980) found alcoholics to be able to learn visual imagery and verbal rehearsal strategies, which facilitate encoding and retrieval on paired associate learning tasks. In another study, recovering alcoholics were able to utilize strategies involving memory notebooks and signs to enhance memory functioning (Laverty, 1975). Moreover, Laverty concluded that alcoholics can learn and likewise benefit from treatment, although alcoholics with documented brain damage are routinely excluded from rehabilitation protocols. Finally, abstinent alcoholics, followed for five years, continued to improve in verbal and visual memory tasks requiring new learning skills (Brandt and Butters, 1983). This finding of significant improvement in ability to learn over time suggests that a return to cognitive function is not a discrete phenomenon, but a prolonged process.

CONCLUSION

In sum, there are reasons to expect that individuals with premorbid substance abuse histories may benefit from aphasia rehabilitation, but virtually all research to date has ignored the question. Although the well-designed study is difficult to carry out, and as greater numbers of patients with such histories are seen in our large cities, it becomes increasingly urgent to know the extent to which they can benefit from rehabilitation. In fact the difficulties we point to in studying patients with substance abuse histories are similar to those of other special populations who are relatively unstudied as compared to more "standard" aphasics. Patients with Alzheimer's disease or traumatic brain injury, for example, may be expected to have diffuse neuropathology similar in some ways to that of patients with substance abuse histories.

References

Albert, M., Sparks, R. and Helm, N. 1973. Melodic intonation therapy for aphasia. *Archives of Neurology*, 29: 130-1.

Basso, A., Lecours, A.R., Moraschini, S. and Vanier, M. 1985. Anatomical correlations of the aphasias as defined through computerized tomography: Exceptions. *Brain and Language*, 26: 201-229.

Binder, L. and Schreiber, V. 1980. Visual imagery and verbal mediation as memory aids in recovering alcoholics. *Journal of Clinical Neuropsychology*, 2: 71-7.

Brandt, J., Butters, N., Ryan, C. and Bayog, R. 1983. Cognitive loss and recovery in long-term alcohol abusers. *Archives of General Psychiatry*, 40: 435-442.

Carlen, P., Wortzman, G., Holgate, R., Wilkinson, D. and Rankin, J. 1978. Reversible cerebral atrophy in recently abstinent chronic alcoholics measured by computerized tomography scans. *Science*, 200, June, 1076-1078.

Cermak, L. 1975. Imagery as an aid to retrieval for Korsakoff's patients. *Cortex*, 11: 163-9.

Fields, F. and Fullerton, J. 1974. The influence of heroin addiction and neuropsychology functioning. Newsletter for Research in Mental Health

and Behavioral Science, Dept. of Medicine and Surgery, VA, Washington, D.C.
Goodglass, H. and Kaplan, E. 1983. *Assessment of Aphasia and Related Disorders*, Philadelphia: Lea and Febiger.
Grant, I. and Mohns, L. 1975. Chronic cerebral effects of alcohol and drug abuse. *Journal of the Addictions*, 10: 883-923.
Helm-Estabrooks, N., Fitzpatrick, P. and Barresi, B. 1982. Visual action therapy for global aphasia. *Journal of Speech and Hearing Disorders*, 44: 385-9.
Kertesz, A. 1982. *Western Aphasia Battery*, NY: Harcourt Brace Jovanovich.
Laverty, S. 1975. Alcohol and brain damage: Implications for the treatment of the individual patient and alcohol programmes. In J. Rankin (ed.) *Alcohol, Drugs and Brain Damage*, Addiction Research Foundation of Ontario.
Marshall, R. 1986. Treatment of auditory comprehension deficits. In R. Chapey (ed.) *Language Intervention Strategies in Adult Aphasia*, Second Edition, Williams and Wilkins Pp. 297-328.
Naeser, M. and Hayward, R. 1978. Lesion localization in aphasia with cranial computed tomography and the Boston Diagnostic Aphasia Examination. *Neurology*, 28: 545-555.
Ron, M., Acker, W., Shaw, G. and Lishman, W. 1982. Computerized tomography of the brain in chronic alcoholism. *Brain*, 105: 497-514.
Schuell, H., Jenkins, J. and Jimenez-Pabon, E. 1965. *Aphasia in Adults*, New York: Harper and Row.

Conclusion

The rationales behind the various aphasia therapies can be grouped into those that assume that cerebral areas subserving language are intact but not easily accessible, and those that assume that whatever language is lost is not recoverable. In the first instance, retraining is attempted; in the second, compensation is sought by tapping non-grammatical means of communication. Language reactivation and retraining approaches propose stimulation therapies, programmed instruction, symptom-oriented approaches, language structure-based approaches, and deblocking methods. Compensatory approaches seek right-hemisphere involvement, either through the use of pragmatic aspects of language use, paralinguistic means of communication, melodic intonation or laughter therapy, as well as means to increase non- verbal communication skills. The primary aim of rehabilitation is seen to be the improvement of the aphasic person's ability to communicate, if necessary by means other than purely grammatical.

Two pragmatic approaches have been identified. One is a theory-driven perspective, the other a data-driven approach. In theory-driven appoaches, emphasis is on coherence in discourse and on cohesive devices. Data-driven approaches focus on methods of production and interpretation of social interaction. Most aphasic people appear to have intact knowledge of how to use language. What is deficient is the linguistic code, which of course interacts with pragmatic aspects in normal language use. Patients can be taught strategies that will capitalize on their preserved pragmatic abilities to compensate for their loss of linguistic structure.

Methods of aphasia therapy differ in relation to their implicit or explicit underlying theory of language. When a behaviorist standpoint is adopted, language is "trained" in rehabilitation through listening and repetition exercises. When language deficits are identified according to principles of structuralist linguistics, language is generally decomposed into distinct levels of structure. Despite repeated claims of the importance of linguistic theory in aphasia therapy, there has been no systematic application of any specific linguistic theory to a rehabilitation programme for aphasic patients.

There are two ways of approaching the language structure to be treated. One is to observe surface symptoms of a particular problem, and one is to analyze aspects of the deficit in relation to theories of normal language processing. These two approaches entail different tasks in therapy. They either capitalize on what is retained or on what has been impaired. Neither approach has proven very successful.

In recent years, therapy has moved from focussing specifically on linguistic structures to programmes that concentrate on the rehabilitation of the patient's communicative abilities. However, tests for the diagnosis of aphasia continue to use purely linguistic and metalinguistic tasks, removed from the context of natural language use. They differ only as to whether their authors see language as unitary or assume that it is organized in a modular fashion. But in neither do linguistic theories determine their structure or content.

Pharmacotherapy, by acting on neurotransmitters, is a new avenue of research into possible ways of access to language functions whose neural threshold of activation may have fallen below a minimum level.

The assessment of the efficacy of various methods for various types of aphasic patients is fraught with inherent methodological problems. Comparison of treatments designs suffer from a number of shortcomings. In particular, it cannot be used to demonstrate whether either of the compared treatments is efficacious. It will only indicate whether one treatment is the same or better than another. On the other hand, while treatment vs. no- treatment designs do provide some evidence about the efficacy of treatment, it does not indicate which patients respond best to which types of treatment. An answer to this question may come from employing single subject alternating treatment designs with replications to test specific treatments with specific types of aphasic patients.

Preliminary results seem to indicate that the linguistic structure approach, when aimed at specific impairments of linguistic knowledge is more favorable in chronic aphasic patients than a purely stimulating or communicatively activating approach. It may be that metalinguistic knowledge, in the absence of (access to) implicit linguistic competence, can serve as a useful compensatory strategy, allowing patients—as it allows beginning foreign language learners—to consciously construct their sentences (as opposed to generating them automatically). If it is the case that procedural memory (here, implicit linguistic competence) is what is impaired in aphasia, while declarative memory (here, conscious metalinguistic knowledge) remains intact (Cohen, 1984), then learning metalinguistic facts through declarative memory may allow patients to compensate for their lack of access to automatic linguistic competence by substituting for it the controlled use of such declarative knowledge.

Two ways of approaching the language structure to be treated have been used. One is to observe surface symptoms of a particular problem, and one is to analyze aspects of the deficit in relation to theories of normal language processing. These two approaches entail different tasks in therapy. They either capitalize on what is retained or on what has been impaired. Neither approach has proven very successful.

Considering the extensive efforts employed, there has been a relative lack of success of language therapy, irrespective of the method, in retraining patients in the use of lost (or inaccessible) linguistic competence—i.e., the lost ability to produce acceptable sentences automatically. This has led researchers to look for alternative compensatory strategies that would circumvent the patients' linguistic impairment, principally in two ways: (1) by focussing on the capacities of the intact right hemisphere (pragmatic aspects of language use, intonation therapy, etc.) and (2) by the use of intact declarative memory to provide patients with the metalinguistic knowledge that will allow them to substitute controlled production for their unavailable automatic linguistic competence. Some methods are overtly metalinguistic (Springer and Weniger, 1980). That is, patients are trained to make conscious decisions about phonemic, graphemic, lexical and grammatical structures.

Traditional language therapy that has focussed on the reactivation (reacquisition, relearning) of linguistic parameters of verbal communication has had very limited success (Prins, Schoonen, and Vermeulen, 1989). Indeed, its success has generally been inversely proportional to the severity of the aphasia, and most often can be attributed to spontaneous recovery. What seems to have been more successful, even under circumstances in which strict linguistic aspects of language were stressed (e.g., DiCarlo, 1980), and is promising as a successful compensatory strategy, is the use of pragmatic aspects of language to compensate for the loss of access to linguistic structure.

Byng and Lesser (Chap. 10) rightly suggest that the limited effects of many therapies must be taken seriously, so that a different approach to remediation can be found that will meet the needs of the patients.

The general impression is that all the approaches to therapy are only loosely connected to a theory base, be it linguistic, psychological, cognitive, or neurological. As has been argued in Chapter 10, there is as yet no theory underlying the application of any therapy, that is, no theory about *how* or *why* certain tasks might bring about change. While information processing models (as discussed in Chapters 7-10) pinpoint the origin and nature of the deficit, they do not specify what techniques are to be used in the treatment of this identified deficit. Thus, while we have a refined method to diagnose the exact processing level of a particular deficit, there still does not seem to be any specific theoretical basis for the use of precise techniques. Thus, what is still needed is a theory that will serve as a foundation for the choice of actual rehabilitation procedures.

The result of this search for theoretical foundations of aphasia rehabilitation turns out to be the discovery that in fact, there are none. To be sure, there are broad theoretical frameworks within which language therapy develops. And there are (albeit sometimes conflicting) linguistic and information-processing theories to locate the source of the deficit. But what is still lacking is a valid theory of how to remediate these newly refined deficits. These theories specify what to focus on, but not what to do about it. We are still in need of a rationale for specific methods of intervention.

Reference

Cohen, N. (1984) Preserved learning capacity in amnesia: Evidence for multiple memory systems. In L. R. Squire and N. Butters (Eds.), *The neuropsychology of human memory*. New York: Guilford Press. Pp. 83-103.

Di Carlo, L.M. 1980. Language recovery in aphasia: Effect of systematic filmed programmed instruction. *Archives of Physical Medicine and Rehabilitation*, 61: 41-4.

Prins, R.S., Schoonen, R. and Vermeulen, J. 1989. Efficacy of two types of speech therapy for aphasic stroke patients. *Applied Psycholinguistics*, **10**: 85-123.

Springer, L. and Weniger, D. 1980. Aphasietherapie aus logopädisch-linguistischer Sicht. In Böhme, G., (ed.), *Therapie der Sprach-Sprech- und Stimmstrungen*, Stuttgart: Fischer.

Author Index

Abidi, R. 416, *419*
Acker, W. 422,*426*
Ackerman, N. 264, 265, *288*, 329, 336, *361*
Ades, H.W. 367, *374*
Ahern, M.B. 105, *133*, 177
Ahlsen, A. 225, *242*
Albert, M. 4, 8, 10, 18, 21, 24, *25*, *26*, *27*, *30*, 43, 48, 56, 61, 67, *74*, 80, *95*, 113, 115, 116, 119, *129*, *131*, 153, 221, *244*, 266, *287*, 373, 380, *407*, 421, *425*
Alexander, M.P. 8, *25*, 266, *287*
Allard, L. 221, *245*
Allport, D.A. 252, *259*
Anderson, J. 5, *28*, 35, 63, *73*
Anzola, G.P. 380, *407*
Araki, G. 372, *376*
Argentino, C. 366, *374*
Arsenault, J.K. 7, *28*, 40, 162
Astrup, J. 365, *374*
Aten, J.L. 4, 16, *25*, *26*, 49, 50, 61, 65, 71, 75, 77, *97*, 120, 121, 125, *129*
Auerbach, S. 370, 371, *376*
Austin, J. 215, 217, *242*

Bachman, D.L. 21, *25*, *26*, 56, 57, *129*
Bachy-Langedock, N. 251, *259*, 277, 278, *287*, 306, 307, *314*
Badecker, W. 321, *360*
Bader, E. 18, 27, 51
Bagnara, S. 379, *407*
Baguley, T. 263, *288*
Bainton, D. 63, *72*, 77, *95*
Baker, E. 265, *288*
Barber, J. 94, *96*
Barnes, N.S. 61, 65, 71, 75, 77, *97*
Barresi, B. 4, 14, 15, *27*, 33, 51, 61, *73*, 77, *96*, 100, *100*, 109, 122, 127, *131*, 154, 421, *426*

Basso, A. 5, *26*, 31, 61, 64, *72*, 77, 95, 250, 257, *259*, 272, *287*, 369, *374*, 421, *425*
Bastard, V. 77, *97*, 113, 114, 115, *134*, 182, 273, *292*
Bates, E. 221, *242*
Bayer, J. 296, *315*
Bayog, R. 425, *425*
Beaumont, J.G. 379, 392, *407*
Beauvois, M.F. 267, 277, *288*, 294, 297, 301, 309, 310, *314*
Behrmann, M. 101, 113, 114, 117, *129*, 135, 234, *242*, 293, 299, 301, *314*
Bellugi, U. 20, *29*
Benenson, M. 63, *73*, 77, *96*
Benjamins, D. 297, *316*
Benson, D.F. 20, 22, *26*, 371, *374*, *376*
Berko-Gleason, J. 221, *242*
Berlin, C. 128, *129*
Berlucchi, G. 379, 384, *407*, *408*
Berman, M. 10, *26*
Berndt, R.S. 251, *260*, 297, *314*, 321, 325, 331, 337, 347, 348, 358, *360*,*361*
Berrini, R. 379, *407*
Bertolini, G. 380, *407*
Bester, S. 229, *244*
Beukelman, D. 236, *243*
Bever, T. 332, *360*
Beyn, E.S. 9, *26*, 41, 331, 355, *360*
Bierwisch, M. 14, *30*, 99, *100*
Binder, L. 424, *425*
Black, M. 321, 323, 331, 334, 353, 354, *360*,*362*
Black, S.E. 293, *314*, 372, *375*
Blanken, G. 280, 282, 283, *288*
Blavier, A. 276, *288*
Blomert, L. 227, *242*
Bloom, L.M. 8, *26*, 37
Boles, D.B. 379, *407*
Boller, F. 380, 389, 391, *408*
Bollier, B. 104, *130*, 144
Bollinger, R.L. 6, *26*, 37

Author Index

Bond, S.L. 416, *418*
Bourgeois, M.S. 17, *27*, 329, 340, 341, 346, 347, *360*
Brain, R. 380, *407*
Brandt, J. 425, *425*
Brannegan, R. 61, 65, 71, *75*, 77, *97*
Brazil, D. 217, *242*
Brennen, T. 263, *288*
Bresnan, J.W. 321, *360*
Bright, J. 263, *288*
Brizzolara, D. 379, *407*
Broida, H. 4, *26*, 31
Brookshire, R.H. 61, 63, 64, 65, 71, 74, *75*, 77, *97*
Brown, J.W. 99, *100*, 119, *131*, 149, 267, *289*
Brown, R. 262, 264, *288*
Brownell, H. 216, 221, *243, 246*
Bruce, C. 11, *26*, 42, 116, *129*, 136, 263, 270, *288*
Bruce, V. *288*
Bruckert, R. 270, 274, 280, 283, *289*
Brunner, A. 77, *95*
Bryan, K. 5, *28*, 35, 63, *73*
Bryden, M.P. 379, *407*
Bub, D.N. 279, *288*, 293, 297, *314*
Buchtel, H.A. 380, *407*
Buck., D. 369, *376*
Bucy, P.C. 367, *374*
Bugbee, J. 239, *244*
Bunel, G. 276, *290*
Burton, A. 107, 115, *129*, 136
Burton, E. 107, 115, *129*, 136
Butfield, E. 63, *72*
Butler, J. 63, *73*, 77, *96*
Butters, N. 425, *425*
Byng, S. 13, *26*, 46, 77, *95*, 101, 112, 113, 117, *129*, 137, 300, 301, 302, 303, *315, 318*, 321, 323, 325, 330, 331, 334, 353, 354, 355, 358, *360, 362*, 416, *418*

Caligiuri, M.P. 4, *26*, 50
Cambier, J. 371, *374*
Cameron, D. 222, 224, *245*
Cannito, M. 262, *291*
Canter, G.J. 262, *292*
Capitani, E. 5, *26*, 31, 61, 64, *72*, 369, *374*

Caplan, D. 201, *206*, 322, *360*
Capon, A. 372, *374*
Cappa, S.F. 372, *374*
Caramazza, A. 257, *259*, 279, 284, *288, 289*, 294, 296, 297, 301, 314, *314, 318*, 321, *360*
Cardebat, D. 276, *288, 290*
Carey, P. 203, *206*
Carlen, P. 422, *425*
Carlomagno, S. 298, 310, 311, 313, *315*
Carrol, V. 4, *29*
Carroll, J.B. 196, 198, *207*
Carter, J. 61, 65, 71, *75*, 77, *97*
Casadio, P 313, *315*
Cauthen, J.C. 380, *408*
Cegla, B. 19, *28*, 55
Cermak, L. 424, *425*
Chapey, R. 4, *26*, 125, *129*, 235, *242*
Chapman, S. 220, 221, *242, 245*
Chassin, G. 77, *97*, 113, 114, 115, *134*, 182, 273, *292*
Chedru, F. 390, 391, *407*
Chertkown, H. 293, *314*
Cheshire, J. 224, *243*
Chin Li, E. 113, 114, 115, 117, 118, 129, *130*, 139
Chlenov, L.G. 413, 414, *418*
Chomsky, N. 195, *206*, 321, *360*
Clark, H.H. 125, *130*, 225, *242*
Clerebaut, M. 88, *95*
Coates, R. 370, *375*
Code, C. 4, *26*, 227, *242*
Coderre, L. 330, *361*
Coelho, C.A. 20, *26*, 54, 122, *130*, 141
Cohen, N. 1, 2
Cohen, R. 270, 271, *288*
Collins, C. 269, *289*
Collins, M.J. 63, 64, *74*, 77, *97*
Colombo, A. 313, *315*
Coltheart, M. 77, *95*, 250, *259*, 294, 302, 303, *315, 318*, 321, 416, *418*
Consolini, T. 13, *26*, 38, 105, 106, 127, *130*, 142
Conway, N. 225, *242*
Cook, K. 255, 256, *259*
Copeland, M. 216, 218, *242*

Corkin, S. 269, *289*
Coslett, H.B. 295, *315*
Coulthard, R. 217, 218, *242*, *245*
Coyette, F. 88, *95*
Crockford, C. 229, *242*
Crystal, D. 103, *130*, 141
Cubelli, R. 13, *26*, 38, 105, 106, 127, *130*, 142
Culton, G.L. 101, 102, 109, 110, *130*, 143
Cummings, J.L. 371, *374*, *376*
Czopf, J. 371, *374*

Dabul, B. 104, 105, *130*, 144
Dallal, G.E. 92, *95*
Dans, J. 238, *246*
Darley, F.L. 62, 70, *72*, 102, 113, 114, 116, 121, 125, 126, *133*, *134*, 175, 268, 276, *291*, 373
David, R.M. 63, *72*, 77, *95*
Davidson, J. 217, *245*
Davies, C.L. 101, 102, 106, 109, *130*, 144
Davis, A. 254, *260*, 275, *291*
Davis, G.A. 4, 12, 15, 16, 25, *26*, *27*, 47, 61, *72*, 77, *95*, 101, 102, 108, 109, 110, 120, 122, 124, 125, *130*, 145, 216, 235, *242*, *243*, *246*, 327, 334, 343, *360*
Davis, L. 61, *74*, 78, 88, *97*
Dax, M. 367, *374*
De Bleser, R. 25, *27*, 296, *315*
De Partz, M.P. 107, 251, *259*, 270, 277, 278, *287*, 296, 298, 303, 306, 307, 308, 309, *314*, *315*, *318*
Deal, J.L. 71, *74*
Dean, E. 212, 217, 226, *246*
Deck, J.W. 18, *29*, 53, 121, 123
DeGiovanni E. 265, *290*
DeGirolami, J. 365, *377*
Déjerine, J. 293, *315*
Della Sala, S. 379, *407*
Deloche, G. 77, *97*, 105, 106, 113, 114, 115, 126, *130*, *134*, 182, 183, 273, 276, *288*, *290*, *292*, 301, 302, *315*
Demeurisse, G. 372, *374*

Denès, P.-A. 416, *418*
Dennis, S. 371, *375*
DeRenzi, E. 340, *360*, 380, 389, 390, 391, *407*
Derouesné, J. 277, 279, *288*, *291*, 294, 297, 301, 309, 310, *314*
DeWeerdt, C.J. 366, *374*
Di Carlo, L.M. 7, *27*, 39, 100, *100*
Dietens, E. 272, *292*
Dordain, M. 276, *288*, *290*
Douglass, E. 369, *375*
Doyle, P.J. 17, *27*, 329, 340, 341, 346, 347, *360*
Duckworth, M. 23, *29*, 60
Duffy, J.R. 4, *27*, 32, 202, *206*
Durieu, C. 415, *418*
Dusatko, D. 113, 129, *130*, 139

Edelman, G. 88, *95*
Edgington, E.S. 84, 91, *96*
Edwards, S. 119, 120, 121, 122, 125, *130*, 146, 221, *243*
Elbard, H. 229, *244*
Elghozi, D. 371, *374*
Elia, E. 262, *291*
Ellis, A.W. 252, *259*, 280, *289*, 295, *315*
Emanuelli, S. 313, *315*
Emery, P. 4, 10, *27*, 43, 113, *131*, 153
Enderby, P. 63, *72*, 77, *95*
Engel, D. 270, 271, *288*
Engl, E. 12, *27*
Exner, S. 293, *315*

Faglioni, P. 77, *95*, 380, 389, 390, 391, *407*
Fair, C. 369, *376*
Falk, K. 416, *418*
Fazzini 16
Feeney, D.M. 366, *374*
Fennell, E.B. 305, *317*
Ferguson, P.A. 101, 110, *130*, 143
Ferrand, I. 276, *288*, *290*, 301, 302, *315*
Ferreira, A. 239, *244*
Feyereisen, P. 88, *95*, 267, *288*
Fields, F. 422, *425*
Finlayson, A. 229, *244*
Fisher, M. 365, *377*

Fitzpatrick, P.M. 4, 14, 15, 27, 33, 51, 61, 73, 77, 96, 100, *100*, 109, 122, 127, *131*, 154, 421, *426*
Flanagan, J. 238, *243*
Flavell, J.H. 204, *206*
Flowers, C. 239, *243*
Fodor, J. 203, *206*
Foldi, N. 216, *243*
Foresti, A. 13, *26*, 38, 105, 106, 127, *130*, 142
Forst 264
Fox, P.T. 251, *260*
Franklin, S. 77, 96, 114, 117, 118, *131*, 156, 158, *259*, 271, 272, 274, 282, *289*, 297, *316*
Frazier, C.H. 63, 72
Fredman, M. 415, *419*
Friden, T. 63, 64, *74*, 77, 97
Friedman, R.B. 293, 304, 305, *315*, *316*
Friedrich, F.J. 262, *291*
Fritsch, G.T. 367, *374*
Frost, R. *288*
Fullerton, J. 422, *425*
Funnell, E. 295, 301, 312, *315*

Gainotti, G. 262, *288*, 380, *407*
Garcia-Bunuel, L. 61, 65, 71, *75*, 77, 97
Gardner, H. 216, 221, *243*, *246*, 267, *288*
Garman, M. 221, *243*
Garrett, K. 236, 237, *243*
Garrett, M.A. 321, *361*
Gazzaniga, M.S. 368, *374*
Gelmers, J.J. 366, *374*
Gerber, S. 226, 230, 231, 232, *243*
Gerstman, L.J. 119, *131*, 149, 267, *289*
Geschwind, N. 372, *375*
Gildston, H. 23, *27*, 59, 119, *130*
Gildston, P. 23, *27*, 59, 119, *130*
Gleason, J.B. 263, *290*, 329, 336, *361*
Gleber, J. 220, *244*
Glindemann, R. 25, *27*, *30*, 88, *97*, 236, *243*
Gloning, I. 372, *375*
Gloning, K. 369, 372, *375*

Glosser, G. 226, *243*, 293, *316*
Godfrey, C.M. 369, *375*
Goldberg, T. 297, *316*
Goldberger, M.E. 367, *375*
Goldblum, M.-C. 416, *419*
Goldblum, N. 264, *288*
Goldfarb, R. 18, 27, 51
Goldstein, H. 10, 17, *27*, *30*, 45, 329, 340, 341, 346, 347, *360*
Goldstein, K. 99, *100*, 366, *375*
Goodglass, H. 11, *25*, *27*, 62, *73*, 108, *131*, 195, 200, 203, 204, 205, *206*, 221, *242*, 251, *259*, 261, 263, 264, 265, 266, 267, 268, 269, 281, 282, *287*, *288*, *289*, *290*, *291*, *292*, 329, 336, *361*, 368, *375*, 421, *426*
Goodman, N.J. 18, 19, *29*, 120, 126, *133*
Goodman, R. 52, 61, 65, 71, *75*, 77, 97
Gordon, W.P. 105 *132*, 165
Gorter, K. 366, *374*
Graddol, D. 224, *243*
Grant, I. 422, *426*
Green, E. 221, *242*, 329, 336, *361*
Green, G. 120, 121, 122, 123, 124, *131*, 147, 235, 236, 238, 239, 241, *243*
Greenbaum, H. 61, 65, 71, *75*, 77, 97
Grice, H. 125, *131*, 212, 213, *243*
Grizzle, J.E. 77, 96
Gross, M.M. 379, *407*
Grunwell, P. 101, *130*, 144
Guilford, A. 216, *243*
Guillotte, N. 308, 309, *316*
Gurland, G. 226, 230, 231, 232, *243*

Haag, E. 80, 97
Haas, G. 5, *29*, 36, 104, *133*, 172, 327, 337, *362*, 370, 371, *376*
Haferkamp, G. 21, *28*
Hagen, C. 64, 73
Hakes, D.T. 200, *206*
Hamby, S. 221, *242*
Hamre, C.E. 103, *133*, 170
Hanlon, R.E. 119, 128, *131*, 149, 267, *289*
Hannequin, D. 276, *288*, 290

Hanson, W.R. 25, 27, 372, *376*
Harlock, W. 370, *375*
Harris, E.H. 105, *133*, 177
Hart, J. 297, *314*
Hartman, J. 63, *73*
Hatfield, F.M. 12, 28, 77, 94, 96, 99, *100*, 101, 102, 103, 104, 109, 113, 115, 117, *131*, 150, 151, 298, 299, 307, 309, *316*, 359, *361*
Haub, G. 372, *375*
Haviland, S.E. 125, *130*
Hawkins, P. 214, 218, 226, *243*
Hay, D.C. 379, *407*
Hayward, R.W. 386, *408*, 421, *426*
Heilman, K.M. 295, *315*, 380, 390, *408*
Heiss, W.D. 369, *375*
Helm, N.A. 8, 18, 24, 25, *30*, 48, 61, 67, *74*, 80, *95*, 266, *287*, *289*, 421, *425*
Helm-Estabrooks, N.A. 4, 10, 14, 15, 21, *25*, *27*, 33, 43, 51, 56, 61, *73*, 77, *96*, 100, *100*, 109, 113, 115, 116, 119, 122, 127, *129*, *131*, 152, 153, 154, 221, *244*, 250, *260*, 320, 323, 329, 336, 337, *361*, 363, *363*, 370, 371, *376*, 421, *426*
Helmick, J.W. 119, *131*, 155
Hemphill, R.E. 414, *419*
Henaff Gonon, M.A. 270, 274, 280, 283, *289*
Henin, D. 371, *374*
Henschen, S.E. 371, *375*
Hermand, N. 77, *97*, 113, 114, 115, *134*, 182, 273, *292*
Heron, W. 384, *407*
Hécaen, H. 293, *316*, *317*, 380, *407*
Hier, D.B. 279, *289*
Higo K. 273, *292*
Hildebrandt, N. 201, *206*
Hillis, A.E. 251, *259*, 275, 279, 283, 284, 285, 286, *288*, *289*, 294, 296, 314, *314*
Hilton, L.M. 414, *419*
Hilton, R. 5, *28*, 35, 63, *73*
Hines, D. 379, *408*
Hiorns, R.W. 272, *291*

Hitzig, E. 367, *374*
Hocki, T. 21, *28*
Hofmann, R. 21, *28*
Holgate, R. 422, *425*
Holland, A.L. 4,6, 15, 17, *26*, *27*, *28*, 50, 61, 65, 71, *73*, *75*, 77, 80, *97*, 102, 109, 123, 125, *131*, *134*, 187, 218, 225, 226, 227, 228, 231, 234, 236, 238, *243*, *244*, *245*, 272, *289*, 327, 328, 339, 340, *361*
Holtzapple, P. 63, 64, *74*, *97*
Horner, J. 23, *29*
Howard, D. 11, 12, *26*, *28*, 42, 77, *96*, 114, 117, 118, *129*, *131*, *132*, 136, 156, 158, 251, 253, *259*, 268, 270, 271, 272, 274, 279, 280, 281, 282, *288*, *289*, 297, *316*, 359, *361*
Howell, J. 372, *375*
Howes, D. 380, 389, 391, *408*
Hubbard, D.J. 63, 64, *74*, 77, *97*
Huber, W. 11, 23, 25, *28*, *29*, *30*, 63, *73*, 77, 81, 88, *96*, *97*, 219, 220, 236, 237, *243*, *244*, *245*
Huff, F.J. 269, *289*
Hunt, J. 108, *131*
Huntley, R.A. 10, *28*, 44
Hyde, M. 221, 242, 263, *290*, 329, 336, *361*
Hyman, R. 384, *407*

Ingham, S.D. 63, *72*

Jackson, J.H. 371, *375*
Jacob, A. 330, *361*
Jarecki, J. 262, *291*
Jefferson, G. 223, 225, *245*
Jenkins, J.J. 4, *29*, 61, *74*, 80, *97*, 99, *100*, 195, 196, 198, 201, 202, 205, *207*, 369, 377, *426*
Jimenez-Pabon, E. 4, *29*, 61, *74*, 80, *97*, 195, 201, 202, 205, *207*, 369, *377*
Job 321
Johannsen-Horbach, H. 19, *28*, 55
Johnson, M.G. 266, *289*
Johnson-Laird, P. 219, *244*

Jones, A.C. 64, 70, 71, *73*, 77, *96*
Jones, C. 94, *96*
Jones, E.V. 77, *96*, 99, *100*, 108, 110, 112, 117, *132*, 159, 323, 329, 330, 338, 346, 350, 352, 359, *361*
Jones, K.F. 203, *206*

Kabe, S. 372, *376*
Kamhi, A.G. 204, *206*
Kaplan, E.F. 62, *73*, 195, 200, 204, 205, *206*, 226, *243*, 251, *259*, 264, 265, 266, 269, *288*, *289*, 329, *361*, 368, *375*, 421, *426*
Kaplan, R.M. 321, *360*
Kashiwagi 312, 313
Katz, R.C. 10, *28*, 44, 61, 65, *73*
Kay, J. 280, *289*, *318*
Kean, M. 227, *242*
Kearns, K.P. 8, *28*, 39, 111, *132*, 160, 320, 323, 328, 329, 334, 349, 350, 358, *361*
Keelean, P.D. 264, *291*
Kelter, S. 270, 271, *288*
Kenhlies, M. 21, *28*
Kerschensteiner, M. 77, *96*, 220, *245*
Kertesz, A. 24, *28*, 61, 62, 65, 71, *73*, *74*, 77, *97*, 272, 279, *288*, *289*, 297, *314*, 368, 369, 370, 371, 372, *375*, *376*, 421, *426*
King, F.A. 380, *408*
Kinsbourne, M. 371, *375*
Kinsey, C. 101, *132*, 161
Kintsch 219
Kirchner, D. 211, 212, 214, 216, 226, 229, *245*
Kitselman, K. 113, 129, *130*, 139
Klassen, A.C. 372, *376*
Klein, H. 203, *206*
Klima, E.S. 20, *29*
Knopman, D.S. 272, *289*, 370, 371, 372, *376*, *377*
Koenig, L.A. 204, *206*
Koenigsknecht, R.A. 77, *97*, 113, *135*, 193, 273, *292*
Koff, E. 372, *376*
Kohlert, P. 220, *244*
Kohlmeyer, K. 365, *376*

Kohn, S.E. 7, *28*, 40, 113, *132*, 162, 261, 263, *289*, *290*
Koskas, E. 261, 262, 279, *290*
Koster, C. 227, *242*
Kotten, A. 11, 12, *27*, *28*, 105, 106, 113, 115, 116, *132*, 163
Kraat 15
Krapf, E.E. 414, *419*
Kremin, H. 261, 262, 265, 276, 279, 280, 281, 285, *288*, *290*, 293, 295, 296, 297, 309, *316*, *317*
Kuhl, D.E. 372, *376*
Kurtzke, J.F. 61, 63, 64, 65, 71, *74*, 75, 77, *97*
Kushner, D. 5, *28*, 34, 101, 102, 119, *132*, 164
Kutas, M. 379, *408*

Lacorte-Pi, T.M. 384, *408*
Laine, M. 293, *317*
Landau, W.M. 64, *73*
Landis, T. 293, *317*, 371, *376*
LaPointe, L.L. 8, *28*, 61, 65, 67, 70, 71, *74*, *75*, 77, *97*
Larroque, C. 276, *288*, *290*
Lashley, K.S. 366, 367, 370, *376*
Laughlin, S.A. 5, *29*, 36, 104, 105, *132*, *133*, 165, 172, 327, 337, *362*
Laverty, S. 425, *426*
Le Bohec, I. 265, *290*
Le Dorze, G. 265, *290*, 330, 352, *361*
Leblanc, M. 390, 391, *407*
Lebrun, Y. 414, *419*
Lecours, A.R. 421, *425*
Leech, G. 213, *244*
Lemme, M.L. 105, *133*, 177
Lendrem, W. 64, 70, 71, *73*, 77, *96*
Lesser, R. 5, 15, *28*, *29*, 35, 63, *73*, 77, *96*, 103, 104, 105, 107, 112, 113, 116, 117, 118, 120, 123, 127, *132*, 165, 167, 218, 226, 231, 232, *244*, 252, 254, 256, 258, 260, 261, 278, *290*
Levine, D.M. 371, *376*
Levine, H.L. 371, 372, *374*, *376*
Levinson, S. 211, 213, 223, *244*
Levita, E. 369, *377*

Levitsky, W. 372, *375*
Levy, C. 327, 339, *361*
Lhermitte, F. 279, *291*, 390, 391, *407*
Li, E.C. 269, *291*
Lieberthal, T. 101, 113, 114, 117, *129*, 135
Lincoln, N.B. 64, 70, 71, *73*, 77, *96*
Linebarger, M.C. 112, *133*, 322, 330, *362*
Linke, D. 415, *419*
Lishman, W. 422, *426*
List, G. 270, 271, *288*
Lomas, J. 229, *244*, 370, *376*
Love, R.J. 269, 271, *291*
Loverso, F.L. 332, 337, 338, 339, *361*, *362*
Low-Morrow, D. 236, *243*
Lubinski, R. 121, 122, 123, 124, 125, *132*
Lucas, D. 107, *129*, 136
Ludlow, C. 369, *376*
Luria, A.R. 5, 24, *28*, 70, *73*, 293, *317*, 368, *376*
Luzzatti, C. 296, *315*

MacFarlane, F.K. 23, *29*, 60
MacMahon, M.K.C. 104, 108, 109, *132*, 168, *206*
Marcie, P. 293, *317*
Maretsis, M. 415, *419*
Margolin, D.I. 262, *291*
Marin, O.S.M. 295, *317*
Marks, L.E. 384, *408*
Marks, M. 63, *73*
Marshall, J.C. 253, 254, *260*, 272, 274, 275, *291*, 294, *315*, *317*, *318*
Marshall, R.C. 61, 65, 71, *75*, 77, *97*
Martin, N. 252, *260*
Marttila, R. 293, *317*
Marzi, C.A. 379, *407*
Masterson, J. 416, *418*
Matovitch, V. 63, 64, *74*, 77, *97*
Mayer, I. 77, *96*
Maylor, E.A. 263, *291*
Mazurski, P. 5, *29*, 36, 104, *133*, 172, 327, 337, *362*
McBride, K.E. 266, 292, 415, *419*

McCabe, P. 71, *73*, 272, *289*, 369, 371, *375*
McConkey, R. 250, *260*
McCrae Cochrane, R. 322, 327, 350, 356, 358, *361*
McGlone, J. 372, *376*
McGuirk, E. 64, 70, 71, *73*, 77, *96*
McNeil, M.R. 103, *133*, 170
McNeill, D. 262, 264, *288*
McReynolds, L.V. 7, *30*, 41, 110, 111, *134*, 188, 189, 250, *260*, 323, 328, 334, 344, 345, 358, *362*
McTear, M. 211, 218, *244*
Meikle, M. 63, *73*, 77, *96*
Menn, L. 263, *290*
Methé, S. 236
Metter, E.J. 25, 27, 372, *376*
Metz-Lutz, M.N. 276, *288*, *290*
Micelli, G. 262, *288*, 295, 301, *314*, *317*
Michel, F. 270, 274, 279, 280, 283, 289, *291*
Michelow, D. 221, *243*
Milianti, F.J. 61, 65, 71, *75*, 77, *97*
Miller, N. 103, 104, 105, 106, 120, *132*, 169, 239, 240, *244*, 262, *291*
Milroy, L. 218, 224, 225, 226, 231, 232, 239, *244*
Milton, S. 322, 327, 350, 356, 358, *361*
Mintum, M. 251, *260*
Mitchell, J.R.A. 64, 70, 71, *73*, 77, *96*
Mitchum, C.C. 251, *260*, 325, 331, 347, 348, 358, *361*
Moger, U. 19, *28*, 55
Mohns, L. 422, *426*
Mohr, J.P. 279, *289*, 371, *376*
Monakow, C. von 367, 368, *376*
Montgomery, J. 101, 102, 115, *132*, 170
Moore, W.H. 24, *29*
Moraschini, S. 421, *425*
Morgan, A. 21, *25*, 26, 56, 57, *129*
Morley, G.K. 63, 64, *74*, 77, *97*
Morton, J. 77, *96*, 114, 117, 118, *131*, *132*, 156, 158, *259*,

271, 272, 274, 278, 282, 289, *291*, 295, 297, 303, *317*
Moss, S. 305, *317*
Moulard, G. 105, 106, *134*, 183
Mulhall, D. 63, *73*, 77, *96*
Mulley, G.P. 64, 70, 71, *73*, 77, *96*
Muma, J.R. 103, 120, 123, *133*, 170
Munk, H. 368, *376*
Müller, D.J. 4, *26*

Naeser, M.A. 5, *29*, 36, 101, 102, 104, 105, 109, *132*, *133*, 165, 171, 172, 327, 337, *362*, 370, 371, 372, *376*, 386, *408*, 421, *426*
Nagata, K. 372, *376*
Nakles, K. 329, 340, 341, *360*
Naud, E. 276, *288*, *290*
Naydin, V.L. 5, 24, *28*, 293, *317*
Negrao, M. 264, *291*
Nelson, D.L. 264, *291*
Nespoulous, J.-L. 265, *290*
Nettleton, J. 254, 256, *260*
Neumann, N. 77, *95*
Neville, H.J. 379, *408*
Newcombe, F. 272, *291*, 294, *317*
Newhoff, M. 239, 240, *244*
Niccum, N. 272, 289, 370, 371, 372, *376*, *377*
Nicholas, M. 250, *260*
Nichols, M. 221, *244*
Nickels, L. 321, 323, 331, 353, 354, *360*, *362*
Nicol, J. 203, *207*
Nielsen, J.M. 371, *376*
Niemi J. 293, *317*
Nolan, K.A. 296, *318*
Noll, J.D. 269, *292*

O'Conner, J. 216, *243*
Obler, L. 221, *242*, *244*
Ohlendorf, I. 12, *27*, 280, *290*
Orchard-Lisle, V. 77, *96*, 114, 118, *131*, 156, 158, *259*, 268, 271, 272, 274, 279, 280, 281, 282, *289*
Otomo, E. 366, *377*

Paradis, M. 1, 2, 237, 416, *419*

Parlato, V. 298, 310, 311, *315*
Pate, D.S. 262, *291*
Paternot, J. 372, *374*
Patterson, K. 42, 77, *96*, 114, 117, 118, *131*, *132*, 156, 158, *259*, 271, 272, 274, 282, 289, *291*, 294, 295, 297, 303, 305, *315*, *316*, *317*, *318*
Pease, D.M. 267, *291*
Peele, L.M. 10, *26*
Peizer, E. 239, *243*
Penn, C. 212, 214, 216, 226, 231, 233, 234, 235, 236, 238, 241, *242*, *244*
Perkins, L. 15, *29*, 224, 225, 231, 232, 238, 239, 241, *244*, *245*, 258
Perrier, D. 265, 276, *288*, *290*
Petersen, S.E. 251, *260*
Peuser, G. 415, *419*
Phelps, M.E. 372, *376*
Pichard, B. 276, *290*
Pichard, D. 276, *288*
Pickard, L. 229, *244*
Pieniadz, J.M. 372, *376*
Pierce, J. 63, 64, *74*, 77, *97*
Pierce, R.S. 108, 111, 125, 126, *133*, 173, 174, 262, *291*
Pindzola, R.H. 10, *28*, 44
Pirozzolo, F.J. 392, *408*
Podraza, B.L. 101, 113, 114, 116,, *133*, 175, 268, 276, *291*
Poeck, K. 11, 23, *28*, *29*, 63, *73*, 77, 81, *96*, *97*, 120, *134*, 192, 220, *245*
Poffenberger, A.T. 380, *408*
Poizner, H. 20, *29*
Polk, M. 371, 372, *375*
Popham, W.J. 86, *97*
Porch, B.E. 63, 64, *74*, 77, *97*, 196, 206, 337, *362*
Poser, E. 12, *27*
Posner, M. 251, *260*
Potter, R.E. 18, 19, *29*, 52, 120, 126, 128, *133*
Pound, C. 253, 254, *260*, 274, 275, *291*
Prescott, T.E. 332, 337, 338, 339, *361*, *362*

Pring, T. 250, 253, 254, *260*, 274, 275, *291*
Prins, R.S. 4, 12, *29*, 48, 65, *73*, 103, 104, 108, 111, 114, *133*, 176
Prinz, P. 216, *245*
Prior M. 416, *418*
Prutting, C.A. 25, *29*, 211, 212, 214, 216, 226, 229, *245*
Purell, C. 271, *291*

Quatember, R. 369, 372, *375*
Quint, S. 276, *288*

Raab, D.H. 367, *374*
Raichle, M.E. 251, *260*
Rainey, C.A. 379, *407*
Ramsberger, G. 27, 33, 109, *131*, 154, 320, 323, 329, 336, 337, *361*
Rankin, J. 422, *425*
Rayner, K. 392, *408*
Razzano, C. 313, *315*
Regard, M. 293, *317*
Reinvang, I. 24, *29*
Resurreccion, E. 61, 63, 64, 65, 71, 74, 77, *97*
Riddoch, J. 416, *418*
Riege, W.H. 25, *27*
Riley, L. 339, *361*
Risser, A. 196, *207*
Rizzolatti, G. 379, 380, 384, *407*, *408*
Robinson, A.J. 71, *74*
Robinson, S.R. 304, 305, *315*
Roman, M. 216, *246*
Romani, C. 301, *314*
Ron, M. 422, *426*
Rosch, E. 113, *133*
Rosen, T.J. 269, *289*
Rosenbek, J.C. 67, 70, 74, 105, *133*, 177
Rosenberg, J. 369, *376*
Rosnet, E. 276, *288*
Ross, A. 103, 104, 114, 115, 116, 121, 123, 127, *133*, 178
Ross, P. 297, *318*
Rothi, L.G. 295, 305, *315*, *317*
Rothi, L.J. 23, *29*
Rousselle, M. 105, 106, *134*, 183

Rubens, A.B. 8, *25*, 266, 272, *287*, *289*, 370, 371, 372, *376*, *377*
Rusk, H.A. 63, *73*
Ryan, C. 425, *425*

Sacchetti, M.L. 366, *374*
Sacks, H. 223, 225, *245*
Saffran, E.M. 112, *133*, 252, *260*, *317*, 295, 322, 330, *362*
Sakai, F. 366, *377*
Salmon, S.J. 111, *132*, 160, 320, 323, 328, 329, 334, 349, 350, 358, *361*
Sands, E. 63, *74*, 102, *133*, 180, 369, *377*
Sarno, M.T. 63, *74*, 102, 109, *133*, 180, 272, *291*, 369, *377*
Sartori 321
Sasanuma, S. 273, *291*, 312, 313, *318*, 414, 415, *419*
Saunders, P. 238, *246*
Schaefer, E. 225, *242*
Schegloff, E. 223, 225, *245*
Schienberg, S. 225, *245*
Schlanger, B.B. 22, 25, *29*, 58, 122, 125, *133*
Schlanger, P.H. 22, 25, *29*, 58, 122, 125, *133*
Schmidt, A. 379, *408*
Schoonen, R. 4, 12, *29*, 48, 65, *73*, 176
Schreiber, V. 424, *425*
Schuell, H.M. 4, *29*, 61, *74*, 80, *97*, 99, *100*, 195, 196, 198, 201, 202, 203, 205, *207*, 340, *362*, 369, *377*
Schwartz, H.D. 380, *408*
Schwartz, M.F. 112, *133*, 295, *317*, 322, 330, *362*
Scott, C. 300, 301, *318*
Scotti, G. 390, 391, *407*
Searle, J. R. 215, 217, *245*
Sefer, J.W. 102, 107, 108, 109, 115, *133*, *134*, 180, 181
Seidenberg, M. 252, *260*
Selinger, M. 332, 337, 338, 339, *361*, *362*
Selnes, O.A. 272, *289*, 370, 371, 372, *376*, *377*

Seron, X. 6, *29*, 77, 88, *95*, *97*, 105, 106, 113, 114, 115, 126, *130*, *134*, 182, 183, 267, 276, 273, 288, 292, 303, *315*
Serrat, A. 293, *317*
Shallice, T. 295, 296, *318*
Shankweiler, D. 63, *74*, 369, *377*
Shapiro, L. 201, *207*
Shaw, G. 422, *426*
Shaw, R. 102, *134*, 181
Shewan, C.M. 61, 65, 71, *74*, 77, *97*, 109, 110, 129, *134*, 184, 185, 320, 326, 334, 342, 343, *362*
Shewell, C. 77, *96*, 99, *100*, 101, 102, 103, 109, 113, 115, 117, *131*, 151
Shokhor-Trotskaya, M.K. 9, *26*, 41, 331, 355, *360*
Shoonen, R. 103, *133*
Siegel, S. 86, *97*
Siesjo, B.K. 365, *374*
Signoret, J.L. 371, *374*
Silveri, M.C. 238, *245*, 262, *288*
Silverman, M. 102,*133*, 180, 369, *377*
Simion, F. 379, *407*
Sinclair, J. 217, *245*
Singer, W. 23, *29*
Sirian, S. 415, *419*
Skinner, C. 212, 217, 226, *245*, *246*
Sloan, L.L. 384, *408*
Smith, A. 391, *408*
Smith, K.L. 7, *28*, 40, 113, *132*, 162, 281, 282, *292*
Sommers, L. 262, *291*
Sonderman, J.C. 328, 340, *361*
Sparks, R.W. 18, 24, *25*, *30*, 48, 53, 61, 67, *74*, 80, *95*, *97*, 102, 109, 110, 121, 123, 128, *134*, 187, 421, *425*
Spinelli, C. 113, 129, *130*, 139
Spinnler, H. 379, *407*
Spreen, O. 196, *207*
Springer, L. 8, 9, 14, 23, 25, 27, *28*, *30*, 77, 80, 88, *96*, *97*, 120, *134*, 192, 236, *243*
Srinivasan, M. 370, 371, *376*
Stachowiak, F.-J. 11, *29*, 220, *245*
Stavraky, G.W. 367, *377*

Stern, G. 63, *73*, 77, *96*
Sterzi, R. 379, *407*
Stevens, M.K. 366, *377*
Stimley, M.A. 269, *292*
Stockel, S. 20, *30*, 57
Stout, C.E. 6, *26*, 37
Street, B.S. 4, *29*
Stubbs, M. 218, *245*
Stuss, D.T. 268, *288*
Subirana, A. 372, *377*
Suzuki, A. 372, *376*
Swann, J. 224, *243*
Swinney, D. 203, *207*
Symon, L. 365, *374*

Tan, L.L. 110, *130*, 145, 327, 334, 343, *360*
Tanemura, J. 273, *292*
Taylor, M. 63, *73*, 211, 231, *245*
Taylor, T. 222, 224, *245*
Tazaki, Y. 366, *377*
Testvetkova, L.S. 5, 24, *28*
Thiery, E. 272, *292*
Thomas, J. 213, *244*
Thompson, C.K. 7, *30*, 41, 110, 111, 120, *134*, 188, 189, 190, 250, *260*, 323, 328, 334, 344, 345, 358, *362*
Thompson, I. 217, *245*
Toni, D. 366, *374*
Trappl, R. 369, *375*
Tsveskova, L.S. 293, *317*
Tupper, A. 63, *73*, 77, *96*
Tyler, L.K. 252, *260*

Ulatowska, H. 220, 221, 242, *245*
Umiltà, C. 379, 384, *407*, *408*
Uno, A. 273, *292*

Valenstein, E. 380, *408*
Vallar, G. 372, *374*, 379, *407*
Van Der Borght, F. 267, *288*
Van Der Linden, M. 303, *315*
Van Dijk, T. 217, 219, *245*
Van Mier, H. 227, *242*
Vance, C. 111, *134*, 189
Vandereeken, H. 272, *292*
Vanier, M. 421, *425*
Vendrell-Brucet, J.M. 384, *408*

Vendrell-Gomez, P. 384, *408*
Vercruysse, L. 266, *289*
Verhas, M. 372, *374*
Vermeulen, J. 4, 12, 29, 48, 65, *73*, 103, *133*, 176
Vignolo, L.A. 5, *26*, 31, 61, 64, 72, *74*, 77, *95*, 340, *360*, 369, *374*
Villa, G. 262, *288*, 301, *314*
Vinarskaya, E.N. 5, 24, *28*
Virnarskaya, E.N. 293, *317*
Vish, E. 415, *419*
Vogel, D. 61, 65, 71, 75, 77, *97*
Voinescu, I. 415, *419*

Wagner, C.M. 125, 126, *133*, 174
Wald, I. 413, 414, *419*
Waller, M.R. 121, 125, 126, *134*
Walsh, M.J. 371, *374*
Wambaugh, J.L. 111, 120, *134*, 190
Wapner, W. 221, *243*
Wärhborg, P. 22, *30*, 58
Warrington, E.K. 295, 296, *318*
Wasterlain, C.G. 372, *376*
Watamori T. 414, 415, *419*
Watson, R.T. 380, *408*
Webb, W.G. 269, 271, *291*
Webster, E. 238, *246*
Wechsler, E. 63, *73*, 77, *96*
Weidner, W.E. 10, *28*, 44
Weigl, E. 13, 14, *30*, 99, *100*
Weiner, M. 226, *243*
Weintraub, S. 221, *242*, 264, 265, *288*, *289*
Weisenburg T.H. 266, *292*, 415, *419*
Weismann 25, *27*
Weiss, D.G. 61, 63, 64, 65, 71, 74, 75, 77, *97*
Weniger, D. 9, 23, *28*, *29*, *30*, 77, *97*, 120, 121, 122, *134*, 192
Wepman, J.M. 4, *30*, 63, *74*, 235, *246*

Wernicke, C. 371, *377*
Wertz, R.T. 61, 63, 64, 65, 67, 70, 71, *73*, *74*, *75*, 77, *97*, 105, 129, *133*, 177
West, J.A. 63, 64, *74*, 77 *97*, 328, 340, *362*
West, R. 20, *30*, 57
Weylman, S. 216, *246*
Whitaker, H.A. 99, *100*
White-Thomson, M. 253, 254, *260*, 274, 275, *291*
Whitney, J.L. 10, *30*, 45
Wiegel-Crump, C. 77, *97*, 113, 114, 115, 116, *135*, 193, 273, *292*
Wier, C.S. 366, *374*
Wiezer, H.J.A. 366, *374*
Wilcox, M. 4, 15, 16, 25, *27*, 61, 72, 78, 88, *95*, 120, 122, 124, 125, *130*, 216, 235, *242*, *243*, *246*
Wilkinson, D. 422, *425*
Williams, S.E. 113, 115, *130*, 139, 262, 269, *291*, *292*
Willmes, K. 25, *30*, 63, *73*, 77, 80, 81, 86, 88, *96*, *97*, 236, *243*
Wingfield, A. 263, 281, 282, *290*, *292*
Winitz, H. 5, *28*, 34, 101, *132*
Wipplinger, M. 119, *131*, 155
Wirz, S. 212, 217, 226, *245*, *246*
Wortzman, G. 422, *425*

Yaksh, T.L. 366, *377*
Young, A.W. 252, *259*, 392, *408*
Yunoki, K. 372, *376*

Zangwill, O. 63, *72*
Zegiser, P. 303, *318*
Zivin, J.A. 365, *377*
Zoghaib, C. 229, *244*
Zurif, E. 203, *207*, 221, *242*

Subject Index

AAC (Augmentative and Alternative Communication) 15, 236, 237
AAT (Aachen Aphasia Test) 81, 82, 89, 90, 93
adverb 9, 42, 55, 117, 147, 234, 355, 356
after-effect 81, 82, 84
agrammatism 7, 9, 14, 33, 34, 41, 50, 109, 112, 117, 137, 145, 146, 151, 154, 155, 159, 166, 167, 178, 179, 188, 189, 190, 257, 308, 309, 320, 321, 323, 327, 328, 329, 330, 331, 337, 352, 353, 354, 355, 356
alcohol abusers 411, 421, 422, 424, 425
amnesic aphasia 193, 265
anatomical asymmetries 372, 373
anomia 11, 178, 253, 257, 262, 270, 272
anomic aphasia 35, 44, 139, 140
APPLS (Assessment Protocol of Pragmatic-Linguistic Skills) 226, 230, 231
apraxia 22, 25, 40, 52, 123, 141, 144, 175, 255, 272
  speech 32, 38, 54, 86, 90, 94, 104, 177, 178
argument 110, 112, 113, 117, 160, 321, 343, 352, 353
ASL (American Sign Language) 20, 54, 55
auxiliaries 9, 42, 111, 160, 161, 299, 328, 331, 334, 348, 349, 350, 352
axonal sprouting 366
BAT (Bilingual Aphasia Test) 416, 417
BDAE (Boston Diagnostic Aphasia Examination) 147, 148, 163, 195, 199, 200, 203, 204, 205, 303, 329, 336, 337
Behaviorist 37

Behaviourist 6, 8, 17, 37, 99, 101, 125
binomial model 86
bottom-up 115, 128, 212, 226, 230, 231
Broca's aphasia 8, 17, 35, 40, 42, 46, 57, 61, 67, 68, 69, 70, 81, 90, 116, 136, 138, 139, 140, 145, 146, 149, 151, 154, 157, 158, 159, 160, 161, 165, 167, 184, 185, 189, 190, 191, 221, 237, 296, 301, 320, 321, 323, 326, 327, 328, 329, 330, 340, 369, 370, 371, 373, 401, 405, 414, 417
buffer 252, 255, 284, 301, 302
CA (Conversation Analysis) 212, 224, 225, 226, 230, 231, 232, 233, 237, 241
CADL (Communicative Abilities in Daily Living) 50, 148, 227, 228
catecholamine 366
ceiling effect 369
cerebral atrophy 370, 373, 422
cerebral blood flow 365, 366, 372
cerebral hemispheres 2, 3, 15, 16, 18, 19, 20, 25, 61, 122, 123, 126, 128, 143, 146, 149, 152, 162, 167, 187, 293, 363, 367, 371, 372, 373, 379, 380, 381, 384, 386, 389, 390, 391, 392, 393, 396, 400, 401, 405, 406, 421, 423
CET (Communicative Effectiveness Test) 148, 229, 230
circumlocution 50, 89, 127, 139, 178, 234, 235, 303, 424
cognition 128, 149, 152, 235, 366, 422
cognitive neuropsychology 5, 46, 77, 105, 107, 118, 231, 235,

443

241, 247, 249, 250, 251, 257, 258, 293
coherence 212, 219
cohesion 121, 167, 219, 220, 221, 222
collateral sprouting 366, 367
communication 3, 4, 6, 7, 8, 14, 15, 16, 17, 19, 22, 23, 25, 88, 89, 90, 94, 100, 119, 120, 121, 122, 123, 124, 125, 126, 127, 204, 219, 225, 228, 229, 232, 233, 234, 235, 236, 237, 238, 239, 240, 241, 255, 256, 258, 312, 319, 320, 322, 339, 341, 358, 369, 411, 415, 424
compensation 23, 24, 77, 93, 117, 142, 367, 368, 371, 372, 406
compensatory strategies 1, 17, 233, 234, 235, 238, 239, 424
competence 1, 2, 99, 100, 101, 119, 120, 124, 125, 147, 150, 171, 179, 181, 182, 212, 234, 236, 249, 415
Compliance 71, 72
comprehension 61, 65, 80, 82, 89, 90, 102, 104, 108, 110, 111, 112, 113, 119, 121, 122, 123, 125, 126, 127, 135, 137, 138, 143, 144, 146, 147, 148, 152, 157, 162, 163, 164, 167, 172, 173, 174, 175, 176, 177, 178, 180, 188, 192, 196, 198, 199, 200, 202, 203, 204, 205, 214, 215, 216, 220, 225, 226, 227, 233, 235, 240, 247, 254, 262, 274, 275, 280, 283, 285, 296, 299, 300, 301, 302, 303, 304, 305, 306, 307, 308, 309, 310, 319, 322, 327, 328, 329, 330, 331, 332, 333, 334, 337, 339, 340, 347, 352, 353, 354, 355, 370, 371, 373, 415
computer 61, 65, 107, 116, 136, 137, 161, 162, 179, 184, 236, 276, 300, 339, 381, 383, 386, 392, 406

conduction aphasia 7, 13, 35, 38, 40, 43, 105, 106, 117, 127, 139, 140, 142, 153, 157, 158, 162, 261, 264, 265, 269, 276, 299
consonant clusters 105, 184, 198, 202, 311
context 22, 58, 59
  communicative 61, 88, 119, 148, 159, 211
  situational 7, 39, 50, 78, 80, 115, 179, 192, 262
Conversational Coaching 17, 236, 238
conversational exchange 125, 167
conversational prompting 327, 356
copula 111, 160, 161, 191, 328, 334, 349, 350, 352
cortical retina 368
Critical Incident Technique 238
CT (computerized tomography) 368, 370, 372, 373, 386, 389, 401, 421, 422
cueing 3, 10, 11, 25, 44, 100, 105, 114, 115, 116, 117, 118, 121, 125, 136, 139, 140, 141, 144, 145, 146, 153, 155, 158, 159, 166, 172, 175, 179, 183, 184, 185, 227, 228, 235, 237, 238, 240, 255, 256, 264, 266, 267, 268, 269, 270, 271, 273, 274, 275, 276, 278, 279, 280, 281, 282, 283, 284, 286, 303, 327, 330, 337, 338, 339, 342, 343, 348, 354, 356, 416, 424
cues 5, 6, 7, 10, 11, 12, 14, 16, 18, 33, 36, 38, 39, 42, 43, 44, 45, 46, 47, 50, 53
cytoskeleton 366
Deblocking 13, 14, 33, 152, 264, 266, 267, 268, 269, 270, 271, 273, 279, 280, 283, 306, 308, 413
deep structure 103, 109, 111, 175, 204, 415
dementia 373
denervation 367
determiners 160, 343, 352

Subject Index

diaschisis 367, 368
diffuse neurologic condition 421
direct effect 81, 84, 86, 336, 422
discourse 78, 79, 94, 100, 120, 121, 126, 147, 148, 167, 174, 192, 204, 205, 212, 214, 215, 217, 218, 219, 220, 221, 223, 224, 225, 226, 231, 233, 234, 237, 238, 240, 326, 329
dysarthria 38, 52, 169, 175, 201, 369
dysgraphia (see also *writing*) 107, 297, 298, 299, 300, 301, 303, 308, 311
dyslexia (see also *reading*)
    deep 107, 127, 166, 278, 293, 294, 296, 297, 298, 306, 308
    surface 165, 166, 285, 294, 297, 300, 302, 303, 304, 305, 307, 309, 310
    visual 294
Edinburgh Functional Communication Profile 217, 218, 226, 228
effect size 71
efficacy 2, 21, 25, 50, 56, 57, 61, 62, 63, 64, 65, 66, 67, 68, 69, 71, 77, 81, 128, 129, 185, 186, 192, 217, 236, 276, 345, 347
elicitation 227, 228, 229, 322, 343, 348
ELT (Everyday Language Test) 227, 228
embedding 94, 226, 231, 233, 234, 327, 352
error analysis 199, 261
ethnomethodology 222
ETOH (Adults with histories significant for alcohol abuse) 421, 424
explicit 1, 2, 11, 101, 102, 105, 106, 108, 109, 110, 112, 113, 142, 151, 170, 195, 196, 220, 224, 225, 235, 241, 277, 282, 283, 287, 322, 330, 338, 343, 347, 348, 352, 354, 355, 356, 357, 358
facilitation 5, 13, 14, 24, 65, 69, 70, 116, 118, 129, 144, 147, 149, 156, 157, 158, 159, 263, 264, 267, 268, 269, 270, 271, 273, 274, 275, 276, 278, 280, 282, 283, 284, 285, 322, 349
FCP (Functional Communication Profile) 148, 229
FCT (Functional Communication Therapy) 16, 49, 121
focal lesion 262, 421
functional impairment 294
gender 12, 108, 182, 372
generalisation 254, 255, 257, 320, 326, 327, 328, 329, 331, 334, 339, 340, 341, 342, 343, 344, 345, 346, 347, 348, 349, 350, 358
generalization 7, 8, 15, 20, 39, 40, 41, 45, 46, 54, 55, 69, 93, 111, 113, 114, 122, 135, 143, 146, 156, 159, 161, 177, 182, 185, 188, 189, 190, 191, 192, 193, 194, 273, 276, 277, 282, 283, 284, 285, 286, 301, 302, 303, 304, 305, 314, 348
generative grammar (see also *transformational grammar*) 99, 103, 110, 111, 119, 151, 219
gesture 7, 15, 16, 20, 22, 39, 50, 51, 52, 58, 70, 88, 100, 115, 119, 121, 122, 127, 128, 139, 141, 146, 148, 149, 178, 179, 183, 194, 234, 237, 239, 240, 252, 255, 266, 267, 310, 424
global 55
global aphasia 5, 8, 15, 16, 18, 19, 24, 34, 35, 36, 37, 51, 55, 56, 61, 89, 94, 111, 122, 127, 135, 149, 152, 165, 173, 176, 183, 186, 187, 276, 279, 283, 370, 371, 381, 394

grammar 1, 20, 49, 54, 55, 99, 100, 101, 102, 111, 119, 126, 164, 168, 181, 217, 320, 363
group design 66, 69, 70, 71, 72
  no treatment 65
  single 63, 64
group designs 62
  single 63
handedness 70, 272, 372, 373
HELPSS (Helm Elicited Language Program for Syntax Stimulation) 14, 33, 34, 152, 154, 155, 167, 179, 329, 336, 346, 347, 359
hemianopsia 368
history of substance abuse 423
holistic processing 401
home treatment 354, 356
homophones 265, 281, 283, 299, 300, 301, 302, 306, 307, 308
hyperosmolar agents 365
hypometabolism 372
imageability 115, 140, 255, 256, 299, 338
imitation 51, 54, 55, 141, 161, 170, 171, 180, 253, 269, 327, 349, 356
implicature 103, 171, 212, 213, 218
implicit 1, 2, 101, 106, 109, 113, 115, 117, 170, 265, 322, 363
indirect requests 214, 216
information processing 19, 52, 118, 129, 278, 279, 283, 287, 293, 294, 295, 297, 298, 300
insula 370, 371, 373
ischemia 365, 366
ischemic penumbra 365
IVDA (Adults with histories significant for heroin abuse) 421, 424
kana 312, 313
kanji 312, 313
L-dopa 21
language comprehension 24, 34
laterality 372
lesion site 172
level of difficulty 18, 37, 102, 182, 395
lexicon 7, 15, 19, 39, 49, 55, 56, 100, 104, 105, 107, 112, 113, 114, 116, 117, 118, 119, 127, 135, 136, 140, 150, 156, 158, 165, 166, 169, 171, 176, 177, 183, 188, 198, 202, 204, 252, 254, 255, 256, 261, 278, 279, 280, 282, 283, 284, 285, 295, 296, 300, 301, 302, 303, 304, 305, 312, 417, 418
linguistically structured learning approach 79, 80, 81, 82, 84
listening 34
localisation 251
localization 168, 178, 199, 203, 287, 293, 297, 370, 371, 380, 393
LOT (Language Oriented Therapy) 65
Macrostructure 219, 220
Mapping 13, 46, 47, 108, 112, 113, 117, 137, 138, 159, 160, 166, 295, 310, 321, 330, 331, 334, 352, 353, 354
mass effect 370, 380
mastery 6, 86, 99, 103, 144, 151, 177
matching
  picture 12, 46, 47, 48, 51, 136, 162, 176, 193, 379, 392, 394, 395, 397, 399, 400, 405
  picture/word 114, 118, 138, 156, 157, 158, 159, 166, 176, 177, 180, 254, 271, 272, 275, 282, 284, 285, 286, 306, 307, 339, 392, 393, 394, 395, 396, 397, 399, 400, 401, 403, 405
  word 394, 395, 397, 399, 400, 401, 405
maxims of conversation 125, 213, 214, 226
memorization 119
memory 1, 19, 23, 52, 54, 59, 80, 94, 99, 102, 103, 113, 119, 127, 145, 150, 152, 179, 181, 193, 198, 202, 205,

235, 252, 284, 300, 301, 305, 332, 359, 422, 425
Meprobamate 20, 21, 57, 58, 373
metalinguistic 1, 2, 9, 38, 88, 100, 106, 142, 201, 203, 204, 205, 206
microcomputer 10, 11, 15, 42, 43, 44, 45
MIT (Melodic Intonation Therapy) 2, 18, 49, 53, 67, 68, 69, 70, 105, 110, 121, 123, 128, 152, 165, 167, 179, 187, 188
mitochondria 366
modularity 99, 200, 203, 204, 205, 206, 258, 287, 293
morphology 12, 101, 103, 104, 107, 108, 111, 113, 119, 150, 151, 168, 198, 204, 336, 341, 347, 413
MRI (magnetic resonance imaging) 368
MTDDA (Minnesota Test for the Differential Diagnosis of Aphasia) 195, 196, 198, 199, 200, 201, 202, 203, 204, 205, *208*, 340
naming 6, 13, 21, 35, 42, 43, 44, 56, 57, 88, 89, 101, 104, 105, 113, 114, 116, 117, 118, 119, 128, 129, 136, 139, 140, 149, 153, 155, 156, 157, 158, 159, 163, 164, 165, 166, 167, 172, 175, 182, 183, 193, 194, 196, 198, 199, 203, 204, 235, 236, 247, 251, 252, 253, 254, 255, 256, 257, 261, 262, 263, 264, 265, 266, 267, 268, 269, 270, 271, 272, 273, 274, 275, 276, 277, 278, 279, 280, 281, 282, 283, 284, 285, 286, 287, 296, 299, 305, 306, 308, 310, 346, 347, 348, 415
narrative 175, 219, 220, 221, 299, 330, 332, 334, 348, 353, 354, 355, 357
neglect 363, 380, 390, 391
neural model 368

neurotransmitter 21, 56, 57, 366, 367, 373
Noun 111, 117, 139, 140, 143, 147, 155, 162, 164, 175, 180, 190, 199, 220, 221, 269, 295, 297, 308, 309, 323, 329, 330, 331, 339, 343, 351, 354, 355, 356
nouns 19, 34, 55
optic aphasia 267, 309
oral start-up 268, 269, 270, 271, 273, 279, 280, 281, 283, 284, 285, 286
outcome measures 69, 70
PACE (Promoting Aphasics' Communicative Effectiveness) 2, 16, 25, 78, 88, 89, 90, 93, 94, 122, 124, 125, 129, 139, 146, 147, 179, 235, 236, 240
paraphasia 56, 199, 279, 331
  numeric 40
  phonemic 9, 38, 105, 127, 166, 178, 225, 235, 237, 241, 253, 269, 280, 306, 310
  semantic 89, 178, 235, 253, 255, 261, 266, 268, 278, 279, 280, 284, 303, 306, 307, 308, 313
pars opercularis 370
pathway 107, 126, 128, 294, 295, 301
pathways 16, 19, 44, 53
performance 10, 11, 16, 18, 21, 37, 43, 50, 52, 99, 102, 113, 128, 205, 206, 227, 228, 229, 230
permutation test 84
PET (Positron Emission Tomography) 372
phonetics 101, 103, 104, 107, 119, 150, 169, 178, 417
phonology 12, 15, 101, 103, 104, 106, 108, 111, 112, 113, 119, 150, 151, 167, 168, 169, 172, 176, 177, 178, 196, 198, 200, 201, 204, 254, 255, 282, 294, 313, 417

picture description 139, 221, 334, 336, 353, 355
planum temporale 372, 373
plasticity 14, 367, 422
practitioner-researcher 249, 250, 251, 258
pragmatics 14, 15, 18, 52, 108, 119, 120, 126, 167, 178, 179, 211, 212, 217, 219, 226, 227, 236
preposition 78, 80, 81, 82, 86, 112, 142, 144, 147, 160, 174, 177, 299, 352, 353
   local 82, 86
   local/locative 82
   temporal 78, 82, 84, 86
prepositions 13
   locative 46
production 1, 5, 7, 9, 12, 13, 33, 34, 38, 39, 40, 41, 44, 46, 48, 49, 50, 51, 54, 55, 56, 94, 102, 104, 105, 112, 113, 119, 121, 122, 127, 137, 138, 141, 142, 143, 145, 146, 147, 149, 160, 161, 164, 165, 166, 172, 178, 179, 180, 187, 188, 189, 190, 191, 199, 202, 203, 204, 216, 220, 221, 222, 223, 225, 226, 227, 233, 234, 235, 237, 239, 240, 241, 247, 263, 267, 268, 271, 272, 275, 277, 278, 279, 280, 282, 283, 284, 285, 286, 287, 293, 294, 295, 296, 297, 298, 299, 301, 303, 304, 305, 309, 311, 313, 314, 319, 320, 322, 323, 327, 328, 329, 330, 331, 332, 333, 334, 335, 336, 337, 338, 339, 343, 344, 345, 347, 348, 349, 350, 352, 353, 354, 355, 356, 357, 359, 414, 416
profile 204, 205, 216, 218, 219, 229, 265, 421
profiles 49
prognosis 1, 21, 31, 48, 104, 186, 272, 373, 391
pronominal co-reference 220
pronunciation 102, 107, 181, 182, 297, 298, 304, 305, 306, 312
psycholinguistic model 127, 165, 167, 251, 253, 254, 255, 256, 257, 258, 320
psycholinguistics 104, 129, 152
random assignment 64, 71, 72
randomization test 91
RCSST 6
reaction time 20, 57, 379, 380, 386, 388, 391, 392, 393, 395, 396, 397, 399, 400
reading (see also *dyslexia*) 6, 14, 31, 35, 38, 42, 61, 65, 80, 81, 105, 106, 107, 122, 127, 136, 144, 162, 163, 165, 166, 167, 176, 180, 196, 198, 199, 200, 202, 204, 229, 247, 254, 273, 274, 275, 278, 279, 283, 284, 285, 286, 293, 294, 295, 296, 297, 298, 299, 300, 301, 302, 303, 304, 305, 306, 307, 308, 309, 310, 311, 312, 313, 319, 328, 339, 344, 363, 391, 400, 401, 403, 405, 406
recognition 54, 55, 118, 141, 157, 160, 175, 176, 177, 178, 180, 198, 229, 230, 265, 271, 278, 281, 282, 287, 294, 295, 297, 298, 300, 302, 306, 314, 379
recovery 1, 12, 20, 23, 24, 25, 31, 38, 47, 48, 53, 57, 63, 64, 65, 156, 163, 167, 170, 176, 177, 180, 185, 186, 187, 193, 202, 257, 258, 272, 276, 277, 293, 303, 304, 327, 343, 350, 356, 358, 363, 365, 366, 367, 368, 369, 370, 371, 372, 373, 379, 413, 414, 415, 416, 417, 422, 423
redundancy 44, 110, 111, 121, 124, 173, 174, 220, 367, 370
regrowth 366
relay 256, 257, 277, 278, 298, 299, 303, 304, 306, 307, 309, 310, 311, 312

repair 223, 224, 225, 226, 227, 230, 231, 232, 238, 239, 241, 305, 356, 365, 366
repetition 6, 7, 14, 33, 35, 40, 43, 53, 67, 68, 69, 99, 102, 105, 113, 117, 127, 136, 142, 143, 144, 146, 154, 155, 156, 158, 162, 163, 164, 181, 187, 193, 199, 200, 201, 202, 204, 219, 224, 234, 235, 255, 265, 269, 273, 279, 283, 285, 286, 299, 303, 308, 310, 328, 339, 343, 346, 349, 356, 359, 424
replication 7, 66, 67, 68, 69, 70, 72, 90
request 159, 191, 215, 216, 217, 218, 232, 234, 235, 329, 340, 341, 342, 424
requests 17, 41
right brain damage 216, 221, 227
Role Play 22, 23, 25, 58, 122, 227, 228
route 107, 116, 127, 128, 166, 179, 186, 254, 277, 279, 295, 296, 300, 305, 310, 311
routes 13, 14
sample size 48, 69, 71, 72
semantics 39, 100, 101, 103, 104, 108, 110, 111, 112, 113, 116, 119, 135, 157, 159, 167, 168, 171, 176, 178, 198, 204, 215, 252, 254, 282, 294, 295, 300
sentence 5, 6, 7, 10, 13, 14, 15, 18, 31, 33, 34, 35, 36, 38, 40, 42, 46, 47, 48, 51, 53, 55
Sentence Construction Board 12, 47, 108, 109, 110, 145
sentence processing 109, 110, 111, 112, 113, 137, 138, 167, 320, 321, 326, 329, 330
single case studies 179, 323, 417
single-subject design
    A-B-A comparison 52
    A-B-A, withdrawl design 62
    alternating treatments design 62, 66, 67, 68, 69, 70, 71, 72
    changing criterion design 62
    cross-over design 62, 77, 78, 89, 94
    multiple baseline design 62, 66
    two-period cross-over design 78, 81, 82
SLAC (Sentence Level Auditory Comprehension) 5, 36, 104, 153, 167, 172, 173, 337
speech act 212, 214, 215, 216, 217, 218, 226, 227, 229, 341
spelling 102, 107, 108, 166, 167, 181, 182, 184, 298, 299, 302, 303, 305
spontaneous .i.recovery 6, 10, 13, 24, 32, 33, 35, 38, 43, 47, 49, 51
spontaneous speech 10, 14, 82, 89, 127, 154, 161, 162, 163, 189, 190, 192, 254, 255, 296, 299, 305, 308, 312, 326, 327, 330, 332, 333, 334, 336, 339, 346, 349, 350, 352, 355, 357, 358, 373
SSP (Syntax Stimulation Program) 14, 33, 154
STACDAP (Systematic Therapy Program for Auditory Comprehension Disorders in Aphasic Patients) 12, 48, 104, 108, 111, 176, 177
stimulation 2, 3, 4, 5, 6, 7, 12, 14, 22, 24, 25, 31, 32, 33, 34, 35, 36, 37, 41, 42, 48, 49, 51, 65, 69, 70, 78, 79, 80, 81, 82, 84, 86, 94, 110, 114, 118, 119, 121, 129, 136, 139, 142, 146, 148, 152, 154, 155, 164, 167, 170, 175, 176, 177, 178, 181, 182, 185, 188, 189, 193, 202, 235, 256, 268, 269, 276, 327, 328, 329, 336, 343, 344, 345, 352, 355, 367, 381, 384, 391, 416
strategies 4, 6, 13, 17, 24, 38, 47
striatal cortex 368
subject selection 423

substitution 2, 7, 9, 13, 23, 24, 39, 108, 127, 142, 144, 261, 299, 306, 308, 363, 367, 371, 373
substratum 366
supramarginal gyrus 370
syllable 7, 10, 13, 31, 32, 40, 44, 45, 103, 104, 105, 106, 107, 113, 116, 140, 142, 144, 162, 165, 166, 171, 178, 184, 194, 196, 202, 262, 263, 264, 265, 267, 269, 270, 274, 276, 283, 294, 301, 302, 306, 308, 311, 312, 313, 393
synaptogenesis 366
syndrome 127, 155, 165, 167, 197, 271, 294, 295, 305, 320, 321, 337, 368, 421, 422
syndromes 8, 9, 10, 11, 20, 52
syntactic structure 19, 55, 106, 109, 110, 111, 113, 124, 126, 137, 169, 174, 192, 195, 196, 198, 199, 240, 321
syntax 12, 13, 15, 33, 39, 100, 101, 103, 104, 108, 109, 110, 111, 112, 113, 117, 119, 137, 144, 145, 150, 151, 152, 154, 159, 160, 163, 167, 168, 169, 171, 172, 174, 176, 177, 178, 182, 188, 189, 190, 195, 196, 198, 202, 204, 234, 328, 329, 330, 336, 339, 417, 418
tacit knowledge 262, 263, 264, 265, 266
tactile aphasia 277
TAP (Treatment of Aphasics' Perseveration) 10, 43, 119, 153, 154
temporal gyrus 370, 373
temporal lobe 128, 164, 371
text grammar 212, 217, 219, 220
thematic roles/relations 13, 46, 47, 112, 113, 117, 137, 138, 330, 331, 334, 338, 351, 353, 354, 355
theoretical model 170, 283, 287
therapy
  direct 82, 230, 232, 233, 235, 241
  indirect 225, 232, 233, 237
  phonological 159, 255, 257
  semantic 253, 254, 255, 257
thrombolysis 365
Token Test 328, 337, 340
top-down 113, 120, 212, 217, 222, 226, 228
topic 16, 121, 129, 147, 178, 214, 230, 231, 234, 237, 238, 239, 262, 264, 272, 277, 329, 340, 341
TOT (Tip-of-the-Tongue) 262, 263, 264, 274
transcoding 107, 127, 165, 295, 298, 301, 306, 308, 311, 312, 313
transcortical motor aphasia 21, 56, 57, 149, 373
transcortical sensory aphasia 43, 153, 303
transfer 10, 18, 25, 51, 60, 94, 180, 183, 184, 236, 286, 304, 311, 371, 406, 413, 414, 415, 416, 417, 418
transformational grammar (see also *generative grammar*) 47, 103, 108, 109, 110, 205
turn-taking 125, 147, 167, 178, 223, 225, 227, 230, 232
universal grammar 415
VAT (Visual Action Therapy) 16, 52, 122, 127, 152
verb 108, 109, 111, 112, 113, 117, 119, 121, 137, 139, 140, 144, 147, 149, 155, 159, 160, 161, 162, 167, 172, 175, 177, 182, 183, 185, 192, 193, 199, 204, 216, 255, 269, 307, 308, 309, 321, 323, 325, 328, 329, 330, 331, 332, 337, 338, 339, 342, 343, 346, 347, 348, 351, 352, 353, 354, 355, 356
verbs 12, 19, 33, 40, 55
vicarious functioning 366, 367
visual hemifield 379, 380, 381, 386, 387, 389

Wernicke's aphasia 5, 18, 24, 35, 36, 69, 81, 89, 122, 139, 140, 146, 149, 167, 173, 221, 225, 261, 264, 265, 268, 276, 327, 370, 371, 381, 394, 417
Wh-question/interrogative 7, 33, 40, 41, 78, 81, 111, 188, 190, 191, 328, 334, 337, 339, 344, 345, 346
word class 19, 169, 299, 301, 310, 358
word frequency 170, 171
word retrieval 90, 94, 109, 114, 115, 116, 136, 158, 185, 193, 201, 237, 262, 263, 264, 269, 282, 326, 331, 343, 359
word-finding 5, 9, 10, 11, 19, 24, 32, 34, 42, 44, 45, 52, 90, 114, 116, 117, 118, 127, 136, 158, 175, 182, 183, 231, 251, 252, 253, 255, 262, 300, 303, 305, 329, 330
word-retrieval 114, 115, 116, 151, 178, 193, 225, 230, 247, 249
writing (see also *dysgraphia*) 6, 16, 17, 31, 35, 42, 50, 61, 105, 106, 107, 127, 147, 153, 159, 163, 180, 184, 195, 196, 198, 199, 200, 204, 229, 234, 237, 247, 252, 273, 274, 275, 284, 285, 286, 293, 294, 295, 296, 297, 298, 299, 300, 301, 302, 303, 304, 305, 306, 308, 309, 310, 311, 312, 313, 325, 328, 338, 339, 414, 424